Fundamentals of
Weed Science

Fundamentals of
Weed Science

Robert L. Zimdahl

Department of Bioagricultural Sciences and Pest Management
Colorado State University
Fort Collins, Colorado

ACADEMIC PRESS

San Diego   London   Boston   New York   Sydney   Tokyo   Toronto

Academic Press
a division of Harcourt Brace & Company
525 B Street, Suite 1900, San Diego, California 92101-4495, USA
http://www.apnet.com

Academic Press
24-28 Oval Road, London NW1 7DX, UK
http://www.hbuk.co.uk/ap/

Library of Congress Catalog Card Number: 98-85114

International Standard Book Number: 0-12-781062-5

PRINTED IN THE UNITED STATES OF AMERICA
98 99 00 01 02 03 EB 9 8 7 6 5 4 3 2 1

This book is dedicated to the memory

of

Ann Osborn Zimdahl

Contents

Chapter 3

Weed Classification   41

Chapter 4

Ethnobotany   55

Chapter 5

Weed Reproduction and Dispersal   71

Chapter 6

Weed Ecology   111

Chapter 12

Properties and Uses of Herbicides   297

Chapter 15

Herbicide Application   389

Chapter 16

Herbicide Formulation   401

Chapter 17

Herbicides and the Environment   411

Chapter 20

Weed Science: The Future   497

Preface

Not too many years ago there were only a few weed science textbooks; each is still useful, and revised and new texts have appeared. Most available texts lack an ecological-management perspective on the rapidly developing science of weeds and their control. This book does not ignore chemical weed control, but strives to include it as one management technique among many rather than the method of choice for most weed problems.

Scientists, including weed scientists, eagerly accepted credit for many of the advances since World War II that were deemed to be benefits of science. The public regarded these advances, which included herbicides and other pesticides, as desirable and benign. Now, however, science is held responsible for many problems that have arisen from its linkage with technology. Herbicides are not regarded as benign but often as threats to humans and the environment, and they are seen by many as an undesirable creation of science. Public attitude toward science and scientists is a mingling of awe and fear. The practice of science is constrained because, even though it claims to be an end in itself, it is supported and tolerated for its utility, and its practical value. Weed science is not atypical and, because of its close identification with chemical herbicides, may be regarded with more fear than other areas of agricultural science. The public's lack of understanding or its misunderstanding of what weed scientists do will not lessen the need for action and increases the responsibility of weed scientists and educators to be clear about the problem of weeds and proposed solutions.

This book addresses herbicides[1] and their use as an important aspect of modern weed management, but strives to place them in an ecological framework. Any text that purports to discuss the present state of the practice (and art) of weed management would be of little consequence and limited

---

[1]Common names of herbicides will be used throughout the text, except in some tables where they may be paired with one or more trade names.

value to students and others who wish to know about weed management as it is now practiced if the text omitted discussion of herbicides. Many weed scientists believe agriculture is a continuing struggle with weeds— without good weed control, good and profitable agriculture is impossible. Each agricultural discipline sees itself as central to agriculture's success and continued progress, and weed science is no exception. While not denying the importance of weed management to successful agriculture, this book places it in a larger ecological context. The roles of culture, economics, and politics in weed management are mentioned, but are not strong themes.

This, the second edition, has been improved in several significant ways, while trying to maintain an overall ecological framework. Some references in the first edition have been omitted and 301 new references have been added, 100 of which are from works published since 1992. Chapter 1 includes a description of the text's organization and addresses the question: Why study weeds? Chapter 1 will guide student discussions on why weeds are important. Chapter 2 begins with an expanded discussion of the definition of the term *weed,* a concept central to weed science. Material on the costs of weeds in Chapter 2 has been improved by the addition of several estimates of regional costs and the most current information on national costs. Chapter 3 presents weed classification, including basic phylogenetic relationships. More information on the role and nature of parasitic weeds, one of the world's great unsolved weed problems, has been included. The primary intent of Chapter 4, ethnobotany, remains the same: to convince students that weeds are part of nature and not always to be regarded as undesirable. The chapter has new information on uses of weeds for medicinal purposes and human food. Chapter 5 now includes more data on the number of seeds produced by weeds, including conflicts in available data. A new section on seed mimicry as an evolutionary technique has been added. Information on the effect of weed-seed dispersal on ecosystem processes and new data on weed-seed burial studies from Mississippi, Nebraska, and Arkansas have been added to complement older studies.

The ecology of weeds remains a central focus of the book. Chapter 6 has been improved by the addition of 48 references and new information on the possible influence of global climate change on weed growth and management, plus new material on the effects of light and temperature on growth and competition. Discussion of the critical period in weed competition has been modified to reflect new thinking about its importance. A major emphasis in weed management research is the development of mathematical decision-aid models. A new section has been added to Chapter 6 to discuss the purposes, value, and uses of such models. Allelopathy is presented in Chapter 7, where the emphasis on its natural importance has been increased. Work on use of allelopathy in turf from Iowa State University has been added. Chapter 8, the introduction to plant competition, has not been changed, but Chapter 9 has been amended with over 50 new references to new information on tillage techniques and the roles of companion cropping, crop rotation, crop interference, and fertility manipulation in weed management. Chapter 10 on biological control has not been changed except for the addition of more examples of success and many new references.

Chapter 11 introduces chemical control and has not been changed in any significant way, but Chapter 12 on properties and uses of herbicides has been updated to reflect the most current thinking on herbicide classification into families. Several new herbicides and all new herbicide families have been added to the text. Chapter 13 has been modified to include more recent examples from the weed science literature on interactions of herbicides and plants. Coverage of soil–herbicide interactions (Chapter 14) now includes information on mass transfer and diffusion as determinants of dissipation in soil. Chapter 15 on herbicide application and Chapter 16 on formulations have been revised slightly. The title of Chapter 17 has been modified to reflect a stronger environmental emphasis, and material on herbicide resistance has been expanded in significant ways. Pesticide registration and legislation have not changed in major ways since the first edition was published, and the chapter has not been modified significantly.

Development of weed management systems is a current and good goal for weed scientists. Chapter 19 incorporates what is known and provides complete information on the components of weed management systems for several cropping and weed management challenges. Chapter 20 now includes a major section on biotechnology and the development of herbicide-resistant crops. This chapter is the most personal and includes the most conjecture of all chapters. It is intended to provoke thought, but whether it is a reliable prediction of the future remains to be seen.

A strong, growing trend in weed science is away from the exclusive study of annual control techniques toward understanding weeds and the systems in which they occur. Control is important but understanding endures. Herbicides and weed control are important parts of the text, but it is hoped that understanding the principles of management and the biology and ecology of the weeds to be managed will emerge as dominant themes.

Study of weeds, weed management, and herbicides is a challenging and demanding task that requires diverse abilities. Weed science involves far more than answering the difficult question of what chemical will selectively kill weeds in a given crop. Weed science includes work on selecting methods for controlling weeds in a broad range of crops, on non-crop lands, in forests, and in water. Weed scientists justifiably claim repute as plant physiologists, ecologists, botanists, agronomists, organic and physical chemists, molecular biologists, and biochemists. Lest the reader be intimidated by that list of disciplines, however, I hasten to add that this text will emphasize general principles of weed science and not attempt to include all applicable knowledge. It is tempting, and would not be much more difficult, to incorporate extensive, sophisticated knowledge developed by weed scientists. Although this knowledge is impressive and valuable, it is beyond the scope of an introductory text.

It is hoped that the book conveys some of the challenges of the world of weeds and their management, as well as the importance of solving weed problems to agriculture and society and to meeting the demand to feed a growing world population. The book is designed for undergraduate weed science courses, with the aim of including most aspects of weed science, without exhaustively exploring each. Readers should note that in nearly all cases I have used the units of measure in the original reference rather than changing all to one measurement system.

Several colleagues provided helpful suggestions on the first edition of *Fundamentals of Weed Science.* I thank all of them, even though some comments were difficult to hear. The first edition had several errors of fact, and those have been corrected in this edition. I thank the following colleagues for suggestions and critical review of portions of the manuscript: Dr. Kenneth A. Barbarick, Dr. K. George Beck, Dr. Gregory L. Orr, Dr. Philip Westra, and Dr. Scott J. Nissen of Colorado State University; Dr. William W. Donald USDA/ARS, University of Missouri; Dr. David L. Mortensen, University of Nebraska; Dr. Robert F. Norris, University of California; Dr. Alan R. Putnam, Gallatin Gateway, Montana; Dr. Albert E. Smith Jr., University of Georgia; and Dr. Malcolm D. Devine, University of Saskatchewan. This book has been improved because of their efforts. Errors of interpretation or fact remain solely my responsibility.

Chapter 1

Introduction

Weeds have never been respected. The fact that many people earn a living and serve society by working to control and manage them is often greeted with amusement, if not outright laughter. Even colleagues who work in esoteric disciplines find it hard to believe that another group of scientists could be concerned exclusively with what is perceived as a mundane topic.

It is not surprising, therefore, that student's who take a course about weeds often wonder why the course is recommended or, perhaps, required, and what it is about. Students who enroll in chemistry or English have a reasonably good idea what the class will be about and how it fits into their curriculum. This is not usually true for students of weed science. Of course, students from farms and ranches know about weeds and the problems they cause, but they do not always comprehend the complexities of weed management. Therefore, it is important to our pursuit, that the nature of the subject be established and that the subject matter be related to students' knowledge of agricultural, biological, and general science. From the beginning, the text, the instructor, and the student should strive to establish relationships among weed science, agriculture, and society.

A brief review will lead to the conclusion that the story of agriculture is a story of struggle. Formidable obstacles have been placed between humans and a continuing food supply. These include physical constraints such as lack of good highways and transportation, economic constraints such as lack of credit and operational funds, environmental constraints such as too much or too little water or too short a growing season, and biological constraints such as problems with fertility, varieties, soil pH, or salinity. One of the most formidable environmental constraints has been pests.

Surveys by the Food and Agriculture Organization of the United Nations (1963, 1975) showed that in the 1960s and 1970's more than onethird of the potential annual world food harvest was destroyed by pests. The $75 billion lost was equivalent to the value of the world's grain crop (about $65 billion) and the world's potato[1] crop (about $10 billion). This means that insects, plant diseases, nematodes, and weeds deprived humans of food worth more than the entire world crop of wheat, rye, barley, oats, corn, millet, rice, and potatoes. These losses were only up to harvest and do not include damage during storage, another large sum. Current, less complete estimates, show that losses due to pests of all kinds have increased since the first FAO estimates were made.

History is filled with examples of human conflicts with pests, from biblical to modern times. Examples include locusts (*Melanoplus* spp.), which still plague the world, and late blight [*Phytophthora infestans* (Mont.) D. By.], which caused the Irish potato famine of the 1840's. The continuing world-wide presence of Colorado potato beetles (*Leptinotarsa decemlineata* Say), and the more recent epidemic of Southern corn leaf blight (*Helminthosporium maydis* Nisik and Miyake), show that the battle has not ended. In fact, the battle has become more intense as agriculture has changed, with the introduction of chemical pesticides and as more people have created demand for ever greater quantities of high-quality food.

One must respect the prescience of Swift (1677–1745; see Williams, 1937) who said:

> "Hobbes clearly proves that every creature
> Lives in a state of war with nature,
>
> ......
>
> So, Nat'ralists observe, a Flea
> Hath smaller Fleas that on him prey;
> And these have smaller Fleas to bite 'em:
> And so proceed *ad infinitum*."

DeMorgan (1850), who probably had read, but did not cite, Swift's poem, expressed the ubiquity of pests several years later:

> "Great fleas have little fleas upon
> their backs to bite 'em,
> And little fleas have lesser fleas,
> and so ad infinitum,
> And the great fleas themselves, in
> turn, have greater fleas to go on;
> While these again have greater still,
> and greater still, and so on."

The subject of this book is weeds, visible but unspectacular pests, whose presence is obvious nearly everywhere, but whose effects are not. Weeds have always been with us and are included in some of our oldest literature:

> "Cursed is the ground for thy sake; in sorrow shalt thou eat of it all the days of thy life; thorns and thistles shall it bring forth to thee; and thou shalt eat the herb of the field."
>
> —Genesis 3:17–18
>
> "Ye shall know them by their fruits. Do men gather grapes of thorns, or figs of thistles?"
>
> —Matthew 7:16
>
> "And thorns shall come up in her palaces, nettles and brambles in the fortresses thereof. . . ."
>
> —Isaiah 34:13

Weeds are also mentioned in the parables of Jesus (Matthew 13:18–23). The Biblical thistles, thorns, and brambles are common weeds and have been identified as such by biblical scholars (Moldenke and Moldenke, 1952). They were and are serious threats in the continuing battle to produce enough food. The tares in the parable (Matthew 13:24– 30), are the common weed, poison ryegrass, a continuing problem in cereal culture.

> "The kingdom of heaven is likened unto a man which sowed good seed in his field: But while he slept, his enemy came and sowed tares among the wheat, and went his way. But when the blade was sprung up, and brought forth fruit, then appeared the tares also."

The Greek word tares is translated as darnel—a weed that grows in wheat. It is a grass resembling wheat or rye, but with smaller, poisonous seeds. The weed called tares in Europe today is a different species.

No agricultural enterprise or part of our environment is immune to the detrimental effects of weeds. They have interfered with human endeavors for a long time. In much of the world, including my garden, weeds are controlled by hand or with a hoe. A person with a hoe may be as close as we can come to a universal symbol for the farmer. For many, the hoe, and the weeding done with it, symbolize the practice of agriculture. The battle to control weeds, waged by people with hoes, is the farmer's primary task in much of the world.

> "Bowed by the weight of centuries he leans
> Upon his hoe and gazes on the ground,
> The emptiness of ages in his face,
> And on his back the burden of the world.
> Who made him dead to rapture and despair,
> A thing that grieves not and that never hopes,
> Stolid and stunned, a brother to the ox?

Who loosened and let down this brutal jaw?
Whose was the hand that slanted back this brow?
Whose breath blew out the light within this brain?
. . . .
O masters, lords and rulers in all lands,
How will the future reckon with this man?
How answer his brute question in that hour
When whirlwinds of rebellion shake all shores?
Excerpt from "The Man with the Hoe"
—Edwin Markham (1899)

There have been four major advances in agriculture that have significantly increased food production.

1. The first was the introduction of mineral fertilizer. Early work on plant nutrition and soil fertility proceeded directly from the pioneering studies of Justus von Liebig (see Liebig, 1942), who questioned prevailing theories of plant nutrition.

2. A second major advance was rapid mechanization that began in the United States with development of Whitney's cotton gin (1793), McCormick's reaper (1834), and Deere's moldboard plow (1837).

3. Understanding and using genetic principles in plant and animal production was an enormous advance for agriculture. The obscure Austrian monk, Gregor Mendel, pursued his studies quietly and in seclusion. He

The introduction of mineral fertilizer increased food production.

Mechanization has increased agricultural productivity.

had no goal of pragmatic application or economic gain. The discoveries made from his beginning, most notably in development of plant hybrids, have had huge, generally positive, effects on our ability to produce food. De Vries, Correns, and Tschermak independently rediscovered Mendel's 1865 report in 1900 when searching the literature to confirm their own discoveries.

4. A fourth major advance in agriculture has been use of pesticides and plant growth regulators. These moved beyond mechanization to the chemicalization of agriculture and led to weed science. Weed science did not develop exclusively because of herbicide development, nor is its continued development dependent on herbicides, although they are an important part of knowledge concerning weeds and their management.

Weed science is vegetation management–the employment of many techniques to manage plant populations in an area. This includes dandelions in turf, poisonous plants on rangeland, and johnsongrass in soybeans. Weed science might be considered a branch of applied ecology that attempts to modify the environment against natural evolutionary trends. Natural evolutionary or selection pressure tends toward the lower side of the curve in Figure 1.1 (Shaw *et al.,* 1960), toward what ecologists call climax vegetation, the specific composition of which will vary with latitude, altitude, and environment. A climax plant community does not presently provide the kind or abundance of food humans want or need. Therefore, we attempt to modify nature to grow high-value crops for food and fiber.

In the beginning there were no weeds. If one impartially examines the composition of natural plant communities, or the morphology of weed flowers, one can find beauty and great aesthetic appeal. The flowers of wild onion, poison hemlock, dandelion, chicory, or tall morning-glory are beautiful and worthy of artistic praise for symmetry and color. By what right do we humans call plants with beautiful flowers, "weeds"? Who has the right to say some plants are unwanted? By what authority do we so easily assign the derogatory term "weed" to a plant and say it interferes with agriculture, increases costs of crop production, reduces yields, and may even detract from quality of life?

Nature knows no such category as weed. One widely accepted definition is a plant growing where it is not desired (Buchholtz, 1967). It is important that students note the anthropocentric dimension of this definition. Desire is a human trait, and therefore a particular plant is a weed only in terms of a human attitude. Ecologists speak of weedy plants, but often their discussion is affected by preconceptions of the vegetation on a particular site. People say that a plant, in a certain place is not desirable, and therefore

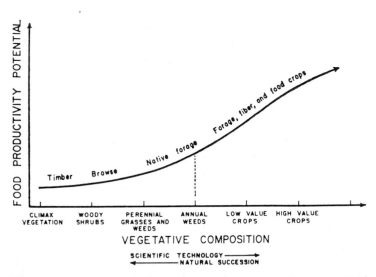

Figure 1.1   The food productivity potential of vegetation (Shaw *et al.,* 1960).

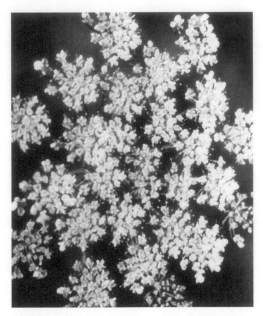

The flowers of many weeds are beautiful and have great aesthetic appeal. This is the flower of wild carrot or Queen Anne's Lace.

arbitrarily assign it the derogatory term "weed." We make it the lowest of the plant kingdom not because it is "naturally" harmful, but because it is harmful to us.

It is homeowners and neighbors who say that dandelions are unacceptable in lawns. Does grass really care what other plants live in the neighborhood? It is hay fever sufferers who say that ragweed or perhaps big sagebrush in the Western United States are unacceptable. It is those allergic to poison ivy who say it is unacceptable in their environment and who want to get rid of it. You decide what plants are weeds and when, where, and how they will be controlled.

This book will discuss many aspects of weeds, their biology, and their control. Chapters are arranged in a logical progression to organize the large, complex subject. The second chapter pursues the discussion of the definition of weed begun in this chapter and presents the characteristics and harmful aspects of weeds. Chapter 2 concludes with a discussion of what weeds cost. Chapter 3 classifies weeds in several ways, and Chapter 4, unique among weed science texts, discusses uses of weeds or ethnobotany. Reproduction and dispersal of weeds are presented in Chapter 5. Chapter 6 is one of the most important, as it presents the fundamental ecological basis of weed science. It is followed by a discussion of allelopathy; a subject included as a minor point in other weed science texts. Having established the ecological base of weed science, the significance of weedcrop competition is presented in Chapter 8.

Chapter 9 begins the large topic of weed management. For many this is what weed science is all about – how are weeds controlled? Weed problems are created, and those who wish to control need to ask why the weed is there as well as how to manage or control it. Key concepts of prevention, control, and management are presented in Chapter 9, followed by presentation of mechanical, non–mechanical, and cultural control techniques. Chapter 10 continues discussion of control, but by biological means. Chapter 11 introduces important concepts related to chemical control of weeds, and Chapter 12, one of the longest and most difficult, classifies herbicides based on how they do what they do–their mechanism of action, and their chemical family. Students will have to work to understand the information in Chapter 12. Chapters 13 and 14 are central to understanding of the interactions of herbicides, plants, and soil. Application and formulation of herbicides are presented in Chapters 15 and 16.

Chapter 17 returns to the ecological theme, but this time with information on the interaction between herbicides and the environment. A central, and intentionally unanswered, question is how one balances and judges the potential harmful and beneficial aspects of herbicide use. Chapter 18 presents the legislative decisions that address some of the questions raised in Chapter 17.

Chapter 19 brings things together by discussing weed management systems, many of which are largely conceptual and not yet prescriptive. They are evolving and improving with time. The last chapter (20) presents a view of the future of weed science. It is meant to provoke thought and discussion. It is not an infallible prediction of what will be.

One can establish a relationship between pesticide use and agricultural yield. Perhaps a better way to put it is that one can find a relationship between good pest control (regardless of how it is accomplished) and agricultural yield. We should not equate good pest control with pesticide use. Good pest control depends on cultural knowledge: what a good farmer or plant grower knows. Cultural knowledge is different from the scientific knowledge that leads to herbicide development and successful use. Both kinds of knowledge–scientific to tell us what can be done, and cultural to tell us what we ought to do–are essential to good weed management. One can also postulate a relationship between the way pests are controlled, the practice of pest management, and a nation's food supply. Figure 1.2 shows the world's tropical and sub-tropical areas, their major crops, and the percent of the world's total crop grown in each area. The region's ability to

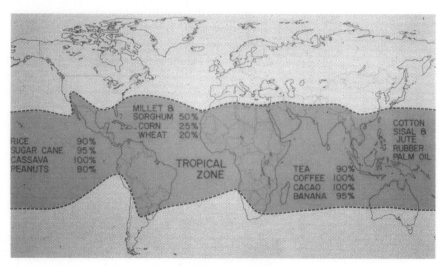

Figure 1.2 Crop production in the world's tropics (Holm 1971).

control weeds is shown in Figure 1.3, with data for the world and four major areas. Each segment in Figure 1.3 is divided into good, moderate or acceptable, low, and very poor weed management. The world's tropical and sub–tropical regions (Figure 1.2) are home to a majority of the world's

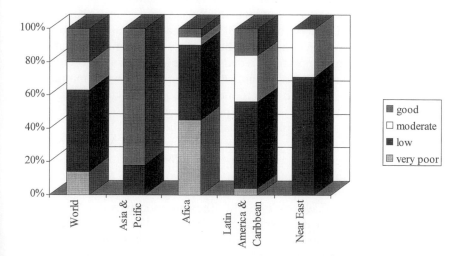

Figure 1.3 The level of weed control practices in the world and four regions (Labrada, 1996).

population, produce most of some of the world's important crops, and, even with all the progress from 1971 to today, still suffer from underdevelopment of weed science and other agricultural technology (Figure 1.3). Seventy countries in Asia, Africa, and Latin America were included in a recent UN/FAO survey (Labrada, 1996). The countries had 43.7% of the world's arable land and 65.8% of the world's people and, of most importance to our subject, inadequate weed control technology and knowledge.

The founder of Latin prose, Cato the Elder, reminds us in his work on agriculture that "it is thus with farming: if you do one thing late, you will be late in all your work." We are late in implementing advanced weed management techniques in much of the world, and agriculture will not progress to its full potential without them.

The agricultural productivity of the developed world is not an accident. U.S. agriculture and that of other advanced nations grew out of a propitious combination of scientific advancement, industrial growth, and abundant resources of soil, climate, and water. One should not regard it as just good fortune or God's benevolence that we, in the United States, can say that after the food bill is paid we have more left over than most other folks in the world. For most Americans (although, unfortunately, not all) this is true. It not only is true, it is regarded as so common that it is treated as a right rather than as something that was created and must be maintained.

Weeds are controlled in much of the world by hand or with crude hoes. The size of a farmer's holding and yield per unit area are limited by several things, and paramount among them is the rapidity with which a family can weed its crops. More human labor may be expended to weed crops than on any other single human enterprise, and it is likely that much of that labor is expended by women. Weed control in the Western world and in certain other areas of the world is done by sophisticated machines and by substituting chemical energy for mechanical and human energy. There is a relationship between the way farmers control weeds and the ability of a nation to feed its people. Weed science is part of that relationship. Good weed management is one of the essential ingredients to increase food production.

The early flights of the Apollo spacecraft gave those of us bound to earth a view of the whole planet, floating in the great, black sea of space. Many had imagined but never seen such a picture before. Space exploration opened exciting new vistas and opportunities for exploration, but, for now, we are confined to this planet. About 1965, world food production began to lose the race with an expanding population, as Rev. T. R. Malthus (1798) predicted it would. Each year the Malthusian apocalypse he predicted is prevented, but it is a daily specter for many in the world. The world's population will soon exceed 6 billion, and it will continue to grow, albeit at a slower rate. More than 85% of the world's people will live in poor, developing countries, and about 95% of the population growth will occur in those countries. As world population expands, food production is barely keeping pace, and often slipping behind. About 10% of the world's 33 billion acres of land are arable, and although the area devoted to productive agriculture can be expanded, the cost will be great. One must also recognize that the world may lack the social and political will to handle the complex problems expansion onto previously untilled land will bring. Such expansion is certainly part of the solution to the world food dilemma, but an equally important solution is use of appropriate, available technology and development of new technology. If all the world's people are going to enjoy higher standards of living and be able to watch their children mature without fear of debilitating disease, malnutrition, or starvation, we must use intelligently all present agricultural technology, and we must continue to develop better, safer technology. Shared technology and knowledge will permit our neighbors in this world to farm in ways that create opportunities to realize full agricultural and human potential.

Weed science is not a panacea for the world's agricultural problems. The problems are too complex for any simple solution, and students should be suspicious of those who propose such solutions to complex problems.

The dandelion is called a weed by many.

In fact, the hope should be not to solve but to diminish, not to cure but to alleviate, and to at least anticipate the "brute question" and have some answers when "whirlwinds of rebellion strike all shores." The work of the weed scientist is fundamental to solving problems of production agriculture in our world. Weeds have achieved respect among farmers who deal with them every year in each crop. Weeds and weed scientists have achieved respect and credibility in academia and the business community. The world's weed scientists are and will continue to be in the forefront of efforts to feed the world's people.

LITERATURE CITED

Ashton, F. M., and A. S. Crafts. 1981. 2nd ed. *Mode of Action of Herbicides.* J. Wiley and Sons, Inc. New York. 525 pp.

Buchholtz, K. P. 1967. Report of the terminology committee of the Weed Science Society of America. *Weeds* **15:**388–389.

Cato, the Elder. De Agri Cultura. 2nd Cent. B.C.

Corbett, J. R., K. Wright, and A. E. Baillie. 1984. 2nd ed. *The Biochemical Mode of Action of Pesticides.* Academic Press, New York. 382 pp.

De Morgan, A. C. 1850. A Budget of Paradoxes. p. 453, item 662.1 *International Thesaurus of Quotations,* 1970. Compiled by R. T. Tripp.

Fedtke, C. 1982. *Biochemistry and Physiology of Herbicide Action.* Springer–Verlag. Berlin. 202p.

Food and Agriculture Organization of the United Nations. 1963. Production Yearbook.

Food and Agriculture Organization of the United Nations. 1975. Production Yearbook.

Hassell, K. A. 1990. *The Biochemistry and Uses of Pesticides: Structure, Metabolism, Mode of Action and uses in Crop Protection.* VCH Publishers, New York. 536p.

Hatzios, K., and D. Penner. 1982. *Metabolism of Herbicides in Higher Plants.* Burgess, Pub. Minneapolis, MN. 142p.

Holm, L. 1971. The role of weeds in human affairs. *Weed Sci.* **19:**485–490.

Labrada, R. 1996. Weed management status in developing countries. Second Int. Weed Cont. Congress. **2:**579–589.

Liebig and After Liebig. 1942. Washington, D.C. American Association for Adv. of Sci. 111 pp.

Malthus, T. R. 1798. *An Essay on the Principle of Population as It Affects the Future Improvement of Society, with Remarks on the Speculations of Mr. Godwin, M. Condorcet, and Other Writers.* Macmillan, 1966 edition. 396 pp.

Markham, Edwin. 1899. "The Man with the Hoe". pp. 303–304 in: *The Pocket Book of Verse,*. 1940. Pocket Books, Inc., New York.

Moldenke, H. N., and. and A. L. Moldenke. 1952. *Plants and the Bible.* Dover Publications, Inc. New York. Pages 70–72, 133–134, and 153.

Moreland, D. E., J. B. St. John, and F. D. Hess (Eds.) 1982. *Biochemical Responses Induced by Herbicides,*. Vol. 181, Amer. Chem. Soc. Symp. Series. 274p.

Shaw, W. C., J. L. Hilton, D. E. Moreland, and L. L. Jansen. 1960. Herbicides in plants, pp. 119–133 in: *The Nature and Fate of Chemicals Applied to Soils, Plants and Animals.* USDA, ARS., 20–9.

Williams, H. 1937. *The Poems of Jonathan Swift,* Vol. II. *On Poetry: A Rhapsody.* pp. 639–659. Specific reference on p. 651.

Chapter 2

Weeds–The Beginning

FUNDAMENTAL CONCEPTS

- The most basic concept of weed science is embodied in the term *weed*.
- Weeds are defined in many ways, but most definitions emphasize behavior that affects humans.
- Weeds share some characteristics.
- There are at least nine separate ways in which weeds express their undesirability.
- There are no completely accurate estimates of what weeds cost. In the United States, losses due to weeds probably exceed $8 billion per year.

OBJECTIVES

- To understand why weeds are defined as they are.
- To know the characteristics that weeds share.
- To understand how weeds cause harm.
- To appreciate how estimates of the cost of weeds are made and what the magnitude of costs is.

> . . . and nothing teems
> But hateful docks, rough thistles, kecksies, burs,
> Losing both beauty and utility.
> And as our vineyards, fallows, meads, and hedges

Defective in their natures, grow to wildness;
Even so our houses, and ourselves, and children,
Have lost, or do not learn, for want of time,
The sciences that should become our country.

W. Shakespeare, *King Henry V,* Act V, Scene 2. The Duke of Burgundy address-
ing the Kings of France and England. First performed 1599; first printed in First
Folio, 1623.

I will go root away
The noisome weeds, which without profit suck the soil's fertility
from wholesome flowers.

W. Shakespeare, A gardener speaking to a servant in the Duke of York's garden,
Richard II, Act III, Scene III.

I. DEFINITION OF THE TERM WEED

To be fully conversant with a subject it is mandatory that one understand
its basic concepts. The most basic concept of weed science is embodied in
the term "weed." Each weed scientist has a clear understanding of the
term, but there is no universal definition shared by all. The Weed Science
Society of America, in a widely accepted definition, defines a weed as a
plant growing where it is not desired (Buchholtz, 1967). The European
Weed Research Society defines a weed as "any plant or vegetation, exclud-
ing fungi, interfering with the objectives or requirements of people"
(EWRS, 1986). These definitions are clear but leave the burden, and responsibility for final definition, with people. People determine when a particular
plant is growing in a place where it is not desired. The *Oxford English
Dictionary* (1973) defines a weed as an "herbaceous plant not valued for
use or beauty, growing wild and rank, and regarded as cumbering the
ground or hindering the growth of superior vegetation." The human role
is again clear because it is we who determine use or beauty and which
plants are to be regarded as superior. It is important that weed scientists and
weed managers remember the importance of definitions as determinants of
our attitudes.

How one defines something, in large measure, determines one's attitude
toward the thing defined. For the weed scientist and weed manager, the
definition determines which plants are weeds and thus undesirable. Weeds,
like other plants, lack consciousness and cannot enter the court of public
opinion to claim rights. Humans can assign rights to plants and serve as
their counsel to determine or advocate their rights or lack thereof in our
environment. Our attitude toward weedy plants need not always be shaped
by another's definition. We humans do not always agree about definitions.

Once in a golden hour,
I cast to earth a seed.
Upon there came a flower,
The people said a weed.

Read my little fable:
He that runs may read
Most can raise the flowers now,
For all have got the seed.

And some are pretty enough,
And some are poor indeed:
And now again the people
Call it but a weed.

A. Tennyson
"The Flower"

Not all people agree about what a weed is or what plants are weeds. Harlan and deWet (1965) assembled several definitions (partially reproduced herein) to show the diversity of definitions of the same or similar plants. The array of definitions emphasizes the care weed scientists and weed managers must take in equating how something is defined with a right or privilege to control.

Definitions of weeds by plant scientists (Harlan and deWet, 1965):

Blatchley	1912	"A plant out of place or growing where it is not wanted."
Georgia	1916	"A plant that is growing where it is desired that something else shall grow."
Robbins et al.	1942	"These obnoxious plants are known as weeds."
Muenscher	1946	"Those plants with harmful or objectionable habits or characteristics which grow where they are not wanted, usually in places where it is desired that something else should grow."
Harper	1960	"Higher plants which are a nuisance."
Salisbury	1961	"A plant growing where we do not want it."
Klingman	1961	"A plant growing where it is not desired; or a plant out of place."

Definitions of weeds by enthusiastic amateurs (Harlan and deWet 1965):

Emerson	1912	"A plant whose virtues have not yet been discovered."
King	1951	"Weeds have always been condemned without a fair trial."

Definitions of weeds by the ecologically minded (Harlan and deWet 1965):

Bunting	1960	"Weeds are pioneers of secondary succession, of which the weedy arable field is a special case."
Blatchley	1912	"A plant which contests with man for the possession of the soil."
Pritchard	1960	"Opportunistic species that follow human disturbance of the habitat."

Salisbury   1961   "The cosmopolitan character of many weeds is perhaps a tribute both
                       to the ubiquity of man's modification of environmental conditions
                       and his efficiency as an agent of dispersal."

Godinho (1984) compared the definition of the French words *d'aventice* and *le mauvaise herbe* with the English *weed* and the German *unkraut*. No single definition was found for weed and unkraut because both have two meanings:

1. In the ecological sense, weed, unkraut, and d'aventice mean a plant that grows spontaneously in an environment modified by man.
2. In the weed science sense, weed, unkraut, and *malherbe* (Italian) or mauvaise herbe mean an unwanted plant.

In some languages weeds are just bad (mal) plants: in Spanish, *mala hierba* or *malezas* and in Italian, malherbe. One must agree with Godinho (1984), Fryer and Makepeace (1977), and others (Anderson, 1977; Crafts and Robbins, 1967) that the word weed is not easy to define.

Aldo Leopold made the point well in a 1943 article critical of a 1926 bulletin, "Weeds of Iowa." Many of the native plants of Iowa are included in the bulletin and Leopold noted that these plants, in addition to their inherent beauty, have value as wildlife food, for nitrogen fixation, or as makers of stable plant communities. He admits that many plants others call weeds are frequent in pastures, but says that soil depletion, overgrazing, and needless disturbance of advanced successional stages often create the need for their control. Leopold (1943) argues that the definition of weed is part of the problem because not all plants that some call weeds "should be blacklisted for general persecution."

About 3,000 of the 350,000+ recognized plant species have been or are cultivated, and one should not assume the rest are weeds. That this is wrong is easily recognized, but when specific, unknown, and noncultivated plants are considered, one's objectivity is often not as clear.

The ulterior etymology of "weed" is unknown, but an exposition of what is known was provided by King (1966). He traced the word to Germanic, Romance-language, and Oriental roots, but concluded that weed is an "example of language as an accident of usage." He was unable to find a common word or words, in ancient languages, for the collective term "weed." The ultimate etymology of weed is unknown.

It is logical to assume that even if one cannot define weed it should be possible to identify the origin of individual species and determine certain characteristics of weeds. They come from native and naturalized flora. These immigrant plants succeeded as weeds because they were able to evolve forms adapted to disturbed environments more readily than other species. Baker's (1965) definition, repeated below, emphasizes success in disturbed environments.

> . . . [A] plant is a "weed" if, in any specified geographical area, its populations grow entirely or predominantly in situations markedly disturbed by man (without, of course, being deliberately cultivated plants). Thus, for me, weeds include plants which are called *agrestals* by some writers of floras (they enter agricultural land) as well as those which are *ruderals* (and occur in waste places as well as along roadsides). It does not seem to me necessary to draw a line between these categories and accept only the agrestals as weeds (although this is advocated by some agriculturally oriented biologists) because in many cases the same species occupy both kinds of habitat. Ruderals and agrestals are faced with many similar ecological factors, and the taxa which show these distributions are, in my usage, "weedy."

If one considers weeds in the Darwinian sense of a struggle for existence, they represent one of the most successful groups of plants that have evolved simultaneously with human disruption of areas of indigenous vegetation and habitats and creation of disturbed habitats (King, 1966).

Aldrich (1984) offered a definition that does not deny the validity of others, yet introduces a desirable ecological base. For Aldrich a weed is "a plant that originated in a natural environment and, in response to imposed or natural environments, evolved, and continues to do so, as an interfering associate with our crops and activities" (Aldrich, 1984). His definition provides "both an origin and continuing change perspective" (Aldrich, 1984). Aldrich wants us to recognize weeds as part of a "dynamic, not static, ecosystem." His definition departs from those that regard weeds as enemies to be controlled. Its ecological base defines weeds as plants with particular, perhaps unique, adaptations that enable them to survive and prosper in disturbed environments. Navas (1991) also included biological and ecological aspects of plants and effects on man in his definition. A weed was defined as "a plant that forms populations that are able to enter habitats cultivated, markedly disturbed or occupied by man, and potentially depress or displace the resident plant populations which are deliberately cultivated or are of ecological and/or aesthetic interest."

Although all do not agree on precisely what a weed is, most know they are not desirable. Those who want to control weeds must consider their definition. When the term weed is borrowed from agriculture and applied to plants in natural communities, a verification of negative effect on the natural community should be a minimal expectation. Simple yield effects are not acceptable, but effects of the presumed weed in a natural community can be estimated in terms of a management goal such as establishment of presettlement conditions, preserving rare species, maximizing species diversity, or maintaining patch dynamics (Luken and Thieret, 1996). Many recognize the human role in creating the negative, often deserved, image. Weeds are detrimental and often must be controlled, but only with adequate justification for the site and conditions.

II. CHARACTERISTICS OF WEEDS

Why is a particular plant a weed? What is it that makes some plants capable of growing where they are not desired? Why are they difficult to control? What are their modes of interference and survival? The most consistent trait of weedy species is not related to their morphology or taxonomic relationships. It is, as Baker (1965), noted their ability to grow well in habitats disturbed by human activity. They are plants growing where someone does not want them, and often that is in areas that have been disturbed or changed intentionally. Gardens, cropped fields, golf courses, and similar places are the places weeds grow well. Their ability to grow in habitats disturbed by man makes them a kind of ecological Red Cross. They rush in to disturbed places to restore the land.

Two nonindigenous species, kudzu and purple loosestrife, illustrate the ability of weeds to spread to new areas and habitats. Both are introductions to the United States and both now exist across large areas (Fig. 2.1) (U.S. Congress, 1993).

Not all weeds have all possible undesirable characteristics, but in addition to growing in disturbed habitats, all have some of the following characteristics (some from Baker, 1965):

1. Weeds have rapid seedling growth and the ability to reproduce when young. Redroot pigweed can flower and produce seed when less than 8 inches tall. Few crops can do this.
2. Weeds have quick maturation or only a short time in the vegetative stage. Canada thistle can produce mature seed 2 weeks after flowering. Russian thistle seeds can germinate very quickly between 28 and 110° F in late spring (Young, 1991). It would spread more, but the seed must germinate in loose soil because the coiled root unwinds as it pushes into soil and is unable to do so in hard soil.
3. Weeds may have dual modes of reproduction. Most weeds are angiosperms and reproduce by seed. Many also reproduce vegetatively.
4. Weeds have environmental plasticity. Many weeds are capable of tolerating and growing under a wide range of climatic and edaphic conditions

Kudzu (*Pueraria lobata*) 1990

Purple Loosestrife (*Lythrum salicaria*) 1985

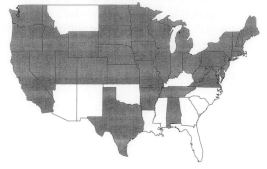

Figure 2.1 U.S. distribution of kudzu and purple loosestrife (U.S. Congress, 1993; Thompson *et al.,* 1987; Anonymous, 1990).

5. Weeds are often self-compatible, but self-pollination is not obligatory.
6. If a weed is cross-pollinated, this is accomplished by nonspecialized flower visitors or by wind.
7. Weeds resist detrimental environmental factors. Most crop seeds rot, if they do not germinate shortly after planting. Weed seeds resist decay for long periods in soil and remain dormant.
8. Weed seeds exhibit several kinds of dormancy or dispersal in time, to escape the rigors of the environment, and germinate when conditions are most favorable for survival. Many weeds have no special environmental requirements for germination.
9. Weeds often produce seed the same size and shape as crop seed, making physical separation difficult and facilitating spread by man.

10. Some annual weeds produce more than one seed crop per year, and seed is produced for as long as growing conditions permit.
11. Each generation is capable of producing large numbers of seeds per plant, and some seed is produced over a wide range of environmental conditions.
12. Many weeds have specially adapted long- and short-range seed dispersal mechanisms.
13. Roots of some weeds are able to penetrate and emerge from deep in soil. While most roots are in the upper foot of soil, Canada thistle roots routinely penetrate 3 to 6 feet, and field bindweed roots have been recorded over 10 feet deep. Roots and rhizomes are capable of growing many feet per year.
14. Roots and other vegetative organs of perennials are vigorous, with large food reserves, enabling them to withstand environmental stress and intensive cultivation.
15. Perennials have brittleness in lower stem nodes or in rhizomes and roots and, if severed, vegetative organs will quickly regenerate a whole plant.
16. Many weeds have adaptations that repel grazing, such as spines, taste, or odor.
17. Weeds have great competitive ability for nutrients, light, and water and can compete by special means (e.g., rosette formation, climbing growth, allelopathy).
18. Weeds are ubiquitous. They exist everywhere we practice agriculture.
19. Weeds resist control, including herbicides.

III. HARMFUL ASPECTS OF WEEDS

Definitions of weeds usually include trouble with crops, harm to people, or harm to animals. Most people do not consider plants to be bad. They are assigned the descriptive, derogatory term "weed" because of something they do to us or to our environment. If they were benign we wouldn't be so concerned about them because there would be no detrimental effects. The nature of weeds' harmful effects will be explored in this section. That harmful effects exist is not questioned. It is important to understand specific effects so appropriate action can be taken.

A. PLANT COMPETITION

From an agricultural perspective we are concerned about weeds because they compete with crop plants for nutrients, water, and light. If they did not, those who grow things would be more willing to tolerate their presence. Weed–crop competition will be discussed in Chapter 7.

B. Added Protection Costs

Weeds increase protection costs because they harbor other pests. A partial listing of diseases, insects, and nematodes that use weeds as alternate hosts is in Tables 2.1, 2.2, and 2.3. Weeds harbor a wide range of organisms, thereby increasing opportunities for those organisms to persist in the environment and reinfest crops in succeeding years.

Weeds that exist on the edges of crop fields serve as hosts when crops are not present and as sources of reinfestation. Volunteer wheat is a primary host of wheat streak mosaic virus. Its presence can be seen in disease transmission up to $\frac{1}{4}$ mile from a stand of volunteer wheat. A virus carried by wheat curl mite (*Aceria tulipae*) causes the disease, and volunteer wheat must be controlled 3 weeks before planting to eliminate the mites and prevent crop infection. This is a complex management problem in which a disease, an insect, and a weed host interact. Another illustration is spread of potato blackleg disease (*Erwinia carotovora* var. *atroseptica*) and potato soft rot (*Erwinia carotovora* var. *carotovora*) by *Erwinia* bacteria via enduring infestations of common lambsquarters, redroot pigweed, or black nightshade that harbor the disease organisms (Cooper and Harrison, 1973).

In addition to direct attack on crops, insects are a primary means of dispersal for many pathogenic organisms. The cause of aster yellows virus is carried by the leafhopper *Macrosteles fasifrons* from lettuce to broadleaf plantain after lettuce emerges and during lettuce harvest. Several aphids carry pepper veinbanding mosaic virus and potato virus Y from weeds to crops (Broadbent, 1964). Fungal spores such as the conidia of *Claviceps purpurea* (the cause of ergot in rye) are transported by fungal gnats. The insects are attracted to sticky substances secreted by wounds. The fungal disease caused by the spores infects a wide range of grasses, including wild species. Piemiesel (1954) found that leafhoppers and curly top virus of sugarbeets used weeds as breeding grounds to increase inoculum density for later crop infection.

A classic case of a weed serving as a host for a pathogen is the heteroecious stem rust fungus (*Puccinia graminis* var. *tritici*) of wheat which uses European barberry as an alternate host. King (1966) estimated that wheat yield losses from this fungal disease were over 600 million bushels per year in the early 1960s.

Over 20 years from the 1970s to the 1990s, wheat rust has caused $100 million in crop losses annually. Eradication of barberry and related species dramatically reduced stem rust and consequent epidemics. Several U.S.

Table 2.1

Plant Diseases Harbored by Specific Weeds

Plant disease	Weed host	Crop infested	Reference
Blackleg	Black nightshade Common lambsquarters Mare's tail Redroot pigweed Smartweed	Potato	Dallyn and Sweet, 1970
Wilt diseases	Netseed lambsquarters Purslane, Common Redroot pigweed	Potato, alfalfa	Oshima *et al.,* 1963
Stem canker	Netseed lambsquarters	Potato, beans	Oshima *et al.,* 1963
Soft rot	Annual sowthistle Dayflower Common lambsquarters	Chinese cabbage and other vegetables	Kikumoto and Sakamoto, 1969
Powdery mildew	Wild oats	Wheat, oats, barley	Eshed and Wahl, 1975
Stripe mosaic virus	Common lambsquarters	Barley	ARS, 1966
Leaf curl virus	Common lambsquarters	Sugarbeet	ARS, 1966
Cucumber mosaic virus	Black nightshade	Several	ARS, 1966
Potato virus X and leaf roll virus	Redroot pigweed	Potato	ARS, 1966
Maize dwarf mosaic virus Maize chlorotic dwarf virus	Johnsongrass	Corn	Bendixen *et al.,* 1979
White rust Early blight Leaf spot Vascular wilts Cottony rot White mold Watery rot	Redroot pigweed	Potato, tomato, annual vegetables and flowers, beans, cabbage, carrot, peanut	Commers, 1967
Leaf Spot and Leaf blight Stalk rot Vascular wilt Damping off Soft rot	Tall morning glory	Sugarbeet, celery, pea, peanut, corn, tobacco, bean, fruits and vegetables	Commers, 1967
Stem rust Leaf spot and Leaf blight	Cocklebur	Wheat, barley, rye, celery, beets, tomato, soybean	Commers, 1967
White rust Banana leaf spot Takeall Stem rust Rusts	Canada thistle	Crucifers, banana, wheat, rye, barley, legumes, bean, pea, fava bean	Commers, 1967

Table 2.2

Insects Harbored by Specific Weeds

Insect	Vector of	Weed host	Crop infested	Reference
Cabbage maggot	Blackleg	Common lambsquarter	Potato	Bonde, 1939
Colorado potato beetle		Black nightshade Buffalobur	Potato	Brues, 1947
Beet leaf hopper, corn borer	Curly top	Russian thistle	Sugarbeet	Brues, 1947

states joined in an effort which was estimated to have saved farmers well over $30 million per year (Stakman and Harrar, 1957).

Russian thistle (Table 2.2) is an alternate host for the curly top virus of sugarbeets and tomatoes (Young, 1991) and the beet leafhopper (*Circulifer tenellus*) (Goeden, 1968). Goeden (1968) points out that hosting a potentially damaging insect may not be a sufficient reason to control a weed. Russian thistle hosts 32 economically important insects from 5 different orders. These are not all harmful, because some may be entomophagous enemies of harmful insects both of which are hosted by Russian thistle.

Crested wheatgrass is widely planted in the western United States for soil conservation. It and other species of *Agropyron* harbor the Russian wheat aphid (*Diuraphis woxia*) an important wheat pest (U.S. Congress, 1993).

Johnsongrass, a major weed in the southern United States, can hybridize with cultivated sorghum to produce the annual weed shattercane. Thus, a weed produces another weed (Mack, 1991).

C. REDUCED QUALITY OF FARM PRODUCTS

Most growers are familiar with weed seed in grain crops and resultant decreases in quality and losses from dockage and cleaning. Weed seed in grain crops perpetuate the problem when the crop seed is replanted. A particularly bad problem is wild onion or wild garlic in wheat. Seeds and aerial bulblets of these weeds are similar in size to wheat grains and difficult to separate. They impart an onion flavor to flour made from grain and an onion odor to milk after cows have grazed them or eaten feed containing them.

Table 2.3

Nematodes Harbored by Specific Weeds

Nematode	Weed host	Crop infested	Reference
Criconemoides onoensis	Nutsedges Junglerice	Rice	Hollis, 1972
Ditylenchus dipsaci	9 weeds from 7 genera	Soybean, snapbean, pea	Edwards and Taylor, 1964
Heterodera glycines	Bittercress Common foxglove Common pokeweed Oldfield toadflax Purslane Rocky Mountain beeplant Spotted geranium	Soybean	Riggs and Hamblen, 1966
Heterodera marioni	47 weeds from 42 genera	Pineapple	Godfrey, 1935
Heterodera schachtii	Black nightshade Lambsquarters Mustards Purslane Redroot pigweed Saltbush	Sugarbeet	Altman, 1968
Hoplolaimus columbus	Henbit Johnsongrass Purple nutsedge Yellow nutsedge	Soybean, cotton	Bendixen *et al.*, 1979 Bird and Högger, 1973 Högger and Bird, 1974
Meloidogyne incognita	Chickweed Johnsongrass Purple nutsedge Yellow nutsedge	Soybean, cotton	Bird and Högger, 1973 Högger and Bird, 1974
Pratylenchus sp.	Johnsongrass	Corn	Bendixen *et al.*, 1979
Trichodorus spp.	19 weeds from 18 genera	Potatoes	Cooper and Harrison, 1973

Wild oats, which affect the quality of bread and other wheat products, infest many acres of small grains, most notably spring wheat. Wild oats also infest barley used for feed and for malting, and any brewer will verify that wild oats make bad beer.

Weeds reduce the quality of seed crops. Purchasers of hybrid or certified seed expect to receive a high-quality product that will give high yields and not be infested with weed seed. This necessitates weed control in seed crops, and failures lead to high cleaning costs before sale.

Weeds cause loss of forage and reduce the carrying capacity of pastures and rangeland. Surveys in the 1990s by the Nebraska Department of Agriculture showed over 2 million acres infested with musk thistle and over 400,000 with leafy spurge. Rangeland and pasture were the dominant sites, and carrying capacity (number of animals supported by the land) was reduced 8 to 100% by musk thistle and 10 to 70% by leafy spurge.

D. REDUCED QUALITY OF ANIMALS

Many acres of western U.S. range land are infested with larkspur, which causes cattle deaths because cattle like it and often eat it selectively. In early growth stages, as little as 0.5% of an animal's weight ingested as larkspur can, within an hour, lead to toxicity, and 0.7% may be fatal (Kingsbury, 1964). Locoweeds and crazyweeds are important poisonous range weeds. All ruminants are susceptible to loco poisoning, but only when large amounts are consumed over weeks or even months. Horses are also poisoned, and symptoms appear at lower levels of intake for shorter periods of time than is true for ruminants (Kingsbury, 1964).

Halogeton grows on arid, alkaline soils and is found in many parts of the world, including the western United States. It is especially toxic to sheep because of its high oxalate content. Photosensitization or excessive sensitivity to light by cattle can be caused or aggravated by St. John's wort and mock bishopsweed (Anonymous, 1977).

Weed science usually emphasizes the negative effects of weeds on animals grown for profit and human food, but game animals are also affected by weeds. In western Montana, elk use of rangeland decreased as spotted knapweed increased. On native bunchgrass sites, 1575 pellet groups were on each acre. On sites infested with spotted knapweed there were only 35 pellet groups per acre (Hakim, 1975).

Poisonous plants may contain one or more of hundreds of toxins from nearly 20 major chemical groups, including alkaloids, glucosides, saponins, resinoids, oxalates, and nitrates (Kingsbury, 1964). There is no way to determine if a plant is poisonous by noting where it grows, when it grows, or how it changes during growth.

Because poisonous plants can occur in many habitats, one must learn to know the important ones in each area. There are no good antidotes after ingestion of poisonous plants by humans or animals. Signs of poisoning differ in intensity depending on the species, its stage of growth, when it is eaten, the soil the plant grew in, the amount of other food eaten with or before the poisonous plant, and each individual's tolerance. Poisonous weeds can be managed, once recognized. A few of the common poisonous weeds found in the United States, their toxic principles, the plant sources, and some clinical signs of poisoning are shown in Table 2.4.

Weeds can affect animals by providing an inadequate diet or a diet that is unpalatable because of chemical compounds in the weed. They can directly reduce the quality of animal products by affecting milk production and fleece or hide quality. Reproductive performance is affected through toxins that cause abortion or kill animals (Table 2.4).

In addition to direct poisoning, weeds cause mechanical damage to grazing animals. Sharp spines on seed-bearing burs of puncture vine and sandbur are strong enough to penetrate tires and shoe leather. Someone who encounters a seed bur in bare feet will quickly recognize the pain and damage it could do to tender mouth tissue. Seed burs of these weeds and those of common cocklebur and burdock also become entangled in sheep's wool, decreasing cleanliness and salability.

E. INCREASED PRODUCTION AND PROCESSING COSTS

We are concerned about weeds because they do things to us or our products and increase production costs. Any weed control operation, from hand hoeing to herbicide application, costs money. These costs are often necessary to prevent greater crop loss or even crop failure and can be perceived as necessary to gain a profit. However, if the weeds were not there, there would be no control cost. Unfortunately, the complete absence of weeds is rare and the costs of their competition and control must be included when calculating profit or loss. Costs of control are relatively easy to calculate if hourly labor, equipment, fuel, and herbicide costs are known. It has been estimated that the cost of tilling cultivated land may equal as much as 15% of a crop's value. Tillage may be required on some soils for crop production, and there are sound agronomic reasons for tillage, including seedbed preparation, trash burial, soil aeration, promotion of water infiltration, and, of course, weed control. Prior to herbicides, an experiment to investigate effects of tillage was always confounded by weeds and the need to control them by tillage. However, the ascendancy of minimum and no-tillage farming and availability of appropriate herbicides have brought many traditional tillage practices into question. Experiments with herbicides, in many soils, have shown little benefit from tillage other than weed control.

There are other, less obvious costs associated with weeds. Wild oat seed in wheat or barley or black nightshade fruit in beans both lead to increased costs due to the necessity of cleaning. Failure to remove these can lead to

Table 2.4

Characteristics of Some Poisonous Weeds (Evers and Link, 1972; Kingsbury 1964)

Name	Toxic principle	Source	Signs
Arrowgrass	Hydrocyanic acid	Leaves	Nervousness, trembling, spasms or convulsions
Bouncing bet	Saponin	Whole plant seeds are most toxic	Nausea, vomiting, rapid pulse, dizziness
Bracken fern	Unknown	Fronds	Fever, difficulty in breathing, salivation congestion
Buffalobur	Solanin	Foliage and green berries	Most serious in nonruminants
Buttercup	Proto-anemonin	Green shoots	Loss of condition, production drops; milk is often red; diarrhea, nervousness, twitching, labored breathing
Chokecherry and other cherries	Glucoside-amygdalin, a cyanogenic compound	Leaves	Rapid breathing, muscle spasms, staggering, convulsions, and coma
Cocklebur	Hydroquinone	Seeds and seedlings	Nausea, depression, weakness esp. in swine
Corn cockle	A glucoside githagin and a saponin	Seeds	Poultry and pigs are most affected; inability to stand, rapid breathing, coma
Horsetail	Thiaminase activity—an alkaloid	Shoots	Loss of condition, excitability, staggering, rapid pulse, difficult breathing, emaciation
Indian tobacco	Alkaloids similar to nicotine	Leaves and stems	Ulcers in mouth, salivation, nausea, vomiting, nasal discharge, coma
Jimsonweed	Alkaloids	All parts	Rapid pulse and breathing, coma
Larkspur	Alkaloid	All parts	Staggering, nausea, salivation, quivering, respiratory paralysis
Nightshade	Solanine—a glycoalkaloid	Foilage and green berries	Most cases in sheep, goats, calves, pigs, and poultry, anorexia, nausea, vomiting, abdominal pain, diarrhea
Ohio buckeye	Alkaloid	Sprouts, leaves and nuts	Uneasy or staggering gait, weakness, trembling

(*continues*)

Table 2.4 (*continued*)

Name	Toxic principle	Source	Signs
Water hemlock	Cicutoxin	Young leaves and roots	Convulsions
Whorled milkweed	A resinoid—galitoxin	Shoots esp. near top	Poor equilibrium, muscle tremors, depression, and then nervousness, slobbering, mild convulsions

loss in quality, dockage losses at the point of sale, or even loss of the crop if it should heat and spoil in storage because of unripened weed seed. If a harvested crop has large amounts of weed seed in it, one can assume that some of the crop was lost in the field from weed competition and that some additional quality was lost due to weeds at harvest and consequent harvest difficulty. Another cost of weeds at harvest is wear and tear on machinery. The extra bulk of weedy plants that pass through mechanical harvesting systems is bound to cause machinery to break down more frequently and wear out sooner. These kinds of things are not usually attributed to weeds because they are not recognized as contributors to increased costs of machinery breakdown, repair, and replacement. Weeds also cost money when they remain in the field and interfere with harvest (Table 2.5).

F. WATER MANAGEMENT

Weeds interfere with water management in irrigated agriculture. Water is consumed and flow is impeded by weeds growing in and along irrigation ditches. Weeds consume water intended for crops, cause water loss by

Table 2.5

Soybean Harvest Losses from Two Weeds (Nave and Wax, 1971)

Weed	Header	Threshing and separating	Total
		Percent loss	
Redroot pigweed	5.35	0.73	6.08
Giant foxtail	1.55	0.81	2.36

seepage, transpire water, and reduce water flow in irrigation ditches, leading to more evaporation.

Terrestrial criteria for assessing weed competition cannot be employed in aquatic environments. There are no known appraisals of direct crop losses due to aquatic weeds. However, Timmons reported in 1960, nearly four decades ago, that man-made lakes above dams across major rivers in Africa, Asia, and Central and South America had become so badly infested with weeds within 5 to 10 years of construction that their usefulness for power development, boat transport, and irrigation was greatly reduced. Aquatic weeds quickly reduced designed flow of some irrigation canals in India by 40 to 50% and in others up to 80% (Gupta, 1973). Submerged weeds retard water flow up to 20 times, whereas floating weeds only retard it 2 times (Gupta, 1976). Decreased flow reduced the possibility of irrigating distant fields, and accelerated opportunities for leakage and evaporation. In addition to agricultural concerns, those who use water for recreation or enjoy the aesthetic appeal of aquatic habitats are often disturbed by weeds. Aquatic weeds are often ugly, but the more important problem is that their presence and inevitable decay hasten eutrophication. There is probably more concern about aquatic weeds in recreational waters than in agricultural waterways.

G. Human Health

Those not associated with agriculture may often think of weeds as plants that impair human health. One who has not experienced the runny nose, sneezing, and watery eyes of hay fever cannot fully appreciate the animosity sufferers may develop toward plants. The pollen that causes hay fever often comes from weedy plants. Ragweed and goldenrod are common causes in many parts of the United States. Sagebrush is a leading cause in the western United States. While hay fever may be an obvious weed menace to some people, others would choose poison ivy as the worst weed. Swelling and itching, after contact with poison ivy, are always bothersome and can lead to serious discomfort. The rash can be caused by contact with any portion of live plants or with smoke from fire in which plants are burned. Most people are quick to put poison ivy or poison oak in the category of unwanted plants after one or the other has disturbed their picnic or camping trip.

Many plants that poison when consumed are common household plants that can be especially hazardous to children. Some weedy species can lead to aberrant behavior or death when consumed by people. Examples of household plants that are poisonous when consumed include narcissus, oleander, lily-of-the-valley, and iris.

H. Decreased Land Value and Reduced Crop Choice

Perennial weeds (field bindweed, johnsongrass, or quackgrass) or the annual parasitic weed dodder can lead purchasers to discount offers to buy or bankers to reduce the amount of a loan, because each recognizes a loss of productive potential. They also recognize the costs required to restore otherwise valuable land to full productivity. These weeds reduce salability of land because they restrict crop choice. Severe infestations of almost any perennial or parasitic weed will reduce yield of most crops and dodder may completely eliminate successful growth of some crops.

I. Aesthetic Value

Weeds in recreation areas often must be controlled. Weeds are fire hazards around power substations and equipment, oil, or chemical storage areas. A very practical need for weed control exists near traffic intersections where, in addition to being aesthetically unappealing, weeds reduce visibility and may contribute to vehicular accidents.

IV. COST OF WEEDS

There are no completely accurate estimates of the total cost of weed control and losses in agriculture due to weed competition, although several attempts have been made. One of the first estimates was reported at the 1969 United Nations Food and Agriculture Organization (FAO) International Conference on Weed Control. For example, U.S. losses due to weeds in potatoes were estimated to be $65,000,000 in 1969 (Dallyn and Sweet, 1970).

In 1967, weeds caused an estimated 8% loss of potential U.S. agricultural production (Irving, 1967). In 1967, Cramer summarized losses attributed to pests of all kinds in the world's major crops. He calculated that 9.7% of potential world crop yield was lost because of weeds. Parker and Fryer (1975) used Cramer's data and calculated that weeds eliminated 14.6% of the world's potential crop production. They estimated weeds eliminated 11.5% of world crop production in 1975 (Table 2.6). A comparison made in 1980 (Ahrens *et al.*, 1981) for wheat and rice shows losses were still about 10%, despite developments in control. Combellack (1989) estimated the total cost of Australian weeds to be $2 billion in 1986, of which $137 million was for herbicides.

An estimate of loss in crop yield from weeds in Canada in 1935 was $69 million (Hopkins, 1938). In 1949, the cost had risen 2.7 times to $186.2

Table 2.6

Estimated Food Losses Caused by Weeds in Three Classes of Crop Production

Class of crop production	Total cultivated area (%)	Relative production per unit area	Total food production (%)	Loss to weeds (%)	Loss as % of world food supply	Estimated food loss per year (metric tons × million)
A. Most highly developed	20	×1.5	30	5	1.5	37.5
B. Intermediate	50	×1.0	50	10	5.0	125.0
C. Least developed	30	×0.67	20	25	5.0	125.0
Total	100		100		11.5	287.5

Estimates in this table are not based on any firm statistical data, but are approximations suggested by the authors. Where food loss is estimated in terms of metric tons, this is based on an approximate world total food production of 2,500,000,000 metric tons per year (Parker and Fryer, 1975).

million (McRostie, 1949) and it was $255 million in western Canada alone (Wood, 1955). By 1956 the total loss was estimated to be $468.6 million, 150% increase over 1949 (Anderson, 1956).

Friesen and Shebeski (1960) estimated that the annual loss due to weeds in Manitoba grain fields was $32.3 million in 1959. Renney and Bates (1971) estimated losses due to weeds in British Columbia at $72 to $78 million per year in 1969. Their study showed that 38 to 42% of weed-caused yield losses in British Columbia were due to yield reduction of agricultural crops, increased insect and disease problems, dockage, harvest losses, and costs of control. If forest weeds were included, losses in yield and costs of control accounted for an additional 45 to 49% of total loss. By 1984, Canadian losses were estimated to be $911.7 million per year (722.6 + 189.1, Table 2.7) in 36 crops, or nearly double what they had been in 1956.

A U.S. soybean loss survey (Anonymous, 1971) found weed competition caused an estimated 3.3 bu/A yield reduction in 28 states. Weeds were responsible for a 12% crop loss, each year. Chandler (1974) summarized other estimates and concluded that weed competition in some southern U.S. states caused as much as 20% soybean yield loss. For the entire country, 5% was regarded as an optimistically low level of loss, except on perhaps half of the most intensively farmed acreage.

Peanut farmers in the southeastern U.S. spend about $50 per acre for weed control. Annual losses from weeds were estimated to be $20 million

Table 2.7

Estimated Average Annual Losses Due to Weeds in Several Commodity Groups (Chandler *et al.*, 1984)

Commodity group	Average annual loss		
	United States ($ × 1,000)	Western Canada ($ × 1,000)	Eastern Canada ($ × 1,000)
Field crops	6,408,183	616,331	69,647
Vegetables	619,072	20,972	29,956
Fruits and nuts	441,449	8,418	—
Forage seed crops	37,400	75,661	—
Hay	—	—	89,507
Total	7,506,104	722,634	189,110

in Alabama, $8 million in Florida, and $72 million in Georgia in 1991 (Dowler, 1992). There are good herbicide choices for peanut weed control, so the reasons for the large losses are of concern to farmers and weed scientists.

A U.S. Department of Agriculture report for the 1950s (ARS, 1965) estimated annual losses due to reduced crop yield and quality and costs of weed control in the United States were $5.1 billion. This value, an educated guess, became enshrined in early weed science textbooks. While the estimate was never proven wrong, changes in the value of crops and inputs, as well as methods employed to arrive at such figures, have increased the loss due to weeds. In 1954, it was estimated that weeds caused an annual loss of >$2 billion in 11 major U.S. agronomic crops (Anonymous, 1962).

In the 1970s poisonous plants alone may have caused a $118 million loss to livestock producers in the Great Plains area of the United States (Deloach, 1976). In the same year, Shaw (1976) estimated that weeds caused a loss of 10% of the value of food, feed, and fiber crops and ornamental plantings. The total annual loss was >$6 billion. He also projected that the $2.7 billion was spent for cultural, ecological, and biological control and another $2.3 billion for chemical control. The total cost of weeds was estimated to be $11 billion per year. For 1980, Shaw (Shaw, 1982) raised the total annual loss to $18.2 billion, with $12 billion due to competitive loss, $3.6 billion for chemical control, and $2.6 billion for other controls.

For 1975 to 1979 the competitive loss due to the weeds in U.S. agriculture for 64 crops was estimated to be $7.5 billion per year (Table 2.7) (Chandler *et al.*, 1984). In a separate publication Chandler (1985) estimated total losses of $14.1 billion with $8 billion due to weed competition, $2.1 billion to herbicide cost, and $4 billion for equipment and labor.

The most recent estimate of the cost of weeds in the U.S. was made for 1989–1991 (Bridges, 1992). The report covered all U.S. states except Alaska and 46 crops, including field crops, vegetables, fruits, and nuts. Research or extension weed scientists from each state estimated the percent yield loss from weeds competing in crops where the current best management practices were employed. The same scientists also estimated losses with best management practices without herbicides. The loss was $4.2 billion annually just in field, nut, and fruit crops, with best management strategies, and 82% of the total was lost in field crops. Without herbicides the loss rose to $19.6 billion. Total losses with best management practices were $6.19 billion and costs of control were above $9 billion for a total loss of $15.2 billion per year.

By any measure, this is a large amount of money, greater than the 1984 estimate. All estimates (by definition) are not absolutely accurate; they are the best information available. Because they are estimates (educated guesses) rather than quantitative experimental data, they cannot be regarded as absolutely true.

Regional or more local estimates are often more accurate, but extrapolation to other areas, while tempting, is often unwarranted. For example, leafy spurge now occupies more than 150 million acres of rangeland in the northern Great Plains of the United States. Direct livestock production losses and indirect economic effects approached $110 million in 1990 (Bangsund and Leistritz, 1991). In North Dakota, losses of income by cattle producers due to leafy spurge were $8.7 million and the producers reduced personal spending $14.4 million. That translates to reduced income for merchants who sell to cattlemen.

In 1990 leafy spurge reduced cattle carrying capacity about 580,000 animal unit months, or by 63,100 cows over a 7.5-month grazing season. The total annual direct grazing land losses were $23.1 million. Indirect grazing land losses were $52.2 million and wildland losses were $2.9 million. A 40% leafy spurge infestation reduced rangeland carrying capacity by 50%, and leafy spurge can reduce carrying capacity 75%. Because of leafy spurge alone, North Dakota lost $87.3 million and 1,000 jobs in 1980 (Leistritz *et al.*, 1992).

World literature concerning domestic and international food production leaves no doubt that weeds cost money: lots of money. They are ubiquitous and their effects on yield create large losses borne by producers and by consumers. Present worldwide inflation and lack of a world or country data base for each crop make it unproductive to attempt more accurate estimates of world, country, region, or crop losses due to weeds, even through present estimates lack precision.

Weed costs are calculated in dollars and associated with commodities. There are other ways to estimate costs and associated benefits of weed management. One is to examine the number of acres of crops treated for weed control. This estimates the value of weed management to farmers and is an accurate estimate of the extent of market penetration by herbicides (Table 2.8). These data do not estimate the use of other weed management techniques. Table 2.9 shows losses due to weeds by comparing weeded and unweeded crops in the Philippines and other Asian countries (Mercado

Table 2.8
Percentage of Crop Acreage Treated with Herbicides and Total Herbicide Use in the United States in 1971 and 1982 (Chandler 1985)

Commodity	Proportion of hectares treated with herbicide (%)		Herbicide applied (Million kg ai)	
	1971	1982	1971	1982
Row crops				
Corn	79	95	45.8	110.4
Cotton	82	97	8.9	7.8
Sorghum	46	59	5.2	6.9
Soybeans	68	93	16.6	56.8
Peanuts	92	93	2.0	2.2
Tobacco	7	71	0.1	0.7
Total	71	91	78.6	184.8
Small grain crops				
Rice	95	98	3.6	6.3
Wheat	41	42	5.3	8.2
Other grain	31	45	2.5	2.7
Total	38	44	11.4	17.2
Forage crops				
Alfalfa	1	1	0.2	0.1
Other hay	[a]	3	[a]	0.3
Pasture and range	1	1	4.8	2.3
Total	1	1	4.0	1.7
TOTAL	17	33	94.0	204.7

[a]Included in alfalfa.

Table 2.9

A Comparison of Yield in Weeded and Unweeded Crops (Mercado, 1979).

Crop	Yield (T/ha)		% increase from weeding
	Weeded	Unweeded	
Lowland rice			
Transplanted	3.9	2.9	34
Direct-seeded	4.1	1.0	310
Upland rice	2.8	0.6	367
Corn	5.1	0.53	862
Soybean	1.15	0.48	140
Mung Bean	0.75	0.57	32
Transplanted tomato	9.2	5.5	67
Direct-seeded tomato	5.1	1.5	240
Transplanted onion	10.8	0.44	2355

1979). The percent increase in yield due to weeding is an impressive statement about the value of weeding, regardless of the technique by which it is done. Similar data are shown in Table 2.10 for studies done on several crops in India, where improved methods may mean only better cultivation and are not to be interpreted as a recommendation for all modern technology.

Table 2.10

Benefits from Weed Control at Various Dryland Centers in India, 1971–81

Location	Crop	Crop yield (ka/ha) with		Increase (%)
		Traditional weed control	Improved weed control	
Varanasi	Upland rice	1,700	2,700	59
Dehra Dun	Maize	1,760	4,600	161
Hyderabad	Sorghum	1,500	3,740	149
Sholapur	Pearl millet	180	950	428
Dehra Dun	Soybean	920	1,840	100
Bangalore	Peanut	420	1,910	355

Unpublished data from: Friesen, G.—Manitoba.

THINGS TO THINK ABOUT

1. What commonalities and differences can be found in the several definitions of the word weed?
2. How does the way we define something determine our attitude toward it?
3. What taxonomic, biological, morphological, and physiological traits do weeds share?
4. What is the best estimate of what weeds cost in the United States?
5. How are cost estimates obtained?
6. What are the problems with estimates of the cost of weeds?

LITERATURE CITED

Agricultural Research Service. 1965. A survey of extent and cost of weed control and specific weed problems. Agric. Res. Serv. Rpt. 23-1. USDA, Washington, D.C. 78 pp.

Agricultural Research Service. USDA. 1966. Plant pests of importance to North American agriculture. ARS Handbook No. 307.

Ahrens, C., H. H. Cramer, M. Mogk, and H. Peschel. 1981. Economic impact of crop losses, pp. 65–73 in Proc. 10th Cong. Plant Protection. Brit. Crop Prot. Council.

Aldrich, R. J. 1984. *Weed-Crop Ecology: Principles in Weed Management.* Breton, N. Scituate, Massachusetts, pp. 5–6.

Altman, J. 1968. The sugar beet nematode. *Down to Earth* 23(4):27–31.

Anderson, E. G. 1956. What weeds cost us in Canada. *Proc. California Weed Conf.* 8:34–45.

Anderson, W. P. 1977. *Weed Science: Principles.* 1st Ed. West, New York, p. 1.

Anonymous, 1962. A survey of extent and cost of weed control and specific weed problems. U.S. Dept. Agric. Agric. Res. Serv. and Fed. Ext. Serv. ARS 34-23. 65 pp.

Anonymous, 1971. Weed losses in soybeans. 1971 National Soybean Weed Loss Survey. Elanco Products Co., Chicago, Illinois.

Anonymous, 1977. Texas weed makes cattle supersensitive to sun. *Chem. & Eng. News* 55:44.

Anonymous, 1990. Scourge of the South may be heading north. *National Geographic* 178(1), July, p. 5.

Baker, H. G. 1965. Characteristics and modes of origin of weeds, pp. 147–172 in *Genetics of Colonizing Species,* Proc. First Int. Union of Biol. Sci. Symp. on Gen. Biol. (H. G. Baker and G. L. Stebbins, Eds.) Academic Press, New York.

Bangsund, D. A., and F. L. Leistritz, 1991. Economic impact of leafy spurge on grazing lands in the northern Great Plains. Agric. Econ. Rpt. No. 275-5. Dept. Agric. Econ. N. Dakota State University, Fargo, North Dakota.

Bendixen, L., D. A. Reynolds, and R. M. Riedel. 1979. An annotated bibliography of weeds as reservoirs for organisms affecting crops. Ohio Agric. Res. and Ext. Center, Wooster, OH. Res. Bull. 1109. 64 pp.

Bird, G. W., and C. Högger. 1973. Nutsedge as host of plant parasitic nematodes in Georgia cotton fields. *Plant Dis. Rep.* 57:402–403.

Bonde, R. 1939. Comparative studies of the bacteria associated with potato blackleg and seed piece decay. *Phytopath.* **29:**831–851.

Bridges, D. C. (Ed.). 1992. Crop losses due to weeds in the United States—1992. Weed Sci. Soc. America, Champaign, Illinois, 403 pages.

Broadbent, L. 1967. Control of plant virus diseases pp. 330–364. In *Plant Virology* (M. K. Corbett and H. D. Sisler, Eds.). Univ. of Florida Press, Gainesville. 346 pp.

Brues, C. T. 1947. *Insects and Human Welfare: An Account of the More Important Relations of Insects to the Health of Man, to Agriculture, and to Forestry,* rev. ed. Harvard Univ. Press, Cambridge, Massachusetts. 154 pp.

Buchholtz, K. P. 1967. Report of the terminology committee of the Weed Science Society of America. *Weeds* **15:**388–389.

Chandler, J. M. 1974. Economic losses due to weeds. Res. Rpt. 27th Ann. Mtg. Southern Weed Sci. Soc., pp. 192–214.

Chandler, J. M. 1985. Economics of weed control. Pages 9–20 in (A. C. Thompson, ed.) *In* Chemistry of Allelopathy. Am. Chem. Soc. Symposium Series No. 268. Am. Chem. Soc. Washington, D.C. pp. 9–20.

Chandler, J. M., A. S. Hamill, and A. G. Thomas. 1984. Crop losses due to weeds in the United States and Canada, May 1984. Weed Sci. Soc. of America, Champaign, Illinois. 22 pp.

Combellack, J. H. 1989. Resource allocations for future weed control activities. Proc. 42nd New Zealand Weed and Pest Cont. Conf., pp. 15–31.

Commers, I. L. 1967. An annotated index of plant diseases in Canada and fungi recorded on plants in Alaska, Canada and Greenland. Pub. No. 1251 Res. Branch. Canada Dept. of Agriculture. 381 pp.

Cooper, J. I., and B. D. Harrison. 1973. The role of weed hosts and the distribution and activity of vector nematodes in the ecology of tobacco rattle virus. *Ann. Appl. Biol.* **73:**53–66.

Crafts, A. S., and W. W. Robbins. 1967. *Weed Control.* McGraw-Hill, New York, pp. 1–2.

Cramer, H. H. 1967. *Plant Protection and World Crop Production* (English translation by J. H. Edwards). Pub. as *Pflanzenschutz-Nachrichten* by Bayer, A. G. Leverkusen, W. Germany. 524 pp.

Dallyn, S., and R. Sweet. 1970. Weed control methods, losses and costs due to weeds and benefits of weed control in potatoes, pp. 210–228 in FAO Int. Conf. on Weed Control, Davis, California.

Deloach, C. J. 1976. Considerations in introducing foreign biotic agents to control native weeds of rangelands, pp. 39–50 in 4th Int. Symp. on Biol Cont. of Weeds. Gainesville, Florida.

Dowler, C. C. 1992. Weed survey—southern states, broadleaf crops subsection. *Proc. So Weed Sci. Soc.* **45:**392–407.

Edwards, D. I., and D. P. Taylor. 1964. Host range of an Illinois population of the stem nematode (*Ditylenchus dipsaci*) isolated from onion. *Nematologica* **9:**305–312.

Eshed, N., and I. Wahl. 1975. Role of wild grasses in epidemics of powdery mildew on small grains in Israel. *Phytopath.* **65:**57–62.

Evers, R. A., and R. P. Link. 1972. Poisonous plants of the midwest and their effects on livestock. Spec. Pub. 24. Univ. Illinois College of Agric. 165 pp.

EWRS. 1986. Constitution Eur. Weed Res. Soc. 15 pp.

Friesen, G., and L. H. Shebeski. 1960. Economic losses caused by weed competition in Manitoba grain fields. I. Weed species, their relative abundance and their effect on crop yields. *Can. J. Plant Sci.* **40:**457–467.

Fryer, J. D., and R. J. Makepeace. 1977. *Weed Control Handbook.* Blackwell, Oxford, p1.

Godfrey, G. H. 1935. Hitherto unreported hosts of the root-knot nematode. *Plant Dis. Rep.* **19:**29–31.

Godinho, I. 1984. Les de'finitions d'aventice et de mauvaise herbe. *Weed Res.* **24**:121–125.

Goeden, R. D. 1968. Russian thistle as an alternate host to economically important insects. *Weed Sci.* **16**:102–103.

Gupta, O. P. 1973. Aquatic weed control. *World Crops* **25**:185–190.

Gupta, O. P. 1976. Aquatic weeds and their control in India. *FAO Plant Prot. Bull.* **24**(3):76–82.

Hakim, S. E. A. 1975. M.Sc. Thesis. U. Montana, Missoula.

Harlan, J. R. and J. M. J. de Wet. 1965. Some thoughts about weeds. *Econ. Bot.* **19**:16–24.

Högger, C. H. and G. W. Bird. 1974. Weed and covercrops as overwintering hosts of plant parasitic nematodes in soybean and cotton fields in Georgia. *J. Nematology* **6**:142–143.

Hollis, J. P. 1972. Competition between rice and weeds in nematode control tests. *Phytopath.* **62**:764.

Hopkins, E. S. 1938. The weed problem in Canada. Proc. Fourth Mtg. Assoc. Comm. On Weeds, East Div., pp. 11–15.

Irving, G. W. 1967. Weed control and public welfare. *Weed Sci.* **15**:296–299.

Kikumoto, T. and M. Sakamoto. 1969. Ecological studies on the soft-rot bacteria of vegetables. VII. The preferential stimulation of the soft-rot bacteria in the rhizosphere of crop plants and weeds.

King, L. J. 1966. *Weeds of the World—Biology and Control,* Chapter 1, pp. 1–24. Interscience Publishers, New York.

Kingsbury, J. M. 1964. *Poisonous Plants of the United States and Canada.* Prentice Hall, New Jersey, 626 pp.

Leistritz, F. L., D. A. Bangsund, N. M. Wallace, and J. A. Leitch. 1992. Economic impact of leafy spurge on grazing land and wildland in North Dakota. Dept of Agric. Econ. Staff Paper Ser. AE-92005. N. Dakota State Univ., Fargo. 14 pp.

Leopold 1943. What is a Weed? *The Weed Flora of Iowa,* Bulletin No. 4. Pp 306–309. *In* S. L. Hader and J. B. Callicoh (Eds). 1991. The River of the Mother of God and other essays by Aldo Leopold. Univ. of Wisconsin Press. Madison, WI.

Little, W., H. W. Fowler, and J. Coulson. 1973. *The Shorter Oxford English Dictionary on Historical Principles,* 3rd Ed., revised and edited by C. T. Onions with etymologies revised by G. W. S. Friedrickson. Clarendon, Oxford. 2 volumes, 2672 pp.

Luken, J. O. and J. W. Thieret. 1996. Amur honeysuckle, its fall from grace. *Biosci.* **46**:18–24.

Mack, R. N. 1991. Pathways and consequences of the introduction of non-indigenous plants in the United States. Rpt. to Office of Technol. Assessment. September.

McRostie, G. P. 1949. Losses from weeds. *Agric. Inst. Rev.* **4**:87–90.

Mercado, B. L. 1979. *Introduction to Weed Science.* Southeast Asian Regional Center for Graduate Study and Research in Agriculture, Laguna, Philippines. 292 pp.

Navas, M. L. 1991. Using plant population biology in weed research: a strategy to improve weed management. *Weed Res.* **31**:171–179.

Nave, W. R. and L. M. Wax. 1971. Effect of weeds on soybean yield and harvesting efficiency. *Weed Sci.* **19**:533–535.

Oshima, N., C. H. Livingston, and M. D. Harrison. 1963. Weeds are carriers of two potato pathogens in Colorado. *Plant Dis. Rpt.* **47**:466–469.

Parker, C. and J. D. Fryer. 1975. Weed control problems causing major reductions in world food supplies. *FAO Plant Prod. Bull.* **23**:83–95.

Piemeisel, R. L. 1954. Replacement control: Changes in vegetation in relation to control of pests and diseases. *Botan. Rev.* **20**: 1–32.

Renney, A. J. and D. L. Bates. 1971. The cost of weeds. *Canada Weed Comm. Western Sect.* **24**:40–49.

Riggs, R. D. and M. L. Hamblen. 1966. Additional hosts of *Heterodera glycines. Plant Dis. Rpt.* **50**:15–16.

Shaw, W. C. 1976. Weed control technology for protecting crops, grazing lands, aquatic sites, and noncropland. U.S. Dept. Agric., Agric. Res. Serv. ARS-NRP-20280. 185 pp.

Shaw, W. C. 1982. Integrated weed management systems technology for pest management. *Weed Sci.* **30**(Supp.):2–12.

Stakman, E. C., and G. J. Harrar. 1957. *Principles of Plant Pathology.* Ronald Press, New York. 581 pp.

Thompson, D. Q., R. L. Stuckey, and E. B. Thompson. 1987. Spread, impact, and control of purple loosestrife (*Lythrum salicaria*) in North American wetlands. U.S. Dept. Interior, Fish and Wildlife Service, Washington, D.C.

Timmons, F. L. 1960. Control of aquatic weeds, pp. 357–386 in FAO Int. Conf. on Weed Cont., Davis, California.

US Congress, Office of Technology Assessment. 1993. Harmful non-indigenous species in the United States, OTA-F-565 U.S. Govt. Printing Office, Washington, D.C.

Wood, H. E. 1955. Herbicides used agriculturally in Western Canada for the control of Weeds. Mimeo, Manitoba Weeds Comm. Winnipeg, Manitoba, Nov. 10.

Young, J. A. 1991. Tumbleweed. *Scientific American,* March, pp. 82–87.

Chapter 3

Weed Classification

FUNDAMENTAL CONCEPTS

- The order in the world of weeds is recognized through systems of classification.
- Weeds can be classified in at least four ways. The most important and oldest system is based on phylogenetics or evolutionary ancestry.

OBJECTIVES

- To learn the fundamentals of weed classification based on phylogenetics or ancestral relationships.
- To learn why and how other weed classification systems are used and why they are important to weed management.
- To understand the unique habitat and role of parasitic weeds.
- To know the major groups of parasitic weeds.
- To understand the importance of a plant's scientific name.

One of the great, often unspoken, hypotheses of modern science is that there is order in the world. With careful study, scientists believe they can discover and describe the order. With each discovery and consequent description science will improve our understanding of how our world functions. Among those who study the order in the natural world are taxonomists, who describe and classify species. Although not everyone agrees on whether or not a particular plant is a weed or exactly what a weed is, most weeds have been classified, as members of the plant kingdom, by plant taxonomists.

There are at least 450 families of flowering plants and well over 350,000 different species. Only about 3,000 of them have been used by humans for food. Fewer than 300 species have been domesticated and of these, there are about 20 that stand between us and starvation. About 15 plants have provided most of our food for many generations. There are at least 100 species of great regional or local importance, but only a few major species dominate our food.

Twelve plant families include 68% of the 200 species that are the most important world weeds (Holm, 1978). These weeds share some characteristics including the following:

1. Long seed life in soil,
2. Quick emergence,
3. Ability to survive and prosper under the disturbed conditions of a cropped field,
4. Rapid early growth, and
5. No special environmental requirements for seed germination.

They are also competitive and react similarly to crop cultural practices. Weeds are usually defined only by where they are and how that makes someone feel about them. The fact that they may have shared characteristics means we may be able to define and classify them based on what their genotype enables them to do. Some characteristics that weeds share are discussed in Chapter 8.

Table 3.1 shows 12 families that include 68% of the world's important weed problems. The Poaceae and Cyperaceae account for 27% of the world's weed problems, and when the Asteraceae are added, 43% of the world's worst weeds are included. Nearly half of the world's worst weeds are in only three families, and any two of these include over a quarter of the world's worst weeds. The Poaceae is the family with the most weedy species, yet it includes wheat, rice, barley, millet, oats, rye, corn, sorghum, and sugar cane: all important crops.

About two-thirds of the world's worst weeds are single-season or annual weeds. The rest are perennials in the world's temperate areas, but in the tropics they are accurately called several-season weeds. The categories annual and perennial do not have the same meaning in tropical climates, where growth is not limited by cold weather but may be limited by low rainfall. About two-thirds of the important weeds are broadleaved or dicotyledonous species. Most of the rest are grasses, sedges, or ferns. The United States has about 70% of the world's important weeds, and they are classified in many ways.

Table 3.1

Families of the World's Worst Weeds (Holm, 1978)

Family	Number of species[a]	Percent		
Poaceae	44			
Cyperaceae	12	27%		
Asteraceae	32		43%	
Polygonaceae	8			
Amaranthaceae	7			
Brassicaceae	7			68%
Leguminosae	6			
Convolvulaceae	5			
Euphorbiaceae	5			
Chenopodiaceae	4			
Malvaceae	4			
Solanaceae	3			

[a]47 other families have three species or less.

I. PHYLOGENETIC RELATIONSHIPS

Weeds are classified by taxonomists and weed scientists in the same way all other plants are: based on phylogenetic (Gk. *phylo = phulon* = race or tribe plus Gk. *gen* = be born of, become) relationships, or a plant's ancestry. All good identification manuals include a key to the species, and all keys are based on a classification developed over many years and, for plants, brought near its present form by the Swedish botanist Carl von Linné (Linnaeus, 1707–1778) who established the binomial system of nomenclature (Genus + species) that is based, primarily, on floral characteristics, especially the presence, number, and characteristics of stamens and pistils.

Phylogenetic keys to plant species, based on ancestry and ancestral similarity, include division, subdivision, class, family, genus, and species. A brief description of a plant key for weed species follows:

Division I: Pteridophyta
   Description   Fernlike, mosslike, rushlike, or aquatic plants without true flowers. Reproduce by spores.

   Representative families
      Salviniaceae
      Equisetaceae
      Polypodiaceae

Division II: Spermatophyta
 Description   Plants with true flowers with stamens, pistils, or both.
 Reproduce by seed containing an embryo.

Subdivision I: Gymnospermae
 Description   Ovules not in a closed ovary. Trees and shrubs with
 needle-shaped, linear, or scalelike usually evergreen leaves.

 Representative families   Pinaceae, Taxaceae.
 Almost no weedy species.

Subdivision II: Angiospermae
 Description   Ovules borne in a closed ovary that matures into a
 fruit.

Class I: Monocotyledoneae
 Description   Stems without a central pith or annular layers but
 with woody fibers. Embryo with a single cotyledon and early
 leaves always alternate. Flower parts in threes or sixes, never
 fives. Leaves mostly parallel veined.

 Representative families
  Poaceae
  Cyperaceae
  Juncaceae
  Liliaceae
  Commelinaceae

Class II: Dicotyledoneae
 Description   Stems formed of bark, wood, and pith with the wood
 between the other two and increasing with annual growth.
 Leaves net-veined. Embryo with a pair of opposite cotyledons.
 Flower parts mostly in fours and fives.

 Representative families
  Polygonaceae
  Chenopodiaceae
  Convolvulaceae
  Asteraceae
  Solanaceae

All classified plants have a genus and specific name. By convention the genus is always capitalized (e.g., *Amaranthus*) and is commonly written in italics or underlined. The species name is not capitalized.

II. A NOTE ABOUT NAMES

The first question one asks or is asked about a weed is: What is it? The expected and best answer is its name. But what name? Most plants have several names. Each has its own, distinctive scientific name plus one to several common names. Common names vary between languages and between regions that share a language. For example, *Zea mays* is the plant Americans call corn, but the British, and most of the rest of the world's people, call the plant maize or (in Spanish) maíz. In England, wheat and other small grains are known as corn.

Reluctantly, and for the reader's convenience, common names have been used throughout this book. The scientific name for a plant (see Appendices A and B) is the name that is known throughout the world, or, at least, is the name that will clear up the confusion that often occurs when just the common name is used.

Students resist learning scientific names because they are regarded as useless, boring, and perhaps even nonsense words designed to confuse them. The arguments against learning them are manifold. The first defense is that the names are difficult because they are in Latin, which after all is a dead language. Outside of the Roman Catholic Church, few speak it at all, and knowing Latin certainly doesn't score many points with one's peers. Besides, the argument continues, common names are widely accepted and convey real meaning.

Latin is difficult, but difficulty should be dismissed as an objection not worthy of one engaged in higher education. Like most worthy goals, obtaining an education will not be achieved without some effort. Latin is dead, but therein lies its advantage as a medium to name things. A dead language doesn't evolve and assume new forms as daily usage modifies it and introduces variation. The rules are fixed, and while the language can be manipulated, it is not pliable as is a living language (Zimdahl, 1989).

As opposed to common names, scientific names have a universal meaning. Those who know scientific names will be able to verify a plant's identity by reference to standard texts or will immediately know the plant in question when the scientific name is used. Those who do not share the same native language can make use of Latin, an unchanging language, to share information about plants.

Scientific plant names have been derived from a vocabulary that is Latin in form and usually Latin or Greek in origin. Other peculiarities that make scientific nomenclature difficult are the frequent inclusion of personal names, Latinized location names, and words based on other languages. Taxonomists have developed and accepted rules for name creation that provide latitude for imagination and innovation, but not license for their neglect (Zimdahl, 1989).

III. CLASSIFICATION METHODS

Other common, and less systematic classification methods for weeds are based on life history, habitat, morphology, or plant type. Knowledge of classification is important because a plant's ancestry. length of life, the time of year during which it grows and reproduces, and its method or methods of reproduction provide clues about management methods most likely to succeed.

A. TYPE OF PLANT

The type of plant or general botanical group is an essential bit of knowledge, but not very useful as a total classification system. It is important that we know whether a weed is a fern or fern ally, sedge (Cyperaceae), grass (monocotyledon), or broadleaved (dicotyledon). One should not even begin to attempt control or try to understand weedy behavior until this information is available. However, when one knows the general classification, other questions about habitat or life cycle must be answered to acquire necessary understanding to control the weed or to create a weed management system.

B. HABITAT

1. Cropland

The first, and most important, weedy habitat is cropland, where many annual and perennial weeds occur. Although it is important to know the crop and whether it is agronomic or horticultural, that knowledge is not particularly useful. It tells us where the weed is, but it doesn't tell us much about it. It is not a precise way to classify because there is so much overlap among crops. There are few, if any, weeds that occur exclusively in agronomic or in horticultural crops or in just one crop. Redroot pigweed, velvetleaf, Canada thistle, and quackgrass are commonly associated with agricultural crops. Others such as crabgrass, common mallow, prostrate knotweed, dandelion, and wood sorrel commonly associate with horticultural crops. Each can occur in many different crops and environments.

2. Rangeland

Some weeds are almost exclusively identified with rangeland, a dry, untilled, and extensive environment. Sagebrush and gray rabbitbrush are rarely weeds in corn or front lawns. Only the worst farmer or horticulturalist would create an environment in which these weeds could thrive. Range weeds include those shown in Table 3.2. Although the list is not exhaustive, it shows that rangeland weeds are commonly perennial and include many members of the Asteraceae. There are poisonous weeds such as locoweed and larkspur on rangeland, and many others including thistles (of several species), dandelion, groundsel, buttercup, and vetch, but these also occur in other places.

3. Forests

There are over 580 million acres of forest in the United States, and in addition to common herbaceous annual and perennial weeds, there are others unique to the forest environment. The woody perennials such as alder, aspen, bigleaf maple, chokecherry, cottonwood, oaks and sumac, and the herbaceous perennial bracken fern (common in the acidic soils of Pacific Northwest Douglas fir forests) are unique forest weeds.

Red alder was eliminated by herbicides from Douglas fir forests in the 1970s. Red alder can fix atmospheric nitrogen, and in soils quite deficient in nitrogen, Douglas fir will grow better with than without red alder. In the 1990s, red alder wood increased in value, and some companies now

Table 3.2

Rangeland Weeds

Weed	Life cycle	Family
Big sagebrush	Perennial	Asteraceae
Sand sagebrush	Perennial	Asteraceae
Fringed sagebrush	Perennial	Asteraceae
Broom		
snakeweed	Perennial	Asteraceae
Gray rabbitbrush	Perennial	Asteraceae
Yucca	Perennial	Liliaceae
Greasewood	Perennial	Chenopodiaceae
Halogeton	Annual	Chenopodiaceae
Mesquite	Perennial	Leguminosae
Locoweed	Perennial or annual	Leguminosae
Larkspur	Perennial	Ranunculaceae

plant it. Some weeds do so well they become crops! Red alder has been the target of biological control with a fungus (Dorworth, 1995).

4. Aquatic

Agriculture is the largest user of fresh water in the world, and crops are sensitive to supply variation. Most of the world's major cities are located on a lake, an ocean coast, or a major river. Water, a finite resource, has been, and will continue to be, essential for urban and agricultural development. Aquatic weeds (Table 3.3) interfere with crop growth because they impede water flow or use water before it arrives in cropped fields. They can interfere with navigation, recreation, and power generation. Free-floating plants (e.g., waterhyacinth) attract attention because their often massive infestations are so obvious. They move with wind and floods and some have stopped river or lake navigation. They float free and never root in soil. Submersed plants (e.g., Hydrilla) complete their life cycle beneath the water. Emersed aquatic weeds (e.g., cattail) grow with their root system anchored in bottom mud and have leaves and stems that float on water or stand above it. They grow in shallow water, but all can impede flow, block boat movement, clog intakes of electric power plants and irrigation systems, and hasten eutrophication.

5. Environmental Weeds

This category includes plants particularly obnoxious to people, such as poison ivy and poison oak, both of which cause itching and swelling for

Table 3.3
Aquatic Weeds

Growth habit	Weed	Life cycle	Family
Free-floating	Waterhyacinth	Perennial	Ponterderiaceae
	Salvinia	Annual/perennial	Salviniaceae
	Waterlettuce	Perennial	Araceae
	Duckweed	Annual	Lemnaceae
Submersed	Hydrilla	Annual/perennial	Hydrocharitaceae
	Elodea, Western	Perennial	Hydrocharitaceae
	Pondweed	Perennial	Potamogetonaceae
	Eurasian watermilfoil	Perennial	Haloragaceae
	Coontail	Perennial	Ceratophyllaceae
Emersed	Cattail	Perennial	Typhaceae
	Alligatorweed	Perennial	Amaranthaceae
	Arrowhead	Perennial	Olismataceae

many people who touch them. Other plants in the environmental group are goldenrod, ragweed, and big sagebrush, primary causes of hay fever.

6. Parasitic Weeds

Parasitic weeds are often placed in other sections in weed science texts. They are here because theirs is a particular and peculiar habitat. Phanerogamic parasites, a word that comes from the Greek *phaneros* = visible and *gamos* = marriage, include more than 3,000 species distributed among 17 families, but only 8 families include important parasitic weeds. The economically important species that damage crop and forest plants are all dicotyledons from five families (Table 3.4) (Sauerborn, 1991). Parasitic weeds from four families will be discussed briefly. Those who want more detailed information are directed to the 1993 book by Parker and Riches.

The Cuscutaceae, dodders, are noxious in all U.S. states except Alaska and are distributed throughout the world's agricultural regions. A mature dodder plant, a true parasite, is a long, fine, yellow branching stem. A single stem of field dodder, one of the most important species, can grow up to 10 cm in one day. It is non-specific regarding hosts and coils and twines on many plants. Dodder is a flowering plant that reproduces by small, sticky seeds. Haustoria penetrate a host's cortex to the cambium and the fine stems dodder (tremble) when the wind blows. Dodder emerges from as deep as 4 feet in soil as a rootless, leafless seedling. The fine, yellow stem, 1 to 3 inches long, emerges as an arch, straightens and slowly rotates in a counterclockwise direction (called circumnutation) until it contacts another plant, which must be within about 1 1/4 inches. Seeds have sufficient resources to search for a host for 4 to 9 days, after which they die (Sauerborn,

Table 3.4
Important Families of Parasitic Weeds

Family	Genera	Species
Cuscutaceae	*Cuscuta*	Dodder
Loranthaceae/Viscaceae	*Loranthus*	Mistletoe
	Arceuthobium	Mistletoe
	Viscum	
Orobanchaceae	*Orobanche*	Broomrape
	Aeginetia	
Schrophulariaceae	*Striga*	Witchweed
	Alectra	

1991). After contact and attachment, the soil connection withers and dodder lives as an obligate stem parasite.

The most important parasite in the Loranthaceae is mistletoe. Mistletoes occur in two families: the Loranthaceae and the Viscaceae. Some taxonomies combine both families in the Loranthaceae. Dwarf mistletoe is a photosynthetic, flowering plant that parasitizes ponderosa pine in the southwestern United States. It occurs on the trunk and branches as a dense tangle of short, brown to yellow-brown stems. Seeds are dispersed by birds or by explosion of seed pods and expulsion of sticky seeds that adhere to adjacent trees. Seeds can travel up to 60 mph over 45 feet. The seeds are usually dispersed in August or early September in the southwestern United States.

Witchweed (*Striga asiatica*) is one of three weedy, hemiparasitic species of the Scrophulariaceae in the world. It is called witchweed because it damages crop plants before it is even visible above ground. There are 35 species of Striga; 23 are found in Africa, and at least 11 parasitize crops (Parker and Riches, 1993). Other most important Striga species are *S. hermonthica,* which parasitizes sorghum, millet, and corn in Africa, and *S. gesnerioides,* the only one that parasitizes dicots, which is important on cowpeas and groundnut in East and West Africa and Asia.

The desert locust (*Schistocerca gregaria*) gains a great deal of publicity when it swarms in Africa, and massive efforts are made to combat it. However over the years and in any one year, *Striga* species causes more crop losses in Africa than the desert locust. The genus has the narrowest host range of the important parasitic weeds and a narrower range of distribution than the dodders. Witchweed is a root parasite on corn, sorghum, and other grasses in Africa, India, and the far East. In the United States it is limited to parts of North and South Carolina. *Striga* species are widely distributed in the world's tropical and subtropical regions. Secretions from corn (and some other grass) roots encourage germination of seed. After parasitization, corn is stunted, yellow, and wilted because of loss of nutrients and water. Many weeds, including crabgrass, serve as alternate hosts. Witchweed seeds are small (about 0.2 by 0.3 mm); 1000 to 1500 seeds, placed end to end, would be only 1 foot long. They survive up to 14 years in soil, and one *S. asiatica* plant can produce up to 58,000 seeds. It parasitizes corn and its 90- to 120-day life cycle is similar to corn's. One corn plant can (but usually does not) support up to 500 witchweed plants. Witchweed seed will not germinate in soil, in the absence of the host-excreted stimulant. But it may be induced to germinate with the artificial stimulant ethylene gas. USDA regulations currently have witchweed under quarantine in North and South Carolina, to prevent its spread throughout the United States. The quarantine has been successful and the area is decreasing.

The Orobanchaceae (from Latin *orobos* = bitter vetch and Latin *anchein* = to strangle) or broomrapes include over 100 species, five of which are important, obligate root holoparasites (lacking all chlorophyll) that attack carrots, broadbeans, tomatoes, sunflowers, red clover, and several other important, small-acreage crops in more than 58 countries (Parker and Riches, 1993; Sauerborn, 1991). The broomrapes have the broadest host range of the parasitic families. They cause major yield losses and often complete crop loss in many developing countries where control is not possible. Broomrape is found in California but is not important in most of the United States. It is important in South and East Europe, West Asia, and North Africa. Seed of some species lives in soil for up to 10 years. One plant can produce up to 200,000 seeds that are as small as Striga seed; 1 gram of seed contains as many as 150,000. Similar to witchweed, germination of Orobanche seed is stimulated by secretions from the host's root or from roots of nonhost plants. Germination will not occur in the absence of host-excreted chemical stimulants. Most damage from root parasites occurs before the parasite emerges, and only 10 to 30% of attached parasites emerge (Sauerborn, 1991).

An important aspect of parasitic weeds is the present inability to manage them with other than sophisticated chemical technology or extended fallow periods. Many of the world's people live in areas where food is scarce and agricultural technology is not sophisticated. These are the same places where parasitic weeds cause the greatest yield losses. Fields have been taken out of production, and production area of some crops has been reduced severely.

C. Life History

Another way to classify weeds is based on their life history. A plant's life history determines what kinds of cropping situations it might be a problem in and what management methods are likely to succeed. All temperate weeds can be categorized as annual, biennial, or perennial. These groups are easy to define and observe and are very useful in temperate zone agriculture. The concept of perennation really is not useful in tropical agriculture, where seasons do not change as they do in temperate zones. It is more accurate to refer to short-season and many-season plants in the tropics than to annuals and perennials.

1. Annuals

An annual is a plant that completes its life cycle from seed to seed in less than 1 year or in one growing season. They produce an abundance of seed, grow quickly, and are usually, but not always, easy to control. Summer annuals germinate in spring, grow in summer, flower, and die in fall, and thus go from seed to seed in one growing season. Many common weeds such as common cocklebur, redroot pigweed and other pigweeds, crabgrass, wild buckwheat, and foxtails are annuals. The typical life cycle of an annual weed is shown in Figure 3.1. Weed ecologists are working to quantify many of the steps in this cycle. The sequence of events is qualitatively accurate, but neither rates nor quantities are defined for most annual weeds. For example, it is known that not all seeds produced by a weed survive in soil. Some die from natural causes at an unknown rate. Others suffer predation by soil organisms or enter the soil seedbank where their life may be prolonged by dormancy. Quantitative understanding of the steps in a weed's life is essential to wise management.

Winter annual weeds germinate in fall or early winter and flower and mature seed in spring or early summer the following year. Downy brome, shepherd's-purse, pinnate tansymustard, and flixweed are winter annual weeds. They are particularly troublesome in winter wheat, a fall-seeded crop, and in alfalfa, a perennial.

Some parts of the world (southern European and north-African Mediterranean countries) have a winter rainy season, rarely with snow or subfreezing temperatures. This is followed by a long dry period. Crops are planted in the fall when, or just before, the rains begin. The crops and their weeds

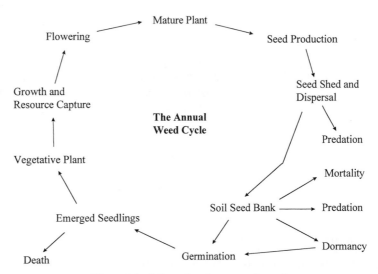

Figure 3.1 Life cycle of an annual weed.

begin to grow with the rain. Because the rains do not begin until late fall, the annual weeds live into the next calendar year and their life cycle fits part of the definition of a winter annual. They are however, best regarded and managed as annuals because their growth is continuous and not interrupted by a cold period when plants live but do not grow.

2. Biennials

Biennials live more than 1, but not over 2 years. They should not be confused with winter annuals, which live during 2 calendar years but not for more than 1 year. Musk thistle, bull thistle, and common mullein are biennials. It is important to know that one is dealing with a biennial rather than a perennial. Spread of a biennial can be prevented by preventing seed production, but this is not true for creeping perennials.

3. Perennials

Perennials are usually divided into two groups—simple and creeping. Simple perennials spread only by seed. If the shoot is injured or cut off, it may regenerate a new plant vegetatively, but the normal mode of reproduction is seed. Simple perennials include dandelion, buckhorn and broadleaf plantain, and curly dock. Creeping perennials reproduce by seed and vegetatively. Vegetative reproductive organs include creeping above-ground stems (stolons), creeping below-ground stems (rhizomes), tubers, aerial bulblets, and bulbs. The life cycle of a typical perennial plant is shown in Figure 3.2. Some of the more important creeping perennials and their mode of vegetative reproduction are shown below:

Canada thistle	Creeping root stocks that give rise to aerial shoots
Johnsongrass and quackgrass	Rhizomes
Field bindweed	Rhizomes
Yellow nutsedge	Rhizomes and tubers
Leafy spurge	Rhizomes
Bermudagrass	Rhizomes and stolons
Russian knapweed	Adventitious shoots from creeping roots

THINGS TO THINK ABOUT

1. How are weed classification systems used?
2. What classification system is most likely to be used by horticulturalists, agronomists, and weed scientists?

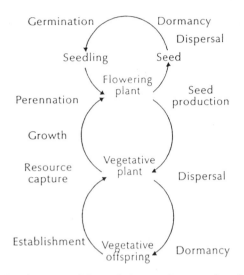

Figure 3.2 Life cycle of a perennial weed that produces seed and vegetative progeny (Grime, 1979).

3. Why are parasitic weeds such difficult problems and where do they exist?
4. If parasitic weeds are not important problems in most developed countries, why do we bother to study them?
5. Why bother to learn the scientific names of plants?
6. How are the scientific names of plants created?

LITERATURE CITED

Holm, L. 1978. Some characteristics of weed problems in two worlds. *Proc. West. Soc. of Weed Sci.* **31**:3–12.

Dorworth, C. C. 1995. Biological control of red alder (*Alnus rubra*) with the fungus *Nectria ditissima*. *Weed Technol.* **9**:243–248.

Grime, J. P. 1979. Plant Strategies and Vegetation Processes. J. Wiley and Sons, New York, p. 2.

Parker, C. and C. R. Riches. 1993. *Parasitic Weeds of the World: Biology and Control.* CAB International, Wallingford, U.K. 332 p.

Sauerborn, J. 1991. Parasitic flowering plants in agricultural ecosystems of West Asia. *Flora et Vegetatio Mundi* **9**:83–91.

Zimdahl, R. L. 1989. *Weeds and Words: The Etymology of the Scientific Names of Weeds and Crops.* Iowa State Univ. Press, Ames, Iowa. 125 pp.

Chapter 4

Ethnobotany

FUNDAMENTAL CONCEPTS

- Many weeds are also useful for food, feed, or medicine.
- Some weeds may be useful for fuel or insulation.
- The same plant can be a weed in one place and a beneficial crop in another. One plant could be a crop and a weed.

OBJECTIVES

- To understand the many ways weeds can be used.
- To encourage thought about the importance of considering uses for weeds.

Ethnobotany includes study of the uses of plants by man and the relationship between man and vegetation. It examines our dependence on plants and our effects on them. If weeds are plants out of place, and regarded as useless by humans, is it possible they could also be useful? Can a single species be weedy and useful? James Russell Lowell provided a clue:

> One longs for a weed here and there, for variety;
> A weed is not more than a flower in disguise,
> Which is seen through at once, if love give a man eyes.
>
> —A Fable for Critics (Lowell, 1848)

I. FOOD FOR HUMANS

There is interest in the possible uses of weeds. It is a sobering thought that we have rediscovered only one new food crop in this century—the soybean, first domesticated in China over 1,000 years ago. What would be the benefit of another food crop like the soybean? Would it be worth the price of a mission to the Moon or of a Trident submarine? I suggest it would be, and that it might be profitable to study the potential value of weeds. There are sources of information to assist the search that range from an article in the *Reader's Digest* (1974) to books on edible native plants (Harrington, 1967), edible weeds (Duke, 1992) potential dietary uses of wild and cultivated plants (Hylton, 1974), and many articles in the scientific and popular literature.

The tradition of using indigenous plants for human food is regarded by some as vegetarianism or food faddishness, but if the tradition is viewed in its historical context, a long history of potentially useful food sources may be discovered. For example, of 158 weed species collected from rice fields in two districts of West Bengal, India, 124 had economic importance to farmers (Vega, 1982). Young pigweeds may be eaten as salad greens, and pigweed seeds can be eaten raw or parched. Several species of amaranth grow rapidly and contain abundant, high-quality protein (Hauptil and Jain, 1977). Leaves of some species contain up to 33% protein and their seeds have 16 to 19% protein (Hauptil and Jain, 1977). Young leaves of shepherd's-purse are eaten as greens, and dried roots can be eaten as a substitute for ginger or candied by boiling in a sugared syrup.

Instead of agonizing over dandelions in turf, why not learn to love and use them? Vineland, New Jersey, proclaims itself the world's dandelion capital. A recipe book for things ranging from dandelion Jell-O to dandelion soup is available from the Mayor's office (Anonymous, 1979). Dandelions are harvested and sold for conversion to dandelion corn chowder, wine, or Italian dandelion casserole. Some say dandelion flowers are quite good when dipped in batter and deep-fat fried. Dandelion roots make a caffeine-free coffee substitute. The leaves are rich in vitamins A and C. More than 100,000 pounds of dandelion are imported to the United States annually for use in patent medicines (Duke, 1992). The root contains a diuretic, and an old European common name is piss-a-bed.

Common purslane contains high levels of fatty acids, vitamin E at six times the level of spinach, and other nutrients. Europeans eat it in salads and it could be developed as a new vegetable crop (Anonymous, 1992). Omega-3 fatty acid has been linked in some studies to reduced heart disease, and purslane contains more than any other green, leafy vegetable (Anonymous, 1992). It is well adapted and could be an alternative crop in arid areas.

Barnyardgrass seeds may be eaten dry or parched, and they have been ground into flour. Some thistles' seedlings may be eaten raw in salads if the spines are removed. Young Canada thistle roots can be peeled and the pith-like interior eaten raw or as a condiment in some cooked dishes. There are recipes for thistle-leaf tea. Seeds of wild oat can be ground into flour, and one can make an artificial fly for fishing from wild oat seeds. Seeds of crabgrass, green foxtail, wild oat, and the common reed have been eaten whole.

Iroquois Indians ate burdock leaves as greens and used its dried roots in soup (Duke, 1992). Martin (1983) reported, and Duke (1992) agreed, that eating raw burdock stems has been reported to stir lust and improve sexual virility. Duke (1992) reported its sale in Japan as an herb for sexual problems.

Wild mustard leaves have a hot, spicy flavor that blends well in salads with lettuce and dandelion. Wild onion has been used as a relish, to flavor cooked foods, and to cover the taste of gamey meat (Ross, 1976).

Martin (1983) reported that the Japanese like kudzu root. It is ground to a fine powder and used as a condiment. The leaves are also eaten. Kudzu was promoted extensively by the U.S. Dept. of Agriculture in the 1930s to stabilize eroding land. The Chinese have long relied on simple kudzu root extract to stop human craving for alcohol. It is sold as an over-the-counter drug in China and is 80% effective when taken for 2 to 4 weeks. The extract is being evaluated in the United States and may gain approval for treatment of alcoholism. Weed scientists will not welcome fields of kudzu, but other priorities may prevail.

Before hops were used, leaves of ground ivy, also called gill-over-the-ground (gill from the French *guiller* = to brew), were added to the brew to clarify and add flavor (Martin, 1983).

Duckweed is one of the world's tiniest flowering plants, but it has potential in the fight against world hunger. Rich in protein, containing high levels of all essential amino acids save one, duckweed is nutritious and abundant. It is found in temperate and tropical regions growing in thick green masses on surfaces of ponds and lakes. People in Thailand have eaten duckweed for generations. It can be harvested every 3 to 4 days and eaten in soups or stir-fried with other vegetables and meat. Duckweed could become a valuable livestock feed as well. Ten acres of duckweed could supply 60% of the nutritional needs of 100 dairy cows for one year. Considering that more than 100 million people each year suffer from severe protein/calorie malnutrition, the food potential of duckweed should be studied carefully.

N. W. Pirie of Rothamsted Experiment Station in England conducted experiments on juice pressed from a random collection of jungle plants.

He was able to extract a juice with 50 to 75% protein that, when coagulated, made a tasteless product that could be textured to resemble cheese.

Note: Readers are cautioned that these examples are intended to be illustrative of the range of potential uses for plants and are not a recipe book or set of recommendations. Because of the danger of poisoning or digestive upset, specific references should be consulted before casual experiments lead to problems.

Increased agricultural production has relied on low cost energy and rapid genetic improvement for several decades (Boyer, 1982). These have allowed farmers to use dense plant populations adapted to high production on soil amended with purchased resources. Weeds grow well in cropped fields and in some environments with limited resources. Weeds have been self-selected to do well with limited and abundant resources, and their genetic abilities may be important resources for plant breeders and crop producers.

A natural stand of giant ragweed in Champaign county, Illinois, had an above-ground biomass similar to that of corn and greater than that of soybeans. Its seed biomass was lower than those of corn or soybeans, but equal to the average soybean grain yield in the United States in 1975 (1610 Kg/ha). Giant ragweed is not a food crop and will not become one, but its high productivity with low inputs provides a valuable lesson for the future of food crops (Boyer, 1982) as energy and water resources decline or are directed away from agriculture.

II. FEED FOR ANIMALS

Weed seed screenings are used in many U.S. states as animal feed; a practice with some disadvantages (see Chapter 5).

Some rangeland plants are weeds, but studies have shown that cattle grazing on native blue grama and buffalo grass range achieve better gains when weeds and shrubs constitute 10 to 70% of total vegetation. One should not neglect the contribution of sagebrush and other weedy range species to the diet of browse animals such as deer, elk, and antelope.

Cattle ranchers in the western United States often use kochia hay as feed. When immature its protein content can be 17%, equal to that of alfalfa. However, it becomes woody as it matures, and because it is an annual, it will not reseed when harvested immature for hay. An important warning is that it can accumulate high amounts of nitrates, and cattle may be intoxicated and lose weight when kochia is 90% or more of the diet.

The forage and nutritional value of many weed species is equal to that of cultivated forage crops. Marten and Anderson (1975) and Temme *et al.* (1979) reported that the annual broadleaved weeds redroot pigweed, common lambsquarters, and common ragweed had digestible dry matter, fiber, and crude protein concentrations about equal to those of good alfalfa hay when alfalfa and weeds were harvested at the same growth stage. Giant foxtail, Pennsylvania smartweed, shepherd's-purse, and yellow foxtail all had lower nutritional value than alfalfa. Dutt *et al.* (1982) concluded that yellow rocket reduced the feeding value of alfalfa hay, but white campion and dandelion did not. Yellow rocket reduced nutritive value index, animal intake, and digestibility.

The perennial, quackgrass is a serious weed problem in the perennial crop alfalfa. It invades alfalfa and decreases hay consumption by cattle. Although its nutritional value is high, its palatability is lower than that of alfalfa or smooth bromegrass (Marten *et al.,* 1987). A biotype, selected for broad leaves, was equal or superior to smooth bromegrass and equal to alfalfa in palatability in Minnesota (Marten *et al.,* 1987). Marten *et al.* (1987) investigated the forage value of nine perennial broadleaved and grass weeds compared to alfalfa and smooth bromegrass. Smooth bromegrass and quackgrass consistently had more neutral detergent fiber, less crude protein, and an in-vitro digestibility similar to that of alfalfa. Jerusalem artichoke, Canada thistle, dandelion, and perennial sowthistle had crude protein and in-vitro digestibility equal to or greater than those of alfalfa (Marten *et al.,* 1987). Broadleaved species generally had lower palatability than alfalfa or smooth bromegrass. Jerusalem artichoke, Canada thistle, curly dock, and hoary alyssum were completely rejected by grazing lambs and are therefore always weedy species in sheep pastures.

Weed forage and hay quality are correlated with plant maturity. More mature plants have lower forage quality. Marten *et al.* (1987) showed this to be true for nine perennial species in Minnesota, where forage quality, measured as digestible dry matter, fiber, or crude protein, declined with maturity. Crude protein of curly dock declined 22% from the vegetative to the mature seed stage (Bosworth *et al.,* 1985). In hay crops, the decision to control weeds must be site specific and is dependent on the weeds present and their growth stage when hay is to be cut. Alfalfa stands often become weedy because of death of alfalfa plants, not because weeds crowd them out (Sheaffer and Wyse, 1982). Control may reduce hay yields and produce hay of lower quality if all weeds are controlled just because they are perceived to be weeds and therefore undesirable. As illustrated earlier, some weeds make good pasture and forage.

III. MEDICAL USES

Plants and plant extracts have been used to treat almost every ailment known to humans, ranging from venereal disease and rheumatism to colds and bleeding. Plants with the word *officinale* (or its derivatives) were at one time included on an official drug or medicinal list. Wort (e.g., St. John's wort), a common suffix in common plant names, means healing. A plant with bane added to its common name (e.g., henbane) was probably used for medicinal purposes. Readers can probably think of plants that fit in one or more of these categories.

Roots of yucca can be chopped and soaked in water to extract a soapy substance which western U.S. Indians used for washing and cleaning. One must assume this is a source of the common name soapweed. The next time you are stung by a bee, hope you are standing near a curly dock plant. Quickly rub some leaves between your hands and press them, with their juice, against the sting. Within 10 to 20 minutes the stinging sensation will be gone. Curly dock has more vitamin C than oranges, and extracts of its yellow root have been used to treat jaundice. Common plantain has similar properties. Curly dock leaves have also been boiled in vinegar to soften the fibers and then combined with lard to make an ointment for treatment of inflammations.

A persistent human problem is the common head cold. If one boils a few ounces of sunflower seed in a quart of water, adds honey and gin, and takes the mixture 3 or 4 times daily, irritating mucus will be discharged from the nose and mouth. One must, however, question whether the mixture's efficacy is due to sunflower seed extract or gin (also a plant product).

Yarrow and big sagebrush have been used as a tea to relieve the fever that accompanies a cold. Yarrow leaves were chewed by Western pioneers to settle an upset stomach. Extracts were also used to regulate menstrual flow and to stop blood flow from a wound. Modern medicine has confirmed its efficacy (Martin, 1983).

Salicylic acid was discovered in meadowsweet (spiraea) and later combined with acetic acid to create acetylsalicylic acid, the effective painkiller and still one of the world's most widely used analgesics, aspirin. Aspirin was first synthesized by Felix Hoffmann in Germany in 1897, patented in February 1899, and marketed the same year, as aspirin, by Bayer Chemical Co.

Dandelion has been used as a laxative, and Shepherd's purse and St. John's wort have been used for control of diarrhea. Common burdock has been used to make a tonic and diuretic. Young shoots and the pith of young leaf stalks can be eaten raw with salt or after boiling in salt water. Water-lettuce has been used to cure coughs and heal tubercular wounds. The Chinese use its leaves as external medicine for boils; American Indians used the leaves to cure hemorrhoids (Harrington, 1967). Extracts or preparations of several weeds have been used as sedatives, including poison hemlock, jimsonweed, poppies, and marijuana. The last has achieved rather widespread use and popularity for its sedative and relaxant properties; it is well known to most, and experienced by many, people.

Healers used to grate the dried root of wild carrot and apply the grated material to soothe burned skin. Modern science has shown the roots contain carotin, which, when mixed with oil, heals burns (Martin, 1983). Modern science has also proven the utility of extracts of bouncing bet to treat jaundice and liver problems (Martin, 1983). Backpackers and campers should know bouncing bet. When torn or bruised leaves are added to cold water, a bubbly lather ensues. This source of soap has been known since the middle ages (ended about 1450 A.D.). Its unusual name comes from the white, reflexed petals that someone thought resembled the posterior view of a washerwoman (named Bet?) with her petticoats pinned up.

Drury (1992, based on Gerard, 1597) reported that sprigs of common tansy were placed in beds and bedding to discourage vermin. Tansy tea tastes terrible, but has been used to treat a variety of illnesses and to cure rheumatism and intestinal worms (Martin, 1983).

Some of the trilliums (usually not considered weeds) have been used as aids in childbirth. Scientists in the Philippines have studied antifertility and abortive characteristics of the common weed called sensitive plant.

IV. AGRICULTURAL USE

Kochia, as mentioned earlier, is a good source of protein for ruminant animals. It is a self-seeding, high-yielding, water-efficient plant with no serious disease or insect pests. It is a serious annual weed in many crops and common in many parts of the United States. Kochia can accumulate high levels of nitrate and will escape from cultivation and become a weed. It may cause photosensitization in cattle. In some experiments, cattle have lost weight and some have died when fed only kochia.

At least one farmer has used kochia as a cover crop to suppress wild proso millet (Cramer, 1992). A thick stand of unirrigated kochia grew through the summer and was mowed before seed set. It was hard to plow because of all the biomass, but millet was suppressed effectively the next year.

Farmers in southeastern Mexico classify plants as crops or non-crops (Chacón and Gliessman, 1982). The latter are classed according to potential use and their effects on soil or crops. Chacón and Gliessman (1982) argue that local farmers understand the contribution of non-crop plants to agriculture. The authors contrast the farmer's view with the dominant view in developed countries that a weed is any plant other than the crop (see Chapter 2).

Weeds have practical, but often unappreciated, value when used as ground cover for wildlife or for prevention of soil erosion on sites that cannot be cropped or otherwise managed by humans. In some situations, weeds can conserve nitrogen. They have been introduced in many places because someone thought they would be useful (see Chapter 6 for other examples). Cogongrass was introduced into the United States in Grand Bay, Alabama, and McNeil, Mississippi (Tabor, 1952). At Grand Bay in 1912, bare-root satsuma orange plants were boxed for shipping with cogongrass and then the grass was discarded. The McNeil introduction was part of a search for a superior forage. Cogongrass is now a weed in many southern U.S. states. Catclaw mimosa was introduced to Thailand from Indonesia in 1980 as a green manure cover crop in tobacco plantations and for control of ditchbank erosion (Thamsara, 1985). It was successful for both purposes, but spread to become a weed problem. The aggressive, weedy annual paragrass was introduced into Africa from several tropical countries for fodder, pasture, and as a cover crop in banana plantations. There are many examples of plant introductions that someone thought would be useful, only to find they became a serious weed.

V. ORNAMENTALS

Many species of weeds have been used as ornamentals, and several species that are now weedy were first imported into the United States for ornamental purposes. One U.S. study (Williams, 1980) documented 33 imported species that became weedy. Of these, 2 were imported as herbs, 12 as hay or forage crops, and 16 as ornamentals. Henbane was imported for its potential medicinal value. One was imported for use in aquaria (hydrilla), one as a fiber crop (hemp or marijuana), and one privately, just for observation (wild melon). Imported plants including bermudagrass, jimsonweed, kochia, musk thistle, johnsongrass, and waterhyacinth have become some of our most detrimental weeds. Some people use weeds as ornamentals in spite of, or in ignorance of, their weedy nature.

Several forbs, grasses, shrubs, and trees have been and still are used in gardening and landscaping, reclamation, or restoration. Some are widely acknowledged as weeds and others may become weedy because of their ability to invade and dominate. All have the ability to escape their intended habitat. A few are shown in Table 4.1. Not everyone may agree that the plants in this table are weeds or could become weedy. At present there is no civil or criminal penalty for planting any of them. The choice of what to plant is the landowner's. When an escape occurs, everyone pays part of the price if the species becomes weedy.

Table 4.1

Plants to Avoid in Gardening, Reclamation, and Restoration

Type	Name	Problem
Forb	Purple loosestrife	Displaces native wetland or marsh plants.
Forb	Mediterranean sage	Forms monoculture and outcompetes native plants.
Forb	Common toadflax	Displaces native vegetation.
Grass	Timothy	Competes with native plants in arid areas.
Shrub	Buckthorn	Competes with native vegetation in riparian areas.
Shrub	Scotch broom	Problem on West Coast of the United States. Displaces native vegetation.
Tree	Tamarisk or salt cedar	Uses large amounts of water and displaces native vegetation.
Tree	Russian olive	Seed dispersed by birds. It displaces native plants.

VI. INSECT OR DISEASE TRAPS

A disadvantage of weeds is that they can shelter insects and disease organisms. They can also be used intentionally in agriculture as traps for insect or disease pests (Table 4.2). They do this in one of three ways:

1. As hosts for adult insect parasites,
2. As hosts for non-economic insects that serve as alternate hosts or food for parasites or predators

Table 4.2

Weeds and Control of Other Pests

Cropping system	Weed species	Pest regulated	Reason
Beans	Goosegrass Red sprangletop	Leafhoppers (*Empoasca kraemeri*)	Chemical repellency and masking
Vegetable crops	Wild carrot	Japanese beetle (*Popilla japonica*)	Increased activity of the parasitic wasp *Tiphia popilliavora*
Corn	Giant ragweed	European corn borer (*Ostrinia nubilalis*)	Provision of alternate host for the tachinid parasite *Lydella grisesens*
Cotton	Common ragweed	Boll weevil (*Anthonomus grandis*)	Provision of alternate hosts for the parasite *Eurytoma Tylodermatus*

3. By increasing effectiveness of biological control organisms and thereby reducing damage to crops.

Johnsongrass is an alternate host of the sorghum midge (*Contarinia sorghicola* Coquillet) an important pest of grain sorghum. Larvae develop and feed in the sorghum spikelet and prevent normal seed development. Johnsongrass maintains the first two or three generations of the insect until grain sorghum flowers are available. Time and duration of johnsongrass flowering (which can be determined by management) may affect the sorghum midge population (Holshouser and Chandler, 1996)

Showy crotalaria, a legume weed in Hawaii, is used in macadamia nut orchards to attract Southern green stink bugs (*Nezara viridula* L.) away from macadamia trees. Showy crotalaria was introduced to Florida in 1921 as a green manure crop because, as a legume, it fixes nitrogen. However, the foliage and seed are toxic, especially to poultry. It is a weed in soybeans in the southern United States where, because of the toxicity of its seeds, contaminated soybean seed cannot be sold.

In parts of California, wild blackberries are grown with grapes as hosts of a non-economic leafhopper that hosts a parasite of the grape leafhopper (*Erythroneura elegantula* Osborn). Japanese farmers graft tomato scion (shoot or bud tissue) onto rootstock of some weedy members of the *Solanaceae* to avoid root diseases. Other examples of this kind of use can be found in Altieri (1985) and Zandstra and Motooka (1978).

Chapter 2 described how weeds serve as hosts for damaging insects and diseases. It is important to realize that not all insects or microorganisms damage other plants. If one plant harbors harmful organisms, it is only logical to assume that other plants may harbor beneficial organisms. The preceding examples verify this, and Altieri (1985) provides other examples.

The agricultural quest for high-yielding monocultures has reduced plant diversity to the point where beneficial insects have been reduced in crop fields. One way to regain a desirable diversity in crop fields is to manipulate the abundance and composition of the weed flora. Weed borders, occasional weedy strips, or weeds at certain times in the crop growth cycle are all possibilities. Weed scientists and farmers may even want to consider planting weeds in attempts to optimize plant protection and crop yield while striving to minimize other inputs.

VII. POLLUTION CONTROL

In addition to the foregoing uses, which most weed scientists would readily acknowledge, weeds have other, less well known, perhaps esoteric, but interesting and potentially valuable uses that a few creative minds have explored.

Star chickweed has been used as a vegetable and is a good source of vitamins A and C. In Elizabethan England it was used to reduce fever (Martin, 1983). Martin (1983) reported it has been used to predict rain. If it blooms fully, there will be no rain for at least 4 hours. If blossoms shut, rain is on the way. Perhaps a look skyward would yield the same prediction.

The waterhyacinth will remove the heavy metals selenium, manganese, and chromium from water and may be useful for detecting them. It concentrates heavy metals up to 2,000 times the level found in water. Waterhyacinth can be used in what is called bioremediation, to remove nutrients from water and reduce eutrophication (Rogers and Davis, 1972). One hectare of waterhyacinth, growing under optimum conditions, could absorb the average daily nitrogen and phosphorus waste of more than 800 people if maximum uptake and plant growth for a whole year were assumed. The hectare would contain 1.6 million plants, and its capacity would be reduced to 300 to 400 people if less than year-round growth was achieved. Under optimum growth conditions, one hectare of waterhyacinth produces 8 to 16 tons of plant material per day that can be dried, ground, and added to corn silage for cattle feed. The supplementary feed value is comparable to that of cotton seed meal or soybean oil meal. Anaerobic fermentation of the plant residue produces methane gas that can be used for heating or light. One pound of dried plants yields up to 6 cubic feet of methane or up to 2 million cubic feet of gas per acre of plants per year. In India and Indonesia researchers have made paper products (blotting paper, cardboard) from waterhyacinth mixed with rice straw. There are problems because waterhyacinth does best in warm water and warm climates, and cold weather kills it (fortunately). There are questions about the cost of establishing and maintaining a processing facility. An obvious problem, if waterhyacinth is to be used for bioremediation, is disposal of waterhyacinth plants to prevent eutrophication of ponds, when no use for the plant residue has been developed.

Because waterhyacinth can be used for bioremediation when there is heavy metal pollution, others have looked at it as a way to harvest valuable heavy metals. Limited research indicates that an acre of waterhyacinth could yield 0.45 kg of silver every 4 days, and work on gold harvest has been done (Anonymous, 1976).

The bulrush has been identified as a way to remove pollutants from water (Zandstra and Motooka, 1978). Sudanese tribesmen have used it cheaply and effectively. Muddy water from the Nile river is stored in jars containing bulrush, and soon one has clean, pure water. A German company has designed a municipal water treatment facility using bulrushes to take up pollutants such as phenols, cyanide, phosphates, and nitrates. Commercialization may not be possible, but we should be cognizant of potential uses for plants we so easily call weeds. The Germans have also experimented with Sakhalin knotgrass, which takes up cadmium and lead without self-injury. They hope it will be useful to reclaim soil treated with metal-contaminated sewage sludge so crops can be grown.

During the 1970s there was great interest in developing systems to use plants to process sewage. Jewell (1994) reported on a hydroponic or nutrient film technique originally developed in England. The technique does not require deep water or a growth-supporting medium. Most terrestrial plants can be grown in a nutrient film system. Cattails, a common weedy species, have been a good choice for the initial stages of sewage treatment in a nutrient-film system (Jewel, 1994).

VIII. OTHER

The common water reed, ground into powder, can be used as a home heating fuel according to Swedish scientists (Bjork and Graneli, 1978). One kilogram of dry reeds will yield 5 kilowatts of energy. About 10 times more energy can be obtained from the powder than is required to cultivate, harvest, grind, and transport it. Cultivation of the weed could greatly increase production per unit area and may have the added advantage of preserving and using some portions of wetlands now threatened by development. Preservation of such lands has positive environmental benefits in terms of habitat for marsh animals and waterfowl.

Scientists at the University of Arizona have compressed Russian thistle to make fireplace logs (Tumble Logs™) with an energy value equal to that of lignite. Scientists are also investigating the biomass potential of mesquite, saltbush, and johnsongrass for energy production.

During World War II, Allied forces lost the World's Far Eastern sources of natural rubber. The war could not be fought without rubber for tires, and there was great activity in the early 1940s to develop alternative sources or substitutes for natural rubber. Gray rabbitbrush and guayule (wy-oolee) were among the plants studied. Gray rabbitbrush contains a high grade rubber called chrysil that vulcanizes well (Ross, 1976). One-fourth of guayule's entire weight is natural rubber. Guayule can be grown on land not suited for many other crops and can be mechanically harvested. There is interest in guayule and other plants as sources of hydrocarbons for replacement of expensive and increasingly scarce petroleum oil. There are several latex-bearing plants from the Euphorbiaceae (spurge) and Asclepiadaceae (milkweed) families. Many are common weeds, or perhaps just plants that are not even sufficiently noticed or bothersome to be raised to the defined category of weed.

For a brief period, the U.S. military used "down" from mature cattail heads (actually part of the female flower) to fill life jackets. It has also been used to insulate clothing and as stuffing for quilts and pillows. Western Indians ate young shoots, roots, stem bases, and seeds. The same "down," the pappus from female flowers, was used to make dressings for burns, for padding, and in talcum powder.

If none of these ideas interest you and you like plants and are a speculator, buy a few acres of land and plant milkweed, a hardy perennial that competes well with most plants and, once established, should thrive with care and pest control. Its produce, in the form of the seed pod, can be harvested, carded to remove seeds, and dried, and the pappus or floss can be used as an acceptable substitute for expensive goosedown in jackets, sleeping bags, and other things designed to trap air and keep us warm (Lione, 1979). Natural Fibers, Inc., of Ogallala, Nebraska, has made great strides toward commercial production of milkweed fiber for use in down comforters and pillows. Their advertisements proclaim that "Nothing warms you up like Ogallala down." Further information about the company can be found on pages 118 to 123 of the 1992 U.S. Department of Agriculture yearbook.

THINGS TO THINK ABOUT

1. How many uses can you think of for a plant you thought was just a weed?
2. Are there situations where we ought to encourage weed growth in crops?
3. Should genetic engineering be used to create useful weeds?

LITERATURE CITED

Altieri, M. A. 1985. Agroecology: *The Scientific Basis of Alternative Agriculture,* 2nd ed., p. 112–113. Div. of Bio. Cont., University of California, Berkeley.

Anonymous. 1976. Roots of water hyacinth may be harvested for gold. *Weed, Trees and Turf* 15(2):50.

Anonymous. 1979. *Dandelion Recipe Book.* Mayor's special events office. Vineland, New Jersey. 12 pp.

Anonymous. 1992. Quarterly report of selected research projects. USDA/ARS. July 1 to September 30.

Bjork, S., and W. Graneli. 1978. Energy reeds and the environment. *Ambio* **7:**150–156.

Bosworth, S. C., C. J. Hoveland, and G. A. Buchanan. 1985. Forage quality of selected cool–season weed species. *Weed Sci.* **34:**150–154.

Boyer, J. S. 1982. Plant productivity and environment. *Science* **218:**443–448.

Chacón, J. C., and S. R. Gliessman. 1982. Use of the "non-weed" concept in traditional tropical agroecosystems of south-eastern Mexico. *Agroecosystems* **8:**1–11.

Cramer. C. 1992. Healthy harvest. *New Farm Magazine* Nov.–Dec., 13–16.

Daniel, J. 1974. Take a weed to lunch. *The Reader's Digest,* pp. 213–215. Condensed from *American Agriculturalist* and the *Rural New Yorker.*

Drury, S. 1992. Plants and pest control in England circa 1400–1700: A preliminary study. *Folklore* **103:**i,103–106.

Duke, J. A. 1992. *Edible Weeds.* CRC Press, Boca Raton, Florida. 232 pp.

Dutt, T. E., R. G. Harvey, and R. S. Fawcett. 1982. Feed quality of hay containing perennial broadleaf weeds. *Agron. J.* **74:**673–676.

Gerard, J. 1597. *The Herbal or Generale Historie of Plants,* 1636 ed., p. 115.

Harrington, H. D. 1967. *Edible Native Plants of the Rocky Mountains.* Univ. of New Mexico Press. 392 pp.

Hauptil, H., and S. Jain. 1977. Amaranth and meadowfoam: two new crops. *Calif. Agric.* (Sept.):6–7.

Holshouser, D. L., and J. M. Chandler 1996. Predicting flowering of rhizome johnsongrass (*Sorghum halepense*) populations using a temperature dependent model. *Weed Sci.* **44:**266–272.

Hylton, W. H. 1974. *The Rodale Herb Book.* Rodale, Emmaus, Pennsylvania. 653 pp.

Jewell, W. J. 1994. Resource–recovery wastewater treatment. *American Sci.* **82:**366–375.

Lione, A. 1979. Make a milkweed down jacket. *The Mother Earth News* (Sept.–Oct.):104–105.

Lowell, J. R. 1894. A fable for critics. In *The Complete Poetical Works of James Russell Lowell,* pp. 117–148. Riverside Press, Cambridge, Massachusetts.

Marten, G. C., and R. N. Andersen. 1975. Forage nutritive value and palatability of 12 common annual weeds. *Crop Sci.* **15:**821–827.

Marten, G. C., C. C. Sheaffer, and D. L. Wyse. 1987. Forage nutritive value and palatability of perennial weeds. *Agron. J.* **79:**980–986.

Martin, L. C. 1983. *Wildflower Folklore.* The East Woods Press, Charlotte, North Carolina. 256 pp.

Murray, C. 1976. Weed holds promise for pollution cleanup. *Chem. Eng. News* (March) **22:**23–24.

Rogers, H. H., and D. E. Davis. 1972. Nutrient removal by water hyacinth. *Weed Sci.* **20:**423–428.

Ross, R. L. 1976. Wild edible and medicinal plants. Coop. Ext. Serv. Montana State Univ. Circular 1183. 7 pp.

Sheaffer, C. C., and D. L. Wyse. 1982. Common dandelion (*Taraxacum officinale*) control in alfalfa (*Medicago sativa*). *Weed Sci.* **30:**216–220.

Tabor, P. 1952. Comments on cogon and torpedo grasses. A challenge to weed workers. *Weeds* **1:**374–375.

Temme, D. G., R. G. Harvey, R. S. Fawcett, and A. W. Young. 1979. Effect of annual weed control on alfalfa forage quality. *Agron. J.* **71:**51–54.

Thamsara, S. 1985. *Mimosa pigra* L. Proc. 10th Asian–Pacific Weed Sci. Soc. Conf., pp. 7–12.

Vega, M. R. 1982. Crop production in the total absence of weeds. Paper presented at the FAO/IWSS Expert Consultation on Weed Management Strategies for the 1980s for the LDCs. FAO, Rome, Italy. Sept. 6–10, 1982.

Williams, M. C. 1980. Purposefully introduced plants that have become noxious or poisonous weeds. *Weed Sci.* **28:**300–305.

Zandstra, H. B., and P. S. Motooka. 1978. Beneficial effects of weeds in pest management–A review. *PANS* **24:**333–338.

Chapter 5

Weed Reproduction and Dispersal

FUNDAMENTAL CONCEPTS

- The soil seedbank in most agricultural soils includes millions of weed seeds/acre and is the primary source of yearly weed problems.
- There are many complex methods for dispersal of weed seeds in space. These involve purely plant mechanisms, man-aided, mechanical, water-aided, and animal-aided systems.
- Continued development of understanding of the process of seed germination and the physiological and environmental factors that affect it is essential to development of good weed management systems.
- Seed dormancy is the way weed seeds are dispersed in time.
- Vegetative reproduction creates some of the most difficult weed management problems because vegetative reproductive organs are hard to reach with available control measures.

OBJECTIVES

- To learn the size and role of the soil seedbank and aerial seed sources.
- To understand how seeds are dispersed in space.
- To understand how seeds are dispersed in time via seed dormancy.
- To understand the causes, classification, and role of weed seed dormancy.

71

• To know the methods of vegetative reproduction and understand its role in weed management.

Weed biology is a part of weed science devoted to study of growth, development, and reproduction of weeds. Although this is not a textbook of weed biology, biological knowledge is essential to understanding the fundamentals of weed science and to development of appropriate weed management systems.

This chapter is divided into four sections dealing with reproduction and dispersal of weeds. The first section discusses seeds and their production, the second includes dispersal of seeds in space, and the third deals with seed germination and dispersal of seeds in time or seed dormancy. The last section covers vegetative or asexual reproduction.

I. SEED PRODUCTION

A seed is a mature fertilized ovule or plant embryo that has stored energy reserves (sometimes missing), and a protective coat or coats. It is a small plant packed for shipment. Survival of many flowering plants depends on production of a sufficient number of viable seeds. This is especially true for annual weeds that reproduce by seed, and, therefore, prevention of seed production is the key to elimination of future problems. Failure to prevent production of weed seed results in increasing numbers of seed in soil and weeds in each crop.

Many weed seeds are small. For example, broadleaf plantain has over 2 million seeds per pound and shepherd's-purse nearly 5 million. Small seeds are easily dispersed by wind and water, and their size precludes easy detection until they germinate and a plant emerges above the soil surface.

Some weeds can produce viable seed by apomixis (nonsexual reproduction, e.g., dandelion) and others are wholly self-fertile (e.g., shepherd's-purse). Weather before or during flowering is not important because with apomixis, seeds are set without pollination and with self-fertility, fertilization occurs before flowers open. These plants escape normal photoperiodic effects on flowering.

The number of weed seeds in arable soil is large. Koch (1969) estimated that the average arable soil has 30,000 to 350,000 weed seeds per square meter (300 million to 3.5 billion per hectare, or 120 million to 1.4 billion per acre).

In lowland (paddy or irrigated) rice fields in the Philippines, 804 million seeds from 12 different species (sedges dominated) were found over 1 hectare 6 inches deep (Vega and Sierra, 1970). Samples of soil on Minnesota farms averaged 1,600 seeds per square foot, 6 inches deep, or 70 million seeds per acre (Robinson, 1949). Other estimates range from 10.8 to 332 million seeds per hectare (Klingman and Ashton, 1982).

In a 15-acre field that had been regularly cropped for several years, seven species were 90% of the total weed population. Good weed control reduced this up to 54% in continuous corn (Schweizer and Zimdahl, 1984b) and 26% in rotational crops in 1 year (Schweizer and Zimdahl, 1984a). Redroot pigweed populations declined 99% from 1.07 billion to 3 million seeds per hectare 25 centimeters deep (10 inches) after 6 years of weed control in continuous corn. Lambsquarters declined 94% from 153.6 billion to 8.6 million seeds per hectare 25 centimeters deep. The total number of seeds declined 98% from 1.3 billion to 20.7 million seeds per hectare, 25 centimeters deep (Figure 5.1). This smaller number still equals 192 weeds per square foot of soil if all germinated in 1 year. The seedbank, enormous at the beginning of the experiment, was still large after 6 years of good weed management. It is generally assumed that 2 to 10% of weed seeds in the soil seedbank emerge each year. With 192 weed seeds per square foot, we would expect 4 to 20 plants per square foot: still a weed problem that must be dealt with. Emergence from weed seedbanks from Ohio to Colorado and Minnesota to Missouri showed that for 15 species found on three or more sites, average percent emergence varied from 0.6 for prostrate knotweed to 31.2 for giant foxtail. Six species had greater than 15% emergence in 3 or more years, four had between 5 and 8.5%, and five others had less than 3.5% emergence (Forcella *et al.,* 1997). Between 2% and 10% may still be a reasonable average, but there is large variation among species.

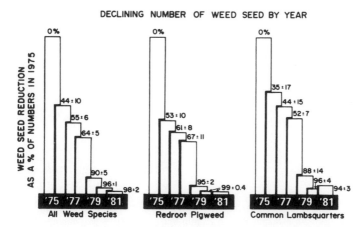

Figure 5.1 Percentage decline in the number of weed seeds for all weed species, redroot pigweed, and common lambquarters after 6 years of continuous corn. Standard errors shown for each weed species and year (Schweizer and Zimdahl, 1984a). Reprinted with permission of Weed Sci. Soc. Am.

In a study with continuous corn (Schweizer and Zimdahl, 1984b), when atrazine was discontinued as the primary herbicide after 3 years, redroot pigweed seed numbers rose to 608 million (Figure 5.2). Common lambs-quarters rose to 22.8 million and the total number of seeds rose to 648.1 million per hectare 10 inches deep. This contrasts with a steady decline with continued weed management (Figure 5.2). The point is that in this system, and in all cropping systems, if weeds are neglected, even briefly, soil seed populations and the annual weed population rebound rapidly.

In rotational crops of barley, corn, and sugarbeets, the total number of weed seeds declined 96.4% from 1.38 billion to 50 million per hectare 10 inches deep after 6 years of weed management (2 rotational cycles) (Schweizer and Zimdahl, 1984a). The number of redroot pigweed seeds declined over the 6-year period, but the percentage of *Chenopodium* species

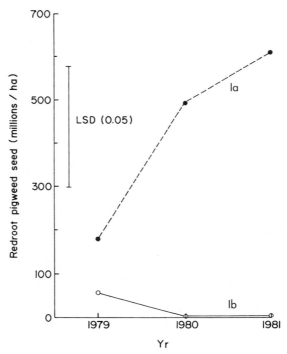

Figure 5.2 Number of redroot pigweed seeds present in soil each spring following conventional tillage and atrazine use in continuous corn. In weed management system la, 2.2 kg/ha atrazine was applied preemergence for 6 consecutive years, beginning in 1975. In weed management system lb, the same rate of atrazine was applied for the first 3 years, and then discontinued (Schweizer and Zimdahl, 1984b). Reprinted with permission of Weed Sci. Soc. Am.

increased because oakleaf goosefoot is more tolerant to cultivation and to the herbicides used than common lambsquarters.

After 1 cropping year, the decline in the number of redroot pigweed and *Chenopodium* species seeds was 34% and 22%, respectively (Figure 5.3). The next significant decline did not occur until after the fourth cropping year. After the sixth cropping year, the decline in the number of redroot pigweed and *Chenopodium* sp. seeds was 99% and 91%, respectively (Schweizer and Zimdahl, 1984a). These data illustrate that weed seed populations can be reduced quickly, but continued attention is required to prevent a rapid increase when a few plants survive.

It is not surprising that a few survivors rapidly increase the number of seeds in the soil bank when one examines the seed-producing capacity of several weed species. Data from single, undisturbed plants are shown in Table 5.1 (Stevens, 1932). The data purport to show maximum seed production and illustrate that production potential is high for many common weeds. These data have been cited in many weed science textbooks and are regarded as accurate, but there are reasons to question their accuracy. Stevens's (1932, 1957) work on 234 species was done with seed collected from diverse habitats in several U.S. states. If the studies were redone with carefully controlled conditions, identified seed sources, and plants grow-

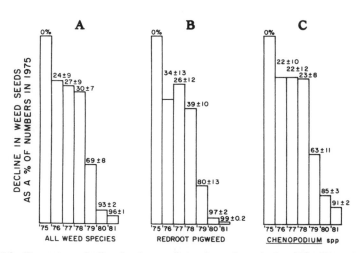

Figure 5.3 Percentage decline in the number of weed seeds for (A) all weed species, (B) redroot pigweed, and (C) *Chenopodium* spp. when averaged over cropping sequence and weed management systems after 6 years of crop rotation. Standard errors shown for each weed species and year (Schweizer and Zimdahl, 1984a). Reprinted with permission of Weed Sci. Soc. Am.

Table 5.1

Number of Seeds Produced Per Plant and Number of Seeds Per Pound for Several Common Weeds (Stevens 1932, 1957. Adapted with permission)

Plant common name	Number of seeds per plant	Number of seeds per pound[a]
Stevens (1932)		
Barnyardgrass	7,160[b,c]	324,286
Black nightshade	8,460	197,391
Buckwheat, wild	11,900	64,857
Charlock	2,700	238,947
Common cocklebur	440	2,270
Common toadflax	2,280	3,242,857
Dock, curly	29,500	324,286
Dodder, field	16,000[c]	585,806
Field bindweed	50	14,934
Foxtail barley	2,420	403,555
Giant ragweed	1,650	26,092
Kochia	14,600	534,118
Lambsquarters	72,450	648,570
Medic, black	2,350	378,333
Mullein	223,200	5,044,444
Mustard, black	13,400	267,059
Nutsedge, yellow	2,420[d]	2,389,484
Oats, wild	250[b]	25,913
Pigweed, redroot	117,400[b]	1,194,737
Plantain, broadleaf	36,150	2,270,000
Primrose, evening	118,500	1,375,757
Prostrate knotweed	6,380	672,593
Purslane	52,300	3,492,308
Ragweed, common	3,380[b]	114,937
Sandbur	1,110[b]	67,259
Shepherd's-purse	38,500[b,c]	4,729,166
Smartweed, Pennsylvania	3,150	126,111
Spurge, leafy	140[d]	129,714
Stinkgrass	82,100	6,053,333
Sunflower, common	7,200[b,c]	69,050
Thistle, Canada	680[b,c]	288,254
Witchgrass	11,400	698,462

(continues)

Table 5.1 (*continued*)

Plant common name	Number of seeds per plant	Number of seeds per pound[a]
Stevens (1957)		
Annual bluegrass	2,050	2,270,000
Catchweed bedstraw	105	59,737
Chicory	4,600	567,500
Common chickweed	600	1,173,127
Common milkweed	600/stem	77,080
Dandelion	12,000	709,375
Giant foxtail	4,030	238,947
Prickly sida	510	142,320
Prostrate knotweed	4,600	504,444
Redroot pigweed	229,175	1,335,294
Toothed spurge	835	97,634
Velvetleaf	4,300	51,885
Venice mallow	58,600	181,600
Wild radish	1,875	53,412

[a]Calculated from the weight of 1,000 seeds.
[b]Many immature seeds present.
[c]Many seeds shattered and lost prior to counting.

ing in isolation with free root growth, seed production would likely be higher.

Barnyardgrass illustrates the point. Stevens (1932) reported that one plant produced 7,160 seeds. Barrett and Wilson (1981) reported 18,000 and Holm, *et al.* (1977) up to 40,000. Research in California (Norris, 1992) predicts that barnyardgrass growing in sugarbeets averages nearly 100,000 seeds/plant and some larger plants produce more than 400,000. Reeves *et al.* (1981) found that wild radish produced 1030 seeds/plant with only one plant per square meter. When there were 247 wild radish plants on each square meter, seed production dropped to 67 per plant. Russian thistle plants typically produce about 250,000 seeds (Young, 1991).

Research in irrigated row crop rotations suggests cropping sequence is the dominant factor influencing species composition of the soil seedbank (Ball, 1992). Herbicides and other cultural techniques vary between crops and shift seedbank composition in favor of less susceptible species. In irrigated row crops dominant species were more prevalent near the surface after chisel as opposed to moldboard plowing (Figure 5.4). The number of species increased more after chisel plowing and there was a greater decrease after moldboard plowing (Ball, 1992).

Forcella, *et al.* (1992) studied weed seedbank size in eight U.S. corn-belt states and found total density ranged from 600 to 162,000 seed/square meter for three annual grasses, redroot pigweed, and common lambsquarters and that 50 to 90% of the total seedbank was dead. Seedling emergence was inversely related to rainfall and air temperature in April and May presumably because anoxia from high water content and high soil temperature induced secondary dormancy or killed the seeds. Forcella, *et al.* (1992) found viable seedlings were less than 1% of the seedbank for yellow rocket to 30% for giant foxtail.

II. SEED DISPERSAL

Weed problems would be much less complicated if weed seed merely fell off plants and gravity determined their destination. One of the most obvious features of many weeds is some structure that gives seed buoyancy in air or the ability to attach to something.

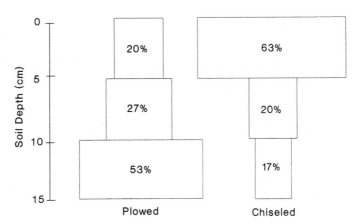

Figure 5.4 Influence of primary tillage on vertical distribution of total weed seed to a 15-cm depth in the soil after a dry bean crop (Ball, 1987, 1992). Reprinted with permission of Weed Sci Soc. Am.

In 1941 a European engineer, George de Mestral (d. 1990) was walking in the Swiss mountains, near Geneva. When he returned, he noticed that he had burdock burrs attached to his pants, jacket, and wool socks, and they were in his dog's fur. Mr. de Mestral must have been a good observer: one of those people who see what they are looking for when it is there, do not see what they are looking for when it is not there, and see what they are not looking for when it is there. Mr. de Mestral did not go walking to collect burdock burrs on his clothing. However, he saw them and observed what he was not looking for when it was there–the unique hook of the burdock burr that allowed it to attach to the wool of his clothing. Following his observation and after 8 years of work he duplicated the grasp of the burr with nylon and invented Velcro™ (from the French *velours,* velvet, and "cro" from the French *croc* or *crochet,* hook). Velcro seems to be everywhere in modern society: children's clothing, airplanes, shoes, and artificial heart valves. It is worth a moment's reflection to consider that it came from a good observer's noticing the way in which one weed species disperses its seed in space.

A. MECHANICAL

The long, slender awn of needle and thread grass moves about easily by attaching to socks and other articles of clothing or to an animal's fur. The hooks on the aggregated cluster of flowers, properly called the capitulum, that form the burlike structure of cocklebur (the fruit) and hold its seeds (achenes) facilitate transport.

Burs of sandbur consist of one to several spikelets surrounded by an involucre of spiny, scabrous bristles. The burs of sandbur and the spines on the fruit of puncturevine penetrate shoe leather and tires. Bicycle riders are familiar with the hazards of puncturevine and sandbur because the bristles so easily penetrate bicycle tires. Seed transport is facilitated by their sharp spines.

The seed pod of unicorn plant is 2 to 4 inches long with a curved beak longer than the body of the pod. At maturity the pod divides into two opposite incurved claws with an inwardly hooked, pointed tip to form an ice-tong-like structure that easily attaches to livestock or equipment. Uni-

The bristly capitulum of cocklebur.

corn plant is a stout, much-branched, bushy plant whose seed pods fall to the ground at maturity, and when they dry and dehisce those incurved hooks can easily grab socks or legs that pass by and enlist them, involuntarily, to move a few seeds.

Another form of mechanical transport of seed is illustrated by curlycup gumweed, a biennial or short-lived perennial that reproduces by seed. The flower heads are bright yellow, $\frac{1}{2}$ to 1 inch across and covered with a sticky resin. The sticky achenes facilitate seed transport.

B. Wind

Many seeds have structural modifications permitting transport by wind. One commonly observed example, although not a weed, is the winged seed of maple trees. Among the weeds, examples of modification for transport by wind include the silky white pappus on dandelion achenes, the white, downy pappus on Canada thistle achenes, and the tuft of silky hairs on the seed of showy milkweed. Many people have seen seed of one or more of these species moving with a summer breeze. The data in Table 5.2 illustrate seed dispersal by wind. Most seed is light (see number per pound in Table 5.1), and can move over great distances with very light winds. On bombed

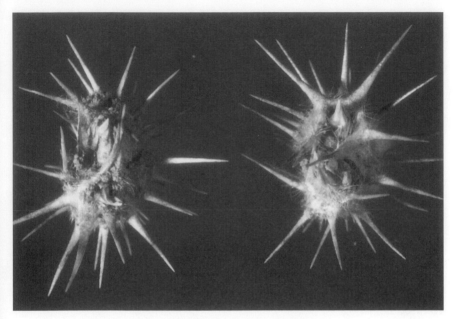

The spikes of puncturevine seed pods.

sites in London after World War II, 140 different species of flowering plants were observed, and of those that established first, about 30% of the total were distributed by wind (Salisbury, 1961).

Another method of wind dispersal is found in Russian thistle or tumbleweed. A mature Russian thistle plant is nearly round. When mature, it

Table 5.2

Rate of Fall of Seeds through Still Air
(Salisbury, 1961)

Plant	Time to fall 10 feet (seconds)	
	Average	Range
Coltsfoot	21.3	14–45
Annual sowthistle	8.5	4.5–12
Groundsel	8.0	6.0–12
Smallflower galinsoga	3.4	2–5

breaks off at the base, and because it is round it tumbles or rolls with the wind. The seeds, held by a series of twisted hairs, are released gradually as the plant rolls and bumps and the hairs break. Other plants that roll to disperse seed include tumble pigweed and witchgrass. In the latter case, only the inflorescence breaks off and rolls.

C. WATER

In the western United States and other areas where irrigation is common, many seeds are dispersed by water. In Nebraska, Wilson (1980) found seed of 77 different plant species in three main irrigation canals over two seasons. He collected a total of 30,346 seeds. Approximately 30% were viable and about 26 times more were found at the end than at the beginning of canals. Most seed floated, and redroot pigweed was 40% of the total seed. He estimated that 120,000 seeds per acre per year entered fields from irrigation water. In the western U.S. alone, surface water irrigates more than 19 million acres each year and is an often unrecognized source of weeds in irrigated fields.

The dandelion seed ready for wind dispersal.

It is not illogical to assume that because seeds are living organisms, they will die quickly when submerged in water. In fact, seeds live a long time under water (Table 5.3). The curly dock fruit is a winged achene and the entire structure floats a long time before sinking. When seeds are deposited in water, the potential problem has not disappeared, it has moved. A Washington study (Comes *et al.*, 1978) found 82 species in irrigation water. Twenty-four species lost viability after storage in water for 12 months or less; however, 27 endured over 12 and some as long as 60 months. After 12 months, 22% of annual monocotyledons germinated and 75% of perennial monocotyledons and annual and perennial broadleaved species still germinated.

D. HUMAN-AIDED

Even though humans have the burden of controlling weeds in our crops, we are still a primary source. We fail to screen or clean irrigation water and facilitate spread by mechanical means. The pattern in which the United States was populated offered a unique opportunity for spread of weeds from the East to the West Coast and from the two coasts inland. Fogg (1966) pointed out the predominance of species of European origin in the United States. Fogg found that about 12.5 % (1051) of the flowering plants

Table 5.3

Germination of Weed Seeds after Storage in Fresh Water (Sources: Bruns and Rasmussen, 1953, 1957, 1958)

Species	Period of storage (months)	Germination (%)
Field bindweed	54	55
Canada thistle	36	About 50
	54	None
Russian knapweed	30	14
	60	None
Redroot pigweed	33	9
Quackgrass	27	None
Barnyardgrass	3	Less than 1%
	12	None
Halogeton	33	Less than 1%
	12	None
Hoary cress	2	5
	19	None

and ferns of the central and northeastern United States and adjacent Canada in Gray's *Manual of Botany* were of foreign origin, and 692 of these were from Europe. About 14% (1200) of the species in Gray's manual are recognized as weedy, and European species dominate. Not all of them are weeds, but 60% of the 1200 plants are from only seven plant families (Table 5.4). The species in the seven plant families are primarily herbaceous (not woody), produce abundant seed, and are aggressive invaders or pioneering plants, that is, they have some of the traits that make weeds successful.

The United States, a major recipient of weeds because of immigration, has also distributed weeds to others. A good example is parthenium ragweed, imported to India from the United States with shipments of grain during the early 1960s. It is an annual that has spread over large areas of southern India and is especially problematic because it contains a toxin that is irritating to human skin. The weed's common name in parts of India is AID weed from its identification with grain distributed by the U.S. Agency for International Development (US/AID). Weeds and their seeds have been imported to countries in forages and feed grains.

Weed seeds are also regularly transported in feed for cattle. Millers usually clean seed received for storage or processing. Screenings can contain weed seed and are routinely transported and used as cattle feed. There is nothing wrong with their nutritional value. Seed viability can be destroyed by cooking, but screenings are of low value and this is usually not done; the seeds are fed whole. Tables 5.5 and 5.6 show examples of seed screenings that have been transported into Colorado. It is obvious that these sources of animal feed can be important sources of weed seed, and similar examples can be found for most places.

Table 5.4

Families of Weed Species Introduced from Europe to the United States (Fogg 1942, 1966; Hill, 1977)

Family	Number of species introduced from Europe
Asteraceae	112
Poaceae	65
Brassicaceae	62
Labiatae	60
Leguminosae	54
Caryophyllaceae	37
Scrophulariaceae	30

Table 5.5

Seed Screenings Analyzed by the Colorado State Seed Laboratory

Seed	Number/lb.	Number in average truckload ($\times$ 10^6)
Lambsquarters	155,700	6,228
Pigweed	9,225	369
Kochia	1,800	72
Russian thistle	900	36
Common sunflower	225	9
Foxtail		
Yellow	225	9
Green	1,575	63
Total noxious	2,700	108

Grinding in a hammer mill does not completely destroy seed viability (Zamora and Olivare, 1994). Less than 1% of spotted knapweed, sulfur cinquefoil, timothy, and alfalfa seeds were intact after passing a 1-mm screen in a hammer mill. Of the four plants, only sulfur cinquefoil failed to germinate.

It is common for farmers to assume that once a crop is harvested and weeds are ensiled (stored in a silo), weed seed can be forgotten about. In general, this is true (Table 5.7). Corn silage pH is between 4.5 and 5.8 and decreases with age. Most seed will completely lose germination after 3 weeks' storage in silage. It is also true that the organic acid content of silage is 1.5 to 2% and silos quickly become anaerobic, both of which lead to seed death (Table 5.7). Downy brome, foxtail barley, and barnyardgrass

Table 5.6

Seed Screenings Analyzed by the Colorado State Seed Laboratory

Item	Number/lb.	Number of seeds in shipment ($\times$ 10^6)
Sample I		
Noxious weed seeds	13,511	540.4
Common weed seeds	142,650	5,706.0
Crop seeds	8,280	
Sample II		
Noxious weed seeds	2,700	113.4
Common weed seeds	279,665	11,745.9
Crop seeds	30,150	1,266.3

Table 5.7
Effect of Ensiling on Viability of Weed Seeds (Tildesley, 1937)

Weed species	Percent germination		
	Month before	2 weeks after	4 weeks later
Quackgrass	99	0	0
Barnyardgrass	61	0	0
Yellow foxtail	20	0	0
Wild buckwheat	64	0	0
Common lambsquarters	82	34	0
Cow cockle	68	0	0
Pennycress	77	0	0
Common mustard	93	0	0

lost all viability after ensiling for 8 weeks or rumen digestion for 24 hours (Blackshaw and Rode, 1991). The same study showed that 17% of green foxtail seed were viable after 24 hours of ruminant digestion. No wild oats survived rumen digestion in the first year, but 88% did in a second year of the study. This was attributed to the different diet in the two years of the study that changed the rumen bacterial population.

There is a story, perhaps apocryphal, that Canada thistle was brought from Canada to the United States to feed the horses in British General John B. Burgoyne's army during the Revolutionary War. The British plan in 1777 was to divide the States by the line of the Hudson River. General Burgoyne was to proceed from Canada by way of Lake Champlain, which forms part of the boundary between northern New York and Vermont. The campaign began in January, and Burgoyne was defeated on October 7 at the second battle of Bemis Heights (near Saratoga, NY). He surrendered his entire force on October 17. Burgoyne had to feed his horses and brought hay contaminated with Canada thistle from Canada. The weed is now ubiquitous in the northern United States. It is worth noting that Canada thistle is called California thistle in Australia, a name indicative of where the Australians think it came from.

E. ANIMAL-AIDED

One might think that if seed is fed to cattle, there is no problem because cattle chew things and rumen digestion is thorough. In one experiment (Beach, 1909), a Jersey cow was fed 6 pounds of flax seed containing 212,912 weed seeds per pound. The seed had 26.4% viability; not atypical for weed seed in feed. The cow voided 40 pounds of feces per day and 1 ounce of feces contained about 1,000 weed seeds, of which 4.5% were viable. Harmon and Keim (1934) confirmed that passage through an animal's digestive tract reduces but does not eliminate weed seed viability, with viability after digestion ranging from 6.4% for sheep to 9.6 % for calves. Chickens destroyed all but 0.2% of viable seed.

Even after weed seeds have been voided in manure, they can reinfest soil. Seeds left in cattle manure in the field had only 3.1% germination, whereas top-dressed manure hauled directly from the barn had 12.8%. Plowing under fresh manure increased seed germination to 23% (Oswald, 1908).

Ridley (1930) listed 124 species whose seeds were dispersed by cattle. In 36 samples of cattle manure from 20 New York state dairy farms, viable seed from 13 grasses and 35 broadleaved species were found (Mt. Pleasant and Schlather, 1994). Four of the farms had cattle manure with no weed seed, while the others averaged 75 to 100 weed seeds/kg of manure. The authors concluded that manure can add seed to fields but the numbers are small compared to the soil seedbank. In contrast, in Iran, sheep manure added 10 million seeds/hectare (ha) each time it was put on soil and was a more important source of new weed seed than the crop seed the farmer planted (182,000 seed/ha) or irrigation water (120 seed/ha) (Dastgheib, 1989).

Other research, in diverse locations, confirms the successful passage of weed seed through cattle (Atkeson *et al.,* 1934; Burton and Andrews, 1948; Dore and Raymond, 1942; and Gardner *et al.,* 1983). Data from a study (Thill *et al.,* 1986) of common crupina, a newly introduced winter annual invader of rangelands in Idaho, show that its seed can be spread by cattle, deer, horses, and pheasants, but achenes were not found in sheep feces. The data support the contention that weeds are spread by game birds, wildlife, and domestic livestock.

The previous data established that many weed seeds can pass through the digestive tract of several different animals without loss of viability. Some seed remains viable even after passage through the digestive tract and storage in manure (Harmon and Keim, 1934) (Table 5.8). These data are further confirmed by studies which show the effect of digestion and

Table 5.8

The Effect of Storage in Cow Manure on the Viability of Weed Seeds (Harmon and Keim, 1934)

Weed	Viability before storage (%)	Viability after storage for			
		1	2	3 (%)	4 months
Velvetleaf	52.0	2.0	0	0	0
Field bindweed	84.0	4.0	22.0	1	0
Sweetclover	68.0	22.0	4.0	0	0
Peppergrass	34.5	0	0	0	0
Smooth dock	86.0	0	0	0	0
Smartweed	0.5	0	0	0	0
Cocklebur	60.0	0	0	0	0

manure storage on germination of seed of several different species (Table 5.9).

Common crupina seed (Thill *et al.*, 1986) passes through the digestive tract of pheasants without loss of viability. Other data (Proctor, 1968) show that viable seed can be retained 8 to 12 hours in the digestive tract of birds. Seed smaller than 1 millimeter in diameter and having a hard seed coat can be retained more than 100 hours. Birds are therefore also agents for weed seed distribution. Still-viable field bindweed, little mallow, and smooth sumac seeds were regurgitated from the digestive tract of killdeer (*Chiradrius vociferus*) after 144, 152 and 160 hours, respectively. Velvetleaf seed was intact for 77 hours. Seed of many species can remain intact and viable in the intestinal tract of some birds long enough to be transported several thousand miles.

F. MACHINERY

An important source of weeds is the farmer's grain drill box. A 1965 study on the western slope of Colorado included 42 drill box samples, obtained by going to the farmer's field during planting. A grain probe removed a sample from the drill box in the field, and analysis of the 42 samples showed that 33% contained prohibited noxious weed seed and 74% contained restricted noxious weed seed. These farmers were planting an average of 2,300 noxious weed seeds per acre. One farmer was planting

Table 5.9

Germination Tests on Weed Seeds before and after Passing through the Digestive Tract of Cattle and after 3 Months' Storage in Manure (Atkeson et al., 1934)

Weed species	Percentage germination before feeding	Percentage germination after 47 hours digestion	Percentage germination after 47 hours digestion plus storage in manure	Percentage decrease due to manure storage and digestion
Redroot pigweed	98	36.0	11.5	88
Common lambsquarters	70	58.0	22.0	69
Alfalfa	86	17.0	80.0	7
Narrowleaf plantain	94	16.0	0.0	100
Curly dock	95	58.0	3.0	97
Green foxtail	21	19.5	0.0	100
Wild oats	74	10.0	0.0	100

13,000 field bindweed seeds and another 14,000 wild oat seed per acre. A second Colorado study included 22 random samples. Fourteen percent of the drill boxes had prohibited noxious weed seed and 77% had restricted noxious weed seed. The average was 6,600 noxious weed seed planted per acre. An Iowa study showed 73% of oat seed was combine-run and had more than 20 weed seeds per pound. Sixty-three percent of all the oats tested contained prohibited or restricted noxious weed seed.[1] In Minnesota, 343 drill box samples averaged 150 weed seeds per pound. One sample of red clover had 24,000 dodder seeds per pound and was, therefore, 10% dodder (Dunham, 1973).

The wide availability of certified seed has reduced this problem, but it still exists. As late as 1988, 31.3 % of wheat, barley, and oat samples taken from grain drills in Utah were infested with an average 313 weed seeds/ pound of grain (Dewey and Whitesides, 1990). The worst sample found in a drill box had 11,118 weed seeds in each pound of grain. Nonnoxious weed seeds were found in 107 samples (23.8%) and noxious weeds were in 76 samples (16.9%). Wild oats were the most common noxious weed seed, occurring in 14% of samples at an average density of 2,136 seed per 100 kg of crop seed (Dewey and Whitesides, 1990). A decline from 52% contaminated drill boxes in a 1958 survey was noted.

The cardinal rule of weed management is "Buy and plant clean seed." Most farmers in the world's developed countries buy seed from a dealer and are confident it is free of weed seed and diseases and has high germinability. This is not true in most of the world's developing countries, where farmers keep harvested grain for planting the next year. Seed is often contaminated at harvest with weed seed and seed-borne pathogens, and the harvested grain may have poor germination. The proper weed management method in these situations is one that emphasizes prevention of the problem before it occurs rather than weed control in the following crop.

One important way people affect weed seed dispersal is through movement of farm machinery, especially itinerant grain combines and accompanying trucks. Spread of many weeds is aided by itinerant combine harvesters that move from field to field, often across large areas of the country. Itchgrass has grown wild in Louisiana sugarcane since the 1920s. It started to migrate when soybean farming expanded. Sugarcane has long been grown in Louisiana, and cane processing machinery is a likely vector, but it rarely leaves a farm. However, soybean harvesting machinery is often itinerant, and itchgrass has spread with itinerant soybean combines.

Some weeds are dispersed by combines because weeds are harvested with the crop and weed seed is dispersed by the combine as straw is spread

---

[1]Personal communication. Colorado State Seed Laboratory, Colorado State Univ.

on the field. Other weeds (e.g., wild mustard and field pennycress) shed seed before harvest in the Northern Great Plains of the United States. Wild oats, downy brome, and Canada thistle shed seed before and during harvest. Green and yellow foxtail, barnyardgrass, quackgrass, redroot pigweed, kochia, wild buckwheat, common lambsquarters, field bindweed, and Russian thistle shed seed during and after harvest and combine harvesting facilitates seed dispersal (Donald and Nalewaja, 1991). These weeds make harvest more difficult by accumulating on the harvester's cutting bar and adding weight and green material to the combine's load and to harvested grain.

Movement and storage in combines is of concern because it has been shown that seed of slimleaf lambsquarters, venice mallow, and curly dock grew better when collected from combines that were harvesting hard red winter wheat than when the seed was harvested by hand from weedy plants in the same field (Currie and Peeper, 1988). Mechanical abrasion or scarification in the combine was the likely cause.

Johnson and Mullinix (1995) suggest that soil tillage distributes weed seed because it affects weed emergence and hence seed production. Crop cultivation, a useful weed management tactic, has been correlated with midseason emergence of Florida beggarweed in peanut (Cardina and Hook, 1989). Mechanical control is discussed more fully in Chapter 9, Section III.

The number of weed seeds in the plow layer can be reduced by repeated tillage (Chancellor, 1985). With optimum rain, 50% of the weed seed in the plow layer of vegetable crop fields germinated within 6 weeks of cultivation (Bond and Baker, 1990). Egley and Williams (1990) increased weed emergence with frequent tillage over 4 years. Subsequent tillage had no effect on emergence, suggesting the seedbank had been depleted. In Minnesota, wild mustard seed in soil was reduced 97% after 7 years of tillage (Warnes and Anderson, 1984). In Alabama, purple nutsedge was eradicated after 5 months of weekly or biweekly harrowing (Smith and Mayton, 1938). Therefore, it is logical to conclude that soil tillage plays an important role in the availability of seed for dispersal by encouraging seed germination and in reducing seed production by destroying emerging seedlings.

G. Mimicry

Gould (1991), discussing the evolutionary potential of crop pests, said, "Of all the crop pests, weeds boast the longest recorded history of adapting to agricultural practices." Weeds have used two mechanisms to survive between cropping seasons: seed dormancy and crop seed mimicry. The second technique is basically one of hiding in crop seed to be planted the next year. It was the easiest technique because it avoided all the perils of remaining in the field, exposed to the environment and to predators. Weedy plants, by evolving to mimic the seed size, shape, or color of the crop they infest, are passed on by humans who plant contaminated seed. Gould (1991) cites mimicry of lentil seeds by common vetch, flax seed by species of falseflax, and rice plants by barnyardgrass. In the last case the mimicry is in plant morphology and growth habit, not seed (Barrett, 1983). Because the plants are very hard to distinguish visually, they are not removed by hand. The foregoing examples are of unrelated plants, but Gould (1991) also cites mimicry in the closely related wild and domestic rices.

H. OTHER

Other sources of weed infestations are associated with human activities. It has been suggested that downy brome first entered California in packing material for glassware shipped from Europe. We also spread weeds growing in nursery stock and ornamentals. Highway construction that demands "fill" soil can easily spread weeds and their seed over wide areas.

I. CONSEQUENCES OF WEED DISPERSAL

It is useful to know that weeds are dispersed in many ways. It is important to know that dispersal of their seeds has real consequences. For example, data from the U.S. Bureau of Land Management show that alien plants (weeds) are expanding their territory by 14% each year or, in other terms, 2,300 acres each day (Culotta, 1994). More than 60% of the 1,350-acre Devil's Tower National Monument in Wyoming has been taken over by leafy spurge, which some regard as the worst of the bad weeds.

Rush skeletonweed, originally from the Balkans, was first spotted near Banks, Idaho, in 1954. In 10 years it had invaded 40 acres and by 1994 was on 4 million acres in Idaho alone (Culotta, 1994). The land occupied by rush skeletonweed now has very low species diversity and high soil erosion.

Weeds affect more than just crop and rangeland. The Sellway-Bitterroot Wilderness in Idaho has prime stream habitat for salmon, but some areas of riverbank are covered with spotted knapweed. Other species do not grow with spotted knapweed, so the soil is bare between the plants. When it rains, erosion increases, soil enters the water, and the quality of salmon spawning area declines (Culotta, 1994).

The weedy tree melaleuca has invaded and taken over more than 450,000 acres of the Everglades and tropical wetlands of south Florida (Schmitz,

1995). Melaleuca is a native of Australia, where it is kept in check by more than 400 insect species. It is expanding its range in Florida by 50 acres a day.

In a small book, Randall and Martinelli (1996) describe 83 foreign invaders and correctly note that they can "change fundamental ecosystem processes such as the frequency of wildfires, the availability of water or nutrients and the rate of soil erosion." Weedy invaders such as melaleuca "change the rules of the game." Invaders that do not change basic ecosystem processes cause other problems. In forests, invading trees and vines can grow into the canopy and shade desirable species. Shrubs can dominate mid-story areas, and herbaceous species can colonize and dominate the forest floor. Prairies and other grasslands across the United States and in other countries are severely infested by nonnative weedy species that are also crop weeds, such as leafy spurge and yellow starthistle. Randall and Martinelli (1996) also point out that on wetlands in the northern third of the United States and southern Canada, purple loosestrife has formed large, dense stands that have displaced native plants and changed or, in many cases, eliminated waterfowl habitat.

III. SEED GERMINATION: DORMANCY

So far we have considered two steps involved in plant reproduction: seed production and seed dispersal in space. The third aspect of reproduction involved in weed management is seed germination. What is really of interest is not the fact that seeds germinate, but the fact that they *do not* germinate because they are dormant. Dormancy is dispersal in time as opposed to dispersal in space. Dormant seed can be dispersed in space without losing their dormancy. Dormancy is not well defined. To be dormant is to be sleeping or inactive. In biology, it is regarded as a state of suspended animation: alive but not actively growing. Thus, dormancy is defined by something that seeds do not do–germinate–rather than by something they do. Scientists have described types of dormancy, but because the basic regulatory processes are unknown, it is difficult to define types of dormancy or to extrapolate from one species to another (Dyer, 1995).

A. CAUSES

Chepil (1946) was the first to report on periodic seed dormancy and germination among weeds. The phenomenon has since been documented for seed of many annual weeds (Karssen, 1982; Dyer, 1995). Dormancy is a highly developed specialization and a complex research problem. Most seeds will germinate when proper environmental conditions exist; however, not all do. Soil disturbance may or may not initiate special germination mechanisms. Changes in soil temperature, soil water content, light, surface drying and wetting, or percent oxygen or carbon dioxide in soil air can create or break dormancy. Soil microflora play a role, as they control oxygen and carbon dioxide content of soil air. One microsite location, a habitable site, may provide appropriate conditions for germination, while a nearby, uninhabitable site may not. There is a range of special requirements for germination and other special conditions that impose dormancy.

If one wants to become famous in weed science, indeed, in agriculture, it could be accomplished by figuring out how to do one of two things. The first is to make most weed seed in soil or those shed from plants dormant forever. The second is to make most weed seed in soil germinate immediately. Because seed dormancy is a complex environmental/physiological/ biochemical phenomenon, it is unlikely that any "magic bullet" solution will ever be found. As weed science moves closer to understanding seed dormancy, it could greatly reduce the need to control annual weeds and perennials that reproduce by seed. It would take time to deplete soil seed-banks, but once that was done, the need to control most annual weeds would decrease. If we could make most seed of annual weeds in the soil seedbank germinate just before frost in the temperate zones, frost alone would kill them. In the tropical dry season, weeds could be managed with tillage. Because weeds have periodicity of germination, timing of tillage and planting is now altered when possible to encourage or discourage weed seed germination (Dyer, 1995; Gunsolus, 1990).

Weeds share many traits with what ecologists call early successional species (Roberts, 1982). Indeed, they are often the same species. For early successional species, seed germination is closely linked to soil disturbance, which ensures the availability of resources for growth. Germination, soon after soil disturbance, reduces the probability of competition with later successional species or crops. Early successional species and many weed seeds usually require light for seed germination, and exposure to light is increased by tillage. Seed germination is favored by fluctuating temperatures and low carbon dioxide concentrations, and may be affected by alternate wetting and drying cycles that tend to break seed coats. All of the conditions favorable for seed germination occur on disturbed sites, and cropped fields are very good examples of disturbed sites.

Seeds of early successional species and many weeds are dormant when shed and can quickly develop secondary dormancy. Induced dormancy (dependent upon environmental interaction) is common. Early successional species grow rapidly above and below the soil and thereby escape the surface zone of maximum environmental variability and stress. Early successional species have a high photosynthetic rate over a wide range of soil water conditions. Photosynthetic rate and environmental resource demand decline quickly with declining soil water potential, permitting survival. Weeds and successful early successional species compress environmental extremes. They are able to maintain constant leaf temperatures to ameliorate stress. They also acclimate rapidly to variable environments because genotypic plasticity facilitates their adaptation.

Seed dormancy is the most important characteristic for perpetuation of annual weed species and perennials that reproduce by seed. Seed of annual weeds germinate under a narrow range of environmental conditions; they are specialists in utilizing their opportunities.

Most weed control techniques treat symptoms rather than the problem itself. Weed control acts on problems either just before they appear (preemergence) or after they have appeared (postemergence). There are no reliable methods for eliminating weed problems by preventing seed dormancy or encouraging germination of dormant seed. Improved weed management depends on better understanding of seven environmental conditions that cause or terminate seed dormancy.

1. *Light.* Of the known causes of dormancy, light may be the most important. At least half of the annual weeds in crops have seed that requires light for germination. This is especially so for small-seeded annual weeds. The length of day and the quality of light are also important. The light requirement is regarded as an evolutionary advantage for small-seeded plants that may not survive germination from lower in soil (Pons, 1991). Light only penetrates 1 or 2 mm in soil, so dormancy can be induced even by shallow burial. Germination of mullein, curly dock, common evening primrose, and buttercup seed is favored by light. Seed of common chickweed, common purslane, johnsongrass, kochia, lambsquarters, prostrate knotweed, and redroot pigweed require light for germination. However, seed germination of wild onion and jimsonweed is favored by darkness. Dormancy of crop seed has been nearly eliminated by breeding light response out of the genome, and most germinate in light or in the dark.

The phytochrome group of photoreceptors, the primary system responsible for light interactions in plants, controls breaking dormancy with light. In a simple but accurate sense, phytochrome exists in two forms: a promoter and inhibitor. The promoting form is favored by red light and the inhibitor by far-red light. This is the same general response and the same pigment that is involved in flowering. The quantity of each form of phytochrome present at a given time is related to light, and more precisely to the ratio of red to far-red light. Sunlight has abundant red light and promotes germination of imbibed seeds. Seeds do not respond to light unless they have taken up water (imbibed).

Light effects on phytochrome and seed germination:

Inactive form (P_r) + red light, 600–680 nm → P_{fr} and germination
promotion

Active form (P_{fr}) + far-red light, 700–760 nm → P_r and germination
inhibition

Light is needed for seed germination of many species, though it is clear that burial in soil will inhibit germination and should be used as a weed control technique. Continued seed burial is encouraged when farmers shift to minimum-tillage and no-tillage agriculture.

Unfiltered light contains a preponderance of the red wavelength that shifts phytochrome to the active (P_{fr}) form and promotes seed germination. Leaf canopies filter red light because chlorophyll absorbs it strongly. A leaf canopy shifts light transmission toward far-red and depresses germination of seeds below.

2. *Immature embryo.* A second cause of dormancy is the presence of an immature embryo. Smartweed and bulrush seeds are typically shed from the plant with an immature embryo and are incapable of immediate germination. This is another example of a mechanism evolved to prevent germination at the wrong time.

3. *Impermeable seed coat.* Seeds from redroot pigweed, wild mustard, shepherd's-purse, and field pepperweed often have seed coats impermeable to water, to oxygen, or both, and the seeds are called "hard." This is another dormancy mechanism. The seed coat can be changed (often referred to as broken) by scarification, action of acids, or microbes. A hard seed coat presents mechanical resistance to germination because the radicle cannot penetrate it. Even though water and oxygen can be absorbed, the hard seed coat prevents germination. In the laboratory, scarification or breaking of a hard seed coat can be accomplished by rubbing on sandpaper, dipping in acid, or pricking with a pin. Such techniques are obviously inappropriate for the field, but the same thing is accomplished by tillage. Anything that stirs or moves soil will inevitably move seeds and abrade seed coats.

4. *Inhibitors.* Some seeds are shed with endogenous (internal) germination inhibitors (e.g., absicisic acid). These varied and complicated chemical inhibitors prevent seed germination until they are removed by leaching with water or by internal metabolic activity. There are also exogenous (external) germination inhibitors that will be discussed in Chapter 7: Allelopathy.

5. *Oxygen.* Partial pressure of oxygen affects seed dormancy. Percent oxygen in soil varies from less than 1% in flooded soil to 8 or 9% in a soil with good tilth, cropped with corn. Soil carbon dioxide content may vary

from 5 to 15%. One of the reasons most seed germinates only near the soil surface is higher oxygen concentration. Soil compaction reduces seed germination, and the mechanism may be reduction of the partial pressure of oxygen.

6. *Temperature.* There is a minimum temperature below which no seed will germinate and a maximum temperature above which germination will not occur. The precise minimum and maximum vary among species, as does the optimum temperature for germination. In late spring, Russian thistle seed germinates readily between 28 and 110°F (Young, 1991). Wild oats will germinate at 35°F (1.7°C), which is a lower temperature than the temperature at which seed of wheat or barley germinate. Temperatures of 40 to 60°F (4 to 15°C) are required for germination of seed of some winter annual weeds. Higher temperatures lead to dormancy. Redroot pigweed seed kept in a germinator at 68°F (20°C) will remain dormant for up to 6 years. It can be induced to germinate, at any time, by alternating storage temperature or by partial desiccation. Germination can also be induced by raising the temperature to 95°F (35°C) for a short time, rubbing the seed, and then lowering the temperature to 68°F.

7. *After-ripening requirement.* There is an after-ripening requirement for some seed. This is not the same as an immature embryo; rather, it is a poorly understood physiological change. A seed's embryo is fully developed but will not germinate, even if oxygen and water are absorbed in the appropriate concentrations. Everything appears to be normal, but the seed will not germinate until it has ripened.

B. CLASSES OF DORMANCY

Dormancy classifications are based on observed seed behavior, not, as mentioned earlier, on complete understanding of the physiology or biochemistry of seed dormancy. Two classification systems will be presented. In the first, a seed dormant when shed from the plant has primary dormancy. All other manifestations of dormancy are secondary. After primary dormancy has been lost, secondary dormancy may be induced by environmental interactions or other special conditions.

The second system of classification includes three types of dormancy (Harper, 1957). The first is *innate* and has three possible causes. It could be an inherent property of the ripened seed based on genetic control when the seed leaves the plant. There may be an after-ripening requirement, perhaps dependent on receipt of a specific environmental stimulus. There could be a rudimentary or physiologically immature embryo, which is not fully developed when seed is shed; an example is smartweed. Innate dormancy can also be caused by impermeable or mechanically resistant seed coats, that is, hard seed. Redroot pigweed, several species of mustard, and all species of wild oats have innate dormancy. A third cause is the presence of endogenous chemical inhibitors. Some species of sumac and fireweed proliferate after forest fires because fire creates permeability in the seed coat and rain leaches out the inhibitor. The amount of an inhibitor is often adjusted to the rainfall of an area. In its simplest form, the presence of an endogenous chemical inhibitor restricts germination to the temperature range where survival is assured. Innate dormancy interacts with the environment because for some species, hot, dry weather during seed maturation yields less dormancy than cool, moist conditions, more favorable to seedling survival.

Sometimes a seed develops dormancy after exposure to specific environmental conditions such as dryness, high carbon dioxide concentration, or high temperature and the acquired dormancy persists after the environmental conditions change. Harper (1957) described this as *induced dormancy.* Seed of winter wild oats and white mustard have induced dormancy that often develops in late spring in temperate climates and persists into fall. Seed buried by tillage may not germinate when brought to the soil surface because of induced dormancy.

Induced dormancy develops because of environmental interaction after seed has been shed from the plant and persists after environmental conditions change. *Enforced dormancy* also depends on environmental interaction, but does not persist when conditions change. In the latter case, dormancy can be caused by lack of water, lack of oxygen, low temperature, etc. When this external limitation is removed, the seed germinates; according to Harper (1957), the seed had enforced dormancy. There is a positive correlation between termination of dormancy and predictable environmental changes.

Wild oats exhibit all three of Harper's classes of dormancy. Harper's system is therefore a classification of mechanism, not of species. In general, termination of dormancy requires exposure to cool, moist conditions, the normal attributes of the temperate zone. Seeds in tropical climates have less and sometimes no seed dormancy.

Figure 5.5 shows how common ragweed succeeds as an early successional plant and a weed. It illustrates integration of Harper's (1957) three dormancy classes (Bazzaz, 1979). Early and late successional environments are different with respect to light intensity and spectral quality. Seed of early successional plants (many are common weedy species) are sensitive to light and seed germination is inhibited by light filtered through plant leaves (P_r form). This is not a problem for weeds that germinate early in the season, before crop leaves filter light. Their germination is favored by fluctuating

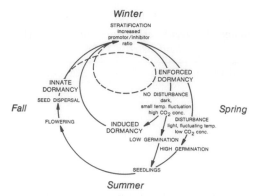

Figure 5.5 Schematic representation of seed germination in common ragweed, a common colonizer in old field succession and a spring annual weed. The dashed line represents seeds that require more than one stratification cycle to germinate and thus assure germination and establishment across a number of seasons. (Bazzaz 1979).

temperatures and low carbon dioxide concentrations in soil. They are not sensitive to soil-water fluctuations and other conditions commonly encountered in cropped fields.

C. Consequences of Weed Seed Dormancy

Dormancy is important because seeds survive for long times in soil and are a continuing source of infestation. Dormancy ensures survival for many years, and the aphorism that 1 year's seeding equals 7 years' weeding is true. One of the first experiments to investigate consequences of seed dormancy was conducted by Duvel near Rosslyn, Virginia (Toole and Brown, 1946). In 1902 seed of 107 different species were buried 8, 22, or 42 inches deep in porous clay flowerpots covered with clay saucers. Samples removed at intervals showed no effect of depth of burial on survival, but a tendency toward longer survival at 42 inches than at 8 inches. His results, summarized in Tables 5.10 and 5.11, show significant seed survival for 38 years. Even after 38 years, 91% of jimsonweed seed was still viable and presumably capable of quickly reinfesting a cropped field.

In 1879, William J. Beal (Darlington, 1951), a pioneer in development of high-yield corn hybrids, began an unusual experiment. He buried 20 pint bottles, each containing 1,000 seeds of 20 weed species, near his lab on the campus of Michigan State University. His aim, in an era before scientific weed control began, was to find out how long seed buried by plowing could

Table 5.10

Number of Weed Species Surviving Burial (Toole and Brown, 1946)

Burial period (years)	Species germinating
1	71
6	68
10	68
20	57
30	44
38	36

survive and thus how long fields had to be left fallow to ensure a weed-free crop when replanted. More than 100 years later, we know the answer to Beal's question: a very long time.

Beal buried bottles upside-down, uncorked, and at an angle so water and oxygen, but not light, could enter. Initially, bottles were dug up every 5 years, and since 1950, every 10 years. Results for 30, 40, 50, and 70 years are reported in Table 5.12. After 70 years, curly dock, evening primrose, and common mullein still germinated. In 1960, three species germinated. In 1970, only one species germinated. In 1980, seeds were planted as usual, in soil sterilized by steam. At first, nothing happened. After several weeks, the first seedling emerged; within 5 months, 29 seedlings had germinated and six died. Of the survivors, 21 were common mullein, one was another species of mullein, and the last was a mallow species that had not germinated

Table 5.11

Germination of Weed Seeds after 38 Years (Toole and Brown, 1946)

91% of jimsonweed
48% of mullein
38% of velvetleaf
17% of evening primrose
7% of lambsquarters
1% of green foxtail
1% of curly dock

Table 5.12

**Results of Beal's Buried Seed Study
(Darlington, 1951)**

Elapsed time (years)	Species still viable
70	Curly dock (8)[a] Evening primrose (14) Common mullein (72)
50	Black mustard Marshpepper smartweed
40	Ragweed Common purslane Redroot pigweed Virginia pepperweed Broadleaf plantain
30	Foxtail Shepherd's-purse

[a]() = % germination.

since 1899. Enough bottles remain to carry on Beal's experiment until 2040. His work shows that fallowing is not a feasible method of weed control for all species.

A study in England (Lewis, 1973), in undisturbed soil, showed that seeds reveal their survival potential during the first 4 years of burial. Rarely did a species that survived 4 years fail to survive 20 years. Seed deterioration occurred more rapidly in acid peat soil than in loam. Seed from the Poaceae were generally short-lived, except timothy. Legumes generally persisted for the full 20 years. The weeds that survived best were common lambsquarters (23%), curly dock (18%), and creeping buttercup (53%).

In Mississippi, seeds of several species were buried in plastic, porous bags to simulate natural conditions and avoid the clay pots of Duvel's experiment and jars of Beal's experiment (Egley and Chandler, 1983). The primary lesson of the Mississippi experiment is that only about six of the species investigated remained viable after 5 years' burial if no new seed was introduced (Table 5.13).

Seed of 41 economically important weed species of the Great Plains region of the United States were buried 20 cm deep (plowing depth) in eastern and western Nebraska in 1976 (Burnside *et al.*, 1996). There were 11 annual grass, 14 annual broadleaved, 4 biennial broadleaved, and 12 perennial broadleaved species. Seeds were exhumed after 1, 9, 12, and 17 years. After 1 year, germination was 57% for all annual grass, 47% for all

Table 5.13

Viability of Weed Seeds after Burial (Egley and Chandler, 1983)

Species	Mean viability after burial for years (%)			
	0	1.5	3.5	5.5
Velvetleaf	99	89	71	30
Purple moonflower	100	84	65	33
Hemp sesbania	100	77	60	18
Common cocklebur	99	27	10	01
Redroot pigweed	96	24	2	01
Common purslane	99	21	2	01
Johnsongrass	86	75	74	48

annual broadleaved, 52% for biennial, and 36% for perennial broadleaved species. Germination dropped steadily with time for each class. After 17 years germination was 4% for annual grasses, 11% for annual broadleaved, 30% for biennial, and 8% for perennial broadleaved species. No explanation was offered for why biennial species survived so well. A conclusion of this study is that after burial at plow depth, germination of annuals will decline rapidly but biennial species will survive well and become problems in crops. The species with the highest survival after 17 years was common mullein, which had 95% germination in western Nebraska. Common mullein was one of the species that survived longest in the Beal (Darlington, 1951) experiment reported earlier. Weed seed germination tended to be greater in the low rainfall and more moderate soil temperatures of western Nebraska.

Soil in Alaska is cold for more of the year than soil in temperate areas. Two studies in Fairbanks, AK (Conn and Farris, 1987; Conn, 1990) showed viability was higher after burial in mesh bags 15 rather than 2 cm deep after 21 months and after 4.7 years. Four of 17 species had 5 to 10% viable seed after 4.7 years, and eight species ranged from 21 to 39%. Viability of American dragonhead did not change during 4.7 years, whereas viability of common hempnettle and quackgrass was zero after 2.7 and 3.7 years, respectively, and viability of two other species was less than 1%.

Taylorson (1970) found that initially nondormant seed of several weed species lost viability after burial sooner and to a greater extent than initially dormant seed. Zorner, *et al.* (1984) found the same thing for kochia seed. Initial rates of loss were much greater in nondormant than in buried dormant populations. After 24 months of burial the number of viable seed remaining and the depletion rates were similar for the two populations.

An important problem in all buried seed studies is the necessity of recovering seed from soil, a complex medium. It is hard to find seed and one must be sure the seed found is the seed that was buried, if longevity is to be estimated. Therefore, all studies use containers. Recent studies use porous, mesh bags that allow transfer of air and water but do not allow other natural processes such as abrasion. However, because seed are concentrated, microbial action and seed interactions may be abnormal. It is generally thought that burial studies overestimate seed longevity. Seed dormancy is a major cause of continuing weed problems, and although a great deal is known about what causes dormancy and how to break it, no one knows how to create it, break it, or use it to manage weeds.

In the laboratory, it is easy to create or break dormancy with a variety of seed treatments (Anderson, 1968). These include abrasive, temperature manipulation, and chemical methods. Abrasive methods include rubbing, dehulling, dipping in sulfuric acid, and alternate wetting and drying to break the seed coat. Temperature manipulation is useful to break dormancy and is common in nature. Alternate freezing and thawing often break dormancy. Stratification or exposure to extremely low temperatures will break dormancy in some seed. Stratification is commonly required to break dormancy in temperate weed species, but rarely works for tropical species. It may act by decreasing the level of an endogenous inhibitor. Finally, chemical methods are used. Leaching with water may remove a chemical inhibitor, and exposure to light will create chemical changes in seed. Chemicals such as potassium nitrate, gibberellic acid, cytokinins, and auxins are all used, and their action is considered to be directed at overcoming the action of or inactivating an inhibitor.

In the field, breaking dormancy on demand is more difficult. Laboratory methods are obviously not suitable to field operations where seed cannot be seen. Plowing soil is a good way to break dormancy and, conversely, not disturbing soil is a good way to maintain dormancy of buried seed. Tillage exposes seed to light (see Chapter 9) and temperature changes. Field methods are nonselective and affect all seeds, so in some species dormancy may be promoted while in others it is released. Weed management will continue to emphasize control until we have obtained a better understanding of weed seed dormancy and developed methods to use it in weed management.

IV. VEGETATIVE OR ASEXUAL REPRODUCTION

Perennial weeds reproduce vegetatively, a most unfortunate aspect of weed management. Simple and creeping perennials also reproduce by seed, but the importance of seed production varies. A good example is water-hyacinth, whose pretty flowers produce seed pods with up to 300 seeds that can live 5 to 15 years submerged in water. But vegetative reproduction can double the size of an infestation in 10 to 15 days and produce floating mats weighing up to 200 tons per acre. Transpired water losses will be 3 to 5 times the loss from an open water surface.

The reproductive organ, the depth to which it penetrates soil, and the importance of seed production for several important perennial weeds is shown in Table 5.14. Seed production is not of great importance for Canada thistle, which is dioecious, and while the pappus is always produced, it does not always have viable seed attached to it. On the other hand, seed production is important for leafy spurge, common nettle, and curly dock.

Many methods of vegetative reproduction are found among weeds. Stolons or creeping above ground stems are found in creeping bent and yarrow. Rhizomes are found in bermudagrass, quackgrass, red top, hedge bindweed, and field horsetail. Bulbs and aerial bulblets are found in wild onion, and wild garlic. Goldenrod has corms–thickened, vertical, underground stems that are reproductive organs. Tubers are produced by yellow and purple nutsedge and Jerusalem artichoke. Reproduction of simple perennials such as dandelion, is from their taproot.

A seedling of a perennial species growing from seed has not yet assumed perennial characteristics (especially the ability to regenerate vegetatively) when it first emerges from soil, and it can be controlled more easily than after it assumes these characteristics. It is generally considered that quackgrass assumes perennial characteristics within 6 to 8 weeks of emergence and johnsongrass after only 3 to 6 weeks. Field bindweed becomes a perennial when it has about 20 true leaves, and yellow nutsedge, 4 to 6 weeks after it emerges from seed. These young plants can be controlled easily before they assume perennial characteristics by tillage or even hoeing.

Seed production of perennials may be unimportant relative to vegetative reproduction (Table 5.14), but it should not be neglected. In April 1990, one field bindweed seed was planted in a small planter, and on April 25 the two-leaf seedling was transplanted to a 4 × 16 × 2 foot box. The plant was harvested on October 19 by opening the box and washing all the soil away with water. The seedling had colonized the entire box, and 197 vertical roots, each about 4 feet long, grew a total of 788 feet. Horizontal root runners from the taproot numbered 34, averaged 4 feet long, and totaled 136 feet of length. They had produced 141 new plants. One little seed can produce a major new weed.[2]

A similar experiment was conducted in Colorado[3] with Canada thistle. One seed was planted in a 2 by 4 by 8 foot box of soil in April 1994. In

[2]Adapted from *Agrichemical Age,* May 1991, p.16.
[3]Westra, P. 1995. Personal communication.

Table 5.14

Characteristics of Important Perennial Weeds (Roberts, 1982)

Species	Reproductive parts and overwintering state	Depth of vegetative reproductive parts[a]	Importance of reproduction by seed
Bermudagrass	Creeping rhizomes, decumbent stems spread laterally	Shallow	Moderately
Bracken fern	Rhizomes; leaves die	Deep	Reproduces by spores
Canada thistle	Creeping roots overwinter; shoots die	Deep	Occasionally produced
Coltsfoot	Rhizomes; leaves die	Very deep	Important
Common nettle	Rhizomes; short green shoots overwinter	Very shallow	Very important
Creeping bent	Aerial creeping stems overwinter	Above ground	Unknown
Creeping buttercup	Procumbent stems; a few leaves overwinter	Above ground	Very important
Curly dock	Taproots; rosette of leaves overwinters	Very shallow, 7 to 10 cm	Very important
Dandelion	Fleshy taproot; few leaves overwinter	Shallow	Important
Field bindweed	Creeping roots overwinter; shoots die	Very deep	Important
Field horsetail	Rhizomes with tubers that overwinter	Deep	Reproduces by spores
Hedge bindweed	Rhizomes overwinter; shoots die	Deep	Rarely produced
Hoary cress	Creeping roots; small rosettes of leaves overwinter	Deep	Important
Japanese knotweed	Rhizomes, dormant underground buds; shoots die	Shallow	None produced
Leafy spurge	Creeping roots overwinter	Very deep	Very
Oxalis sp.	Bulbils, taproots and rhizomes; leaves die	Shallow	Important in some
Perennial sow thistle	Creeping roots; shoots die	Very deep	Important
Quackgrass	Rhizomes with dormant underground buds; shoots overwinter	Shallow	Moderately
Red top	Rhizomes with dormant underground buds; shoots overwinter	Shallow	Very important
Roughstalk bluegrass	Short stolons; a few leaves overwinter	Above ground	Very important
Slender speedwell	Stems creeping on the surface	Above ground	None produced
Wild onion	Offset bulbs and bulbils overwinter	Aerial or very shallow	Rarely produced
Yarrow	Stolons; terminal rosettes of leaves overwinter	Very shallow	Very

[a]Depth varies: Very shallow, 6–10 in.; shallow, 12–18 in.; deep, down to 40 in.; very deep, greater than 120 in.

July 1995, the plant was harvested. If the height of all 142 shoots was added together, the plant would have been 157 feet tall. There were 331 flowers on 60 shoots. Vegetative buds producing new shoots were found up to 4 feet below the soil surface. Total root length was estimated to be 1700 feet.

Tillage can worsen the problem after plants become perennial. Canada thistle spreads by creeping roots, and pieces as small as $\frac{1}{4}$ inch long have produced new plants. Field bindweed spreads by creeping roots, and while they seldom emerge from greater than 4 feet, they can emerge from 20 feet. Pieces as small as 1 inch can produce a new field bindweed plant. Most quackgrass plants, from rhizomes, emerge from the top 12 inches of soil. Deep plowing may therefore be a control method if rhizomes can be permanently buried. Most quackgrass roots are 2 to 4 inches below the surface, and shoots do not emerge from deep in soil. The ability of root segments to produce new plants varies with season and is highest in spring and lowest in fall (Swan and Chancellor, 1976). Many root segments produced shoots, but regeneration of roots was largely from vertical roots.

Leafy spurge roots penetrate up to 20 feet. Over 56% of the total root weight is in the upper 6 inches of the soil profile, and the majority of leafy spurge shoots originate from buds in the top foot of soil. Shoots emerge freely from $1\frac{1}{2}$ feet deep, and some emerge from as deep as 6 feet.

Vegetative buds are not killed by winter freezes. Studies in Iowa on the winter activity of Canada thistle roots showed that buds on horizontal roots continued to develop new shoots until soil was frozen 50 cm deep (Rogers, 1929). When soil finally froze, the shoots were killed but the root bud was not. In January, when soil was still frozen, the latent buds on large roots were larger than they had been in December. By mid-January these buds had developed thick, vigorous shoots up to 20 mm long. By February shoots were 4 to 7 cm long and each had roots 10 to 20 cm long. When the soil thawed, root growth increased rapidly and green shoots appeared by mid-April. Rogers (1929) noted that the cycle of bud and root formation in field bindweed and skeletonleaf bursage was similar to that described for Canada thistle.

THINGS TO THINK ABOUT

1. What is a reasonable range for the number of weed seeds likely to be found in the plow layer of a cropped field?
2. Describe the influence of seed dormancy on weed management.
3. How many different ways are weeds dispersed in space?
4. What are the causes of seed dormancy?

5. What are the classes of seed dormancy? How can these classes be used?
6. How many types of vegetative reproductive organs do weeds possess, and why do they make weeds hard to control?

LITERATURE CITED

Anderson, R. N. 1968. *Germination and Establishment of Weeds for Experimental Purposes.* Weed Sci. Soc. of America, Champaign, IL. 236 pp.

Atkeson, F. W., H. W. Holbert, and T. R. Warren. 1934. Effect of bovine digestion and of manure storage on the viability of weed seeds. *J. Am. Soc. Agron.* **26:**390–397.

Ball, D. A. 1987. Influence of tillage and herbicides on row crop weed species composition. Ph.D. Dissertation, University of Wyoming, Laramie. 157 pp.

Ball, D. A. 1992. Weed seedbank response to tillage, herbicides, and crop rotation sequences. *Weed Sci.* **40:**654–659.

Barrett, S. C. H. 1983. Crop mimicry in weeds. *Econ. Bot.* **37:**255–282.

Barrett, S. C. H., and B. F. Wilson 1981. Colonizing ability in the *Echinochloa crus-galli* complex (barnyardgrass). I. Variation in life history. *Can. J. Bot.* **59:**1844–1860.

Bazzaz, F. L. 1979. The physiological ecology of plant succession. *Ann. Rev. Ecology and Systematics* **10:**351–371.

Beach, C. L. 1909. Viability of weed seeds in feeding stuffs. *VT Agr. Expt. Stn. Bull.* **138:**11–30.

Blackshaw, R. E., and L. M. Rode. 1991. Effect of ensiling and rumen digestion by cattle on seed viability. Weed Sci. 39:104-108.

Bond, W., and P. J. Baker. 1990. Patterns of weed emergence following soil cultivation and its implications for weed control in vegetable crops, pp. 63–68 in Brit. Crop Prot. Council Monograph 45. Organic and Low Input Agric.

Bruns, V. F., and L. W. Rasmussen. 1953. The effects of fresh water storage on the germination of certain weed seeds. I. White top, Russian knapweed, Canada thistle, Morning glory and Poverty weed. *Weeds* **2:**138–147.

Bruns, V. F., and L. W. Rasmussen. 1957. The effects of fresh water storage on the germination of certain weed seeds. II. White top, Russian knapweed, Canada thistle, Morning glory and Poverty weed. *Weeds* **5:**20–24.

Bruns, V. F., and L. W. Rasmussen. 1958. The effect of fresh water storage on the germination of certain weed seeds. III. Quackgrass, Green bristlegrass, Yellow bristlegrass, Watergrass, Pigweed, and Halogeton. *Weeds* **6:**42–48.

Burnside, O. C., R. G. Wilson, S. Weisberg, and K. G. Hubbard. 1996. Seed longevity of 41 weed species buried 17 years in eastern and western Nebraska. *Weed Sci.* **44:**74–86.

Burton, G. W., and J. S. Andrews. 1948. Recovery and viability of seeds of certain southern grasses and lespedeza passed through the bovine digestive tract. *J. Agric. Res.* **76:**95–103.

Cardina, J., and J. E. Hook. 1989. Factors influencing germination and emergence of Florida beggarweed (*Desmodium tortuosum*). *Weed Technol.* **3:**402–407.

Chancellor, R. J. 1985. Tillage effects of annual weed germination. *Proc. World Soybean Res. Conf. III* **3:**1105–1111.

Chepil, W. S. 1946. Germination of weed seeds. I. Longevity, periodicity of germination and vitality of weed seeds in cultivated soil. *Sci Agric.* **8:**307–346.

Comes, R. D., V. F. Bruns and A. D. Kelley. 1978. Longevity of certain weeds and crop seeds in fresh water. *Weed Sci.* **26:**336–344.

Conn, J. S. 1990. Seed viability and dormancy of 17 weed species after burial for 4.7 years in Alaska. *Weed Sci.* **38:**134–138.

Conn, J. S. and M. L. Farris. 1987. Seed viability and dormancy of 17 weed species after 21 months in Alaska. *Weed Sci.* **35:**524–529.

Culotta, E. 1994. Meeting briefs: The weeds that swallowed the west. *Science* **265:**1178–1179.

Currie, R. S., and T. F. Peeper. 1988. Combine harvesting affects weed seed germination. *Weed Technol.* **2:**499–504.

Darlington, H. T. 1951. The seventy-year period for Dr. Beal's seed viability experiment. *Am. J. Bot.* **38:**379.

Dastgheib, F. 1989. Relative importance of crop seed, manure, and irrigation water as sources of infestation. *Weed Res.* **29:**113–116.

Dewey, S. A., and R. E. Whitesides. 1990. Weed seed analyses from four decades of Utah small-grain drillbox surveys. *Proc. West. Soc. Weed Sci.* **43:**69–70.

Donald, W. W., and J. D. Nalewaja. 1991. Northern great plains, pp. 90–126 in. W. W. Donald (ed.) *Systems of Weed Control in Wheat in North America.* Monograph #6 Weed Sci. Soc. Am., Champaign, Illinois.

Dore, W. G., and L. C. Raymond. 1942. Viable seeds in pasture soil and manure. *Sci. Agric.* **23:**69–79.

Dyer, W. E. 1995. Exploiting weed seed dormancy and germination requirements through agronomic practices. *Weed Sci.* **43:**498–503.

Egley, G. H., and J. M. Chandler. 1983. Longevity of weed seeds after 5.5 years in the Stoneville 50-year buried seed study. *Weed Sci.* **31:**264–270.

Egley, G. H., and R. D. Williams. 1990. Decline of weed seeds and seedling emergence over five years as affected by soil disturbance. *Weed Sci.* **38:**504–510.

Fogg, J. M., Jr. 1942. The silent travelers. *Brooklyn Botanic Garden Record* **31:**12–15.

Fogg, J. M., Jr. 1966. The silent travelers. *Plants and Gardens* **22:**4–7.

Forcella, F., R. G. Wilson, K. A. Renner, J. Dekker, R. G. Harvey, D. A. Alm, D. D. Buhler, and J. Cardina. 1992. Weed seedbanks of the U.S. corn belt: Magnitude, variation, emergence and application. *Weed Sci.* **40:**636–644.

Forcella, F., R. G. Wilson, J. Dekker, R. J. Kremer, J. Cardina, R. L. Anderson, D. Alm, K. A. Renner, R. G. Harvey, S. Clay, and D. D. Buhler. 1997. Weed seedbank emergence across the corn belt, 1991–1994. *Weed Sci.,* **45:**67–76.

Gardner, C. J., J. G. McIvor, and A. Jansen. 1983. Survival of seeds in the digestive tract and feces of cattle. Annu. Rep. 1982–1983, pp 120-121. CSIRO Div. of Trop. Pastures. Brisbane, Australia.

Gould, F. 1991. The evolutionary potential of crop pests. *Am. Scientist* **79:**496–507.

Gunsolus, J. L. 1990. Mechanical and cultural weed control in corn and soybeans. *Am. J. Alt. Agric.* **5:**114–119.

Harmon, G. W. and F. D. Keim. 1934. The percentage and viability of weed seeds recovered in the feces of farm animals and their longevity when buried in manure. *J. Amer. Soc. Agron.* **26:**762–767.

Harper, J. L. 1957. The ecological significance of dormancy and its importance in weed control. Proc. 7th Int. Conf. Plant Protection, Hamburg, pp. 415–420.

Hill, T. A. 1977. *The Biology of Weeds.* E. Arnold, London.

Holm, L. G., D. L. Plucknett, J. V. Pancho, and J. P. Herberger. 1977. The World's Worst Weeds: Distribution and Biology, pp. 32–40 Univ. Press of Hawaii, Honolulu.

Johnson, W. C. III, and B. G. Mullinix, Jr. 1995. Weed management in peanut using stale seedbed techniques. *Weed Sci.* **43:**293–297.

Karssen, C. 1982. Seasonal patterns of dormancy in weed seeds. In *The Physiology and Biochemistry of Seed Development, Dormancy and Germination* (A.A. Kahn, ed.). Elsevier, New York.

King, L. J. 1966. *Weeds of the World: Biology and Control,* p. 220. Interscience, New York.

Klingman, G. C., and F. M. Ashton. 1982. *Weed Science: Principles and Practices,* 2nd. Ed. Wiley Interscience, New York.

Koch, W. 1969. Influence of environmental factors on the seed phase of annual weeds, particularly from the point of view of weed control. Habilitations-schrift Landw. Hochech. Univ. Hohenheim, *Arbeiten der Univ. Hohenheim* **50:**204.

Lewis, J. 1973. Longevity of crop and weed seeds: Survival after 20 years in soil. *Weed Res.* **13:**179–191.

Mt. Pleasant, J., and K. J. Schlather. 1994. Incidence of weed seed in cow (*Bos* sp.) manure and its importance as a weed source for cropland. *Weed Technol.* **8:**304–310.

Norris, R. F. 1992. Case history for weed competition/population ecology: barnyardgrass (*Echinochloa crus-galli*) in sugarbeets (*Beta vulgaris*). *Weed Technol.* **6:**220–227.

Oswald, E. J. 1908. The effect of animal digestion and fermentation of manures on the vitality of seeds. *MD Agr. Expt. Stn. Bull.* **128:**265–291.

Pons, T. L. 1991. Induction of dark dormancy in seeds: its importance for the seed bank in the soil. *Funct. Ecol.* **5:**669–675.

Proctor, V. W. 1968. Long distance dispersal of seeds by retention in the digestive tract of birds. *Science* **160:**321–322.

Randall, J. M. and J. Martinelli (eds.). 1996. *Invasive Plants: Weeds of the Global Garden.* Brooklyn Botanic Garden Handbook No. 149. 111 pp.

Reeves, T. G., G. R. Code, and C. M. Piggin. 1981. Seed production and longevity, seasonal emergence, and phenology of wild radish (*Raphanus raphanistrum* L.). *Aust. J. Exp. Agric. Anim. Husb.* **21:**524–530.

Ridley, H. N. 1930. *The Dispersal of Plants throughout the World,* pp. 360–368. L. Reeve, Ashford, U. K.

Roberts, H. A. (ed.) 1982. *Weed Control Handbook: Principles,* 7th ed., pp. 32–33. Brit. Crop Protection Council.

Robinson, H. J. 1949. Annual weeds, their viable seed population in soil and their effects on yield of oats, wheat and flax. *Agron. J.* **41:**515–518.

Rogers, C. F. 1929. Winter activity of the roots of perennial weeds. *Science* **69:**299–300.

Salisbury, E. J. 1961. *Weeds and Aliens.* N. N. Collins, London. 384 pp.

Schmitz, D. C. 1995. Diversity disappears in Florida. *Newsweek,* March 13, p. 14.

Schweizer, E. E., and R. L. Zimdahl. 1984a. Weed seed decline in irrigated soil after rotation of crops and herbicides. *Weed Sci.* **32:**84–89.

Schweizer, E. E., and R. L. Zimdahl. 1984b. Weed seed decline in irrigated soil after six years of continuous corn (*Zea mays*) and herbicides. *Weed Sci.* **32:**76–83.

Smith, E. V. and E. L. Mayton. 1938. Nut grass eradication studies: II. The eradication of nut grass, *Cyperus Rotundus* L. by certain tillage treatments. *J. Am. Soc. Agron.* **30:**18–21.

Stevens, O. A. 1932. The number and weight of seeds produced by weeds. *Amer. J. Botany* **19:**784–794.

Stevens, O. A. 1957. Weights of seeds and numbers per plant. *Weeds* **5:**46–55.

Swan, D. G., and R. J. Chancellor. 1976. Regenerative capacity of field bindweed roots. *Weed Sci.* **24:**306–308.

Taylorson, R. B. 1970. Changes in dormancy and viability of weed seed in soils. *Weed Sci.* **18:**265–269.

Thill, D. C., D. L. Zamora, and D. L. Kambitsch. 1986. The germination and viability of excreted common crupina (*Crupina vulgaris*) achenes. Weed Sci. 34:273–241.

Tildesley, W. T. 1937. A study of some ingredients found in ensilage juice and its effect on the vitality of certain weed seeds. *Sci. Agr.* **17:**492–501.

Toole, E. H., and E. Brown. 1946. Final results of the Duvel buried seed experiment. *J. Agr. Res.* **72:**201–210.

Vega, M. R., and J. N. Sierra. 1970. Population of weed seeds in a lowland rice field. *Philipp. Agric.* **54:**1–7.

Warnes, D. D., and R. N. Anderson. 1984. Decline of wild mustard (*Brassica kaber*) seeds in soil under various cultural and chemical practices. *Weed Sci.* **32:**214–217.

Wilson, R. 1980. Dissemination of weed seeds by surface irrigation water in Western Nebraska. *Weed Sci.* **28:**87–92.

Young, J. A. 1991. Tumbleweed. *Sci. Am.* March:82–87.

Zorner, P. S., R. L. Zimdahl, and E. E. Schweizer. 1984. Effect of depth and duration of seed burial on kochia (*Kochia scoparia*). *Weed Sci.* **32:**602–607.

Chapter 6

Weed Ecology

FUNDAMENTAL CONCEPTS

- Weed ecology is the study of the adaptive mechanisms that enable weeds to do well under conditions of maximum soil disturbance.
- There is a strong human influence on weed ecology.
- There are good reasons for the shift toward ecologically based weed management systems.
- Species are products of natural selection, and they interact with their environment to obtain the resources for growth. The rate of supply and amount of those resources determine growth.
- Plant competition occurs when two or more plants seek what they need and the immediate supply is below the combined demand.
- There are characteristics that make some plants more competitive than others.
- The effect of weeds on crop yield is best described by regression analysis that yields a straight-line relationship for lower densities, but a curvilinear relationship over all possible densities.
- Mathematical models are used in research studies of weed–management, but are not yet widely used by farmers to manage weeds. Models will be used more in the future as they are perfected and tested against biological knowledge.

LEARNING OBJECTIVES

- To understand the importance of ecological relationships to weed management systems.

- To know the components of the weed–crop ecosystem.
- To understand weed–environmental interactions.
- To know the factors affecting weed–crop associations.
- To understand the role of fundamental ecological concepts in weed management and weed establishment.
- To be able to define plant competition and to know what resources plants compete for.
- To understand what characteristics make a plant competitive.
- To appreciate the magnitude of crop yield loss from weeds.
- To understand the current and potential role of mathematical models in crop–weed interference research and weed management recommendations.

Plant ecologists study the reciprocal arrangements between plants and their environment. The concern is how climate, soil or edaphic, and biotic factors affect plant growth, development, and distribution. Weed scientists are further concerned with how weed management affects weed and crop growth and development. For many years, ecologists emphasized only natural environmental factors in studies of reciprocal arrangements, plant distribution, and behavior. We now realize the importance of the role people play in ecological interactions and the human role is particularly evident with weeds.

Weed ecology emphasizes the adaptive mechanisms that enable weeds to survive and prosper under conditions of maximum soil disturbance. It is concerned with growth and adaptations that enable weeds to exploit niches in environments disturbed by people. The most successful weed management programs will be developed on a foundation of adequate ecological understanding.

High food production from annual crops requires repression of ecological succession (to the left in Figure 1.1). Production of food for humans from natural vegetation is presumed to be low (Figure 1.1). Fiber and food crops have high food productivity potential, but the fields in which they are grown are disturbed ecological sites. Farming often works against, rather than with the natural order to produce high-value crops. Perennial crops (e.g., coconuts or apples) may create a permanent pre-climax state, and, although this state is more ecologically stable, it will revert to the left side of the curve in Figure 1.1 if not tended.

Rangelands, forests, and other areas of native vegetation present relatively closed habitats. Most weeds are not good invaders of these sites and may not be weedy (i.e., good invaders) in the classic ecological sense. Monocultural cropping seldom utilizes all the moisture, nutrients, or light available in a given field and creates open ecological niches that weeds fill. There are few similar niches in undisturbed prairies and forests.

I. HUMAN INFLUENCES ON WEED ECOLOGY

Farmers try to provide pure culture conditions for crops and limit the incidence and spread of weeds. Their task is complicated because people have carried weed seed across the globe while traveling, in grain, in seed shipments, with armies, and when moving animals. As discussed in Chapter 5, many weeds of temperate Europe and North America migrated west from the early centers of civilization in the eastern Mediterranean region to Europe, and thence to the United States with immigrants from Europe to the New World. That is a major reason many of the dominant U.S. weeds can be traced to Europe rather than Asia. A few examples of people's major, but usually unwitting, influence on weed ecology follow.

When the first settlers came to the great plains of the United States and some of the Pacific Northwest, they found bluebunch wheatgrass and needlegrass dominating the sides of wagon trails. When roads were cut, downy brome invaded and dominated roadsides. When chemical weed control was available, roadsides were sprayed with the triazine herbicide simazine to control downy brome. Simazine worked well and, because it persists in soil, it prevented reinvasion by downy brome and other grasses. It does not control sandbur, which became the dominant species on many western roadsides (Muzik, 1970).

Downy brome had arrived in the interior Pacific Northwest United States by 1889 (Mack, 1981). It was deliberately introduced at least once in the search for new grasses for overgrazed, denuded range. By 1928 it reached its present distribution (Mack, 1981), although not its current density or ubiquity on over 100 million western U.S. acres (Devine, 1993). Devine (1993) calls it the nation's most destructive plant. Downy brome has more than half the 12 features Baker (1974) considered characteristics of the ideal weed (Mack, 1981). It thrived in two human-created ecosystems of the intermountain west: winter wheat and rangeland. It persisted as land was converted from one use to the other.

Russian thistle was introduced in grain imported from Russia to a farm in Bonhomme County, South Dakota, about 1877 (Dewey, 1894). By the 1890s the infestation was so extensive in North Dakota that the value of wheat production lost exceeded the taxes collected by the state (Young, 1991). Agriculture created the conditions for the weed's success and humans aided its dispersal. Without agriculture it would have remained innocuous (Young, 1991). The pioneer farmer's practice of destroying tall and mid-height prairie grass land to plant cereal grains created the right ecological niche to assure Russian thistle's success. Russian thistle, similar to many annual weeds, competes poorly with established plants. It cannot tolerate shade or long periods of high moisture (Young, 1991). But it thrived on dry, disturbed, cereal grain land that often was fertilized.

An examination of early weed science literature reveals that the dominant weed problem in many different crops, and especially in small grain crops such as wheat and barley, was annual broadleaved weeds. After the development and widespread use of selective phenoxy acid herbicides (e.g., 2,4-D, MCPA), there was a gradual shift from annual broadleaved weeds, which are controlled effectively by the phenoxy acid herbicides, to annual grass weeds and broadleaved species not susceptible to phenoxy acid herbicides. This induced ecological change was not intentional but it was inevitable. A similar change has been seen in corn, where the widespread use of triazine herbicides has eliminated many annual broadleaved and grass weeds and created an ecological niche for invasion by annual and perennial grass weeds and yellow nutsedge.

Green revolution cultivars created during the 1960s helped feed the world and avoid starvation in many of the world's developing countries. The new cultivars changed the architecture (the shape) of wheat and rice plants and led to higher grain yields when appropriate production needs (fertilizer and irrigation) were provided. The green-revolution cultivars also changed the harvest index (proportion of the total plant harvested as grain). Snaydon (1984) posits that the main factor leading to an increase in harvest index over several years was selection for shorter straw length, and this has had important effects on weeds. Short stiffer straw (stems) was less likely to lodge with higher rates of nitrogen fertilizer. This same characteristic also opened the plant canopy to more light and changed the light environment for weeds. Thus, a change in plant architecture worsened the weed problem by providing more light to stimulate seed germination and growth of seedling weeds.

Agricultural practice nearly always has changed for reasons of convenience, to save labor, or to increase returns. Rarely has it been changed to intentionally reduce weed growth. The opposite has happened. Indian balsam, a Himalayan wet woodland native, is a showy plant, 2 meters tall, with pink flowers. It found the European climate favorable and escaped but did not become a common weed in Britain until about 1930 when use of artificial fertilizer became widespread. Until then, in attempts to maximize yield, hay was harvested right up to riverbanks where Indian balsam flourishes. When fertilizer was used, yields increased and harvest of riverbanks was no longer necessary. The weed then flourished along previously harvested riverbanks. It has been reported that Indian balsam is invading riverbanks in the Czech republic and moving to wet woodlands where it crowds out the forest's less aggressive species (Anonymous, 1995).

Table 6.1

Components of Production Systems Controlled by Man that Are Relevant to Weed Management[a]

Component	Influence on species composition[b]	Density
Soil tillage	9	9
Water—irrigation	9	5
Nutrient supply—fertilization	9	7
pH—liming	9	5
Date of planting	7	7
Growing period of crop	6	3
Shading period and intensity	6	8
Seed dispersal at harvest	3	5
Seed cleaning before planting	4	2
Weed control	9	9

[a]Source: Koch, W. 1988. Personal communication.
[b]Influence is ranked on a scale of 0 to 10, with 0 indicating no influence and 10 equaling maximum influence.

A shift from annual to perennial weeds has been documented in Japanese rice culture and attributed to the extensive use of herbicides that controlled annual weeds.[1]

United States agriculture has shifted from a mix of crops on a farm to extensive monoculture. Wheat is the dominant crop in the central Great Plains of the United States and soybeans and corn dominate Midwestern states. Cotton and soybeans dominate in the southern U.S. states. These monocultural environments create ecological changes that influence what weeds will succeed.

Conscious introduction, multiplication, and release of parasites and predators for biological control of pests is also ecological change. To date, this is a less important shift in ecological relationships than those mentioned earlier, but the careful weed manager needs to be aware of how such changes affect weeds.

Each agricultural practice has a potential to influence the density and survival of species in a cropped field. The foregoing are a few examples of how human activity influences weeds. Production practices that influence weeds are shown in Table 6.1, with an estimate of the relative importance of each to species composition and weed density.

[1]Itoh, K. Nat. Agric. Res. Center, Isukuba, Ibaraki, Japan, Personal communication.

Ghersa *et al.* (1994) state that "in modern agriculture, social and biological systems have diverged" in their influence. Weed management practices
are more and more uncoupled from biology. They are controlled and designed by social and economic forces that are often devoid of a biological
base. This represents the ultimate human influence on weed ecology because
it is neglect of ecology.

II. THE WEED–CROP ECOSYSTEM

Herbicide use has masked the importance of weed prevention and the
need to understand weed–crop ecology. Understanding weed–crop ecology
will lead to more effective weed prevention, management, and control.
There is a shift toward ecologically based weed management systems for
at least six reasons:

1. Weeds highly susceptible to herbicides have been replaced by
   species more difficult to control.
2. Herbicide resistance has developed in many weed species and
   some weeds are resistant to several herbicides.
3. There are weed problems in monoculture that cannot be solved
   easily with present techniques.
4. New weed problems have appeared in reduced and minimum
   tillage systems.
5. Economic factors have forced consideration of alternative control
   methods.
6. There is increased awareness of the environmental costs of
   herbicides.

Aldrich (1984) diagrammed the weed–crop ecosystem (Figure 6.1). For
too long, weed scientists have focused primarily on weed–crop interactions
and on protecting crops from weeds. Figure 6.1 suggests that weed management must deal with interaction of all factors, rather than just two. There
is a lack of knowledge about these interactions. It is not the intent of this
book to discuss these interactions in depth; other available books do that
well (Aldrich, 1984; Harper, 1977; King, 1966; Radosevich and Holt, 1984;
Cousens and Mortimer, 1995).

Weed science has been dominated by control technology that focused
on how to control weeds in a crop. As weed management systems are
developed, ecological knowledge will be essential, and the complexity
shown in Figure 6.2 must be considered. As complex as Figure 6.2 is, it is

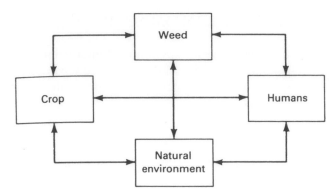

Figure 6.1   The weed-crop ecosystem (Aldrich, 1984).

too simple to represent all factors that affect weed–crop relationships and that should be considered as management systems are developed.

From genes to organisms (individuals) to populations and communities, relationships are the essence of life. The weed–crop system is a product of interactions–its essence is relational as illustrated in Figures 6.1 and 6.2. All levels of life are interdependent, and no level can exist independent of another. The individual cannot survive for long independent of its population, nor can a population survive without individuals. Weed management systems directed only at weeds are founded on error, and while they may succeed temporarily, they are doomed to fail.

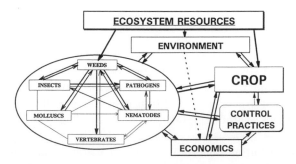

Figure 6.2   The interaction of weeds and other components of the agricultural production system (Norris, 1992).

III. ENVIRONMENTAL INTERACTIONS

There are three important weed–environment interactions: climate, soil, and biota or living organisms. These will be discussed separately but cannot be separated in nature, where life is characterized by interactions.

A. WEEDS AND CLIMATE

The important factors that determine a weed's ecological interactions are light, temperature, water, wind, humidity, and their seasonal aspects–the climate. Yellow nutsedge does well in the subhumid tropics and warm, temperate regions. It does not survive well in temperate areas with prolonged frost. Purple nutsedge thrives in the humid tropics and subtropics with some excursions into subhumid temperate regions. Halogeton thrives under desert conditions of low rainfall and sometimes high alkalinity. Water-hyacinth is an important aquatic weed in the tropics and subtropics, but has not yet invaded temperate waters. Weeds are found in the environment they prefer, and weed control or weed management often may be aided by changing the environment. Irrigation and tillage are major environmental changes that lead to shifts in species composition in the affected areas.

Annual weeds use opportunities to escape environmental stresses. Light intensity, quality, and duration affect weed presence and survival. Photoperiodic responses govern flowering and determine time of seed maturation. If light is too intense or days too long or short, plants will not flower and the species may not endure. Light and temperature response determine a species latitudinal limits. Some weeds tolerate shade well, and their ability to grow under a crop canopy is one reason they succeed. The length of frost-free period or the time soil is frozen affect a weed's ecological relationships. Soil temperature is a primary determinant of seed germination and survival, especially where soil freezes. Freezing also affects winter survival of vegetative reproductive organs. Air and soil temperature are important determinants of species distribution and ecological interactions. Common chickweed survives well in cold climates because it continues to grow in winter without injury (King, 1966). When temperature is below freezing, the weed is often erect and it continues to flower, although the flowers are cleistogamous (without petals and closed). The self-pollinated seeds are fertile.

Seasonal distribution and total supply of water determine species survival. Shortage of water at critical stages is often responsible for reproductive failure, death, or both. The world's arid areas would produce far less food if we did not affect seasonal distribution and total supply of water by irrigation.

Wind can affect water supply through evaporation and an increase of transpiration loss. Wind also affects the microclimate within a plant canopy and the relative concentration of carbon dioxide and oxygen.

Climate may change in the future because of increasing concentration of CO_2 and other triatomic gases that interact with radiant energy. There are reliable scientific data that show the world is warming, and these changes will affect weeds. Agriculture has always been aided and hindered by climate. Crops are vulnerable to unfavorable weather, and weed management may be more difficult during rapid climate change (Patterson, 1995b). It is likely that the negative effects of all agricultural pests will increase with rapid climate change, particularly in less intensively managed production systems. Crops affected by environmental stress will be more vulnerable to attack by insects and diseases and less competitive with weeds (Patterson, 1995a).

B. EDAPHIC FACTORS

Edaphic comes from the Greek *edaphos,* meaning soil or ground. Soil water, aeration, temperature, pH, fertility, fertility source, and the cropping system and associated practices imposed on a soil determine what weeds survive to compete. Many weeds do well in soils too low in fertility for crop production, but others grow only in well-fertilized soil. Few species associate with a soil type. Most weeds can be found in soils differing widely in physical characteristics, moisture content, and pH. This adaptability explains, in part, why they are successful weeds. Some species of Asteraceae and Polygonaceae grow in soils with 1.2 to 1.5% sodium chloride, and although this may not make them better weeds, it illustrates their ability to adapt to diverse environments. Kochia grows well in alkaline or saline soils, but not in acidic soil. Saltgrass can be a weed in turf in alkaline areas where soil pH is 8 or above. Alkaliweed grows only at pH 8 and above. Other species, including common mallow and plantain, are relatively intolerant of alkaline conditions. Crabgrass, a turf weed, grows well on acid soil; bluegrass is sensitive to acidity; and common chickweed is not common in acid soil (King, 1966).

Flooding is a method of weed control. Some water is required for seed germination and plant growth. Too much water changes soil ecology and can control some weeds, as it has in rice for centuries. Several species are adapted to flooding and rice is not free of weeds. No crop plants and few weeds do well in waterlogged soil or compact soil with poor aeration.

Karibaweed is sensitive to salinity and grows only in fresh water. The government of Kuttanad, Kerala, India, wanted to develop rice production. Kuttanad is close to the sea, and salt water decreases or eliminates rice production. Traditionally, farmers had built soil barriers across canals and rivers to prevent incursion of heavier salt water during the growing season. After harvest the barriers were removed to allow incursion of sea water. Experts in the government planning office "knew" that the farmers' practice was not going to be adequate for extensive rice production. Therefore, they built a spillway channel into the open sea and salinity was regulated with a 1400-meter-long regulator channel that checked the advance of salt water. The channel and regulator worked well and the advance of salt water was halted. Invasion of karibaweed was encouraged because salt water no longer invaded the land annually and killed it. Karibaweed stopped rice production because engineers were very good at building structures to stop the sea, but knew nothing about weed ecology. The weed won.

Warm, moist soil conditions are best for germination of weed seed and seedling growth. Dormancy, in temperate regions, is usually associated with cold, freezing conditions.

Soil reaction or pH is an important determinant of what plants grow in an area. However, no generalizations can be made about the influence of soil pH on weeds. LeFevre (1956) reviewed the pH tolerance of 60 weeds and grouped them into basophile (those that love high pH, such as sow thistles, green sorrel, quackgrass, and dandelion), acidophile (those that love acid soil, such as red sorrel and corn marigold), and neutrophile (such as shepherd's-purse, prostrate knotweed, and common chickweed) groups. Some nutrition is essential for plant growth, but most weed species are valueless as indicators of soil reaction or fertility. Luxuriant weed growth does not indicate a highly productive agricultural soil. Weed growth is determined by many factors in addition to a soil's physical and chemical properties These include field cropping history, proximity of sources of infestation, the weed seed population present or supplied to a field, water supply, and growing season conditions. The effects of soil structure, water-holding capacity, and nutrient level are more important than soil type.

C. WEEDS AND BIOTA

Association of weeds and crops is determined largely by the degree of competition offered by a particular crop and weed. It is also determined by cultural operations and rotational practices associated with each crop. The factors contributing to association include the following:

1. *Similarity of seed size.* If a weed seed is similar in size to a crop's seed, it can be a common, unnoticed companion while seeding. It will also be more difficult to clean or separate the weed's seed from the crop's seed. (See Chapter 5.) Weeds have a long record of adapting to agricultural practice. A striking example of seed size mimicry is that of lentil seeds by common vetch (Gould, 1991). Lentil seed is lens-shaped and vetch seed is usually more rounded. In Europe, vetch seed evolved to mimic the shape of lentil seed and made separation nearly impossible.

2. *Time of seed germination and formation.* If weed seed germinates just before or only slightly after a crop's seed, the weed's chances of successful competition are enhanced. If weeds flower and set seed before the crop is harvested, this may ensure their presence in the next crop. These things do not guarantee successful competition, but they are not deterrents to it. A weed whose life cycle is similar to that of the crop will usually be a more successful competitor than one whose life cycle is much shorter or longer than the crop it associates with.

3. *Tillage, rotation, and harvest practices.* Dandelions are common weeds in turf, as are several species of spurge and common chickweed. These are weeds of turf because they are adapted to turf's cultural practices and withstand mowing. The perennial grass quackgrass grows well in the perennial crop alfalfa. The association of wild oats and green foxtail with small grain crops and wild proso millet with corn is related to all of these factors. The weed and the crop have a similar seed size, their time of germination and ripening is nearly identical, and the weeds easily withstand the tillage and harvest practices of the crops. It is unlikely that a weed adapted to survive in plowed fields will do equally well in no- or minimum-till fields (Gould, 1991).

Some weeds germinate after a crop is laid by (after the last tillage has occurred). These include johnsongrass, some of the foxtails, and barnyardgrass. They often germinate later than broadleaved species and elongate rapidly to compete in row crops. Downy brome competes effectively in winter wheat because it germinates in the fall after wheat has been planted, survives over the winter, and develops and sheds seed the next spring, before wheat has completed its life cycle. It is also an effective competitor during the fallow year in a wheat–fallow rotation because it is shallow-rooted, not affected by many cultural operations, and competes for water before the crop has been planted.

Other plants and animals modify the environment: Grazing animals determine weed survival in pastures. Knowledge of crop competition and the relationship of weeds and biota are required to develop better control techniques and management strategies.

Plant environments, and especially cropped fields, are very heterogeneous. A height difference between the top of a furrow and the bottom of only 5 cm may represent a factor of 250 for the smallest weed seeds (Aldrich, 1984). When we irrigate, fertilize, or till a field we perceive uniformity, but across a large or even a small area, weed seeds experience a nonuniform environment. Nonuniform seed distribution in soil is the rule, not the exception. We also know that our management techniques, including herbicides, are not applied uniformly.

There are significant differences in soil temperatures determined by small amounts of litter cover or shading of soil. There are even greater influences on soil moisture and relative humidity of air just above the soil surface, determined by litter and shading. These small environmental differences explain why several different plant species occupy a single environment.

Differences in growth form are often unobserved ecological interactions. They are external expressions of a plant's ability to sample its environment. They are illustrated in Figures 6.3 to 6.8. These differences enable plants to fill different ecological niches. Weeds create and fill ecological niches and change the environment through their germination, growth, and death. They affect moisture, temperature, nutrient supply, and, ultimately, organic matter in soil. Weeds are active, not passive, participants in the agricultural environment.

IV. FUNDAMENTAL ECOLOGICAL CONCEPTS

A. SPECIES

Species is the fundamental biological classification. It is a subdivision of genus and is composed of a number of individuals with a high degree of physical similarity, that can generally interbreed only among themselves, and that show persistent differences from other species. The species *retro-*

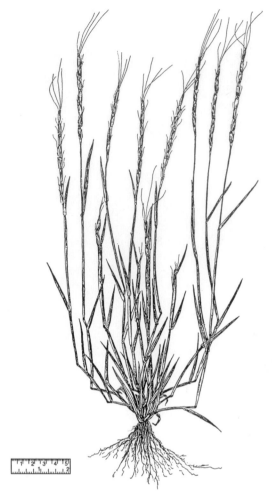

Figure 6.3   The upright, narrow, unbranched leaves of jointed goatgrass.

flexus (redroot pigweed) of the genus *Amaranthus* is consistently different
from the species *spinosus* (spiny amaranth), and they do not interbreed.
Species are products of natural selection and genetic manipulation that
create new gene pools. That is what happens, but the more important, and
more interesting, question is, why does it happen? Organisms are controlled
in nature by the total quantity and variability of the supply of things essential
for growth. All plants have a minimum requirement for various growth
factors and interact with their physical and chemical environment to obtain

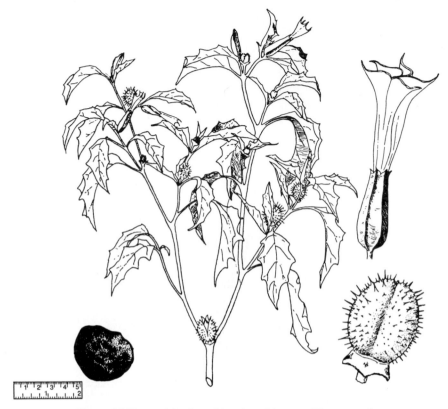

Figure 6.4 The upright, branching, broad leaves of jimsonweed.

them. Plants also have a limit of tolerance to various environmental components. The "why" question is usually answered in terms of rate and amount.
Plant presence and growth are controlled by too little or too much of
the things needed for growth and by the conditions under which they
are available.

Weeds have been continually exposed to conditions that encourage speciation. The major models of speciation are the allopatric and sympatric
models. In allopatric speciation, parent species become physically separated
into daughter populations by geographic separation that restricts or eliminates gene flow between the populations. This occurs because of continual
movement of people and plants from continent to continent or to different
regions within a continent. When weeds are introduced between continents,
species development is a long-term process. Allopatric speciation is the
primary mechanism for development of new species. Charles Darwin's

Figure 6.5 The climbing, twining growth of field bindweed.

Galapagos Island finches are an excellent example of allopatric speciation. Although many weeds may have originated from allopatric speciation, there are no good examples. Weeds have been imported to many places (see Chapter 5 and elsewhere in this chapter), but there has been little to no study of allopatric speciation of weeds. We assume that is what has happened.

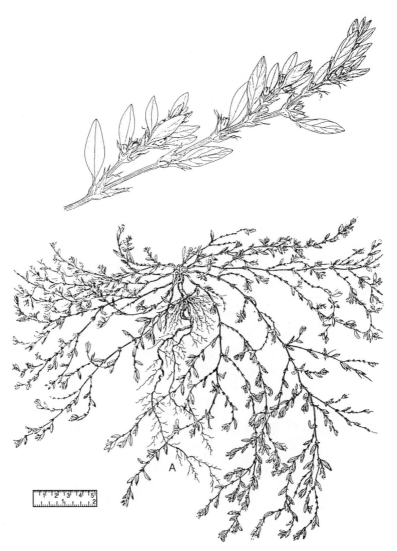

Figure 6.6   The prostrate growth of prostrate knotweed.

In sympatric speciation, a parent species differentiates in the absence of physical restriction on gene flow. Sympatric speciation is a local, short-term process. The continual disturbance of land and changing agricultural practices provide numerous opportunities for hybridization, selection, and

Figure 6.7 The taproot of redroot pigweed.

response to imposed and shifting environmental conditions. Species development has not stopped; weeds in a crop are different today from what they were several years ago, and they will continue to evolve. However, once again there are few examples of weeds that have evolved because of sympatric speciation. One is species of the genus *Passiflora* or passionflower (Harper, 1977). There are about 350 species, and a few are weedy. Nearly

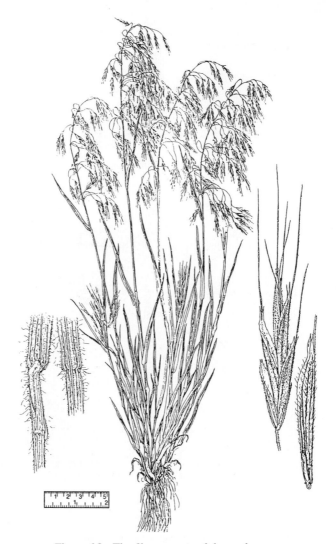

Figure 6.8   The fibrous roots of downy brome.

every species is unique from others. Leaf shape varies enormously, as do leaf surface characteristics. They are identified by feeding habits of the monophagous butterflies of the genus *Heliconius*.

The ready development of ecotypes, or physiological races adapted to various climatic conditions around the world, has occurred in common chickweed and is responsible for its worldwide distribution (Figure 6.9).

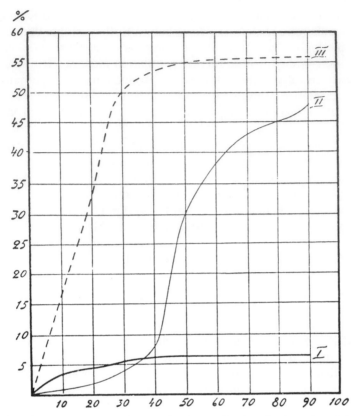

Figure 6.9 The average germination obtained over 100-days for lots of seed from three ecotypes of common chickweed: I, arctic, alpine latitudes; II, oceanic latitudes; III, maritime regions in northern latitudes (Peterson, 1936, cited in King, 1966).

Ecotypes exist in dandelion and in members of many other genera. Their development has implications for weed control and management. Control techniques that work in one place may not work for the same weed in another place because it is not the same weed; it is an ecotype. Ecotype development is sympatric speciation as locally adapted populations are changed. Aldrich (1984) summarized examples of the development of ecotypes for a wide variety of weed species including johnsongrass (Burt, 1974; McWhorter, 1971; McWhorter and Jordan, 1976; Wedderspoon and Burt, 1974), Canada thistle (Hodgson, 1964), common ragweed (Dickerson and Sweet, 1971), yellow nutsedge (Yip, 1978), purslane (Gorske *et al.*, 1979), annual bluegrass (Warwick and Briggs, 1978), and medusahead (Young *et al.*, 1970).

Weeds and other plants have two survival strategies called K and r that define a population's response to disturbance. Both terms are derived from terms in the logistic growth equation where K = environmental carrying capacity and r = the intrinsic rate of population increase. The so-called r reproductive strategy is characterized by production of a large number of seeds (or vegetative reproductive units) and high dispersability. It is the potential rate of increase of a population for a given set of environmental conditions where there is no shortage of resources or any other constraints on growth. The r strategy dominates among annual weeds and is expressed in competitive ability, seed germination, seed dormancy, and seed longevity.

Plants adopting the K reproductive strategy depend on exploitation. They have fewer reproductive units, relatively low dispersability, and strong exploitative ability. K measures an upper size limit beyond which populations cannot go. The limit is determined by available resources and other constraints on population growth. Large-seeded annual weeds (some authors consider sunflower to be a good example, but others disagree) and many perennial weeds generally utilize the K reproductive strategy. Plants with the K strategy are usually not first colonizers.

Some species combine K and r strategies. Canada thistle, for example, is r for vegetative growth because it produces a large number of vegetative buds and its creeping roots disperse plants. At the same time, Canada thistle is K for seed production. It is a dioecious plant and usually produces few seeds that have high dispersability but are not strongly exploitative.

It is important to point out that survival strategy as depicted by K and r reproduction is not equivalent to, and should not be confused with, competitive ability, which is controlled by other factors.

Undisturbed plant communities generally have a large number of a few species and a few individuals of many different species. Undisturbed communities are more complex than disturbed communities. Farmers want fields to be dominated by a single species—the crop and disturb (plow, cultivate, control weeds, etc.) fields to achieve that goal. Crop dominance is favored by weed control. Weed management systems that rely on single control techniques stabilize weed populations and encourage emergence of weeds that are not affected by the control technique.

B. The Community

The crop–weed community is important to weed management because it is the organizational level where change occurs. Change can occur within a species by mutation and ecotype development or by replacement of one species with another. There are at least four reasons (Harper, 1977) why two or more species coexist. They can have the following:

1. Different nutritional requirements, as illustrated by legumes and grass coexisting in pasture and hay fields.
2. Different causes of mortality, observed in pastures where animals selectively graze.
3. Different sensitivity to toxins, including allelochemicals and herbicides.
4. A different time demand for growth factors. Many plants require the same things to grow, but they do not demand them at the same time. This may be the most common reason for coexistence.

C. ECOLOGICAL SUCCESSION

Ecological succession is a natural, orderly, continuous process. In agriculture, it occurs in continually disturbed areas from which the natural community has been removed. Environmental modification is a driving force for succession, and agriculture is conducted by modifying and controlling the environment. Dominance is found in agricultural plant communities, and usually a few (rarely only one) weeds dominate cropped fields in modern agriculture. Their removal creates open niches, and another species will move in, usually not immediately. Therefore, weed control, especially successful weed control, is a never-ending process. Weed management may be designed best when it achieves less than 100% control and is thus not as successful at opening niches and creating an endless process of succession. The best weed management systems may combine techniques to gain the desired level of control, but not a completely open environment that encourages arrival of new weeds that are not controlled by present techniques and thus may be more difficult to control.

V. PLANT COMPETITION

Plant competition is part of plant ecology. To compete comes from the Latin *competere,* which means to ask or sue for the same thing another does.

Imagine yourself having the good fortune to receive free tickets to your favorite football team's next home game. Your tickets are a little way up on the 50-yard line, or, if you are very fortunate, in someone's private box. You know you'll see vigorous competition as the two teams charge up and down the field competing for the ball, for scores, for glory, and perhaps, even for money.

The next time you drive around in the spring or summer, a careful look at most agricultural fields will reveal competition just as vigorous as that you would have seen at the football game. You won't see the plants leaping up and running around and into each other, but they will be competing vigorously for environmental resources. There is no glory, and not as much money is involved, but the competition is real.

Competition is what weed control is all about. If weeds were just there and benign, we wouldn't care much about them. Because they compete (see Chapter 2), we are compelled to care and attempt to control or manage them.

Among the references to weeds, some of the earliest and frequently quoted ones are in the Bible.

> And some fell among the thorns and the thorns sprang up and choked them.
>
> Matthew XIII:7

> Cursed is the ground for thy sake; in sorrow shalt thou eat of it all the days of thy life; thorns and thistles shall it bring forth to thee; and thou shalt eat the herb of the field.
>
> Genesis III:17-18

The Reverend T. R. Malthus, in his essay on the principle of population (1798), said, "The cause to which I allude is the constant tendency in all animated life to increase beyond the nourishment prepared for it." Malthus' concern was the increasing human population and consequent poverty and misery he saw in Liverpool, England. The Malthusian apocalypse, when the human population is greater than the ability of the earth to produce food, has been avoided because of developments in food production technology. The apocalyptic possibility still concerns many.

A. PLANT COMPETITION DEFINED

Clements *et al.* (1929) said competition is a question of the reaction of a plant to the physical factors that encompass it and the effect of these upon adjacent plants. For them, competition was a purely physical process.

> In the exact sense, two plants—no matter how close, do not compete with each other so long as the water content, the nutrient material, the light and heat are in excess of the needs of both.

> Competition occurs when each of two or more organisms seeks the measure they want of any particular factor or things and when the immediate supply of the factor or things is below the combined demand of the organisms.
>
> Clements *et al.*, 1929

This definition makes competition different from the broader term *inter-ference* that includes competition, allelopathy (see Chapter 7), and other environmental modifications. Competition is for something in limited sup-ply. For the physiologist, competition is usually for things. For agronomists and weed scientists, competition is often between individuals (Donald, 1963).

B. Factors Controlling the Degree of Competition

Figure 6.10 illustrates the factors determining the degree of competition encountered by an individual plant. For weeds, density, distribution, and duration (how long weeds are present) are important. For crops, density, distribution (including spacing between rows and spacing in the row), and duration (whether or not thinning is required) are important. These factors, modified by soil (edaphic) and climatic conditions, determine the degree of competition encountered by each plant. The primary things plants com-pete for are nutrients, light, and water. When any one is lessened, others

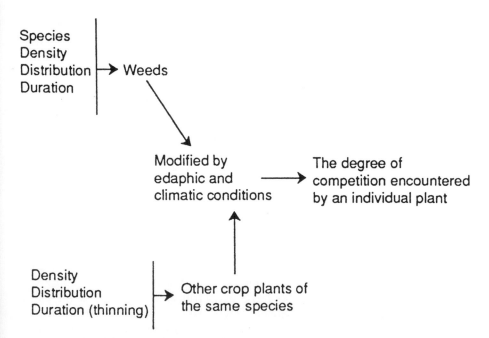

Figure 6.10   Schematic diagram of the competition encountered by a plant (Bleasdale, 1960).

cannot be used as effectively. Plants may compete for heat, but it is difficult to conceptualize how they do so. It is well known that accumulation of degree days enhances plant growth:

Degree day = Daily max. temp. − Daily min. temp. − Threshold temp/2

The threshold temperature differs between species and is the temperature below which the plant does not grow. Because they do not grow well at high temperatures, there is a maximum cutoff temperature, in the range of 30°C, for many plants. Plants grow better when it is warm, but no studies have reported competition for heat.

Yield reductions are generally in proportion to the amount of light, water, or nutrients weeds use at the expense of a crop. A very general rule is that for every unit of weeds grown, there will be one less unit of crop grown. Inconsistent results between weed management experiments in a single year or between years are regularly attributed to environmental (i.e., light, water, nutrient, or climatic) variation. In most cases data are insufficient to define cause and effect.

It is simple and neat to separate the elements of competition (nutrients, light, and water). H. L. Mencken (1880–1956) reminded us that "For every human problem, there is a solution that is simple, neat, and wrong." It is not wrong to separate the elements of competition experimentally, but it is wrong to assume that plants do so and it is nearly impossible to separate the elements of competition in nature. DeWit (1960) was among the first to point out the futility of separating the elements of competition. His work changed the approach to the study of competition. He derived mathematical expressions for competition and advocated consideration of space and what it contained, rather than studies that separated the components of competition. For example, competition for light affects growth, which in turn affects a plant's ability to compete for nutrients and water. Competition will be greatest among similar species that demand the same things from the environment. Those species that best use or first capture environmental factors will succeed.

Only in recent years has research progressed to consider the spatial distribution or where weeds are in a field. Weed scientists have long been concerned with what weeds (the species) and how many weeds (the density) are present in a field. Control has been directed at the dominant weed or weeds. Studies of weed biology have emphasized seed production, seed dormancy and survival, and seedling growth, establishment, and survival. Results of these good studies have been translated into areas (acres or hectares) without considering the patchiness or nonuniformity of weeds in all fields. Control included the usually unstated assumption that weed distribution and density were uniform over the field. Thus, tillage or herbicides are nearly always applied uniformly over the field even though weed scientists agree the weeds are not distributed uniformly. Farmers and others who try to manage weeds have long recognized that weed distribution in a field is not uniform and control practices are unnecessary in some places. Weed distribution is heterogenous, not homogenous (see Chapter 19 on the importance of mapping weed populations). The technology for weed and crop recognition systems that control weeds in parts of fields rather than whole fields is developing. Biological knowledge to define how the seedbank, seed dispersal, plant demography, and habitat interact to determine the stability of weed or weed seed distribution across fields and across time is not as developed (see Mortimer and Cousens, 1995, Chapter 7). There is also a poor understanding of how control techniques affect weed and weed seed distribution over time. As this knowledge develops, weed managers will be able to manage weeds on less than a whole-field basis, and that will lead to reduced need for tillage and herbicides (Mortensen *et al.*, 1996; Johnson *et al.*, 1995). The dynamics of patches, defined as how inherent weed biology interacts spatially with landscape characteristics (Cousens and Mortimer, 1995), is an important area of weed management research. Weed scientists want to understand why weeds are where they are, rather than know only what species are present and use the spatial information as another tool to predict and manage weed populations.

C. COMPETITION FOR NUTRIENTS

Nitrogen, phosphorus, and potassium are primary plant nutrients. One mustard plant needs twice as much nitrogen and phosphorus, four times as much potassium, and four times as much water as an oat plant. Success in gaining nutrients may lead to more rapid growth and successful competition for light and water. Fertilization is used to improve crop growth, but may worsen the weed problem.

Table 6.2 shows the pounds of nutrients required to produce equal amounts of dry matter for three crops and five weeds that frequently compete with the crops. The important point about these data is not that weeds require greater amounts of nitrogen and phosphorus than crops. Consumption of nitrogen and phosphorus for weeds and crops is very similar. The point is that weeds require the same nutrients, at the same time, and are often more successful in obtaining them. Remember, competition occurs when two or more organisms seek what they want or need and the supply falls below the combined demand.

Table 6.3 compares the nutrient content of weed-free corn, corn-free redroot pigweed, and corn grown with redroot pigweed (Vengris *et al.*,

Table 6.2

**Kilograms of Nutrients Required to Produce Equal
Amounts of Dry Matter**

Plant	Nitrogen	Phosphorus
Wheat	5.5	1.2
Oats	4.9	1.7
Barley	8.4	2.6
Common lambsquarters	7.6	1.6
Common ragweed	6.6	1.4
Redroot pigweed	5.1	1.4
Common purslane	3.1	0.8
Mustards	9.8	2.7

1953). When weed-free corn was set at a nutrient content of 100, in all cases except phosphorus, redroot pigweed grown alone contained more of each of the nutrients than did corn. The more interesting data are those in the center row, where the nutrient content of corn infested with redroot pigweed is shown. In every case, nutrient content was reduced. In another study (Vengris *et al.*, 1953), corn was compared with six annual broadleaved weeds and one annual grass (Table 6.4). Weeds contained 1.6 to 7.6 times more of each nutrient. In this study, application of supplemental phosphorus made several weeds more competitive. High fertility did not reduce the detrimental effects of weeds on corn.

A similar study in Poland with wheat, barley, sugarbeets, and rape (Malicki and Berbeciowa, 1986) (Table 6.5) showed that the mineral content

Table 6.3

**Comparison of Nutrient Content of Weed-free Corn, Corn and Redroot
Pigweed, and Redroot Pigweed Alone (Vengris *et al.*, 1955)**

Species	Relative nutrient content				
	N	P_2O_5	K_2O	Ca	Mg
Weed-free corn			100		
Corn infested with redroot pigweed	58	63	46	67	77
Redroot pigweed	102	80	124	275	234

Table 6.4
Mineral Composition of Corn and Weeds (Vengris *et al.,* 1953)

Species	Mean percent composition				
	N	P	K	Ca	Mg
Common lambquarters	2.6	0.4	4.3	1.5	0.5
Common purslane	2.4	0.3	7.3	1.5	0.6
Corn	1.2	0.2	1.2	0.2	0.2
Crabgrasses	2.0	0.4	3.5	0.3	0.5
Galinsoga	2.7	0.3	4.8	2.4	0.5
Pigweeds	2.6	0.4	3.9	1.6	0.4
Ragweeds	2.4	0.3	3.1	1.4	0.3
Smartweeds	1.8	0.3	2.8	0.9	0.6

Table 6.5
Nitrogen, Phosphorus, and Potassium Content of Wheat and Barley (Grain, Straw, and Roots) and Selected Annual Weeds (Malicki and Berbeciowa, 1986)

Species	Percent dry matter		
	N	P	K
Barley	1.5	0.2	1.4
Canada thistle	1.6	0.3	2.0
Common chickweed	2.1	0.6	3.8
Common hempnettle	2.0	0.4	2.3
Common lambsquarters	2.7	0.4	4.1
Corn speedwell	1.5	0.3	1.7
Field bindweed	2.7	0.3	2.7
Hairy vetch	3.0	0.2	1.1
Perennial sowthistle	2.3	0.5	4.0
Shepherd's-purse	1.6	0.3	2.0
Wheat	1.2	0.3	0.8
Wild buckwheat	2.7	0.4	2.5

of most weeds is higher than that of wheat or barley. The authors proposed that common lambsquarters, Canada thistle, field bindweed, wild buckwheat, perennial sowthistle, and common chickweed are dangerous in wheat because of their high nutrient requirement. Rape responded like the grain crops. The percentage of nutrients in roots and leaves of sugarbeets was high, and few weeds exceeded it. This is explained by the high nutrient concentration in the large sugarbeet root.

In a crop heavily infested with weeds, it seems logical that more fertilizer should reduce nutrient competition. If competition does not occur until the immediate supply falls below combined demand, then when supply increases, competition should decrease. Actually, although this seems logical, it is wrong. Fertilizer usually stimulates weed growth to the crop's detriment. With low fertility, competition is primarily for nutrients; with high fertility, however, competition is just as vigorous, and primarily for light. Yields in unweeded, fertilized plots are usually equal to those in weeded, unfertilized plots. Table 6.6 shows that increasing nitrogen reduced flax yield and tended to increase wild oat density and number of seed-bearing stems (Sexsmith and Pittman, 1963). The opposite situation is more common: Nitrogen raises crop yield and then, when nitrogen is in excess, crop yield decreases (Table 6.7) (Okafor and DeDatta, 1976).

Table 6.8 shows similar data on competition of barnyardgrass and barnyardgrass plus the annual broadleaved weed monochoria in rice. It is apparent that increasing nitrogen fertilizer increased yield. It is also apparent that with just barnyardgrass, increasing nitrogen fertilizer from 0 to

Table 6.6

Effect of Form and Timing of Nitrogen Fertilizer on Wild Oats and Flax, (Sexsmith and Pittman, 1963)[a]

	Wild Oats		
Fertilizer	Density	Seed-bearing stems (number/m^2)	Flax yield (kg/ha)
None	96 a	124 a	7.0 a
Ammonium nitrate April 12 (early)	215 ab	254 a	4.2 ab
Ammonium sulfate April 12 (early)	435 bc	444 b	2.4 bc
Ammonium nitrate June 1 (seeding)	476 c	530 b	1.9 c

[a]Means in a column followed by different letters are significantly different at $P = 0.05$.

Table 6.7

**Effect of Nitrogen Fertilizer on Rice Yield and Purple Nutsedge
Competition (Okafor and DeDatta, 1976)**

Nitrogen (kg/ha)	Purple nutsedge (no./m^2)	Rice yield (t/ha)
0	0	1.6
0	750	1.2
60	0	4.4
60	750	2.8
120	0	4.0
120	750	2.4

60 kg/ha decreased yield. Only after a further doubling of nitrogen did yield increase. Even then, yield was lower than the same amount of fertilizer with no weeds. With both weeds, neither level of nitrogen fertilizer increased yield, and both yielded less than the check plot with no fertilizer and no weeds. These data are confirmed by those in Table 6.9, which show nitrogen uptake of rice and barnyardgrass in two trials in Australia (Boerema, 1963).

The influence of fertility treatments for 47 years on weed types and populations was evaluated in Oklahoma (Banks *et al.,* 1976). Plots with lowest weed density were those that had received no fertilizer for 47 years. Highest weed density occurred on plots that received complete fertilizer (N, P, K) and lime (CaCO$_3$). Grass weeds were most abundant with complete fertility, whereas broadleaved species declined.

Table 6.8

**Weed Competition for Nitrogen in Rice
(Moody, 1981)**

Weed(s)	Tons/ha of rice grain with nitrogen fertilizer applied at (kg/ha)		
	0	60	120
None	4.5	5.3	6.6
Barnyardgrass	4.4	4.0	5.5
Barnyardgrass + monochoria	4.1	3.1	3.5

Table 6.9

Nitrogen Uptake of Weeds and Rice in Two Trials (Boerema, 1963)

Species	Trial 1		Trial 2	
	Weeds present	Weeds absent	Weeds present	Weeds few
Barnyardgrass	56.3	0	94.1	1.6
Rice	36.8	99.7	15.5	111.8
Total	93.1	99.7	109.6	113.6

The effects of interactions of soil moisture and fertility on competition between wheat and wild buckwheat were studied in North Dakota (Fabricus and Nalewaja, 1968). Biomass of wheat growing alone increased with increasing fertility. Wheat biomass declined 30 to 37% regardless of soil moisture or fertility when wheat grew with wild buckwheat. The weed also reduced flax growth 47 to 57% when they grew together for 90 days (Gruenhagen and Nalewaja, 1969). There was proportionately greater flax seed loss with higher fertility.

Table 6.10 shows five densities of wild oats with three levels of nitrogen. It is clear that as wild oat density increases, it is less and less profitable to add nitrogen. Wild oats' advantage is due to its higher nitrogen use efficiency (Carlson and Hill, 1986). Increasing fertilizer application rate is not an

Table 6.10

Yield of Wheat Grown in Competition with Wild Oats at Three Levels of Fertilization (Carlson and Hill, 1986)

	Wheat yield with preplant nitrogen			
	0	67	134	Avg.
Wild oat density (plants/m^2)		(kg/ha)		
0	6990	7520	7650	7390
4	6430	6660	6640	6580
8	6460	6100	6140	6230
16	5940	5200	5470	5540
32	5400	4120	3450	4320
Avg.	6240	5920	5870	

economic, agronomic, or energy-efficient way to avoid or reduce crop losses due to weed competition.

In general, weeds have a large nutrient requirement and will absorb as much or more than crops. Nitrogen is the first nutrient to become limiting in most instances of weed-crop competition. The nitrate ion is not held strongly in soil and is highly mobile. Nitrogen depletion zones are likely to be quite wide and similar to those for water. Therefore, rooting depth and root area of plants determine the ability to obtain resources, and relative competitiveness for nitrogen is largely determined by the soil volume occupied by roots of competing species. The amount of nitrogen taken up by plants in any combination is about equal (Table 6.9).

Movement of phosphorus and potassium is slow compared to that of nitrogen, and these minerals move over shorter distances. Smaller depletion zones minimize interplant competition. Competition for phosphorus and potassium is therefore most likely to occur after plants are mature and have extensive, overlapping root development. It is reasonable to assume that competition for phosphorus will be more apparent in perennial crops. Competitiveness of barley cultivars with wild oats varied in response to potassium (Siddiqi et al., 1985) or phosphorus (Konesky et al., 1989) supply. There are few studies of weed-crop competition for phosphorus or potassium.

Although competition for nitrogen can sometimes be overcome by nitrogen fertilization, this is rarely true for phosphorus and potassium. It may be possible to prevent or delay weed invasion of perennial crops by maintaining a vigorous crop with fertilizer.

D. Competition for Water

Water, or its lack, is often the primary environmental factor limiting crop production, and water is probably the most critical of all plant growth requirements (King, 1966). Without irrigation, rainfall determines the geographic limit of crops. The water-use efficiency of nine weeds and nine crops is shown in Table 6.11.

The point of Table 6.11 is not that weeds use a great deal more water or use water more efficiently than crops. They use about the same amount as the crops with which they compete. Weeds effectively explore soil to obtain water (Table 6.12).

Comparison of rooting depth, uptake diameter, and the volume of soil from which resources can be consumed by one sorghum plant and five weeds makes the reason for weed competitiveness clear. Of the five weeds shown, all have a greater rooting depth, and all but pigweed have a larger

Table 6.11

Water Use Efficiency (Dillman, 1931; Shantz et al., 1927)

Plant	Water use efficiency[a]	Transpiration coefficient[b]
Weeds		
Common cocklebur	2.41	415
Common lambsquarters	1.52–2.30	435–658
Common purslane	3.47–3.56	281–288
Foxtail millet	3.65–3.98	251–274
Prostrate knotweed	1.47	678
Redroot pigweed	3.28–3.83	261–305
Russian thistle	3.18–4.46	224–314
Sunflower	1.73	577
Witchgrass	3.94	254
Crops		
Alfalfa	1.15–1.25	798–870
Corn	2.77	361
Cotton	1.76	568
Oats	1.65–1.87	536–605
Smooth bromegrass	1.02–1.28	784–977
Sorghum	3.51–3.73	268–285
Soybean	1.55	646
Sugarbeets	2.65–3.29	304–377
Rape, oilseed	1.40	714

[a]Water use efficiency = mg of dry weight produced per ml of water consumed.
[b]Transpiration coefficient = ml of water transpired per g plant dry weight.

Table 6.12

Soil Water Uptake Patterns of Common Weeds and Grain Sorghum in Summer Fallow (Personal Communication, Adapted from Davis et al., 1967)

Weed species	Rooting depth (m)	Feeding diameter (m)	Root area plant (sq. m)	Plants to consume water/ha (number)
Common cocklebur	2.9	8.5	17.9	704
Grain sorghum	1.7	4.3	6.5	2841
Kochia	2.2	6.7	9.5	1136
Pigweed	2.4	3.6	5.2	3853
Puncturevine	2.6	6.6	10.8	1136
Russian thistle	1.8	5.0	6.5	2149

feeding diameter and volume affected per plant than grain sorghum. All except pigweed have a greater capacity to consume water than grain sorghum does.

The classic work on water requirements of plants was done at Akron, Colorado, in the early 20th century (Briggs and Shantz, 1914; Dillman, 1931; Shantz *et al.*, 1927). Individual crop and weed plants were grown in separate pots, and the grams of water required to produce a gram of plant dry matter were determined. Some of the data are in Table 6.13.

Weeds compete for water, reduce water availability, and contribute to crop water stress. They require just as much, and often more, water than crops and are often more successful in acquiring it. Weedy sunflowers require approximately twice as much water as corn. It takes more water to produce a potato tuber than to produce a common lambsquarters plant. Therefore, if, as is commonly found, common lambsquarters infests potato fields, and water is limiting, fewer and smaller tubers will be produced. About 80 gallons of water are required to produce one pound of dry matter in barnyardgrass, and more than the 60 gallons to produce a pound of

Table 6.13
Water Required to Produce One Pound of Dry Matter (Dillman, 1931; Shantz *et al.*, 1927)

Plant	Kilograms of water
Alfalfa	377
Barley, grain	431
Barley, whole plant	237
Bursage	535
Common lambsquarters	300
Common purslane	128
Common sunflower	338
Corn	159
Mustard	1091
Potato, tuber	430
Potato, vine	150
Redroot pigweed	132–139
Russian thistle	143
Sorghum	283
Wheat	227

wheat. Crabgrass, a common turf weed, requires 83 gallons of water per pound of dry matter.

Many field, laboratory, and greenhouse studies have examined the role of water in weed-crop competition. A few are presented to illustrate water's role.

One of the early studies (Wiese and Vandiver, 1970) compared growth of corn and sorghum with three grass and five broadleaved weeds at three soil moisture levels in the greenhouse. Corn produced the most biomass at all moisture levels. Common cocklebur, barnyardgrass, and large crabgrass normally grow well in humid regions and in irrigated crops and were the most competitive with wet soil conditions. Kochia and Russian thistle, weeds of dry areas, were more competitive with dry soil conditions and grew poorly when soil was wet. Russian thistle produced twice as much growth in dry as in wet soil.

In field experiments in Texas (Stuart et al., 1984), water competition from smooth pigweed reduced leaf water potential and turgor pressure in cotton. Smooth pigweed was affected less by low soil water because it transpires less water and its larger root system draws water from deeper in soil. Smooth pigweed illustrates what may be called water wasting by weeds. In fact, water use is wasteful only from a human perspective or in comparison to another plant, a crop, that uses less water. Each plant uses the water it requires. Stomata in some weeds are less sensitive to declining leaf water potential than those of crops with which they compete (Patterson, 1995). When this is combined with a larger root system (Table 6.12) or better drought tolerance, weeds are formidable competitors for water. High water use by weeds may be ecologically advantageous in weed-crop competition, especially when soil moisture is limiting (Patterson, 1995).

When soybean and velvetleaf competed in Texas, rooting depths were similar at first. After 10 weeks, soybean was able to draw water from greater soil depths and velvetleaf had little effect on soybean's water status (Munger et al., 1987). When the same species competed in Indiana, a wetter, more humid area, velvetleaf reduced soybean growth more in dry than wet years (Hagood et al., 1980).

In Arkansas, soybean had higher leaf water potential than common cocklebur because of stomatal regulation of transpiration. Common cocklebur had lower stomatal resistance and higher transpiration. It is a high water user and exhausts soil water resources rapidly, to soybean's disadvantage (Geddes et al., 1979; Scott and Geddes, 1979).

Patterson (1995) surveyed weed-crop competition studies that included water as a variable and found a slight tendency for decreased water availability to favor crops by reducing weed competition. This reasonable generalization may not always be true, because it will be affected by each crop-weed combination and the cultural and environmental conditions in each crop season or over several seasons.

For example, the influence of season is shown by competition from anyone of three broadleaved weeds that reduced soybean yield more when soil moisture was adequate early followed by drought than when drought was early (Eaton *et al.*, 1973, 1976).

Scientists in arid areas have developed fallow cropping systems. Many arid areas have sufficient rainfall to support crop growth only every other year. Often wheat is grown one year and the land is fallowed (no crop) the next year, and rotated back to wheat in the third year. The primary purpose of this rotation is water conservation. Natural rainfall is not sufficient to grow wheat each year and extensive dryland cannot be irrigated. Therefore, minimum or no-tillage systems, sometimes called "ecofallow," have been developed to conserve water. The data in Table 6.14 show the increase in water stored in the soil profile for a minimum-tillage, ecofallow system compared to a tilled, spring fallow system. The ecofallow system increased soil nitrate, grain protein, and wheat yield.

Water is the least reliable resource for plant growth because we do not know precisely when it will arrive or how much will be received. This is why arid areas are irrigated. Because roots grow more rapidly than shoots early in a plant's life, competition for water and nutrients usually begins before competition for light. Competition for water is determined by the relative root volume occupied by competing plants and will be greatest when roots closely intermingle and crops and weeds try to obtain water from the same volume of soil. Less competition occurs if roots of crops and weeds are concentrated in different soil areas. More competitive plants have faster-growing, large root systems so they are able to exploit a large volume of soil quickly. If plants have similar root length, those with more

Table 6.14

Conventional Tillage vs Ecofallow (Greb and Zimdahl, 1980)

Measurement	Treatment		
	Spring tillage fallow	Ecofallow	Increase
Gain in soil water during fallow (cm)	3.9	5.4	1.5
Gain in soil nitrate during fallow (cm)	51.6	77.4	25.8
Percent gain protein	11.0	11.8	0.8
Wheat yield (bu/A)	34.4	41.8	7.4

widely spreading and less branched root systems will have a comparative advantage in competition for water.

E. COMPETITION FOR LIGHT

The total supply of light is the most reliable of the several environmental resources required for plant growth. But in contrast to water and nutrients, light cannot be stored for later use; it must be used when received or it is lost forever (Donald, 1963).

Light regulates many aspects of plant growth and development. It varies in duration, intensity, and quality. Neighboring plants may reduce light supply by direct interception–shading. Leaves are the site of light competition. Leaves that first intercept light may reflect it, absorb it, convert it to photosynthetic products, convert it to heat, or transmit it. If transmitted, the light is filtered so that when it reaches lower leaves, it is dimmer and spectrally altered. Any time one leaf is shaded by another, there is competition for light.

Light competition is most severe when there is high fertility and adequate moisture because plants grow vigorously and have larger foliar areas. Plants with large leaf area indices (LAIs) have a competitive advantage and normally outcompete plants with smaller leaf area. Leaf area index, a measure of the photosynthetic surface over a given area, is correlated with potential light interception. Successful competitors do not necessarily have more foliage, but have their foliage in the most advantageous position for light interception. Thus, a plant's ability to intercept light is influenced by its angle of leaf inclination and leaf arrangement. Plants with leaves disposed horizontal to the earth's surface are more competitive for light than those with upright leaves disposed more or less perpendicular to the earth's surface. Plants with opposite leaves are probably less competitive than those with alternate leaves. Plants that are tall or erect have a competitive advantage for light over short, prostrate plants. A heavily shaded plant suffers reduced photosynthesis, leading to poor growth, a smaller root system, and a reduced capacity for water or mineral uptake. The effect of shading is independent of direct competition for water or nutrients and entirely under the influence of light (Donald, 1963). Current cropping practices used, at least partially, to manage weeds, such as smother crops and narrow row spacing (see Chapter 9), exploit plant responses to light (Holt, 1995). Most weeds and crops respond to shading in similar ways via morphological and physiological adaptations (Patterson, 1995). This is not surprising because these plants evolved in disturbed habitats where shade adaptation has few selective advantages (Patterson, 1995).

Reports that crops are physiologically and genetically capable of higher productivity and photosynthetic efficiency than are obtainable in the field confirm that intercepted light is a limiting factor in crop canopies (Holt, 1995). Reduced production in low-light-acclimated crop plants is undesirable. Several reviews of responses of weeds and crops to light are available (Holt, 1995; Patterson, 1982, 1985, 1995a; Radosevich and Holt, 1984).

Crops and weeds differ in shade tolerance. Soybean and several of its associated weeds (e.g., Eastern black nightshade, tumble pigweed, and common cocklebur) were most photosynthetically efficient under low growth irradiance (Regnier et al., 1988; Stoller and Myers, 1989). Many other weeds acclimate to low growth irradiance by plastic responses that reduce the growth-limiting effects of shading and allow restoration of high rates of photosynthesis when the plant is exposed to high irradiance (Dall' Armellina and Zimdahl, 1988; Patterson, 1979; Singh and Gopal, 1973).

Bazzaz and Carlson (1982) generated photosynthetic response curves for 14 early, mid, and late successional species grown in full sunlight and 1% of full sunlight. Early successional species, all common annual weeds, had the highest difference in response between sun- and shade-grown plants. The magnitude of photosynthetic flexibility decreased in plants from later successional stages. All species studied were able to change their photosynthetic output in response to light, but the change was larger for early successional annuals (Bazzaz and Carlson, 1982). These findings suggest that weeds are not only adapted to high light, but are more capable of adapting to extreme variation in light, particularly deep shade. Thus, managing the light environment in a crop field to deter weed growth is likely to be difficult (Holt, 1995).

Available light is a major factor in yellow nutsedge competition with corn. More yellow nutsedge grows between corn rows than within the row because less light reaches the soil under plants. Yellow nutsedge density decreases as corn density increases (Ghafar and Watson, 1983); therefore, an acceptable yellow nutsedge management technique is increasing corn population. Increasing corn population density from 66,700 to 133,000 plants per hectare reduced yellow nutsedge tuber production 71%. Reducing corn population from 66,700 to 33,300 plants per hectare increased tuber production 41% (Ghafar and Watson, 1983). Field studies of the effect of artificial shade on yellow nutsedge concluded that rapidly developing crops (e.g., corn or potato) suppressed the weed through competition for light (Keeley and Thullen, 1978). Shading greatly reduced shoot and biomass production and reduced, but did not eliminate, tuber production. Stoller and Woolley (1985) estimated that competition for light caused almost all soybean yield loss in competition with velvetleaf or jimsonweed and half of the yield reduction in soybean competing with cocklebur.

Many studies have quantified effects of light competition between weeds and crops. Cudney *et al.* (1991) showed that wild oats reduced light penetration and growth of wheat by growing taller. When wild oats were clipped to the height of wheat, light penetration in a mixed canopy was similar to that in monoculture wheat. Interference from wild oats planted at low densities reduced light penetration to wheat at later growth stages (Cudney *et al.*, 1991).

Similar height effects were observed in studies of competition between velvetleaf and soybean. Greater light interception by velvetleaf was due to greater height and dry-weight allocation to more upper branches (Akey *et al.*, 1990). Reductions in tomato yield were greater when the tomato plants grew in competition with eastern black nightshade compared to black nightshade, because eastern black nightshade is taller (McGiffen *et al.*, 1992). These studies show that plant architecture, especially height, location of branches, and height of maximum leaf area, determine competition for light and influence crop yield (Holt, 1995).

Interaction of light and water is illustrated in a study of how yield of quackgrass infested soybeans was increased by irrigation when soil moisture was limiting. Soybeans infested with quackgrass yielded less than quackgrass-free soybeans. Quackgrass was nearly the same height or taller than soybeans at all stages of soybean development and competed for light throughout the growing period. Adequate moisture reduced quackgrass competition in soybeans, but did not eliminate it, because quackgrass continued to compete with soybeans for light (Young *et al.*, 1983).

Studies in India (Shetty *et al.*, 1982) have shown that dicots are less shade-sensitive than monocots and help explain why monocots are often important tropical weeds. Broadleaved weeds usually do not appear until after tropical crops are well established. It seems that manipulation of tropical crop canopies could suppress weeds via shading. The height of the dicot weeds celosia and coat buttons was reduced by 90% shade, but that shade level had no effect on height of southern crabgrass. Ninety percent shade reduced height of bristly starbur 50% and purple nutsedge 30%. The effects were most pronounced early in the growing season, and similar reductions in leaf area index and plant dry matter were observed. Slender amaranth's height was not affected by shade, but as light decreased, seed production decreased. For most annuals, 90% shade reduced seed production up to 90% and 40% shade reduced seed production 45%. Shading reduced purple nutsedge tuber production 89%.

F. FACTORS FOR WHICH PLANTS GENERALLY DO NOT COMPETE

Plants that emerge at the same time rarely compete for space, even though plant density may be high. When plants emerge at different times, the first plant that occupies an area will tend to exclude all others and have a competitive advantage and, in this sense, plants compete for space by occupying space first. Occupancy or competitive exclusion can be regarded as competition for the resources in a space.

In general, plants that emerge at the same time and plants that grow together do not compete for space, but rather for what space contains. This may not be true in root crops that are planted closely, but in most cases it is the light, nutrients, and water that space contains for which plants compete.

Plants may compete for oxygen; although there are no studies to document this, it is theoretically possible. In most soils, diffusion of oxygen is rapid enough that adequate supplies are available for all roots. However, oxygen may be limiting for plant growth in very wet soils. Similarly, in most circumstances, carbon dioxide concentrations are always higher than the carbon dioxide compensation point (the light intensity at which there is a balance between carbon dioxide given off by respiration and required by photosynthesis). Competition for carbon dioxide is unlikely to occur under field conditions, but crop yields can be increased by supplemental carbon dioxide. (See earlier comment in this chapter on climate change.) More efficient utilization of carbon dioxide by weeds with high photosynthetic capacities may contribute to their rapid growth and provide a competitive advantage. Therefore, a plant's competitive ability could depend on its capacity to assimilate carbon dioxide and use the photosynthate to extend foliage or increase size. Plants that fix carbon dioxide at high rates are potentially more competitive.

There is no evidence that plants compete for environmental factors such as heat energy or agents of pollination.

VI. PLANT CHARACTERISTICS AND COMPETITIVENESS

In general, it is true that plants possessing one or more of the following characteristics are more competitive than plants that lack them. This list is not in rank order, and it cannot be said that a plant with a certain characteristic will always win over a plant with another. Most competitive plants have the following:

Rapid expansion of a tall, foliar canopy
Horizontal leaves under overcast conditions and obliquely slanting
  leaves (plagiotropic) under sunny conditions
Large leaves;
A C_4 photosynthetic pathway and low leaf transmissivity of light

Leaves forming a mosaic leaf arrangement for best light interception
A climbing habit
A high allocation of dry matter to build a tall stem
Rapid stem extension in response to shading

The most obvious competition among plants is what we see–foliar competition. Competition for nutrients and water takes place beneath soil, where they cannot be seen. The most competitive plants also share some of the following root characteristics:

Early and fast root penetration of a large soil area
High root density/soil volume
High root–shoot ratio
High root length per root weight
High proportion of actively growing roots
Long and abundant root hairs
High uptake potential for nutrients and water

VII. THE RELATIONSHIP BETWEEN WEED DENSITY AND CROP YIELD

Early weed science literature assumed that the relationship shown in Figure 6.11 described the effect of weeds on crop yield.

Figure 6.11 says that with no weeds, crop yield will be maximized and at some large weed density, crop yield will be zero. The intervening relation-

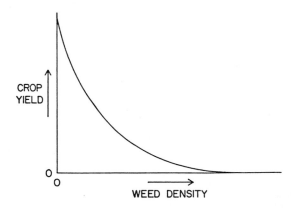

Figure 6.11 A schematic curvilinear relationship depicting the effect of increasing weed density on crop yield (Zimdahl, 1980).

ship is curvilinear, not linear. Such a relationship is supported by data (Figure 6.12) showing the effect of kochia, an annual broadleaved weed, on sugarbeet root yield.

Other data show the curvilinear relationship of Figure 6.13, which is intuitively logical, but also wrong. Some of the data in Table 6.15 show that the relationship is neither linear nor curvilinear. Doubling of weed density does not double crop loss in any of these studies, and even when weed density is increased by a factor of 25, crop loss does not go to zero. Therefore, although the curvilinear relationship is not entirely incorrect, it is not correct, and can be misleading.

Smith (1968) studied the interaction of rice and barnyardgrass density and showed that the appropriate relationship is neither linear nor curvilinear. The curvilinear relationship fails especially because it predicts that a high weed density will reduce crop yield to zero, and that does not happen. Some crop plants always survive, even though they may be very small and the yield is unprofitable. Smith's data show how, as crop density increases, the effect of weed density decreases (Table 6.16).

An interpretation of the relationship between crop yield and weed density has been described by the sigmoidal curve in Figure 6.13 (Zimdahl, 1980). At very low weed densities, there is no effect on crop yield; as weed density increases, although there may be an effect, it is barely discernible. As weed density continues to increase, crop yield drops quickly but never goes completely to zero. Even very high weed densities do not eliminate all crop plants. This represents most weed–crop competition data and provides a picture of what happens, but *it is still not correct.* Its appeal is that it is very difficult to measure the effect of a few weeds in a large area. It may not be wise to attempt to do so. For practical purposes, the effect of 1 weed/acre is zero and that weed has no immediate, measurable economic effect. Yet one weed affects crop plants near it, and it produces seed and can thereby affect future crops.

There are many places in the literature of weed science that state that the relationship between yield loss and weed density is sigmoidal (Figure 6.13), with little or no loss at low weed density, or nearly so. Cousens *et al.* (1987) state unequivocally that the data do not support this. When yields are plotted over a range of weed densities there is no evidence to support a sigmoidal response. The most accurate representation of crop-weed interactions is that created by regression analysis of crop yield and weed density. This is because densities observed in the field and those used in experiments cannot represent the whole range of possible weed densities depicted in Figure 6.13. Multiple regression models must be chosen carefully so they reflect biological reality and not just mathematical convenience.

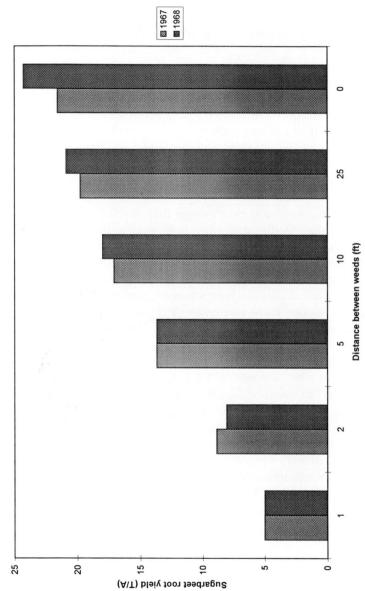

Figure 6.12 Effect of kochia on sugarbeet yield (Weatherspoon and Schweizer, 1971). Each yield bar in each year (not between years) is significantly different than every other bar (yield) for that year.

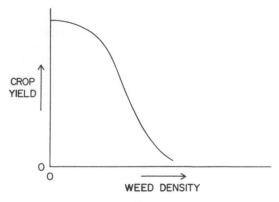

Figure 6.13 A schematic sigmoidal relationship depicting the effect of increasing weed density on crop yield (Zimdahl, 1980).

Table 6.15

The Effect of Weed Density on Crop Yield

Crop	Weed	Weed density	Yield reduction from control (%)	Source
Wheat	Wild oats	$58.5/m^2$	22.1	Bell and Nalewaja (1968)
		$134/m^2$	39.1	
Wheat	Green foxtail	$721/m^2$	20	Alex (1976)
		$1575/m^2$	35	
Cotton	Prickly sida	2/5 cm of row	27	Ivy and Baker (1972)
		4/5 cm of row	40	
		12/cm of row		
Rice	Barnyardgrass	$1/0.09m^2$	57	Smith (1968)
		$5/0.09m^2$	80	
		$25/0.09m^2$	95	
Soybean	Common cocklebur	3297/ha	10	Barrentine (1974)
		6597/ha	28	
		12295/ha	43	
		25989/ha	52	
Corn	Giant foxtail	$\frac{1}{2}$/5 cm of row	4	Knake and Slife (1962)
		1/5 cm of row	7	
		3/5 cm of row	9	
		6/5 cm of row	12	
		12/5 cm of row	16	
		54/5 cm of row	24	

Table 6.16

Interaction of Rice and Barnyardgrass (Smith, 1968)

Rice plants/0.09 m^2	Barnyardgrass	% yield reduction
3	0	0
3	1	57
3	5	80
3	25	95
10	0	0
10	1	40
10	5	66
10	25	89
31	0	0
31	1	25
31	5	59
31	25	79

VIII. MAGNITUDE OF COMPETITIVE LOSS

Tables 6.17, 6.18, and 6.19 show the magnitude of loss in a few studies of weed competition in corn, soybeans, and small grains. This small set of data provides evidence that weeds decrease crop yield, often by a great deal. The data also show that the effect of weeds is not entirely predictable, nor is the effect of a particular density consistent. The data in the tables are shown as they appeared in the original publication in order to make an important point about many studies–the lack of precision of the data. There is no uniform definition of a heavy stand or a small infestation (Table

Table 6.17

Weed Competition in Corn (Zimdahl, 1980)

Location	Density	Yield reduction
Illinois	Heavy stand	55%
Illinois	54 foxtail/ft of row in 4-in. band over row	25%
Iowa	Handweeded	50% greater than unweeded
Iowa	Small infestations of foxtail	6–8 bu/A

Table 6.18
Weed Competition in Soybeans (Zimdahl, 1980)

Location	Density	Yield reduction
Nebraska	86 lbs/A	1 bu/A
Iowa	10–12 weeds/ft of row	7.5–17.1%
Illinois	54 foxtail/ft of row in 4-in. band over row	28%

6.17), and therefore the work is not repeatable. In competition studies, it is important to define precisely the number of weeds and crop plants per unit area (the density).

IX. DURATION OF COMPETITION

It is obvious that a weed present for one day in the life of a crop will probably have no measurable effect on final yield. But what if the weed is present for 2, 20, or 200 days? The question of duration of competition has been asked in two ways. The first kind of study asks, what is the effect when weeds emerge with the crop and are allowed to grow for defined periods of time? After each of these times, the crop is then kept weed-free for the rest of its growing period. These studies define what many call the critical duration of weed competition. The second kind of study asks, what is the effect when the crop is kept weed-free from emergence for certain periods of time, and then weeds are allowed to grow for the rest of the growing season? These define what many call the critical weed-free period.

Table 6.19
Weed Competition in Small Grains (Zimdahl, 1980)

Location	Crop	Density	Yield reduction
Montana	Spring wheat	Canada thistle/sq. ft	
		3–5	4.2 bu/A
		20–25	9.0 bu/A
		40–45	15.3 bu/A
Oregon	Winter wheat	1 fiddleneck/sq. ft	10.0 bu/A
New York	Oats	15 mustard/sq. ft.	11.0 bu/A
Nebraska	Sorghum	15 lbs. of weeds/A	1.0 bu/A

Vega *et al.* (1967) studied the effect of duration of weed control on rice. Weeds grew for no time at all or in intervals of 10 days up to 50 days after rice was planted. They also allowed weeds to compete for 10, 20, 30, 40, or 50 days after planting and then kept the crop weed-free thereafter (Table 6.20).

The data show that yield is reduced when rice is weeded for a short time after planting. When it was weeded for 40 days, yield reached a maximum and there was no benefit from weeding an additional 10 days. In the same way, if weeds were allowed to grow up to 20 days after planting and then removed, there was no effect on yield. Therefore, rice (and many other crops) can withstand weed competition early in the growing season and does not have to be weeded immediately. However, weeds in rice cannot be present more than about 30 days or yield will go down.

Corn must be kept weed-free for 3 to 5 weeks after seeding or 9 weeks after emergence, dependent on location and the weeds (Table 6.21). The opposite study for corn is shown in Table 6.22, which shows the length of early weed competition tolerated by corn. If provided with a weed-free period for 3 weeks after emergence, corn will compete effectively with weeds emerging afterwards. Conversely, corn can withstand weed competition for up to 6 weeks if it is then weeded and kept weed-free.

Table 6.20
The Effect of Duration of Weed Control and Weed Competition on Rice Yield (Vega *et al.*, 1967)

Weed control duration (days after planting)	Yield (kg/ha)
0	46
10	269
20	1544
30	2478
40	3010
50	2756
Weed competition duration (days after planting)	
10	2944
20	3067
30	2752
40	2040
50	1098
Unweeded	55

Table 6.21

**Weed-free Period Required to Prevent Yield Reduction in Corn
(Zimdahl, 1980)**

(Weed-free weeks required after) seeding	emergence	Competing weeds	Location
9		Mixed annuals	Mexico City
	5	Mixed annuals	Vera Cruz, Mexico
	3	Giant foxtail	Illinois

These kinds of data have been used to derive the critical period for weed competition (Table 6.23). A critical period, defined as the period of time between that period after seeding when weed competition does not reduce yield and the time after which weed presence does not reduce yield, has been found for several crops. It is a *time between* the early weed-free period required and the length of competition tolerated (Figure 6.14). It is not a fixed period for a crop because it varies with season, soil, weeds, and location. It is a useful measure because it gives an idea of when to weed. For example, potatoes, if kept weed-free for 6 weeks, will survive the rest of the season without yield reduction, even if weeds grow. If

Table 6.22

Length of Early Weed Competition Tolerated without Yield Loss in Corn (the Critical Duration) (Zimdahl, 1980)

Weeks of competition tolerated after seeding	emergence	Competing weeds	Location
3		Mixed annuals	Vera Cruz, Mexico
	4	Mixed annuals	Mexico City
4		Mixed annuals	Chapingo, Mexico
2–4		Halberdleaf orach and Persian speedwell	England
4		Green foxtail	Ontario, Canada
6		Giant foxtail	Illinois
	6	Redroot pigweed	Oregon
	2–3	Mixed annuals	New Jersey
	8	Itchgrass	Zimbabwe

Table 6.23

Crops with an Apparent Critical Period for Weed Competition (Zimdahl, 1980)

Crop	Weed-free weeks required	Weeks of weed competition tolerated
Bean	5	8
Corn	3	6
Cotton	6	8
Peanuts	4	8
	3	6
Potato	6	9
Rice, paddy	3	6
Soybean	3	8–9

potatoes are weeded 9 weeks after seeding, yield will not be reduced if they are subsequently kept weed-free. Therefore, weeding of potatoes must be done sometime between 6 and 9 weeks after seeding or yield will decrease. Critical-period analyses show that preemergence weed control is not essential, nor is weed control immediately after emergence. The method of weed control dictates when it must be applied, but the lesson of critical period studies is that weed control does not have to be done in the first

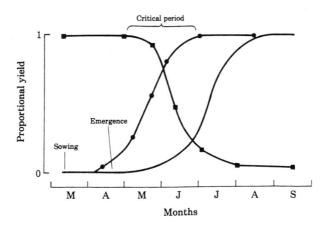

Figure 6.14 The "critical period of competition" illustrated for onions —, Changes in crop dry weight from sowing to harvest; ■, yield response from delaying the start of continuous weed removal; ●, yield response from delaying the termination of weed removal, as adapted by Mortimer (1984) from Roberts (1976).

few weeks after crop emergence. Critical periods have practical weed management value, but Mortimer (1984) points out their limitation. All weeds are considered equally injurious and no distinction is made between the kinds of competition that can occur. Most of us would be injured in a fist fight, but we will be injured less if we get to pick the opponent than if the opponent is the heavyweight boxing champion.

X. ECONOMIC ANALYSES

More economic analyses of weed control are being done. Farmers know weeds reduce yield, and the question they ask is not whether weeds will reduce yield, but how many weeds reduce yield how much? Their question really is, should I control weeds, and what method(s) is best? The farmer's definition of best usually means the method that offers highest profit potential. The farmer knows a few weeds are not of consequence and asks how many weeds are of consequence? The data in Table 6.24 illustrate how the answer might be provided.

The study showed, for three potential wheat yields, what the profit or loss would be for spraying, given a certain value of wheat and a defined spraying cost. For example, if a farmer has $\frac{1}{2}$ weed per square foot, the estimated yield loss is 5%. If the wheat yield is estimated to be 15 or 20 bushels per acre, the cost of controlling the weeds will exceed the benefit to be gained. If, on the other hand, yield will be 30 bushels, then the gain

Table 6.24

Potential Profit or Loss from 2,4-D Application to Control Pinnate Tansymustard in Winter Wheat (Wiese, 1965)

Weeds per sq. ft	Percent estimated yield loss	Potential wheat yield, bu/A		
		15	20	30
			Profit or loss, $[a]	
$\frac{1}{4}$	2.5	−1.03	−0.87	−0.56
$\frac{1}{2}$	5	−0.56	−0.25	0.38
1	10	0.38	1.00	2.25
1	20	2.25	3.50	6.00
4	40	6.00	8.50	13.50

[a]Profit or loss, value of yield loss if weeds are uncontrolled − spray cost; wheat, $1.25/bu; spray cost, $1.50/A.

will exceed the cost and the weeds should be controlled. The values in Table 6.24 are out of date, but the table is provided to illustrate the principle, which remains valid. A similar set of data assist with decisions on controlling wild oats in barley, wheat, or flax. These data (Table 6.25) show the potential yield loss for each crop from a wild oats density that a farmer could determine.

A farmer could calculate control costs and value of yield lost to determine whether or not control should be done. Other studies of decision models have been done (King *et al.*, 1986; Lybecker *et al.*, 1984), but most decisions about what to do are still made by growers with incomplete information. Weed science needs more information on the efficacy of various weed control techniques and weed management systems in different soils and cropping systems. This information must be combined with information on percent emergence of the weed species in the soil seedbank, expected crop yield, weed control cost, and the farm's current economic situation to make wise weed management decisions.

XI. MATHEMATICAL MODELS OF COMPETITION

A large number of experiments have been done to demonstrate that weeds reduce crop yield. This work has demonstrated that some weeds are more detrimental to one crop than another, and that the effect is always modified by environmental interactions. Weed scientists do not need more experiments to establish that weeds are detrimental. In fact, the important

Table 6.25

Yield Loss Caused by Wild Oats in Barley, Wheat, and Flax (Bell and Nalewaja, 1967)

Wild oat seedlings/m^2	Yield reduction in bu/A		
	Barley	Wheat	Flax
10	1.6	1.5	2.0
40	2.7	3.5	5.0
70	4.9	5.2	6.3
100	6.0	5.4	6.9
130	6.2	7.3	7.4
160	7.1	8.7	7.5

questions in weed control and management cannot be answered by experiments to determine yield loss as a function of weed density.

Mortimer (1987) cited four primary issues in weed management:

1. For a given crop management, what is the likelihood of invasion by weeds?
2. Given the presence of weed infestation, how rapidly will the weeds spread and what crop losses will be suffered?
3. How much of any proposed control measure is required to contain the infestation or lead to total eradication?
4. What are the comparative costs of different weed control measures, and what risks are involved in switching weed management strategies?

It is possible to answer these questions with standard field experiments, but it is not desirable to do so, because there is not enough time, money, or weed scientists. Therefore, weed scientists are working to develop models to test experimental hypotheses and complement experimentation. Cousens *et al.* (1987) described four ways models can enhance research:

1. Models can be used as the framework to integrate available information. Critical gaps in research can be pinpointed; incompatibilities and erroneous or abnormal results may become apparent.
2. Mathematics is a formal, rigorous language in which theories and intuition can be expressed. Models can reduce ambiguity and describe complex systems.
3. When used with an experimental program, models can increase the speed with which understanding develops. They can be used to identify critical experiments, thereby making the most economical use of resources.
4. Models can be used to forecast and predict what might be observed under conditions not previously included in experiments.

Models can be empirical, describing data or a response. They can also be mechanistic, attempting to incorporate knowledge of processes that determine response (Cousens *et al.*, 1987). Much modeling effort has been expanded to develop computerized decision-aid software to answer the third and fourth questions posed by Mortimer (1987). Decision-aid models are based on the knowledge that weed effects are population dependent, and all models attempt to predict the biological (weed density) and economic consequences of management decisions (Coble and Mortensen, 1992). Models incorporate the concept of threshold or beginning point for weed effects. At least four kinds of thresholds are used in decision-aid models (Coble and Mortensen, 1992):

Damage–the weed population at which a negative crop yield response is detected.

Economic–the weed population at which the cost of control is equal to the crop value increase from control.

Period–time or times during the crop's life when weeds are most detrimental.

Action–the point when a control measure should be initiated.

Mathematical, computer-based models are not widespread in weed science. As models are developed, perfected, and tested against biological knowledge, they will be used more and more. Models are increasingly able to fulfill the basic requirements for a good weed–crop competition model (Cousens, 1985):

1. Without weeds there is no yield reduction.
2. At low weed densities, the effect of increasing weed density will be additive.
3. Yield loss can never exceed 100%.
4. At high weed densities there is a nonlinear response of crop yield to weed density.

It is beyond the scope or intent of this book to present a more detailed discussion of crop-weed interference modeling in weed science. Readers are directed to the references cited and to current literature for more information on this rapidly expanding research area.

THINGS TO THINK ABOUT

1. Why do plants compete?
2. What do plants compete for?
3. Do plants compete for space?
4. What factors determine weed-crop associations?
5. What makes a plant competitive?
6. How is the critical period of competition developed, and what is it used for?
7. What is the most appropriate description of the relationship between crop yield and weed density?
8. How much yield is lost because of weeds?
9. What must be known about crop-weed competition to make good weed management decisions?

10. How do economic analyses help in making weed management decisions?
11. What is the role of mathematical models in weed science?
12. What kinds of thresholds are used in crop-weed interference models?
13. How can models aid research and weed management?

LITERATURE CITED

Akey, W. C., T. W. Jurik, and J. Dekker. 1990. Competition for light between velvetleaf (*Abutilon theophrasti*) and soybean (*Glycine max*). *Weed Res.* **30:**403–411.

Aldrich, R. J. 1984. Weed–crop *Ecology: Principles in Weed Management*, p. 17. Bretton, N. Scituate, Massachusetts.

Alex, J. F. 1967. Competition between *Setaria viridis* (green foxtail) and wheat at two fertilizer levels. *Res. Rep., Can. Nat. Weed Comm., West Sect.* p. 286–287.

Anonymous. 1995. Czeching the spread of unwanted plants. *The Economist.* Nov. 11, p. 82.

Banks, P. A., P. W. Santlemann, and B. B. Tucker. 1976. Influence of soil fertility treatments on weed species in winter wheat. *Agron. J.* **68:**825–827.

Baker, H. G. 1974. The evolution of weeds, pp. 1-24 in *Ann. Rev. Ecol. and Systematics.* R. F. Johnston, P. W. Frank, and C. D. Michener (eds.), Academic Press, New York

Barrentine, W. L. 1974. Common cocklebur competition in soybeans. *Weed Sci.* **22:**600–603.

Bazzaz, F. A., and R. W. Carlson 1982. Photosynthetic acclimation to variability in the light environment of early and late successional plants. *Oecol.* **54:**313–316.

Bell, A. R., and J. D. Nalewaja. 1967. Wild oats cost more to keep than to control. *N. Dakota Farm Res.* **25:**7–9.

Bell, A. R., and J. D. Nalewaja. 1968. Competition of wild oat in wheat and barley. *Weed Sci.* **16:**505–508.

Bleasdale, J. K. A. 1960. Studies on plant competition, pp. 133–142 in *The Biology of Weeds* (J. L. Harper, Ed.). Blackwell, Oxford. 256 pp.

Boerema, E. B. 1963. Control of barnyardgrass in rice in the Murrumbridge irrigation area using 3,4-dichloropropionanilide. *Aust. J. Exp. Agric. and Anim. Husb.* **3:**333–337.

Briggs, L. J., and H. L. Shantz. 1914. Relative water requirement of plants. *J. Agric. Res.* **3:**1–63.

Burt, G. W. 1974. Adaptation of johnsongrass. *Weed Sci.* **22:**59–63.

Carlson, H. L., and J. E. Hill. 1986. Wild oat (*Avena fatua*) competition with spring wheat: Effects of nitrogen fertilization. *Weed Sci.* **34:**29–33.

Clements, F. E., J. E. Weaver and H. C. Hanson. 1929. Plant competition-an analysis of community function. Publ. No. 398. Carnegie Inst., Washington, D.C. 340 pp.

Coble, H. D., and D. A. Mortensen. 1992. The threshold concept and its application to weed science. *Weed Technol.* **6:**191–195.

Cousens, R. 1985. A simple model relating yield loss to weed density. *Am. Appl. Biol.* **107:**239–252.

Cousens, R., and M. Mortimer. 1995. *Dynamics of Weed Populations.* Cambridge Univ. Press, Cambridge, U.K. 332 pp.

Cousens, R., S. R. Moss, G. W. Cussans, and B. J. Wilson. 1987. Modeling weed populations in cereals. *Rev. Weed Sci.* **3:**93–112.

Cudney, D. W., L. S. Jordan, and A. E. Hall. 1991. Effect of wild oat (*Avena fatua*) infestations on light interception and growth rate of wheat (*Triticum aestivum*). *Weed Sci.* **39:**175–179.

Dall'Armellina, A. A., and R. L. Zimdahl. 1988. Effect of light on growth and development of field bindweed (*Convolvulus arvensis*) and Russian knapweed (*Centaurea repens*). *Weed Sci.* **36:**779–783.

Davis, R. G., A. F. Wiese and J. L. Pafford. 1965. Root moisture extraction profiles of various weeds. *Weeds* **13:**98–100.

Davis, R. G., W. E. Johnson and F. O. Wood. 1967. Weed root profiles. *Agron. J.* **59:**555–556.

Devine, R. 1993. The cheatgrass problem. *Atlantic Monthly* **271** (5):40, 44–48.

Dewey, L. H. 1894. The Russian thistle: Its history as a weed in the United States with an account of the means available for its eradication. Bull. 15, Div. of Botany, USDA, Washington, D.C.

deWit, C. T. 1960. On competition. Verslagen van landbouwkundige onderzoekingen. No. 66.8.

Dickerson, C. T., Jr., and R. D. Sweet. 1971. Common ragweed ecotypes. *Weed Sci.* **19:**64–66.

Dillman, A. C. 1931. The water requirements of certain crop plants and weeds in the Northern Great Plains. *J. Agric. Res.* **42:**187–238.

Donald, C. M. 1963. Competition among crop and pasture plants. *Adv. Agron.* **15:**1–118.

Eaton, B. J., K. C. Feltner, and O. G. Russ. 1973. Venice mallow competition in soybeans. *Weed Sci.* **21:**89–94.

Eaton, B. J., O. G. Russ, and K. C. Feltner. 1976. Competition of velvetleaf, prickly sida and venice mallow in soybeans. *Weed Sci.* **24:**224–228.

Fabricus, L. J., and J. D. Nalewaja. 1968. Competition between wheat and wild buckwheat. *Weed Sci.* **16:**204–208.

Geddes, R. D., H. D. Scott, and L. R. Oliver. 1979. Growth and water use by common cocklebur (*Xanthium pensylvanicum*) and soybeans (*Glycine max*) under field conditions. *Weed Sci.* **27:**206–212.

Ghafar, Z. and A. K. Watson. 1983. Effect of corn (*Zea mays*) population on growth of yellow nutsedge (*Cyperus esculentus*). *Weed Sci.* **31:**588–591.

Ghersa, C. M., M. L. Roush, S. R. Radosevich, and S. M. Cordray. 1994. Coevolution of agoecosystems and weed management. *BioScience* **44:**85–94.

Gorske, S. F., A. M. Rhodes and H. J. Hopen. 1979. A numerical taxonomic study of *Portulaca oleracea*. *Weed Sci.* **27:**96–102.

Gould, F. 1991. The evolutionary potential of crop pests. *Am. Scientist* **79:**496–507.

Greb, B. W., and R. L. Zimdahl. 1980. Ecofallow comes of age in the Central Great Plains. *J. Soil Water Cons.* **35:**230–233.

Gruenhagen, R. D., and J. D. Nalewaja. 1969. Competition between flax and wild buckwheat. *Weed Sci.* **17:**380–384.

Hagood, E. S. Jr., T. T. Bauman, J. L. Williams, Jr. and M. M. Schreiber. 1980. Growth analysis of soybeans (*Glycine max*) in competition with velvetleaf (*Abutilon theophrasti*). *Weed Sci.* **28:**729–734.

Harper, J. L. 1977. *Population Biology of Plants*. Academic Press, New York. 892 pp.

Hodgson, J. M. 1964. Variations in ecotypes of Canada thistle. *Weeds* **12:**167–71.

Holt, J. S. 1995. Plant responses to light: A potential tool for weed management. *Weed Sci.* **43:**474–482.

Ivy, H. W., and R. S. Baker. 1972. Prickly sida control and competition in cotton. *Weed Sci.* **20:**137–139.

Johnson, G. A., D. A. Mortensen, and A. R. Martin. 1995. A simulation of herbicide use based on weed spatial distribution. *Weed Res.* **35:**197–205.

Keeley, P. E., and R. J. Thullen. 1978. Light requirements of yellow nutsedge (*Cyperus esculentus*) and light interception by crops. *Weed Sci.* **26:**10–16.

King, L. J. 1966. *Weeds of the World–Biology and Control,* p. 270. Interscience, New York.

King, R. P., D. W. Lybecker, E. E. Schweizer, and R. L. Zimdahl. 1986. Bioeconomic modeling to simulate weed control strategies for continuous corn (*Zea mays*). *Weed Sci.* **34:**972–979.

Knake, E. L., and F. W. Slife. 1962. Competition of *Setaria faberi* with corn and soybeans. *Weeds* **10:**26–29.

Konesky, D. W., M. Y. Siddiqi, A. D. M. Glass, and A. I. Hsiao. 1989. Wild oat and barley interactions: varietal differences in competitiveness in relation to phosphorus supply. *Can. J. Bot.* **67:**3366–3376.

LeFevre, P. 1956. Influence du milieu et des conditions d'exploration sur le developpement des plantes adventices. Effet particulier du pH et l'etat calcique. *Ann. Agron. Paris* **7:**299–347.

Lybecker, D. W., R. P. King, E. E. Schweizer, and R. L. Zimdahl. 1984. Economic analysis of two weed management systems for two cropping rotations. *Weed Sci.* **32:**90–95.

Mack, R. N. 1981. Invasion of *Bromus tectorum* L. Into western North America: An ecological chronicle. *Agro-Ecosystems* **7:**145–165.

Malicki, L., and C. Berbeciowa. 1986. Uptake of more important mineral components by common field weeds on loess soils. *Acta Agrobotanika* **39:**129–141.

McGiffen, J. E., Jr., J. B. Masiunas, and J. D. Hesketh. 1992. Competition for light between tomatoes and nightshades (*Solanum nigrum* or *S. ptycanthum*). *Weed Sci.* **40:**220–226.

McWhorter, C. G. 1971. Growth and development of johnsongrass ecotypes. *Weed Sci.* **19:**141–47.

McWhorter, C. G., and T. N. Jordan. 1976. Comparative morphological development of six johnsongrass ecotypes. *Weed Sci.* **24:**270–275.

Moody, K. 1981. Weed fertilizer interactions in rice. Int. Rice Res. Inst., Thursday Seminar, July 9.

Mortensen, D. A. , J. A. Dieleman, and G. A. Johnson. 1996. Weed spatial variation and weed management. CRC Press, Boca Raton, Florida.

Mortimer, A. M. 1984. Population ecology and weed science, pp. 363–388 in *Perspectives on Plant Population Ecology* (R. Dirzo and J. Sarukhan, eds.). Sinauer, Sunderland, Massachusetts.

Mortimer, A. M. 1987. The population ecology of weeds–implications for integrated weed management, forecasting and conservation. *Proc. Brit. Crop Prot. Conf.–Weeds,* pp. 936–944.

Mortimer, M., and R. Cousens. 1995. *Dynamics of Weed Populations.* Cambridge Univ. Press. Cambridge, U. K. 332 pp.

Munger, P. H., J. M. Chandler, and J. T. Cothern. 1987. Effect of water stress on photosynthetic parameters of soybean (*Glycine max*) and velvetleaf (*Abutilon theophrasti*). *Weed Sci.* **35:**15–21.

Muzik, T. J. 1970. *Weed Biology and Control,* p. 17. McGraw-Hill, New York.

Norris, R. F. 1992. Have ecological and biological studies improved weed control strategies. *Proc. 1st Int. Weed Cont. Cong.* **1:**7–33. Melbourne, Australia.

Okafor, L. I., and S. K. DeDatta. 1976. Competition between upland rice and purple nutsedge for nitrogen, moisture, and light. *Weed Sci.* **24:**43–46.

Patterson, D. T. 1979. The effects of shading on the growth and photosynthetic capacity of itchgrass (*Rottboellia exaltata*). *Weed Sci.* **27:**549–553.

Patterson, D. T. 1982. Effects of light and temperature on weed/crop growth and competition, pp. 407–420 in *Biometeorology in Integrated Pest Management* (J. L. Hatfield and I. J. Thomason, eds.). Academic Press, New York.

Patterson, D. T. 1985. Comparative ecophysiology of weeds and crops, pp. 101–129 in *Weed Physiology* (S.O. Duke, ed.). CRC Press, Boca Raton, Florida.

Patterson, D. T. 1995a. Effects of environmental stress on weed/crop interactions. *Weed Sci.* **43:**483–490.

Patterson, D. T. 1995b. Weeds in a changing climate. *Weed Sci.* **43:** 685–701.

Peterson, D. 1936. Stellaria-Studien. Zur zytologie, genetik, okologie und systematik der gattung stellaria in besonders der media-gruppe. *Botan. Not.,* pp. 281–419.

Radosevich, S. R., and J. S. Holt. 1984. *Weed Ecology: Implications for Vegetation Management.* Wiley Interscience, New York. 265 pp.

Regnier, E. E., M. E. Salvucci, and E. W. Stoller. 1988. Photosynthesis and growth responses to irradiance in soybean (*Glycine max*) and three broadleaf weeds. *Weed Sci.* **36:**487–496.

Roberts, H. A. 1976. Weed competition in vegetable crops. *Ann. Appl. Biol.* **83:**321–324.

Scott, H. D., and R. D. Geddes. 1979. Plant water stress of soybean (*Glycine max*) and common cocklebur (*Xanthium pensylvanicum*): a comparison under field conditions. *Weed Sci.* **27:**285–289.

Sexsmith, J. J., and U. J. Pittman. 1963. Effect of nitrogen fertilizer on germination and stand of wild oats. *Weeds* **11:**99–101.

Shantz, H. L., R. L. Piemeisel and L. Piemeisel. 1927. The water requirement of plants at Akron, Colorado. *J. Agric. Res.* **34:**1093–1190.

Shetty, S. V. R., M. V. K. Sivakumar, and S. A. Ram. 1982. Effect of shading on the growth of some common weeds of the semi-arid tropics. *Agron. J.* **74:**1023–1029.

Siddiqi, M. Y., A. D. M. Glass, A. I. Hsiao, and A. N. Minjas. 1985. Wild oat/barley interactions: varietal differences in competitiveness in relation to K$^+$ supply. *Ann Bot.* **56:**1–7.

Smith, R. J., Jr. 1968. Weed competition in rice. *Weed Sci.* **16:**252–254.

Snaydon, R. W. 1984. Plant demography in an agricultural context, pp. 389–407 in *Perspectives on Plant Polulation* Ecology (Dirzo, R., and J. Sarvichan, Eds). Sinauer, Sunderland, Massachusetts.

Stoller, E. W., and R. A. Myers. 1989. Response of soybeans (*Glycine max*) and four broadleaf weeds to reduced irradiance. *Weed Sci.* **37:**570–574.

Stoller, E. W., and J. T. Woolley. 1985. Competition for light by broadleaf weeds in soybeans (*Glycine max*). *Weed Sci.* **33:**199–202.

Stuart, B. L., S. K. Harrison, J. R. Abernathy, D. R. Kreig, and C. W. Wendt. 1984. The response of cotton (*Gossypium hirsutum*) water relations to smooth pigweed (*Amaranthus hybridus*) competition. *Weed Sci.* **32:**126–132.

Vega, M. R., J. D. Ona and E. P. Paller, Jr. 1967. Weed control in upland rice at the University of the Philippines College of Agriculture. *Philippine Agric.* **51:**397–411.

Vengris, J., M. Drake, W. G. Colby, and J. Bart. 1953. Chemical composition of weeds and accompanying crop plants. *Agron. J.* **45:**213–218.

Vengris, J., W. G. Colby, and M. Drake. 1955. Plant nutrient competition between weeds and corn. *Agron. J.* **47:**213–216.

Warwick, S. I., and D. Briggs. 1978. The genecology of lawn weeds. 1. Population differentiation in *Poa annua* in a mosaic environment of bowling green lawn and flower beds. *New Phyto.* **81:**711–23.

Weatherspoon, D. M., and E. E. Schweizer. 1971. Competition between sugarbeets and five densities of kochia. *Weed Sci.* **19:**125–128.

Wedderspoon, I. M., and G. W. Burt. 1974. Growth and development of three johnsongrass selections. *Weed Sci.* **22:**319–22.

Wiese, A. F. 1965. Effect of tansy mustard and 2,4-D on winter wheat. Texas Agric. Expt. Stn. Bull. Mp-782.

Wiese, A. F., and W. Vandiver. 1970. Soil moisture effects on competitive ability of weeds. *Weed Sci.* **18:**518–519.

Yip, C. P. 1978. Yellow nutsedge ecotypes, their characteristics and responses to environment and herbicides. *Diss. Abstr. Int. B* **39:**1562–1563.

Young, J. A. 1988. The public response to the catastrophic spread of Russian thistle (1880) and halogeton (1945). *Agric. Hist.* **62:**122–130.

Young, J. A. 1991. Tumbleweed. *Sci. Am.* March:82–87.

Young, J. A., R. A. Evans, and B. L. Kay. 1970. Phenology of reproduction of medusahead. *Weed Sci.* **18:**451-54.

Young, F. L., D. L. Wyse, and R. J. Jones. 1983. Effect of irrigation on quackgrass (*Agropyron repens*) interference in soybeans (*Glycine max*). *Weed Sci.* **31:**720–726.

Zimdahl, R. L. 1980. *Weed-Crop Competition–A Review.* The Int. Plant Prot. Center, Oregon State Univ. 195 pp.

Chapter 7

Allelopathy

FUNDAMENTAL CONCEPTS

- Allelopathy is a form of plant interference that occurs when one plant, through living or decaying tissue, interferes with growth of another plant via a chemical inhibitor.
- Allelopathy may be present in many plant communities.
- Allelopathy has a potential but unknown role in weed management.

LEARNING OBJECTIVES

- To know the definition of allelopathy.
- To understand the complexity of research to discover true allelopathy.
- To understand the complexity of allelopathic chemistry.
- To understand how allelochemicals enter the environment.
- To know the application of an analogous form of Koch's postulates to allelopathy.
- To know some examples of allelopathic interference.

In 1754, Horace Walpole wrote "The Three Princes of Serendip." He based his story on an ancient Persian tale in which the characters make fortunate, unexpected, wonderful discoveries. In his story, the three princes, each vying for the hand of a princess, are assigned impossible tasks by the princess. Each fails to accomplish the assigned tasks, but wonderful, serendipitous, things happen to them anyway. Serendipity is an apparent aptitude for making fortunate discoveries accidentally; unexpected, good things happen. Serendipity may be available to weed science if the presence of allelopathy can be used to control weeds. Organisms from microbes to mammals find food, seek mates, ward off predators, and defend themselves against disease via chemical interactions. Allelopathic interactions are chemical, and discovery of the cause of these interactions may yield a treasure of biological and chemical approaches to control weeds. At least 25% of human medicinal products originated in the natural world or are synthetic derivatives of naturally occurring substances. Many natural interactions are chemical interactions, and some of them could influence the course of weed science.

Interference is the term assigned to adverse effects that plants exert on each others' growth. Competition is part of interference and occurs because of depletion or unavailability of one or more limiting resources. Allelopathy, another form of interference, occurs when one plant, through its living or decaying tissue, interferes with growth of another plant via a chemical inhibitor (Figure 7.1). Allelopathy comes from the Greek *allelo* or "each other," which is similar to the Greek *allelon* or "one another." The second root is the Greek *patho* or *pathos,* which means suffering, disease, or intense feeling. Allelopathy is therefore the influence, usually detrimental (the pathos), of one plant on another, by toxic chemical substances from living plant parts, through their release when a plant dies, or their production from decaying tissue.

There is a subset of allelochemicals known as kairomones (from Greek *kai* = new, and *hormaein* = to set in motion, excite, stimulate) that have favorable adaptive value to organisms receiving them. A natural kairomone from waterhyacinth is a powerful insect attractant for a weevil (*Necochetina eichhorniae*) and the waterhyacinth mite (*Orthogalumna terebrantis*). The kairomone is liberated when waterhyacinth is injured by surface wounding or by the herbicide 2,4-D. The kairomone enhances control of waterhyacinth by attracting large numbers of weevils and mites to the area of the plant's wound (Messersmith and Adkins, 1995). Thus, the kairomone has favorable value to the insects but not to the waterhyacinth. Control of waterhyacinth is enhanced when insect damage is combined with herbicide stress.

For weed management purposes, allelopathy is considered a strategy of control. Corn cockle and ryegrass seeds fail to germinate in the presence of beet seeds. If tobacco seeds germinate and grow for 6 days in petri dishes

Interference = Competition + Allelopathy

Figure 7.1   Components of plant interference.

and then an extract of soil, incubated for 21 days with timothy residue, is added, the root tips of tobacco blacken within 1 hour while radicle elongation is unaffected. If an extract of soil incubated with rye residue is added, the symptoms are reversed (Patrick and Koch, 1958). Residues of timothy, maize, rye, and tobacco all reduce the respiration rate of tobacco seedlings (Patrick and Koch, 1958).

Kooper (1927), a Dutch ecologist, observed the large agricultural plain of Pasuruan on the island of Java, Indonesia, where sugarcane, rice, and maize grew. After harvest, the fallowed fields developed a dense cover of weeds. Kooper observed that the postharvest floristic composition of each community was stable year after year. He found that floristic composition was determined at the earliest stages of seed germination, not by plant survival rate or a struggle for existence, but by differential seed germination. He showed that seeds of other species were present but could not germinate unless removed from their environment. Competition for light, nutrients, or water did not cause the consistent floristic composition. Kooper (1927) concluded that previous vegetation established a soil chemical equilibrium (an allelopathic phenomenon) and determined which seeds could germinate and subsequently which plants dominated.

The word allelopathy was first used by Molisch (1937), an Austrian botanist. He included toxicity exerted by microorganisms and higher plants, and that use has continued. The phenomenon, however, had been observed much earlier by several scientists (Putnam, 1985). A classic example of allelopathy is found in the walnut forests of Central Asia (Stickney and Hoy, 1881). Few other plants survive under the forest plant canopy because of the presence of juglone, a quinone root toxin derived from walnut trees (Massey, 1925). The effect of juglone could not be reproduced in the greenhouse because some plant metabolites, including phenolics, require ultraviolet light for their biosynthesis (Davis, 1928).

Another classic study is the work by Muller and Muller (1964) in California who observed that California chaparral often occurred near, but not intermixed with, California sagebrush. Neither species grew in the zones of contact between the respective communities; other species grew between the communities. They found that terpenes, particularly camphor (a monoterpene ketone) and cineole (a terpene ether) produced by the chaparral, were responsible for the no-contact zones. They concluded that plants, in this case the chaparral, are fundamentally leaky systems. Other studies are described by Rice (1974, 1979) and Thompson (1985).

One plant does not consciously set out to affect another. Rather, the effect occurs as a normal, perhaps serendipitous, ecological interaction with evolutionary implications. Allelopathic species have been selected by evolutionary pressure, because they can outcompete neighbors through energy-expensive biochemical processes that produce allelochemicals. The energy expense is not a waste of resources because no species evolves successfully by wasting resources. Exploration of the phenomena will lead to better understanding of plant evolutionary strategies and, possibly, provide clues for herbicide synthesis and development.

Reviews of allelopathy are found in Putnam (1985, 1994) and the proceedings of the American Chemical Society symposium on the chemistry of allelopathy (Thompson, 1985). Putnam (1985, 1994) lists 50 weeds alleged to interfere with one or more crops (Table 7.1). Allelopathy has also been explored with a number of crops, and there have been attempts to find crop cultivars with a competitive allelopathic edge (Putnam, 1983, 1985; Rice, 1979; Thompson, 1985). Residues of several crops have phytotoxic activity on other plants (Table 7.2, from Putnam, 1994).

I. ALLELOPATHIC CHEMISTRY

Plants produce a myriad of metabolites of no known utility to their growth and development. These are often referred to as secondary plant metabolites and are often defined as compounds having no known essential physiological function. The idea that these compounds may injure other forms of life is not without a logical base. However, proof is questionable because most allelochemical effects occur through soil, a complex chemical matrix. Conclusive studies require extraction and isolation of the active agent from soil. Any allelopathic chemical may be chemically altered prior to or during extraction. That which is extracted, isolated, and studied may not be what the plant produced.

Allelochemicals vary from simple molecules such as ammonia, to the more complex quinones, juglone, and the terpenes camphor and cineole, to very complex conjugated flavonoids such as phlorizin (isolated from apple roots), or the heterocyclic alkaloid caffeine (isolated from coffee) (Putnam, 1985; Rice, 1974; Thompson, 1985). Putnam (1985) lists several chemical groups from which allelopathic agents come: organic acids and aldehydes, aromatic acids, simple unsaturated lactones, coumarins, quinones, flavonoids, tannins, alkaloids, terpenoids and steroids, a few miscellaneous compounds such as long-chain fatty acids, alcohols, polypeptides, nucleosides, and some unknown compounds. Some of the diversity and complexity of allelopathic chemistry is shown Table 7.3. The diversity suggests several mechanisms of action and a multiplicity of effects, and it is one reason for the slow emergence of a theoretical framework. The chemistry of allelopathy is as complex as synthetic herbicide chemistry.

Table 7.1

Common Weeds with Alleged Allelopathic Activity in Agroecosystems (Putnam, 1983, 1994)

Weed	Susceptible species
Beggarticks	Several
Bermudagrass	Coffee
Bluegrass	Tomato
California peppertree	Cucumber, wheat
Canada thistle	Several
Catnip	Peas, wheat
Cogongrass	Several
Common chickweed	Barley
Common lambsquarters	Cucumber, oats, corn
Common milkweed	Sorghum
Common purslane	Peas, wheat
Common ragweed	Several
Corn cockle	Wheat
Diffuse knapweed	Ryegrass
Dock	Corn, pigweed, sorghum
Flaxweed	Flax
Giant foxtail	Corn
Giant ragweed	Peas, wheat
Goldenrod	Several
Heath	Red clover
Italian ryegrass	Oats, brome, lettuce, clover
Jimsonweed	Several
Johnsongrass	Several
Ladysthumb	Potato, flax
Large crabgrass	Several
Leafy spurge	Peas, wheat
Mayweed	Barley
Mugwort	Cucumber
Mustard	Several
Nutsedge, purple	Sorghum, soybean
Nutsedge, yellow	Corn
Parthenium ragweed	Several
Prince's feather	Mustard

(continues)

Table 7.1 (*continued*)

Weed	Susceptible species
Prostrate spurge	Several
Quackgrass	Several
Russian thistle	Several
Spiny amaranth	Coffee
Sunflower	Several
Syrian sage	Wheat
Velvetgrass	Barley
Velvetleaf	Several
Western ragweed	Several
Wild cane	Wheat
Wild garlic	Oats
Wild marigold	Several
Wild oats	Several

There is little doubt that allelopathy occurs in plant communities; but there are questions about how important allelopathic chemicals are in nature and if they can be exploited in cropped fields. Allelopathy has been reported for many crop and weed species (Putnam, 1983, 1985, 1994), but proof of its importance in nature is lacking. Proof will require something similar to the application of Koch's postulates (1912) proposed for plant pathology in 1883 and amended by Smith (1905).

The analogous postulates applied to allelopathy (Aldrich, 1984; Putnam, 1985) are as follows:

1. Observe, describe, and quantify the degree of interference in a natural community.
2. Isolate, characterize, and synthesize the suspected toxin.
3. Reproduce the symptoms by application of the toxin at appropriate rates and times in nature. (Koch's (1912) postulates called for reisolation of the bacterial agent from the experimentally infected plant—an inappropriate criterion for allelopathic research).
4. Monitor release, movement, and uptake, and show they are sufficient to cause the observed effect(s).

These four steps describe difficult and expensive procedures. Rigorous proof has rarely been applied to any ecological interaction, but such proof is vital if allelopathic research is to move from description to causation.

Table 7.2

Some Crops Whose Residues Have Been Reported to Be Phytotoxic (Putnam, 1994)

Crop	Susceptible species
Alfalfa	Alfalfa (autotoxicity)
Apple	Apple (autotoxicity)
Asparagus	Several
Barley	Several
Bean	Pea, wheat
Cabbage	Spinach
Clover, red	Several
Clover, white	Radish
Coffee	Several
Corn	Several
Cucumber	Several weeds
Lentil	Wheat
Oats	Several
Pea	Several
Rice	Rice (autotoxicity)
Rye	Several
Ryegrass	Several
Smooth bromegrass	Several
Sorghum	Several
Sunflower	Several
Wheat	Several

In short, it is insufficient to make an observation and suspect a toxin. It is insufficient to demonstrate that the toxin is produced by one plant. Specific cause and effect must be demonstrated through chemical and plant studies. It may not be necessary to prove that plant X is the source of allelochemical Y. If an allelochemical, effective as a natural herbicide, can be isolated and identified, it might be useful without absolute proof of its plant origin or physiological mode of action. The basic chemistry and biology would remain a scientific challenge, but it might be possible to exploit the activity. Proceeding with partial knowledge is more risky, but not impossible. For example, medical science still does not know exactly how aspirin

Table 7.3

Allelopathic Compounds Isolated from Plants (Putnam, 1983)

Common name	Chemical class	Natural source
Acetic acid	Aliphatic acid	Decomposing straw
Allylisothiocyanate	Thiocyanate	Mustard plants
Arbutin	Phenolic	Manzanita shrubs
Bialaphos	Amino acid derivative	Microorganisms
Caffeine	Alkaloid	Coffee plants
Camphor	Monoterpene	*Salvia* shrubs
Cinnamic acid	Aromatic acid	Guayule plants
Dhurrin	Cyanogenic glucoside	Sorghum plants
Gallic acid	Tannin	Spurge plants
Juglone	Quinone	Black walnut trees
Patulin	Simple lactone	*Penicillium* fungus on wheat straw
Phlorizin	Flavonoid	Apple roots
Psoralen	Furanocoumarin	*Psoralea* plants

relieves pain, and weed science does not know exactly how 2,4-D kills a plant. Both can be used productively and safely.

II. PRODUCTION OF ALLELOCHEMICALS

Production of allelochemicals varies with environment and associated environmental stresses. It can occur in any plant organ (Rice, 1974), but roots, seeds, and leaves are the most common sources. Source becomes important for exploitation of allelochemicals for weed control. For example, an allelochemical found in flowers or fruits would have less potential value than if it were concentrated in roots or shoots (Putnam, 1985). (This is a statement about availability, not allelochemical potency.) For control, soil incorporation of whole plants might create proper distribution regardless of which plant part produced the chemical. The amount is important for control purposes, and if specific effects are to be predicted in the field, total quantity and concentration must be determined (Putnam, 1985).

There is evidence that allelochemical production may be greater when plants suffer from environmental stress (Putnam, 1983, 1985; Rice, 1979). Production is influenced by light intensity, quality, and duration, with a greater quantity produced with high ultraviolet light and long days (Aldrich,

1984). Weeds, commonly understory plants, might be expected to produce lower quantities of allelochemicals because UV light is filtered by overshadowing crop plants. This, of course, assumes that crops provide shade and that shade effectively suppresses allelopathic activity. Quantities of allelochemicals produced are also greater under conditions of mineral deficiency, drought stress, and cool temperatures as opposed to more optimal growing conditions. In some cases, plants affected by growth regulator herbicides may increase production of allelochemicals. Because stress frequently enhances allelochemical production, it is logical to assume that stress accentuates the involvement of allelopathy in weed–crop interference and that competition for limited resources may increase allelopathic potential or sensitivity of the weed, the crop, or both. Thus, weed–crop competition and allelopathy should not be regarded as totally separate, unrelated components of interference in a crop ecosystem.

Allelochemicals enter the environment in a number of ways at different times, and mode and time of entry can alter their effects (Figure 7.2). Although chemicals with allelopathic activity may be present in many species, presence does not mean that allelopathic effects will ensue. Even after a chemical has been isolated and identified, its placement in the environment after plant release or its time of release may preclude expression of potential activity.

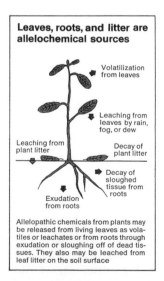

Figure 7.2   Sources of allelochemicals (Putnam, 1983).

Allelochemicals enter the environment through volatilization or root exudation and move through soil by leaching (Figure 7.2). These modes of entry may be regarded as true allelopathy. Toxins also result from decomposition of plant residues. This is properly regarded as functional allelopathy: that is, environmental release of substances that are toxic as a result of transformation after their release by the plant.

Allelochemicals can be produced by weeds and affect crops, but the reverse is also true, although it has not been as widely studied (Putnam, 1994). It is possible that some crop cultivars produce allelochemicals, and these cultivars could be planted to take advantage of their allelochemical potential. It has been suggested that crops with allelopathic potential could be planted as rotational crops or companion plants in annual or perennial cropping systems to exert their allelopathic effect on weeds. Rye and its residues have been shown to provide good weed control in a variety of cropping systems (Barnes and Putnam, 1983). Rye residues reduced emergence of lettuce and proso millet by 58 and 35%, respectively. Rye shoot tissue inhibited lettuce seed germination 52%. It also was phytotoxic to barnyardgrass and cress (Barnes and Putnam, 1986). A final possibility is genetic transfer of the ability to produce a weed controlling allelochemical to a crop plant. Much more physiological and chemical knowledge is required before this can be done successfully, but it is an enticing possibility—a crop that does more of its own weed control because it has a chemical advantage.

III. ALLELOPATHY AND WEED–CROP ECOLOGY

Aldrich (1984) suggested that allelopathy was significant for weed–crop ecology in three ways:

1. As a factor affecting changes in weed species composition
2. As an avenue of weed interference with crop growth and yield
3. As a possible weed management tool

Allelopathy should not always be implicated when other explanations do not suffice, but it should not be overlooked because of the difficulty of establishing causality.

A. EFFECTS ON WEED SPECIES COMPOSITION

Why one species succeeds another is a question that has intrigued ecologists for many years. Weed scientists are interested in the same question, but often only for the lifespan of an annual crop. Weed scientists accept that plants change the environment and are changed by it. Early colonizers succeed by producing large numbers of seeds, whereas late arrivals have greater competitive ability. This is true in old-field succession and in annual crops. Ecologists have shown that successful plants may change the environment to their advantage by subtle means, such as changes in soil nitrogen relationships caused by release of specific inhibitors of nitrogen fixation or nitrification (Putnam, 1985; Thompson, 1985).

B. WEED INTERFERENCE

Weed seeds survive for long periods in soil, and chemical inhibitors of microbial decay have been implicated in their longevity, but specific identification of inhibitors from weed seeds has not been accomplished. Allelochemicals have been implicated in the inability of some seeds to germinate in the presence of other seeds or in the presence of crop residues in soil. Although neither phenomenon has been exploited for weed management, there is little doubt that both occur. Eventual exploitation may depend on discovery of specific chemicals and their mode(s) of action. Because of the mass of plant residue and its volume compared to the volume of seed (even though the number of seeds may be large), the possibility of effects from residues is greater than that of effects from seed.

The problems with replanting the same or different crops in a field have been cited (Putnam, 1985; Rice, 1974) to show the effect of allelochemicals on crop growth. Putnam (1983) showed that the allelopathic potential of sorghum residues has been exploited for weed control in subsequent rotational crops. Although there is little doubt that allelochemicals inhibit crop growth, a problem still exists in separating allelopathic effects from competition. Most greenhouse studies cannot be directly translated to the field because of different climatic, edaphic, and biological conditions, and possible effects of soil volume. Allelopathy awaits adequate experimental methods for independent but related field and greenhouse studies.

C. WEED MANAGEMENT

A living cover crop of spring-planted rye reduced early-season biomass of common lambsquarters 98%, common ragweed 90%, and large crabgrass 42% compared to control plots with no rye (Barnes and Putnam, 1983). Wheat straw has reduced populations of pitted morningglory and prickly sida in no-tillage culture. It was suggested the wheat produced an allelochemical that inhibited emergence of several broadleaved species (Liebl and Worsham, 1983). It is reasonable to assume that many plants have allelopathic potential or some susceptibility to allelochemicals when they are presented in the right amount, form, and concentration at the appropriate time. It is equally reasonable to assume that allelopathy may have no role in the interference interactions of many species. However, enough work has been done to suggest that allelopathy could be utilized for development of new weed management strategies. Field trials in South Dakota showed that fields planted to sorghum had 2 to 4 times fewer weeds the following year than similar fields planted to soybean or corn (Kozlov, 1990). It was proposed, although not proven, that reduced weed seed germination was due to phenolic acids and cyanogenic glucosides given off by sorghum. Suppression of weeds by sorghum has been reported by Guenzi and McCalla (1966) and Hussain and Gadoon (1981). Sunflower has been reported to have an allelopathic effect against grain sorghum (Schon and Einhellig, 1982) and against other weeds (Leather, 1983). Guenzi and McCalla (1966) found allelopathic phenolic acids in oats, wheat, sorghum, and corn residues, and Lodhi *et al.* (1987) discussed the role of allelopathy from wheat in crop rotations. Other sources are available to describe and summarize the major findings of allelopathy research and their application in weed management (Putnam, 1983, 1985, 1994; Rice, 1974, 1979 and Thompson, 1985). A few examples follow to illustrate the research and its potential.

Walker and Jenkins (1986) were the first to demonstrate that sweet potato residues inhibited growth of sweet potato and cowpea. Decaying residues reduced uptake of calcium, magnesium, and sulfur by other plants (Walker *et al.,* 1989). Additional studies showed that after one growing season, shoot dry weight of yellow nutsedge growing with sweet potatoes was less than 10% of the weight when yellow nutsedge was grown alone. Moreover, remaining yellow nutsedge had no effect on sweet potato growth (Harrison and Peterson, 1991). Allelochemicals were present in the tuber periderm that is continually sloughed off during root growth. Proso millet was susceptible to all extracted fractions, but other plants showed differential susceptibility, indicating that several allelochemicals may be present (Peterson and Harrison, 1991).

Another interesting example began with a study of the root parasitic damping off fungus (*Pythium* spp.) in turf. Christians (1991, 1993) wanted to establish the fungus in the soil of a new golf-course green at Iowa State University. *Pythium* was cultured in the laboratory on corn meal, a standard procedure. The culture was placed on field plots and other plots were treated with the same amount of fresh corn meal. The attempt to establish *Pythium* failed, but seeded cultivars of creeping bentgrass did not germinate well on plots that had received fresh corn meal. This was unexpected.

Further study showed potential for selective control of crabgrass in Kentucky bluegrass turf. Liu *et al.* (1994) demonstrated that enzymatically hydrolyzed corn gluten meal was more herbicidally active than corn gluten. Corn gluten hydrolysate completely inhibited germination of crabgrass and creeping bentgrass seed and root emergence of perennial ryegrass seed.

With this kind of compelling evidence, one is inclined to agree with Putnam's (1985) suggestion that not believing in allelopathy, now, is like not believing in genetic inheritance before DNA's structure was known. One area to explore might be testing for suppression of weed seed germination and seedling emergence by potential allelopathic species. Work to date has shown this to be an inconsistent effect and, if developed, it could be used with other methods of weed management. Allelopathy is not, and will never be, a panacea for all weed problems. It is another weed management tool to be placed in the toolbox and used in combination with other techniques. It is not a technique that will finally solve all weed problems or make the hoe obsolete.

The second strategy where allelopathy may be used is weed-suppressing crops. This can be realized by discovering, incorporating, or enhancing allelopathic activity in crop plants. This technique would be most useful in crops maintained in high-density monocultures, such as turf grasses, forage grasses, or legumes.

The third area for allelopathic research and development includes the use of plant residues in cropping systems, allelopathic rotational crops, or companion plants with allelopathic potential. Many crops leave residues that are regarded as a necessary but not a beneficial part of crop production, except as they contribute to soil fertility or tilth. Research (Putnam, 1985, 1994; Rice, 1979) indicates that plant residues have allelopathic activity, but the nature and use of this activity has not been explored sufficiently to permit effective use. Rotation, a neglected practice in many agricultural systems, is being studied because of its potential for weed management through competition and allelopathy. Companion cropping is a new and interesting technique for agricultural systems in developing countries. Multiple cropping is common in many developing countries where allelopathy may be operational without being obvious and defined. These systems may hold valuable lessons for further agricultural development of allelopathy as a useful weed management tool.

Weed scientists need to look beyond the immediate assumption that interference is always competition and see what they may not be looking for—an allelopathic effect. An unexpected, good thing. Perhaps there are expressions of allelopathy before our eyes that we do not see because we are not looking for them. If there are compounds in nature with such great specificity, they should be examined. The patterns of herbicide development

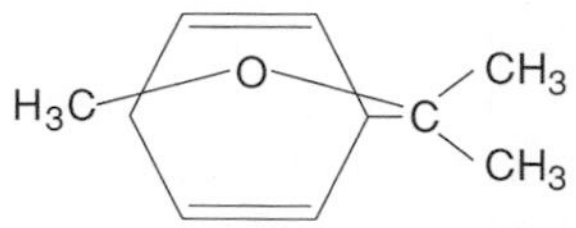

H₃C⟨⟩O⟨⟩=C⟨CH₃ / CH₃⟩

Figure 7.3   Structure of 1,8-cineole.

point to greater specificity, and nature may have solutions if we recognize them, learn how they work, and exploit their capabilities.

One of the first phytotoxic compounds to be implicated in higher plants was 1,8-cineole (Muller and Muller, 1964) (Figure 7.3). Cinmethylin was developed as a herbicide but never used commercially for weed control in crops. It controls many annual grasses and suppresses some broadleaved species. Its structure (Figure 7.4) is similar to the structure of 1,8-cineole, an allelopathic chemical produced by species of sage. Cinmethylin is produced synthetically, but the thought behind it could have been derived from the known phytotoxicity of the allelopathic chemical.

A second and clearer example of a natural herbicide is AAL-toxin, a natural metabolite produced by *Alternaria alternata* f. sp. *Lycopersici,* the pathogen that causes stem canker of tomato (Abbas *et al.,* 1995). The phytotoxic effects of AAL-toxin were tested on 86 crop and weed species (Abbas *et al.,* 1995). Monocots were generally immune to its effects. Black nightshade, jimsonweed, all species of tomatoes tested, and several other broadleaved plants were susceptible at low doses. Other broadleaved species were susceptible but only at higher doses. Abbas *et al.* (1995) proposed that the differential susceptibility of species to AAL-toxin could be exploited for selective weed control. There may be other potentially valuable chemicals hidden from us because we are looking for something else. Promising observations await the good observer.

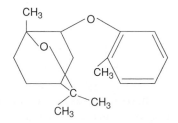

Figure 7.4   Structure of cinmethylin.

THINGS TO THINK ABOUT

1. What is the present role of allelopathy in weed management?
2. What is the potential role of allelopathy in weed management?
3. Why has so little research been done on allelopathy?
4. What are the essential ingredients of a research program to discover allelochemicals?

LITERATURE CITED

Abbas, H. K., T. Tanaka, S. O. Duke, and C. D. Boyette. 1995. Susceptibility of various crop and weed species to AAL-toxin, a natural herbicide. *Weed Technol.* **9:**125–130.

Aldrich, R. J. 1984. Weed–crop ecology—Principles in weed management, Chapter 8 in Allelopathy in weed management. Breton, N. Scituate, Massachusetts.

Barnes, J. P., and A. R. Putnam. 1983. Rye residues contribute to weed suppression in no-tillage cropping systems. *J. Chem. Ecol.* **9:**1045–1057.

Barnes, J. P., and A. R. Putnam. 1986. Evidence for allelopathy by residues and aqueous extracts of rye (*Secale cereale*). *Weed Sci.* **34:**384–390.

Christians, N. E. 1991. Preemergence weed control using corn gluten meal. U.S. Patent No. 5030268, pp. 63–65.

Christians, N. 1993. A natural product for the control of annual weeds. *Golf Course Management,* October:72, 74, 76.

Davis, R. F. 1928. The toxic principle of *Juglans nigra* as identified with synthetic juglone and its toxic effects on tomato and alfalfa plants. *Am. J. Bot.* **15:**620.

Guenzi, W., and T. McCalla. 1966. Phenolic acids in oats, wheat, sorghum, and corn residues and their phytotoxicity. *Agron J.* **58:**303–304.

Harrison, H. F., Jr., and J. K. Peterson. 1991. Evidence that sweet potato (*Ipomoea batatas*) is allelopathic to yellow nutsedge (*Cyperus esculentus*). *Weed Sci.* **39:**308–312.

Hussain, F., and M. A. Gadoon. 1981. Allelopathic effects of *Sorghum vulgare* Pers. *Oecologia* **51:**284–288.

Koch, R. 1912. *Complete Works* **1:**650–660. 10th Int. Medical Congr. Berlin. 1890. George Thieme, Leipzig.

Kooper, W. J. C. 1927. Sociological and ecological studies on the tropical weed-vegetation of Pasuruan (the Island of Java). *Recueil Travaux Botaniques Neerlandais* **24:**1–255.

Kozlov, A. 1990. Weed woes. *Discover* February:24.

Leather, G. R. 1983. Sunflowers (*Helianthus annuus*) are allelopathic to weeds. *Weed Sci.* **31:**37-42.

Liebl, R., and D. Worsham. 1983. Inhibition of pitted morningglory (*Ipomoea lacunosa* L.) and certain other weed species by phytotoxic compounds of wheat (*Triticum aestivum* L.) straw. *J. Chem Ecol.* **9:**1027–1043.

Liu, D. Lan-Ying, N. E. Christians, and J. T. Garbutt. 1994. Herbicidal activity of hydrolyzed corn gluten meal on three species under controlled environments. *J. Plant Growth Regul.* **13:**221–226.

Lodhi, M. A. K., R. Bilal, and K. A. Malik. 1987. Allelopathy in agroecosystems: Wheat phytotoxicity and its possible role in crop rotation. *J. Chem. Ecol.* **13:**1881–1889.

Massey, A. B. 1925. Antagonism of the walnuts (*Juglans nigra* L. and *J. Cinerea* L.) in certain plant associations. *Phytopath.* **16:**773–784.

Messersmith, C. G., and S. Adkins. 1995. Integrating weed-feeding insects and herbicides for weed control. *Weed Technol.* **9:**199–208.

Molisch, H. 1937. *Der Einfluss einer Pflanze auf der Ander–Allelopathie.* G. Fischer, Jena.

Muller, W. H., and C. H. Muller. 1964. Volatile growth inhibitors produced by Salvia species. *Bull. Torrey Bot. Club* **91:**327–330.

Patrick, Z. A., and L. W. Koch. 1958. Inhibition of respiration, germination, and growth by substances arising during the decomposition of certain plant residues in the soil. *Can. J. Bot.* **36:**621–647.

Peterson, J. K., and H. E. Harrison, Jr. 1991. Differential inhibition of seed germination by sweet potato (*Ipomoea batatas*) root periderm extracts. *Weed Sci.* **39:**119–123.

Putnam, A. R. 1983. Allelopathic chemicals. *Chem. & Eng. News* **61**(19):34–43.

Putnam, A. R. 1985. Weed allelopathy, Chapter 5, pp. 131–155, in *Weed Physiology,* Vol. I: *Reproduction and Ecophysiology* (S. O. Duke, Ed.). CRC Press, Boca Raton, Florida.

Putnam, A. L. 1994. Phytotoxicity of plant residues, pp. 285–314 in Managing Agricultural Residues (P. W. Unger, Ed.). Lewis Pubs. (CRC Press), Boca Raton, Florida.

Rice, E. L. 1974. *Allelopathy.* Academic Press, New York. 353 pp.

Rice, E. L. 1979. Allelopathy-an update. *Bot. Rev.* **45:**15–109.

Schon, M. K., and F. A. Einhellig. 1982. Allelopathic effects of cultivated sunflower on grain sorghum. *Bot. Gaz.* **143:**505–510.

Smith, E. F. 1905. Bacteria in Relation to Plant Disease, Vol. 1. Carnegie Inst. of Washington, Washington, D.C.

Stickney, J. S., and P. R. Hoy. 1881. Toxic action of black walnut. *Trans. Wis. State Hort. Soc.* **11:**166–167.

Thompson A. L. (Ed.) 1985. *The Chemistry of Allelopathy—Biochemical Interactions among Plants.* Am. Chem. Soc. Symp. Series No. 268. American Chemical Society, Washington, D.C. 470 pp.

Walker, D. W., and D. D. Jenkins. 1986. Influence of sweet potato plant residues on growth of sweet potato vine cuttings and cowpea plants. *Hort. Sci.* **21:**426–428.

Walker, D. W., T. J. Hubbell, and J. E. Sedberry. 1989. Influence of sweet potato crop residues on nutrient uptake of sweet potato plants. *Agric. Ecosystems and Environ.* **26:**45–52.

Chapter 8

The Significance of Plant Competition

FUNDAMENTAL CONCEPTS

- There is no complete explanation of, or a scientific basis for, plant competition.
- The concept of competitive ability is useful but cannot be precisely defined.
- A proposed biochemical basis for plant competition is based on six factors.
- Plants fix atmospheric carbon dioxide via a C_3 and a C_4 pathway. The latter is generally regarded as more efficient.
- Plants have definable characteristics that make them competitive.

LEARNING OBJECTIVES

- To understand the biochemical basis for plant competition as proposed by Black *et al.* (1969)
- To know the difference between C_3 and C_4 carbon fixation.
- To understand the role of carbon fixation in plant competition.
- To know the arguments against the primacy of carbon fixation in plant competition.
- To know the basis for other explanations of plant competition.
- To know the characteristics that lead to competitiveness.

I. GENERAL CONSIDERATIONS

It is a salutary thought that we do not know—nor have we even given the matter much consideration—what determines the density of population of cereal plants giving maximum yield. Yet until we know this, and especially until we understand the interaction of density with such factors as water and nitrogen, then the development of suitable varieties of plants must depend in the future—as in the past—on empirical plant breeding. We can claim great advances in genetics, and great advances in producing plants with drought escape or disease resistance, fatter pods, or finer flowers. And the breeder can point to varieties which quite apart from these specific virtues, are able under the keen intraplant competition of a commercial crop, to yield more grain, more leaf, more dry matter. Why? The breeder has no idea. Indeed, the answer to such question will often be that it yields more because it has more ears, or more florets, or more fertility, or less abortion, which of course is little more than a paraphrase of the statement that it yields more. Actually what happened was that the breeder selected it because it yielded more, not that it yielded more because it was consciously bred to do so. Why does a modern wheat variety, whether in Greece or New Zealand, yield more than a variety of like maturity and disease resistance of 50 years ago? Because it either (A) fixes more carbon, or (B) has a greater proportion of the carbon in the grain. Why? No one knows. Perhaps it has a different root system, better leaf arrangement and light utilization, more glume surface, or one of many factors affecting growth and photosynthesis. And in particular, it has the desired characteristics when growing under the acute stress conditions of a commercial crop.

—Donald (1963)

These words, from 1963, are still largely true. There is no complete explanation of, and scientific basis for, plant competition, but we are closer. We know that yield and growth are a function of carbon assimilation by photosynthesis, and plant growth is affected by many environmental and physiological factors. It is known that carbon dioxide uptake and fixation are primary determinants of growth, and plant environmental responses are mediated through biochemical reactions. In agriculture, some plants have high yields, grow fast, are competitive, and may be weeds. Black *et al.* (1969) tried to provide a scientific basis for plant competition and weediness. They took data from the work of others, applied unique ideas, and proposed a biochemical basis for plant competition, based on the assumption that the primary determinant of success is the capacity to fix carbon. Their work is not conclusive and has not been included in this book as the one definitive explanation of competition and weediness. It is included because it provides clues about how the process of competition has been studied and about how to think about weed-crop competition.

Black *et al.* (1969) classified plants as efficient or non-efficient on the basis of six factors:

1. Light intensity response
2. Temperature response
3. Response to oxygen
4. Presence or absence of photorespiration

5. Pathway of photosynthetic carbon dioxide assimilation
6. Photosynthetic compensation point level

They developed the hypothesis that efficient plants are often used in agriculture because of their high production *and* because they are often very competitive. Almost all the weeds they examined were efficient.

In efficient plants, carbon dioxide uptake increases with light intensity (Figure 8.1). The ability of non-efficient plants to fix carbon dioxide falls off rapidly as light intensity increases, whereas efficient plants continue to fix carbon as light intensity increases to near full sunlight. The same thing is true for the response of plants to temperature. Non-efficient plants peak in their ability to fix carbon around 20°C (Figure 8.2). Efficient plants continue to fix carbon at higher temperatures, although the curve plateaus. Efficient plants fix carbon at much higher light intensities and temperatures than non-efficient plants.

At normal atmospheric oxygen concentration (21%), photosynthesis in non-efficient plants is inhibited by oxygen. Photosynthesis in efficient plants is not inhibited by oxygen.

In some plants, respiration decreases with increased light: a phenomenon called photorespiration that has not been demonstrated in efficient plants. It is a wasteful, light-stimulated oxidation of photosynthetic intermediates to carbon dioxide and other waste products.

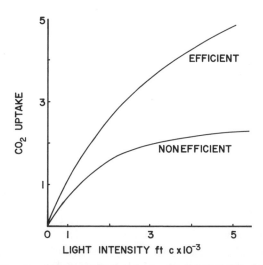

Figure 8.1 The response of photosynthesis to increasing light intensity for efficient and non-efficient plants (Black *et al.*, 1969).

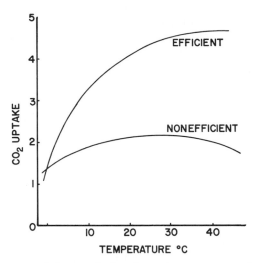

Figure 8.2   The response of photosynthesis to increasing temperature for efficient and non-efficient plants (Black *et al.*, 1969).

The C_3 cycle for carbon fixation is the dominant mechanism in plants. Ribulose diphosphate is the carbon dioxide acceptor. Some plants also fix carbon dioxide in a 4-carbon dicarboxylic acid (malic and aspartic acid) cycle. This is the Hatch-Slack or C_4 cycle in which phosphoenolpyruvate (PEP) is the carbon dioxide acceptor. The efficiency of C_4 fixation results from the fact that phosphoenolpyruvate has a much higher affinity for carbon dioxide than ribulose diphosphate carboxylase (R-uBP), the enzyme responsible for initial fixation in C_3 plants. Black *et al.*'s survey (1969) of a number of plants showed that the presence of the C_4 cycle was characteristic of efficient plants. Plants do not fix carbon dioxide by the C_3 *or* the C_4 cycle. The C_4 cycle, characteristic of efficient plants, supplements, but does not replace, the C_3 cycle. Table 8.1 shows some characteristics usually associated with the C_4 pathway. The table is a quantitative demonstration of the proposed superiority of the C_4 pathway, a superiority that mainly occurs under conditions of high irradiance. It is a reasonable generalization that C_4 plants have higher photosynthetic rates. Corn, a C_4 plant, has a maximum photosynthetic rate (measured as CO_2 fixed in μmol m^{-2} s^{-1}) between 20 and 40, whereas soybean, a C_3 plant, fixes 10 to 20. For further discussion of pathways of carbon fixation, readers are referred to a good textbook of plant physiology, such as Salisbury and Ross (1992).

The sixth characteristic of the Black *et al.* (1969) scheme is the level of carbon dioxide compensation point. Under normal physiological conditions,

Table 8.1

**Some Physiological and Performance Characteristics Associated with the
C_4 Pathway**

Characteristic	Approximate quantitative relationships compared with C_3 species
High temperature optimum for photosynthesis	30–45 vs 15–30°C
High light optimum for photosynthesis	Full sunlight vs 30% full sunlight
High photosynthesis rates per unit leaf area	About twice as much under optimal conditions
High growth rates under optimal conditions for photosynthesis	About twice as much
High dry matter production per unit of water used	Two to three times as much

efficient plants have a carbon dioxide compensation point of 5 ppm or less. Non-efficient plants have a compensation point in the range of 30 to 70 ppm CO_2. The compensation point is the concentration of carbon dioxide below which net carbon assimilation does not occur via photosynthesis. In plants, carbon dioxide released by respiration is used in photosynthesis with no net oxygen evolution at the compensation point. Obviously, plants with a high compensation point fix less carbon because of the inefficiency of their respiration.

A list of efficient and non-efficient plants from Black *et al.* (1969) is shown in Table 8.2. Consistent with their hypothesis, most weeds are efficient and many crops are non-efficient.

An application of this scheme compares Kentucky bluegrass and crabgrass, a common weed in bluegrass turf (Table 8.3). The data illustrate that crabgrass is efficient and will be a good competitor with Kentucky bluegrass, an observation any turf manager can verify.

The data on water requirements of different plant species (Shantz *et al.*, 1927) have been combined with Black *et al.*'s (1969) hypothesis (Table 8.4). They show that some crops and weeds with the C_4 pathway have a low water requirement and others with the C_3 pathway typically have a higher water requirement, lending additional credence to the Black hypothesis.

The Weed Science Society of America composite list of weeds (Anonymous, 1989) contains more than 2,000 species from 500 genera and 125 families. Of that number, 146 species in 53 genera and 10 families have C_4 carbon fixation. This is 17 times higher than the percentage of C_4 plants in the world's plant population.

Table 8.2

A List of Efficient and Non-efficient Plants (Black et al., 1969)

Efficient crops	Non-efficient crops	
Corn	Common bean	Ryegrass
Sugarcane	Soybean	Wheat
Sorghum	Sugarbeet	Oat
	Spinach	Barley
	Tobacco	Kentucky bluegrass
	Cotton	Rice
	Lettuce	
	Orchardgrass	

Efficient weeds	Non-efficient weeds	
Pigweed	Lambsquarters	
Saltbush	Velvetleaf	
Purslane		
Russian thistle		
Barnyardgrass		
Crabgrass		
Foxtails		
Johnsongrass		
Witchgrass		
Nutsedge		

Fourteen of the 18 worst weeds of the world (Holm et al., 1977) are C_4 weeds, and 8 of the top 10 are. Of the 76 worst weeds of the world, 42% are C_4, but only 20% of the 15 major world crops are. The C_3 pathway of photosynthetic fixation dominates among crops. There are many C_4 weeds, but there is an equal number of important weeds that fix carbon by the C_3 pathway.

In the eastern United States, C_3 and C_4 plants are poor competitors and many weeds have C_3 carbon fixation. Baskin and Baskin (1978) proposed that C_4 photosynthesis is less important than other features in determining competitive ability, whereas Black et al. (1969) thought rate of carbon dioxide fixation was the main determinant of competitive ability. This difference is one reason why the Black et al. (1969) hypothesis is presented as a helpful way to think about weed-crop competition, but not as a definitive explanation of competition. With natural temperature and radiation in arid southwest Australia, two C_3 species, rape and sunflower, had higher net

Table 8.3

A Comparison of Kentucky Bluegrass and Crabgrass Using the Scheme of Black
***et al.* (1969)**

Factors	Kentucky bluegrass	Crabgrass
1. CO_2 uptake increases with light intensity	Light saturation at 1000 to 3000 foot-candles	Light saturation at 5000 foot-candles
	Assimilates 15 to 35 mg CO_2/sq dm/hr	Assimilates 50 to 80 mg CO_2/sq dm/hr
2. Temp. optimum	10 to 25°C	30 to 40°C
3. Ps inhibited by O_2	No information	—
4. Photorespiration	Yes	No
5. C fixation cycle	No information	—
6. Ps comp. point	30 ppm CO_2	5 ppm CO_2

Table 8.4

Grams of Water Required to Produce 1 g/lb Dry Matter for Several
Plants (Black *et al.*, 1969; Shantz *et al.*, 1927)

Species	Grams of water required per pound of dry matter
C^4 pathway	
Prostrate pigweed	260
Common purslane	281
Foxtail millet	285
Sorghum	304
Corn	349
Average	296
C^3 pathway	
Wheat	557
Cotton	568
Cowpea	569
Common lambsquarters	658
Prostrate knotweed	678
Rice	682
Beans	700
Prostrate vervain	702
Smooth bromegrass	977
Average	667

assimilation rates and relative growth rates than corn, a C_4 plant (Baskin and Baskin, 1977). Baskin and Baskin (1978) proposed that C_3 and C_4 weeds compete well with crops, but not with climax vegetation. Among successful southeastern U.S. pasture grasses, there is about an even distribution between C_4 and C_3. Bermudagrass, bahiagrass, and pangolagrass are all C_4. Kentucky bluegrass, orchardgrass, and fescue are C_3. Table 8.5 shows the presence of the two pathways in several crops and weeds.

At high temperatures (34°C day/38°C night), redroot pigweed, a C_4 plant, outcompetes common lambsquarters, a C_3 plant, but at low temperatures (18°C day/14°C night), the reverse is true (Pearcy *et al.*, 1981). There is no inherent advantage to C_4 photosynthesis (Baskin and Baskin, 1977). Rate of leaf production or time of emergence may be more important determinants of a weed's competitiveness than rate of photosynthesis. This is not to say that carbon fixation is not important. Life is complex, and explanations of behavior will not be found in single causes.

II. CHARACTERISTICS OF WEEDS

There is a lot to know about why weeds are such good competitors. What makes some plants so capable of growing where they are not desired? Why are weeds such good competitors? What are their modes of competition and survival?

Weeds share some traits (see Chapter 2 on the characteristics of weeds). Not all weeds have all traits, but all weeds have some of the following characteristics, related to growth and physiology (competitive ability), reproduction, and cultural practices (Baker, 1965, 1974; Bazzaz, 1979; Elmore and Paul, 1983).

A. COMPETITIVE ABILITY

Weeds that are most competitive have rapid seedling growth and a high growth rate compared to the crop with which they are interfering. They will also have a short vegetative period to flowering and be able to complete reproduction quickly. They often produce seeds that mature soon after flowering. Canada thistle matures seed within 2 weeks of flowering. Russian thistle seeds held at 80°F will germinate within 90 minutes of wetting. This weed would spread more than it does, except it must germinate in loose soil because the coiled root unwinds as it pushes into the ground; in hard soil the seedling dies before it roots well.

Table 8.5

Photosynthetic Pathways of Some Crops and Weeds (Patterson, 1985)

Crops		Weeds	
C_3	C_4	C_3	C_4
Alfalfa	Corn	Ageratum	Barnyardgrass
Banana	Foxtail millet	Canada thistle	Bermudagrass
Barley	Pearl millet	Catchweed bedstraw	Common purslane
Bean	Sorghum	Cocklebur	Cogongrass
Cassava	Sugarcane	Common chickweed	Crabgrass
Coconut		Milkweed	Dallisgrass
Corn		Docks	Dropseed
Cotton		Field bindweed	Fall panicum
Oats		Hairy beggarticks	Foxtail
Orchardgrass		Jimsonweed	Garden spurge
Peanut		Lambsquarters	Goosegrass
Potato		Largeleaf lantana	Guineagrass
Rice		Morningglory	Itchgrass
Rye		Mustards	Johnsongrass
Soybean		Nightshades	Kikuyugrass
Sugarbeet		Plantain	Kochia
Sweetpotato		Poison ryegrass	Nutsedge, purple and yellow
Tomato		Quackgrass	Pigweeds
Wheat		Ragweeds	Puncturevine
		Sandburs	Russian thistle
		Sensitive plant	Signalgrass
		Sicklepod	Sprangletop
		Sida, prickly	Texas panicum
		Smartweeds	Torpedograss
		Velvetleaf	
		Water hyacinth	
		Wild oats	
		Witchweed	

Weeds with great competitive ability have fast seedling growth and grow tall quickly or gain competitive ability by twining on larger plants. They may also be tolerant of shade, and their highest carbon dioxide assimilation rate may not be in full sunlight.

Consistent with the hypothesis of Black *et al.* (1969), the most competitive weeds have a high photosynthetic rate and rapid partitioning of photosynthate into new leaf production. They have a high light saturation intensity and a low carbon dioxide compensation point.

Competitive weeds develop a large exploitative root system rapidly and have a high tolerance for climatic and soil variations. Their general-purpose genotype frees them from many environmental constraints; they grow well in many places. This genotype enables weeds to grow under adverse conditions, gives them a high ability to recover quickly from resource limitation, lets them acquire resources fast, and ensures that some survive in many different environments.

Many common weeds are not good competitors in the ecological sense. They have evolved to be successful competitors in the intensely managed and regularly disturbed habitats characteristic of cultivated fields. Weeds and crops benefit from things that reduce environmental stress, such as irrigation, fertilization, and pest control. Most weeds lack the ability to tolerate extreme shade and to invade or survive in established vegetation. Few weeds important in crop competition are present in the earliest successional stages following abandonment of cropland. They cannot compete in such a resource-starved environment. Some of the best weeds have the ability to compete by special means such as allelopathy. Other successful weeds have adaptations that repel grazing, such as spines, bad taste, or bad odor.

B. Reproductive Characteristics

The most successful weeds have no special environmental requirements for germination. They may be especially detrimental in crops because their success, after germination, is tied to the same factors that lead to crop success. They succeed in well-fertilized fields, planted at certain times, often with irrigation or regular rainfall.

Successful weeds have a relatively long period of high seed production with favorable growing conditions. Some have almost continuous seed production. Redroot pigweed is able to produce seed as early as when it is 1 to 8 inches tall and for a long time after it first flowers. Good weeds produce some seed under a wide range of environmental conditions.

Weeds have special adaptations for short- and long-distance dispersal of their seeds in space. They are usually self-compatible, but not obligate self-pollinators. Cross-pollination is achieved by nonspecialized flower visitors or wind.

C. CULTURAL PRACTICES

Weed seeds resist degradation in soil and disperse in time via seed dormancy. Even though they produce a large number of seeds per plant, many of which may germinate immediately, they can acquire secondary dormancy. Weed seed often has the same size and shape as many crop seeds, and weed maturation coincides with crop maturity. Morphological and physiological similarity to crop seed makes weed seed hard to detect and clean from crop seed.

Plowing and preparing soil for planting are vigorous practices that disturb plant growth. Most common crop plants, turf, and ornamentals do not survive these practices, but many weeds can. In fact, some weeds, such as nightshades, are dependent on tillage for establishment. Weeds survive and prosper under the disturbed conditions of a cropped field or an environment created to favor human crops or goals. Weeds have the environmental plasticity to do well under these conditions.

If they are perennials, they have vigorous vegetative reproduction with large food reserves in roots. They also may have dual modes of reproduction and do not rely solely on vegetative or sexual reproduction. Perennials normally have brittleness at lower rhizome or root nodes and cannot be pulled. Perennials usually have the ability to regenerate from small root segments (often as small as 1 inch).

Perhaps of greatest importance to the success of many weeds is their resistance to or ability to develop tolerance of different methods of control, including chemical control.

THINGS TO THINK ABOUT

1. What are the six factors included in the Black *et al.* (1969) scheme to explain plant competition?
2. What is the definition of an efficient and a non-efficient plant, and how are the concepts used?
3. Carbon fixation is a logical determinant of plant competitiveness. Why does it fail in some cases?
4. What are the characteristics of plant growth, reproduction, and response to cultural practices that contribute to weediness?
5. What factors contribute to a plant's competitiveness?

LITERATURE CITED

Anonymous. 1989. *Composite List of Weeds,* revised 1989. Weed Sci. Soc. Am., Champaign, Illinois. 112 p.

Baker, H. G. 1965. Characteristics and modes of origin of weeds, pp. 147–172 in *The Genetics of Colonizing Species, Proc. First Int. Union of Biol. Sci. Symp. on Gen. Biol.* (H. G. Baker and G. L. Stebbins, Eds.). Academic Press, New York.

Baker, H. G. 1974. The evolution of weeds, pp. 1–24 in *Ann. Rev. Ecol. and Systematics* (R. F. Johnston, P. W. Frank and C. D. Michener, eds.). Academic Press, New York

Baskin, J. M., and C. C. Baskin. 1977. Productivity of C_3 and C_4 plant species. *Ann. Assoc. American Geog.* **67**:639–640.

Baskin, J. M., and C. C. Baskin. 1978. A discussion of the growth and competitive ability of C_3 and C_4 plants. *Castenea* **43**:71–76.

Bazzaz, R. A. 1979. The physiological ecology of plant succession. *Ann. Rev. Ecol. Systematics* **10**:351–371.

Black, C. L., T. M. Chen, and R. H. Brown. 1969. Biochemical basis for plant competition. *Weed Sci.* **17**:338–344.

Donald, C. M. 1963. Competition among crop and pasture plants. *Adv. in Agron.* **15**:1–118.

Elmore, C. D., and R. N. Paul. 1983. Composite list of C_4 weeds. *Weed Sci.* **31**:686–692.

Holm, L. G., D. L. Plucknett, J. V. Pancho, and J. P. Herberger. 1977. *The World's Worst Weeds–Distribution and Biology.* Univ. Press of Hawaii, Honolulu. 609 pp.

Patterson, D. T. 1985. Comparative ecophysiology of weeds and crops, Chapter 4 in *Weed Physiology,* Vol. I (S. O. Duke, Ed.). CRC Press. Boca Raton, Florida.

Pearcy, R. W., N. Tumosa, and K. Williams. 1981. Relationships between growth, photosynthesis and competitive interactions for a C_3 and C_4 plant. *Oecologia* **48**:371–376.

Salisbury, F. B., and C. W. Ross. 1992. Pp. 253–265 in *Plant Physiology,* 4th ed. Wadsworth, Belmont, California.

Shantz, H. L., R. L. Piemeisel, and L. Piemeisel. 1927. The water requirements of plants at Akron, Colorado. *J. Agric. Res.* **34**:1093–1190.

Chapter 9

Methods of Weed Management and Control

FUNDAMENTAL CONCEPTS

- Weed prevention, control, eradication, and management are different concepts, and each uses or combines technologies differently.
- Prevention is the best strategy to combat weeds.
- Many important weeds in any country are escaped imports.
- Mechanical, nonmechanical, and cultural weed control techniques have their own advantages and disadvantages.
- No weed control method has ever been abandoned. Each new method introduced in large-scale crop culture has reduced the need for human and animal power.
- Cultural weed control is intuitively sensible and widely employed.

LEARNING OBJECTIVES

- To know the definition and relative merits of weed prevention, control, eradication, and management.
- To be familiar with weed seed laws and the federal noxious weed law.
- To understand the importance of planting clean crop seed.
- To know the practices that prevent introduction and spread of weeds.
- To know the advantages and disadvantages of each mechanical, nonmechanical, and cultural weed control technique.

- To know the present role and to consider future weed management roles of living mulches and companion cropping.
- To appreciate the role of minimum and no-tillage in weed management.

I. CONCEPTS OF PREVENTION, CONTROL, ERADICATION, AND MANAGEMENT

When asked, students in weed science classes usually say they are taking weeds or weed control. Those who work on weeds often spend a great deal of time on weed control. However, weed science is not just about weed control. Weed science as practiced by its leaders tries to answer fundamental questions about weeds and weed management. Scientists want to know: Why are weeds problems? Why are some weeds problems in many places? Why do different weed management strategies work differently in different cropping systems? Why are some plants so successful as weeds? Answers to these and similar questions lead to concepts and theories and greater clarity about what ought to be done to manage weeds and why.

The most difficult part of weed management is prevention, defined as stopping weeds from contaminating an area. It is a practical means of dealing with weeds, but it takes time and careful attention to many details, and success is difficult to achieve. Preventive measures include the following:

Isolating imported animals for several days
Not importing weeds or weed seeds in animal feed
Using only clean crop seed free of weed seed
Cleaning equipment between fields and especially between farms
Preventing weed seed production especially by new weeds
Preventing vegetative spread of perennials

Weed control includes many techniques used to limit weed infestations and minimize competition. These control techniques attempt to achieve a balance between cost of control and crop yield loss, but weed control is used only after the problem exists; it is not prevention. Weed control techniques have been adopted widely because control is easier to do than prevention or eradication. Control can be made to work well with short-term economic or cultural planning goals. Prevention and eradication require long-term thinking and planning. Effective preventive techniques may reduce short-term economic gain.

Eradication is complete elimination of all live weeds, weed parts, and weed seed. It could be defined as 100% or complete control. It sounds easy, but is difficult to do. It is usually easy to eliminate live plants, because they can be seen. It is difficult to eliminate seed and vegetative reproductive parts in soil. Eradication is the best program for small populations of perennial weeds, but present technology does not make it easy.

In weed science, as in human medicine, prevention is better than control, but control is required because the weeds arrive before they can be prevented.

Weed management is the combination of the techniques of prevention, eradication, and control to manage weeds in a crop, cropping system, or environment. Weed managers recognize that a field's or area's cropping history, the grower's management objectives, the available technology, financial resources, and a host of other factors must be combined to make good management decisions. Complete weed control in a crop may be the best decision in some cases, but it is not automatically assumed to be the goal. Maintenance of a weed population at some level in a cropping system may be the most easily achievable and financially wise goal for a weed management program.

II. WEED PREVENTION

People want to be and stay healthy. When we become ill, we are pleased to have competent physicians, hospitals, and medical services, yet people would rather remain healthy than have to cure an illness. The same kind of thinking applies to weed management. Weed control cures weeds, but it does not prevent them.

A good weed management program includes vigilance or watchfulness. The good weed manager can identify weed seeds, seedlings, and mature plants and has a management program for each crop and field and appropriate follow-up programs. The good manager is ever watchful for new weeds that may become problems and, whenever possible, emphasizes prevention rather than control. Several preventive practices can be included in management programs:

1. Isolation of introduced livestock to prevent spread of weed seeds from their digestive tract
2. Use of clean farm equipment and cleaning of itinerant equipment, including combines, cultivators, and grain trucks
3. Cleaning of irrigation water
4. Mowing and other appropriate weed control practices on irrigation ditch banks
5. Inspection of imported nursery stock for weeds, seeds, and vegetative reproductive organs

6. Inspection and cleaning of imported gravel, sand, and soil
7. Special attention to fence lines, field edges, rights-of-way, railroads, etc., as sources of new weeds
8. Prevention of deterioration of range and pasture to stop easy entry of invaders such as downy brome (Mack, 1981).
9. Seed dealers and grain handlers must clean crop seed and dispose of cleanings properly.

The first rule for weed prevention and the first step of any good weed management program is the purchase and planting of clean seed. The U.S. Federal Seed Act of 1939 regulates transport and sale of seeds in foreign and interstate (but not intrastate) commerce. The law is enforced by the U.S. Department of Agriculture, which has provided supplementary rules aimed primarily at interstate movement of parasitic plants and noxious weed seeds. The Federal Seed Act and state laws mandate labeling of crop seed to show the kind of seed, its variety, and the state and specific locale where it was grown. Complete labels also show percent pure seed, percent weed seed, percent other crop seed, percent inert matter, percent germination of pure seed, percent hard seeds (those seeds that are viable but not capable of immediate germination), and the date on which the tests were performed. Seed labels also include the name and number per lb of each noxious weed seed.

Each U.S. state has a noxious weed seed law that identifies and regulates sale and movement of crop seed containing what the state law has identified as noxious weed seed. These laws may prohibit importation of crop seed with greater than a certain percentage of specific noxious weed seed and require identification of each noxious weed seed. Noxious weed seed in excess of 1 g in 10 g of the crop seed provide for exclusion in many states. For large seeded crops such as beans, the exclusion is often 1 g in 100 g of crop seed. These laws may also regulate import and sale of crop seed screenings because they contain viable weed seed. State seed laws are designed to protect seed consumers (farmers and other purchasers). These laws do not mean and should not be viewed as implying equal regulation of weedy plants that may be detrimental to agriculture or the environment.

Seed standards are not restricted to the United States. Standards from the Canada Seeds Act of 1987 are shown in Table 9.1. These regulations allow various levels of weed seed to be present depending on the crop and the level of classification desired. The standards apply to barley, buckwheat, lentils, rye, and sanfoin and with minor variation to wheat, canola, flax, and oats. The seller must supply a certificate, on request, that states the number and kinds of weed seed present.

About a third of U.S. states have no limitation on total weed seed in crop seed. Limitations range from 1 to 4%. Most state laws exempt seed

Table 9.1

Maximum Number of Weed Seeds for 500 g of Crop Seed

Certification level		Total weed seeds	Total other crop seed
Canada registered	-1	3	2
Canada registered	-2	6	4
Canada certified	-1	3	4
Canada certified	-2	6	10
Canada	-1	10	25
Canada	-2	20	50

sold by a grower, without advertising. All state laws designate certain weeds as noxious. About 20 states have no limitation on prohibited or restricted noxious weeds. Prohibited noxious weed seeds are usually seed from perennial, biennial, or annual plants that are highly detrimental and especially difficult to control. The presence of these seeds in any amount prohibits sale of crop seed for planting purposes in many states.

A bushel of clover seed weighs 60 lb and was 88% clover, with 35% germination. Therefore, in 1 bushel, there was 18.5 lb of live clover seed, 34.3 lb of dead clover seed, 4.2 lb of weed seed, representing 11 different species, 2.8 lb of inert matter, and 0.2 lb of other crop seed. The purchased seed contained 7,800 Canada thistle seeds per bushel, 5,700 curly dock seeds per bushel, and 114,000 wild mustard seeds per bushel. The bargain seed cost $5.90 per bushel, or $19.14 per 60 lb of 100% viable seed. The same variety of clover could have been purchased for $8.40 per bushel. That bushel had 99.15% purity and 95% germination, or a cost of $8.84 per 60 lb of 100% viable seed. The difference in cash cost ($8.40 − $5.90 per bushel) was $2.50. This cash cost is the only thing some buyers notice. The bargain seed cost $19.14 for 100% good seed versus $8.84 for 100% good seed in the second source: a difference of $10.30 per bushel in favor of the second source (Barnes and Barnes, 1960). Purchasing bargain seed or cheap seed is rarely a good idea and can lead to weed problems.

Restricted noxious weed seeds are seeds of plants that are very objectionable in fields, lawns, or gardens, but can be controlled by good cultural practices.

More than 175 different species are named as noxious weed seeds by the 48 continental United States. An additional 50 species are named in Hawaii. It is important to note that these are legal, not botanical, definitions that are informed by agronomic and horticultural practice.

Most states have state seed laboratories that determine seed quality. One aspect of quality is the number of weed seed or other crop seed in a sample (Tables 9.2 and 9.3). In both cases, too many weed seeds were sown when the purchased seeds were sown. Planting clean seed is an easy method of preventing weeds.

In 1975 weed prevention took a major step forward when the federal noxious weed law empowered the Secretary of the U.S. Department of Agriculture to control import, distribution, and interstate commerce of weeds declared to be noxious. Previous laws regulated just seed, not plants.

Nearly all U.S. states list some prohibited agricultural weeds in addition to those included in the Federal noxious weed law. At present these laws provide some protection, but in most states it is inadequate for agriculture and the environment. The Federal law includes 93 weedy species, yet at least 750 weeds that meet the Act's definition remain unlisted (U.S. Congress, 1993). Many of these are agricultural problems, but some infest other

Table 9.2

Sample Seed Analysis from Colorado State Seed Testing Laboratory

Bromegrass—61% germination	
Seeded @ 4–6 lb/A	136,000 seeds/lb
Redroot pigweed	27,968 seeds/lb
Japanese brome	512 seeds/lb
Stinkgrass	256 seeds/lb
Barnyardgrass	64 seeds/lb
Oldfield cinquefoil	64 seeds/lb
	28,864 seeds/lb
Timothy	448 seeds/lb
Barley	64 seeds/lb
Sweetclover	64 seeds/lb
Sand dropseed	64 seeds/lb
Bentgrass	64 seeds/lb
	704 seeds/lb

Table 9.3

**Sample Seed Analysis from Colorado State Seed
Testing Laboratory**

Alfalfa
224,000 seeds/lb   seeded at 8–10 lb/A
                        alone

Sample 1
79% germination   Dodder   432/lb
                        Mallow   180/lb,
84% live   Groundcherry   90/lb
At 10#/A 4,320 dodder and 2,240,000 alfalfa seeds will be sown per acre.

Sample 2
66% germination   Russian knapweed   9/lb
84% live   Chicory   270/lb
                        Netseed lambsquarters   360/lb
                        Kochia   180/lb
                        Buckhorn plantain   117/lb
                        Other weeds   189/lb
                        Other crop
                        Red clover   6930/lb
At 10 lb seed/A, 1,478,400 alfalfa seeds, 11,250 weed seeds, and 69,300
red clover seeds will be sown/A.

environmental areas such as wetlands and natural areas. Some important
invaders of these areas are (U.S. Congress, 1993):

Purple loosestrife
Brazilian peppertree
Eurasian watermilfoil
Smooth cordgrass

A survey of weed and seed laws in five contiguous western states—Idaho,
Oregon, Utah, Washington, and Wyoming (U.S. Congress, 1993)—showed
the laws provided adequate to inadequate protection based on the likelihood of unlisted weeds causing economic or ecological problems. Specifically, the study found that protection was adequate to inadequate and many
potential threats were omitted (Table 9.4).

Federal and state laws do not include enough weedy plants, and they
regulate only agricultural and vegetable seed. The laws do not cover horticultural seeds, including known sources of weed seed such as wildflower
and native grass mixtures (U.S. Congress, 1993).

In spite of existing laws, regulations are not stringent, and it should not
be surprising that 36 weed species now resident in the United States were

Table 9.4

Survey of Weed and Seed Laws in Five Western States (U.S. Congress, 1993)

State	Number of species listed	Adequacy of protection	Number of potential threats omitted
Idaho	47	Adequate	6
Oregon	67	More than adequate	Few
Utah	23	Inadequate	11
Washington	75	More than adequate	Few
Wyoming	34	Barely adequate	11

imported and escaped to become weeds—in some cases, noxious weeds (Williams, 1980). Of the 36, 2 were imported as herbs, 12 as hay or forage crops, and 16 as ornamentals. Weeds were imported as a windbreak (multiflora rose), for possible medicinal value (black henbane), for use in aquaria (hydrilla), as a fiber crop, just for observation, and as a dye (dyer's woad).

Bermudagrass, a valuable forage species in the southern United States and many other parts of the world, is also an important weed in many areas and was introduced into the United States as a forage crop. In 1849 the U.S. Cotton Office proposed and introduced a new forage grass—crabgrass (Brosten and Simmonds, 1989). More recent introductions of grassy weeds include sorghum-almum, promoted as a hay crop with names such as perennial sudangrass, sorghum grass, and Columbia grass. It was promoted as a drought-resistant emergency forage crop. It is a hybrid between johnsongrass and grain sorghum and was first described and cultivated in Argentina (Brosten and Simmonds, 1989). Wild proso millet was first recognized as a weed in the North Central United States in the early 1970s and now infests several million acres in Wisconsin and Minnesota, as far west as Colorado, and in the midwestern states and Canada. It is the same species as cultivated millet and difficult to control in corn. A major reason it is such a good weed is that its seed germinates throughout the growing season rather than in a short period, and it escapes control by nonresidual herbicides and single cultivations.

The latter two cases are interesting because of their implications for biotechnology. Hybridization of weeds and crops is uncontrolled and may be uncontrollable. Cross-pollination is inevitable when two phenologically similar, outbreeding plants share a small area (exist in an overlapping range). Little research has been done to determine the potential for gene transmittal, in cropped fields, from weeds to crops or vice versa. There is a possibility that a crop genetically engineered for high yield or herbicide resistance will contribute to the generation of new, difficult-to-control weed hybrids (Brosten and Simmonds, 1989).

Two species of toadflax were introduced as ornamentals and became weeds. Jimsonweed and kochia were brought to the United States for use as ornamentals, and kochia was studied as a forage crop. The artichoke thistle escaped to become a weed in artichokes and is a recurring problem on 150,000 acres in California (Brosten and Simmonds, 1989). Waterhyacinth was introduced from South America to the United States by Japanese entrepreneurs as part of a horticultural exhibit at the cotton centennial exposition in New Orleans in 1884 (Penfound and Earle, 1948). It originally came from the Orinoco River in Venezuela, and single plants were given away at the cotton exposition. It has been introduced around the world primarily because its flowers are pretty. At the New Orleans exposition, people liked it so much, they took it home and put it in ponds and gardens, from which it escaped because people discarded it or water flowed out of these places and carried the weed along. It reproduced profusely in ponds and escaped to the St. Johns River in Florida, where it became a major weed problem by clogging the waterway. Waterhyacinth was brought to the Tonkin region of China (now Vietnam) in 1902 as an ornamental. It reached southern China and Hong Kong in the same year. Soon after, it was observed in Sri Lanka and then India, where the sluggish rivers of east Bengal were ideal for its growth. In the 1950s it was discovered in Africa (Vietmeyer, 1975), and in 1958 it had infested more than 1000 miles of the Nile river from Juba in the south to the Jebel Aulia dam in northern Sudan (Heinen and Ahmad, 1964). It is a serious weed problem in all of these places and many others, but not in Venezuela where its spread is controlled by natural enemies.

Cogongrass or alang-alang, a perennial, was introduced at Grand Bay, Alabama, and McNeil, Mississippi (Tabor, 1952). Bare-root orange plants were imported to Grand Bay in 1912, and the cogongrass that lined boxes the plants were shipped in was discarded. In McNeil, scientists were searching for better forage plants; cogongrass escaped from farmers' fields and the experiment station and spread rapidly.

Kudzu, a nonindigenous species, was introduced to the United States at the Philadelphia Centennial Exposition in 1876 (Shurtleff and Aoyagi, 1977). It was promoted by the U.S. Department of Agriculture for erosion control and forage, but it became a major weed and now grows in many areas throughout southeastern United States and has spread to some Midwestern states.

Further evidence of distribution of the world's weeds and the necessity for vigilance to prevent introduction of new species is shown in Table 9.5. Most of our important weeds have come from somewhere else, and vigilance

Table 9.5

Origin and Distribution of the World's Most Serious Weeds (Holm *et al.*, 1977)

Weed	Origin	Distribution (no. of countries)	Associated crops
Purple nutsedge	India	92	52
Bermudagrass	Africa or Indo-Malaysia	80	40
Barnyardgrass	Europe and India	61	36
Junglerice	India	60	35
Goosegrass	China, India, Japan, Malaysia	60	36
Johnsongrass	Mediterranean	63	30
Cogongrass	Old World	75	35
Spiny amaranth	Tropical America	54	28
Sour paspalum	Tropical America	30	25
Tropic ageratum	Tropical America	46	36
Itchgrass	India	28	18
Carpetgrass	Tropical America	27	13
Hairy beggarticks	Tropical America	40	31
Paragrass	Tropical Africa	34	23
	Mexico, West India, tropical S. America	23	13
Smallflower umbrella sedge	Old World tropics	46	1
Rice flatsedge	Old World tropics	22	17
Crowfootgrass	Old World tropics	45	19
Eclipta	Asia	35	17
Globe fringerush	Tropical America	21	(rice)
Witchweed	Europe or South America	35	2
Halogeton	Asia	unk.	rangeland
Russian knapweed	Asia	unk.	>10
Quackgrass	Eurasia	>80	many

is necessary to prevent new problems. Among 300 nonindigenous weeds in the western United States, 8 were former crops, and 28 escaped from horticultural areas (U.S. Congress, 1993).

All is not lost. There is evidence that most imported plants do not become weeds. In the United Kingdom, about 10% of the invaders became established, but only 1% of those became weeds (Williamsen and Brown,

1986). In Australia only 5% of introduced plants became naturalized and only 1–2% of those became weeds (Groves, 1986). Once a plant is naturalized in an area, whether it remains insignificant or becomes a weed problem is dependent on the absence of damaging natural enemies and the presence of suitable soil, crops, and land-use and weed-management practices, and on how the plant responds to the local climate (Panetta and Mitchell, 1991). The few that become weeds can be costly problems. Although the chance is small, the consequences can be great. We can identify areas at risk of invasion, but weediness cannot be predicted as easily (Panetta and Mitchell, 1991).

III. MECHANICAL CONTROL

No weed control method has ever been abandoned completely. New techniques have been added in large-scale agriculture, but old ones are still used effectively, especially in small-scale agriculture. Mechanical methods have a long use history and are the primary weed control technique in many crops.

A. HAND PULLING

Hand pulling is practical and efficient, especially in gardens, but it is hard work. It is very effective for annual weeds, but not for perennials capable of vegetative reproduction, because shoots separate from roots that then produce a new shoot. A disadvantage is that hand pulling does not get the job done when it is most needed. Most of us are too busy or too lazy to go out and weed before weeds become obvious. By the time they become obvious, easy to grab and pull, yield reduction due to weed competition will have occurred.

B. HAND HOEING

Hand hoeing has been used for weed control for many years. It is still the method of choice for most gardens and ornamental plantings and is used regularly in many vegetable crops. Hand hoeing controls the most persistent perennials if it is done often enough. Although efficient and widely used, it takes a lot of time and human energy. Some data on the time required to hand weed some crops in several different places are

Hand pulling weeds on the grounds of the Imperial Palace, Kyoto, Japan

Hand hoeing weeds in rice in the Philippines.

shown in Table 9.6. If human labor is abundant, and labor cost is not high, hand pulling or hoeing is an acceptable method of weed control. If human labor is not abundant and it is expensive, hand methods are expensive and not efficient.

C. TILLAGE

When most people think of mechanical control, the first thing that comes to mind is tillage with an implement to disturb, cultivate, or mix the soil. On arable land, tillage alone, or in combination with cropping or chemical treatment, may be the most economical system of weed control. Tillage turns under crop residue, conditions soil, and facilitates drainage. It controls weeds by burying them, separating shoots from roots, stimulating germination of dormant seeds and buds (to be controlled by another tillage), desiccating shoots, and exhausting carbohydrate reserves of perennial weeds.

Other reasons for tillage include breaking up compacted soil, soil aeration, seedbed preparation, trash incorporation, and crop cultivation. All of

Table 9.6

Time Required for Hand Weeding

Crop	Location	Hours per hectare to hand weed
Soybeans	Peru	360 if 6-hour day
Transplanted tomatoes	Ohio	71 after herbicide, 133 after cultivation
Corn	Zimbabwe	24–48 for 6-hour day
Beans	Wyoming	4.4–15.5 after broadcast herbicide, 32 if no herbicide
Sugarbeet	Washington	2–111 after broadcast herbicide, 141 without herbicide
Vegetables	California	10 after broadcast herbicide
Rice	Several	16–500 depending on location and rice culture
Wheat		101
Sorghum		50
Millet		88–298
Cotton		50–700
Jute		140
Groundnut		102–293
Cassava		115–1,069

Source: Int. Weed Sci. Soc. *Newsletter* 1979 **4** (1).

these are important, but the main accomplishment of most tillage done in the world's developed countries is weed control. The advent of no-till farming and minimum-till farming has shown that tillage is not essential to grow crops and may do no more than control weeds. Too-frequent tillage can increase soil compaction, a disadvantage. Other disadvantages include exposure of soil to erosion, moisture loss, and stimulation of weed growth by encouraging germination of dormant seeds and vegetative buds. In some soils, without tillage, soil can crust and there will be poor water penetration. Decisions about the role of tillage must be made for each soil type and farming system.

Tillage is usually divided into primary and secondary. Primary tillage is initial soil breaking or disturbance. Depth of primary tillage varies from at least 6 (except where primitive tools are used with limited animal power) to as many as 24 inches. The implements of primary tillage include moldboard and chisel plows. These implements cut and invert soil and bury plant and other surface residue. This is often the first step of seedbed preparation. Secondary tillage may be subsequent to primary tillage, or it may be the first tillage operation. Soil is disturbed, often vigorously, but upper layers are usually not inverted. A wide selection of tools is available. Secondary tillage is fast and inexpensive, and its tools are appropriate for large areas. Tools include the double disk, several kinds of harrows, field cultivator, spring tooth harrow, rototillers, rod-weeder (a dryland implement), and the cultipacker (combination of harrow and roller). This diverse group of implements tills soil from a few inches deep to a maximum of 5 or 6 inches. Secondary tillage implements break clods, firm soil, and remove weeds. Many regard secondary tillage implements as both weed-control and seedbed preparation tools.

Primary and secondary tillage is followed, in many row crops, by selective inter-row cultivation or intertillage. Tractor-mounted cultivators or animal-drawn implements move soil between crop rows to loosen it and control weeds. In general, inter-row tillage is just that: It works between crop rows. Some implements prepare inter-row areas for furrow irrigation (where water runs down furrows between crop rows). Implements used for inter-row cultivation include a wide range of tine (long finger-like rods) and flared or straight steel shovel-like tools at the end of solid or flexible (flat steel) shanks that travel through soil at shallow depths (1–2 inches). They break soil crusts and facilitate irrigation, but the main purpose is weed control.

Intrarow tillage implements to control small weeds in the crop row are available. Limited research (Schweizer *et al.,* 1994; VanGessel *et al.,* 1995) has shown that, in corn, intrarow cultivators require early-season weed control (cultivation or herbicide) for optimum efficacy. Intrarow cultivators

Cultivating for weed control in beans.

are more efficient (control more weeds) that inter-row cultivators. Without herbicides, weeds in corn were always controlled better by an in-row cultivator than by the standard inter-row cultivator when each operation was performed at the right time. In-row cultivators have special tools (Figure 9.1) that disturb soil around crop plants and uproot weeds in rows. The tools include spyders (toothed disks that move soil toward or away from crop rows) and torsion weeders and spring-hoe weeders that flex vertically and horizontally to uproot weeds in crop rows. Spinners displace weeds in crop rows. Standard inter-row crop cultivators are most effective on weeds 15 cm tall or less. Inter-row cultivators (Figure 9.1) are most effective on weeds less than 6 cm tall (Schweizer *et al.,* 1994). These cultivators do not work well in corn when weed density is high.

There are situations where plowing and subsequent tillage do not prepare land for planting. These include land heavily infested with perennial sod-forming grasses, often encountered in agriculture in developing countries. Many tillage implements give inadequate results in the crop row after the crop has emerged and begun to grow. Tillage between rows is efficient. Crop cultivation tillage can be done to within a few inches of crop plants, but not as well in the crop row except by moving soil and burying weeds.

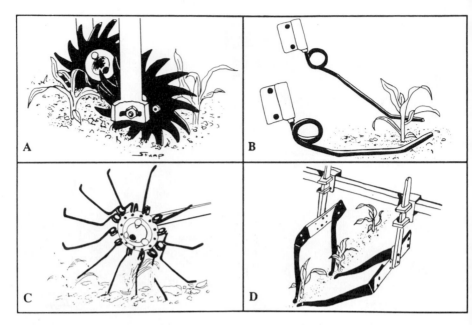

Figure 9.1. Types of tillage implements used for in-row cultivation A. A pair of spyders, B. A pair of torsion weeders, C. A spinner, and D. A pair of spring-hoe weeders (Schweizer *et al.,* 1994). Reproduced with permission.

To maximize tillage benefits, uniform spacing of crop rows, straight rows achieved by precision planting, gauge wheels, and depth guides are needed. Uneven stands and driver error often lead to damage from mechanical cultivation and destruction of some crop plants.

The success of tillage for weed control is determined by biological factors:

1. How closely weeds resemble the crop. Weeds that share a crop's growth habit and time of emergence may be the most difficult to control with tillage, especially when they grow in crop rows. Weeds that emerge earlier or later than the crop are often easier to control.
2. If a weed's seeds have a short, specific period of germination, it is easier to control them by tillage as opposed to those whose seeds germinate over a long time.
3. Perennial weeds that reproduce vegetatively are particularly difficult to control with tillage alone.

Successful mechanical control of weeds is also determined by human factors. Gunsolus (1990) noted that science could explain why certain weed management practices work the way they do. Science develops basic principles to guide action. Human cultural knowledge is different from scientific knowledge, although each may work toward the end of good weed management. Cultural knowledge tells one when and how to do something on a given soil and farm. Tillage is a cultural practice and therefore, by definition, it requires cultural knowledge. It requires the mind of a good farmer who knows the land. Successful mechanical control requires managerial skill (cultural knowledge) that cannot be acquired from science. Such knowledge is acquired by doing and by observing those who have done things well. Cultural knowledge is the art of farming whereby one knows how to select and apply scientific knowledge to solve problems. Successful mechanical control of weeds, regardless of implement, is always related to the timeliness of the operation. Research can determine when to do something, but knowing when on a particular farm is part of the cultural knowledge good farmers have.

The operative principle for use of tillage for control of perennial weeds (point 3 in the list) is carbohydrate depletion. The vegetative reproductive system of perennial weeds is a carbohydrate storehouse. When shoots grow and photosynthesize, eventually the storehouse will be replenished. If shoots are cut off, the plant calls on its reserve to create new growth. When tillage is done frequently, the management assumption is that reserves will be depleted and plants will die because of exhaustion of root reserves and increased susceptibility to other stresses (e.g., frost or dryness). Unfortunately, root reserves are vast and outlast patience and time. Tillage may have to be so frequent that crops cannot be grown. If tillage and destruction of foliage are delayed from a few days to up to a week after emergence, the greatest depletion of root reserves occurs. With most perennial weeds, the great majority of roots and vegetative buds are in the top 6 to 12 inches of soil. Tillage when a crop is growing cannot go this deep without disturbing crop roots: a disadvantage of tillage for control of perennial weeds.

Early work showed that if field bindweed was tilled 12 days after it first emerged, 16 successive tillage operations at approximately 12-day intervals were required to approach eradication. If it was tilled immediately after emergence, about twice as many tillages were needed. The efficacy and impracticality of tillage are also illustrated by a 1938 study that showed that purple nutsedge could be controlled in Alabama by disking at weekly or biweekly intervals for 5 months (Smith and Mayton, 1938). Obviously, no crop can be grown during the 5 months. In more recent work, Buhler *et al.* (1994) demonstrated over 14 years that greater and more diverse populations of perennial weeds developed in reduced-tillage systems than on areas that were moldboard plowed. Practices used to control annual weeds and environmental factors interacted with tillage to regulate (but not eliminate) perennial weeds.

It is often thought, incorrectly, that as long as one tills, it does not matter how or when it is done, as long as the weed is there to be controlled (Schweizer and Zimdahl, 1984). Studies were established in a field where corn had been grown continuously for 6 years. Half of the plots received regular chemical weed control each year, while the other half had herbicides for the first 3 years, then no herbicide and only cultivation for the last 3 years. Plots that received herbicide for 3 years also received optimum supplemental weed management, including cultivation in each of the 6 years. In the plots with herbicide for the first 3 years but only cultivation thereafter, redroot pigweed dominated. At the end of the 6-year experiment, the field was divided in half. One half was plowed in January and disked in April prior to normal spring planting. The other half was disked in January and again in April prior to normal spring planting. More redroot pigweed emerged when the field was disked in the fall than when it was plowed. Where herbicide and optimum weed management had occurred for 6 years, almost no redroot pigweed survived to produce seeds for the last 3 years of the study, and tillage did not make any difference in the redroot pigweed population in the 7th year (see Figure 9.2).

Most disking accomplishes weed control *and* seedbed preparation. (Courtesy of Deere and Co., Moline, Illinois.)

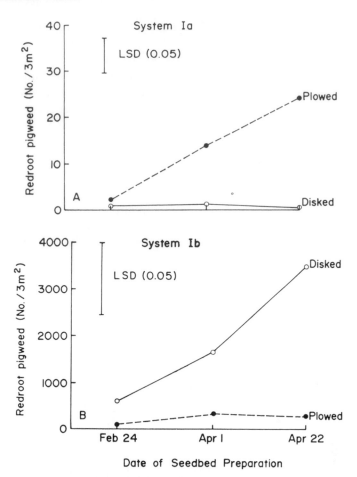

Figure 9.2. Population of redroot pigweed seedlings following several conventional tillage practices and atrazine use in continuous corn. In weed management system 1a, 2.2 kg/ha of atrazine was applied preemergence for 6 consecutive years. In weed management system 1b, the same rate of atrazine was applied for the first 3 years and discontinued thereafter. In the fall, one-half of each system 1a and 1b plot was plowed (hatched line) and the other half disked (solid line) (Schweizer and Zimdahl, 1984).

Disking soil (secondary tillage) in plots that had only cultivation for 3 years enhanced germination of seeds on the soil surface by bringing them nearer the surface. Plowing (primary tillage) buried seeds. Therefore, in the experiment, if weed control had not been good, disking instead of plowing made the weed problem worse. If weed control was good, the kind and time of tillage did not matter (Schweizer and Zimdahl, 1984).

A second example of the importance of tillage timing is from land to be planted to wheat in North Dakota (Donald, 1990). Moldboard plowing 18–20 cm or chisel plowing 9–15 cm deep in the fall (primary tillage) followed by a combined field cultivator-harrow in spring (secondary tillage) controlled established foxtail barley on previously untilled sites. Foxtail barley is a problem only in no-till spring wheat and other spring-sown no-till crops in the northern Great Plains. Often it could be managed by changing tillage practices (e.g., rotating from no-till to primary tillage). If land was chisel plowed in spring and then harrowed, the weed was not controlled (Donald, 1990).

Research to determine the influence of the type of tillage implement and the timing of tillage leads to understanding of how land management and weed control may actually create weed problems. Roberts and Stokes (1965) showed that plowing distributes weed seeds throughout the plow layer. Rotary cultivation leaves 50% of weed seeds in the top 3 inches and 80% in the top 6 inches where they germinate best. The type of tillage implement and tillage timing can determine the weed problem.

Plowing is used to prepare land for planting *and* it controls weeds. (Courtesy of Deere and Co., Moline, Illinois.)

In a rare study of tillage over time, Wicks *et al.* (1971) grew winter wheat annually for 12 years and studied the effect of a sweep plow, one-way disk, and moldboard plowing (all primary tillage implements) after harvest on downy brome. The moldboard plow eliminated the downy brome population after 12 years, compared to 94 plants per square meter for sweep plowing and 24 for the one-way disk. Sweep plows do not bury seed as deeply as moldboard plows. The moldboard buries seed that germinates but cannot emerge. Spread of downy brome is hastened by changing from spring to winter wheat because land is then plowed and prepared for seeding at exactly the right time for the winter annual life cycle of downy brome (McCarty, 1982).

The same kind of evidence about the affect of timing and type of tillage is found in several farming systems. Evidence from rice culture shows that the method and timing of land preparation influences the subsequent weed population. In fields where tractor plowing during the dry season was followed by two harrowings in the wet season, junglerice was over 85% of the weed population in rice, and purple nutsedge was negligible. In the same region, where two plowings and two harrowings occurred in the wet season, junglerice was virtually nonexistent and purple nutsedge was the dominant weed (Pablico and Moody, 1984).

Annual grass weeds are likely to remain a problem with use of minimum cultivation in cereal production, particularly when early planting is practiced (Froud-Williams *et al.,* 1981). Other, previously unimportant, weeds became more prevalent, especially weedy species of brome in winter cereals in the United Kingdom. Buhler and Oplinger (1990) working with spring-sown crops in the United States showed that common lambsquarters density was not influenced by tillage method, but redroot pigweed density was usually higher in chisel plow systems prior to planting soybeans. Moldboard plowing (primary) followed by cultipacking (secondary) always had greater densities of velvetleaf than no-till, and no-till always had more foxtail than plowing. Giant foxtail and redroot pigweed became more difficult to control when tillage was reduced, whereas velvetleaf was less of a problem.

Growers need to be aware of the effect of tillage type and timing on weed populations and, whenever possible, choose a system that contributes to weed control. That is good management, and the integration of techniques will follow. Reduced cultivation encourages establishment of wind-disseminated species and annual broadleaved species decline. In corn, green foxtail density was greater in chisel plow and no-till systems than with moldboard plowing and ridge tillage had lower green foxtail density than all other systems (Buhler, 1992). Common lambsquarters density was nearly 500 plants per square meter after chisel plowing whereas it was only 75 in other tillage systems. Redroot pigweed responded differently to tillage with average densities of 307 and 245 plants per square meter after no-tillage and chisel plowing, vs. only 25 plants per square meter after moldboard plowing or ridge tillage. Weed populations were affected by tillage, but corn yield was not.

Many weed seeds require light to stimulate germination (see Chapter 5). Weed scientists have asked if germination could be reduced if soil tillage or cultivation was done at night. In Oregon's Willamette Valley, cultivating agricultural land during the day increased germination 70 to 400% above levels found after nighttime tillage (Scopel *et al.,* 1994). The effect was attributed to the light to which seeds are exposed during tillage. Buhler and Kohler (1994) showed that tilling soil in absolute darkness can reduce germination of some weed species up to 70%. Night tillage is most effective against small broadleaved species such as pigweed, smartweed, ragweed, nightshade, wild mustard, and common lambsquarters. It is not effective to reduce germination of foxtail or barnyardgrass and it has no effect on large-seeded broadleaved weeds such as velvetleaf, giant ragweed, and cocklebur. Hartmann and Nezadal (1990) were the first to report, after 7 years of study, that tillage between 1 hour after sunset and 1 hour before sunrise reduced weed emergence as much as 80% compared to day tillage. They saw night tillage as a way to manipulate and control weed populations on a purely cultural basis. They also advocated daytime tillage to photostimulate germination of dormant weeds seeds with the goal of diminishing the soil seedbank. They recommended that early primary tillage (plowing) should be carried out in full sunlight to encourage seed germination. Secondary tillage to prepare the seedbed, should be done after dark to destroy emerged seedlings and not encourage germination of seeds. However, do not become too enamored of this idea. Although it is true that exposure to light favors germination of many weeds seeds, some are light insensitive. Light is only one of many environmental factors that affect weed seed germination. Regulating light exposure will favor management of some weeds and enhance chances for success of others.

When undisturbed in soil, most light-sensitive seeds are not photoinduced to germinate by light penetration below 1 cm. Germination stimulation comes from brief (a few seconds or less) exposure to light during mechanical cultivation (soil disturbance) in daylight. This observation is consistent with early work by Wesson and Wareing (1969), who showed that weed-seed germination was dependent on exposure of seeds to light during mechanical cultivation. Most weed seeds germinated within 2 weeks after exposure to light. They also demonstrated that stirring soil for 90 seconds in bright light increased weed-seed germination up to 60%.

Minimum or no-tillage agriculture is practiced for many reasons, including economic ones, and a desire to reduce soil erosion. As emphasized earlier, tillage, including minimum or no tillage, affects the weed population. Any method of weed control that minimizes tillage is potentially of benefit to soil structure. The data in Table 9.7 on eco-farming encourage minimum tillage for production of crops grown under low-rainfall conditions. The point is that minimum-tillage wheat and minimum-tillage grain sorghum yield as well and frequently have lower production costs than more intensive tillage systems. Minimum-tillage, nonirrigated corn does not yield what irrigated corn does, but production costs are lower.

The extent of use and weed-control implications of no or minimum tillage have been reviewed for developing countries (Akobundu, 1982, and Buckley, 1980), and it has been shown that these systems rely on herbicides and may complicate soil management because of the presence of crop residues. With an abundance of weed seed in soil, the best approach may be to use minimum or no tillage and let natural factors deplete the population of buried seed. If weed control fails one year and the soil weed seedbank has been depleted, the best strategy will be to plow deeply and then use minimal tillage thereafter (Mohler, 1993). In the first year after minimum tillage begins, no tillage will have more seedlings than tillage, but in subsequent years, fewer weed seedlings will emerge unless dormancy is high or there is very good survival of seed near the soil surface (Mohler, 1993).

There are important advantages to minimum and no-tillage (Phillips, 1979):

Table 9.7

Yield and Production Costs for Different Cropping Systems in Southwest Nebraska (Klein, 1988)

Crop	Tillage	Average yield (bu/A)	Production cost ($/bu)
Wheat	Clean fallow	37	3.88
Wheat	Stubble mulch	43	3.44
Wheat	Ecofallow-reduced tillage	45	3.30
Sorghum	Conventional	40	3.09
Sorghum	Ecofallow-reduced tillage	65	2.42
Corn	Conventional tillage with center-pivot irrigation	140	2.59
Corn	Ecofallow-reduced tillage	65	2.52

1. Soil erosion is reduced. (A primary disadvantage of tillage is the possibility of increased erosion.)
2. Because of reduced erosion, land subject to erosion may be used more intensively.
3. Reducing tillage saves energy.
4. There is less compaction with decreased travel over soil.
5. Because land is continually covered, soil moisture is not as limiting as it can be on bare soil.
6. Irrigation requirements are lower because post-tillage evaporation of soil moisture is reduced.
7. Less horsepower is required for land preparation and machinery costs can be reduced.

It is generally agreed that reduction or absence of tillage increases problems with perennial weeds. Tillage may increase or decrease weed seedling density (Mohler, 1993); some studies have found more seedlings in tilled plots and others have found more without tillage. The effects of tillage vary between species.

Froud-Williams *et al.* (1981) reviewed changes in weed flora associated with reduced tillage systems. They found several studies where perennial monocot and dicot species increased in the absence of tillage. They suggested that perennial monocot weeds with rhizomes or stolons would be the greatest threat to successful adoption of reduced tillage systems.

There are equally important disadvantages to reducing or eliminating tillage (Akobundu, 1982):

1. Average soil temperature is lower, and this may delay spring planting and subsequent crop emergence.
2. Insect and disease problems may increase because plant residues on the soil surface provide a good environment for insects and disease pathogens (Musick and Beasley, 1978, and Suryatna, 1976).
3. A greater degree of farm managerial skill may be required because:
   (a) Fertilizer requirements and application techniques must be changed.
   (b) Crop establishment may be more difficult because of surface residue.
   (c) Irrigation techniques may have to be modified.
   (d) Weed control is essential, but as species change, methods must change.
   (e) The variety of available herbicides is not great.

Disadvantages have not deterred growers from learning required skills and shifting to no- or minimum-tillage systems. In the United States no-till acreage increased from 10.6 to 32.9 million acres from 1972 to 1980 (Triplett, 1982). Triplett (1982) suggested that 80% of U.S. crop acreage will be planted using some form of reduced tillage and 50% of the acreage will be no-tillage.

Seed burial studies (see Chapter 5) support the contention that the shift to minimum or no-tillage systems of crop production will not eliminate the need for weed management. The need will continue, but the weeds to be managed will change as tillage systems change. Data from seed burial studies show that as tillage is reduced, biennial weeds invade cropland, partially because their seeds survive longer when buried (Burnside *et al.,* 1996). Other annuals, adapted to no-till, will appear in cropping systems. Present federal farm programs promote conservation tillage and require maintenance of plant and residue cover on the soil surface to reduce wind and water erosion.

D. MOWING

Mowing to remove shoot growth prevents seed production and may deplete root reserves of some upright perennials. If repeated often enough, it can be used to control upright perennials in turf. Prostrate perennials such as field bindweed survive mowing.

To maximize the benefits of mowing, it must be done before viable seeds have been produced. Weeds should be cut in the bud stage or earlier. Table 9.8 shows the percentage of germinable seeds produced at various stages of maturity.

Mowing is a useful technique, but it rarely accomplishes much weed control because it is done late. It removes unsightly growth and, if done at the right time, can prevent seed production, which is important in control of annuals and biennials. Its effectiveness for control of the biennial musk thistle is shown in Table 9.9.

The foregoing deals with mowing as an operation performed to control weeds or clean up an area. Mowing is a normal cultural operation for some crops (e.g., turfgrass and hay) and is properly regarded as a potential weed management technique rather than solely a necessary part of producing the crop. Norris and Ayres (1991) showed that cutting interval (but not irrigation timing after cutting) affected yellow foxtail biomass in alfalfa and alfalfa yield. Percent yellow foxtail ground cover was greatest after a 25-day cutting interval and least after a 37-day interval (Figure 9.3). Yellow foxtail biomass was also greatest for the short cutting interval and least for the longest interval. In the 3 years of the study, the 37-day cutting interval

Table 9.8

Germination of Weed Seeds from Plants at Three Stages of Maturity (Gill, 1938)

| Weed | In bud | Cut | | |
		Flowering	Medium ripe	Ripe
Annual sowthistle	0	100		100
Canada thistle	0	0		38
Cat's ear, spotted	0	0		90
Common chickweed		0	56	60
Common groundsel	0	100		100
Curly dock		0	88	84
Dandelion	0	0		91
Meadow barley	0		90	94
Shepherd's-purse	0		82	88
Soft brome	0		18	96
Corn speedwell	0		69	70

always had a higher yield than the 31- or 25-day interval (Table 9.10), thus demonstrating the utility of mowing for weed management.

E. FLOODING, DREDGING, DRAINING, AND CHAINING

These techniques cause ecological change. If a normally dry area is flooded or a normally wet area is drained, the ecological system is changed and weed species will change. These techniques are effective only when

Table 9.9

Seed Production by Musk Thistle (McCarty, 1982)

Time of harvest	Seeds/plant
Full bloom	26
+ 2 days	72
+ 4 days	774
Mature plant	3,580

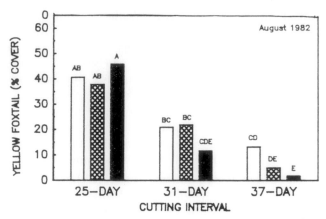

Figure 9.3. Percent yellow foxtail cover in relation to cutting frequency and duration of irrigation delay following cutting. Columns with different letters are different at P = 0.05 according to the LSD (Norris and Ayres, 1991).

an area is immersed or drained for 3 to 8 weeks. Immersion, an anaerobic treatment, is not equally effective on all weeds; lowland or paddy rice fields have weeds such as barnyardgrass and junglerice that survive the flooded conditions of the rice paddy as well as rice does. Flooding does not eliminate all weed problems, just some of them and it creates an environment where other weeds succeed. Weeds found in lowland rice are generally different from those found in upland rice. Purple nutsedge occurs in both systems. Flooding will control established perennials such as silverleaf nightshade, camelthorn, and the knapweeds in arid areas, but the expense of creating

Table 9.10

Alfalfa Dry Matter Yield in Relation to Cutting Interval (Norris and Ayres, 1991)

| Year | Alfalfa yield (tons/acre) with different cutting intervals | | |
	25 days	31 days	37 days
1	9.0	12.8	14.9
2	15.0	21.7	24.0
3	11.0	16.2	20.0

dikes and obtaining water make the practice economically unfeasible (Slife, 1981).

Draining is an excellent control for cattails, bulrushes, and reed canary-grass, which grow best in wet areas. Draining and flooding are not applicable to most agronomic or horticultural environments, but they should not be forgotten when considering weed control on a site.

Chaining is employed on rangelands to destroy emerged vegetation. A large chain similar to a ship's anchor chain is dragged between two bulldozers and uproots sagebrush, rabbitbrush, and other range weeds. Chaining removes emerged growth and completely controls annuals, but not perennials that reproduce vegetatively. The technique is not suited for most cropland. Chains are also used to stop passage of weeds in irrigation channels in many countries. Removing collected weeds from the impoundment created by the chain is a labor-intensive, unpleasant operation.

IV. NONMECHANICAL METHODS

A. HEAT

1. Flaming

Many plant processes are susceptible to high-temperature disruption attributed to coagulation and denaturation of protein, increasing membrane permeability, and enzyme inactivation. Photosynthesis is decreased or stopped. Initial thermal disruption of cellular membranes is followed by dehydration. Heat, short of setting fire to an area, usually does not kill by combustion. Thermal death points for most plant tissue are between 45 and 55°C (113 to 131°F) after prolonged exposure. Flame temperature approaches 2000°F, but flamers may be used selectively when distance from the crop and speed are controlled.

A flamer directs a petroleum-based fuel emitted under pressure and ignited. Plant size at treatment influences efficacy much more than plant density. To achieve 90% control of white mustard with up to 2 leaves required at least 40 kg/ha (36 lb/A) of propane, whereas plants with 2 to 4 leaves required 70 kg/ha (62 lb/A) (Ascard, 1994). Required dose increased with growth stage, and some species of annual weeds are more tolerant than others. The most tolerant species cannot be controlled with one flaming, regardless of dose (Ascard, 1994).

Weeds with unprotected meristematic areas and thin leaves such as common lambsquarters, common chickweed, and burning nettle were completely killed by 20 to 50 kg/ha of propane when they had fewer that 5 true leaves (Ascard, 1995). Shepherd's-purse and pineapple-weed have protected growing points and were killed by flaming only at very early growth states. Annual bluegrass could not be killed with a single flaming, regardless of its size or the propane rate. Plants with up to 4 true leaves were killed by 10 to 40 kg propane/ha, whereas those with 4 to 12 leaves required 40 to 50 kg/ha (Ascard, 1995).

Corn between 2 and 12 inches tall cannot withstand flaming. Before corn is 2 inches tall its meristematic region is underground and will regenerate the plant. After 12 inches, the flame can be directed at the plant's base and used selectively, if the weeds are shorter than the crop. Intensity and duration of exposure are important. If one held a flame on a corn plant for several minutes, the plant would die, so flamers must be kept moving, and speed affects selectivity. Flame has been used selectively in cotton and onions. When cotton stems are 3/16 inches in diameter, or greater, flaming can be used.

Flaming kills green shoots where tillage is impractical, such as along a railroad. Buried weed seeds or perennial plant parts are not affected. Dry seeds withstand high temperatures and rather long exposures because soil protects and insulates. Burning can destroy weed seeds, but only if they are on the soil surface. Even a small layer of soil from shallow burial will protect most seeds. Therefore, flaming is not effective for controlling anything except emerged weeds.

Burning mature weeds destroys debris but does not prevent crop losses from competition. Flaming has no residue: a problem with chemical methods of control. Other than high-rainfall conditions, flaming is not affected by prevailing environmental conditions. It may induce erosion by eliminating vegetation that holds soil. Heat could induce germination of dormant seeds or create conditions favorable for their germination by eliminating emerged, competing plants. This is especially true when brush is burned.

Controlling a flamer's direction eliminates drift, and one can achieve some degree of insect and disease control. As petroleum energy costs increase, flaming's cost may become prohibitive. An additional advantage is immediate observation of results. Flaming is often used to eliminate vegetation along irrigation ditches.

2. Solarization

The fact that mulches (see next section) raise soil temperature makes weed control possible using the heat of the sun in a process called solarization. Weed seed germination is suppressed by high soil temperatures and seedlings are killed. Transparent and opaque polyethylene sheets raise soil temperature above the thermal death point for most seedlings and many seeds.

Solarization uses plastic sheets placed on soil moistened to field capacity and thus heats soil by trapping solar radiation just as a glasshouse does (Horowitz *et al.,* 1983). Its effectiveness for weed control is dependent on a warm, moist climate and intense radiation with long days to raise soil temperature enough to kill weed seeds and seedlings. Moisture increases soil's ability to conduct heat and sensitizes seeds to high temperatures (Horowitz, 1980). Solarization controls soil-borne diseases and increases crop growth because of soil warming.

When different types of plastic were used for 4 weeks in Israel the temperatures under clear plastic exceeded 45°C. Temperatures under black plastic exceeded 40°C about half the time, but did not reach 45°C. UV-absorbing transparent plastic raised temperatures above 50°C. At 5 cm, temperatures increased 9° for black and 19° for clear plastic.

The effects of solarization on weed emergence were apparent for a short time after plastic was removed. During the first 2 months after removal, the number of emerging annuals was less than 15% of an untreated check, and clear plastic was more efficient. Only clear plastic reduced weed populations for 1 year after solarization (Horowitz, 1980). Table 9.11 shows some data on the sensitivity of annual weeds to solarization.

In other work, a month after solarization, field bindweed, annual sowthistle, and prostrate pigweed covered 85% of the soil surface in plots not solarized, compared to only 18% in solarized plots (Silveira and Borges, 1984). A 1-week period of solarization reduced the percentage of buried seeds of prickly sida, common cocklebur, velvetleaf, and spurred anoda in

Table 9.11

Sensitivity of Annual Weeds to Solarization (Horowitz *et al.,* 1983)

Weed	Weeks of solarization to reduce seedling numbers to less than 10%of control
Blue pimpernel	2–4
Bull mallow	>8
Fumitory	6
Heliotrope	4
Horseweed	>8
Pigweeds	2

soil in Mississippi (Egley, 1983). Solarization reduced emergence of all weeds except purple nutsedge. Total weed emergence was reduced 97% 1 week after removal of plastic and up to 77% for the season (Egley, 1983). The major effect of high soil temperature (up to 150°F) is to kill weed seedlings that germinate under the plastic. Solarization has not been employed on a large scale in field crops, but is used effectively in high-value vegetable crops in California's Imperial Valley. Because there is no cold winter season, solarization is used for 6 weeks before crops are planted. The plastic is removed prior to planting and must be disposed of, a problem all by itself, but solarization nearly eliminates use of herbicides. Solarization has potential to improve weed management, but costs, compared to those of other methods, preclude widespread adoption in other than high-value crops.

Research by Campbell's Soup Co. in California has used solarization in a different way (Hoekstra, 1992). The previous comments related to use of plastic mulch to heat soil and kill weeds. D. Larsen of Campbell's Soup has experimented with a solar-powered lens that heats soil and kills weeds. The curved lens is an acrylic sheet made of an array of small lenses. It is cheaper and lighter than glass and not as easily damaged . Lens concentration of solar energy has two primary disadvantages: (1) It does not work on cloudy days, and (2) The lens must be pulled slowly over the field to focus energy sufficiently to kill seedling weeds. Stronger lenses capable of concentrating more energy may enable faster movement.

B. MULCHING

Mulching excludes light and prevents shoot growth. Thick, wide mulches are required to control perennials that creep to the edge of a mulch and emerge. Mulches increase soil temperature and may promote better plant growth. Several different materials have been used for mulch, including straw, hay, manure, paper (first used on sugarcane in Hawaii), and black plastic. It is common to see mulches used in greenhouses where plants grow in a soil floor. Mulches are used most in high-value crops grown on small areas and in crops (e.g., sugarcane) where laying the mulch can be mechanized.

Shredded paper was one of the first mulches used in a crop. It has been replaced by plastic mulch, and its use is rare. Pellett and Haleba (1995) evaluated use of chopped paper in perennial nursery crops over two seasons. Their work showed that paper was an effective mulch that provided weed control over two seasons, especially when the paper was wetted and rolled after application. They applied 2.3 or 3.6 kg/m^2. The higher rate was 15 cm deep. The equivalent rate per hectare was almost 38 tons and the cost of hand application of baled paper, in Wisconsin, was more than $2500 per hectare. The mulch provided good weed control for 2 years and it was possible to rototill paper into soil with power equipment. A tackifier (a substance to make the paper sticky) was important to prevent paper from blowing away or piling due to wind. Cost of the paper and its application prohibits consideration of the use of paper mulch in any but high-value crops.

As the amount of wheat straw mulch increased in a wheat–corn–fallow dryland production system, weed growth decreased (Crutchfield and Wicks, 1983). Others have shown that planting no-till corn into a desiccated green wheat cover crop reduced morningglory biomass 79% compared to a non-mulched, tilled treatment (Liebl *et al.*, 1984). Rye mulch was also successful in reducing biomass of three annual broadleaved species in three crops (Liebl *et al.*, 1984). Rye has been used successfully as a crop mulch in the fall and winter before corn (Almeida *et al.*, 1984): a practice known as green manuring. The rye contributed to weed control in corn because of its allelopathic activity. Its foliage was dense enough that a contact herbicide had to be applied before corn planting.

A synthetic, woven, black cloth is available for mulching. it is sold commercially in rolls about 6 feet wide and can be applied by machine when trees are planted. It is easy to spread and prevents emergence of most annual weed seedlings.

C. SOUND AND ELECTRICITY

Use of high-frequency energy and electricity has been considered since the late 19th century. Ultrahigh-frequency (UHF) fields are selectively toxic to plants and seeds, and the first use of sound for weed control was patented in 1895. UHF fields produce thermal and nonthermal effects, but thermal effects are the chief source of toxicity. There is a linear and positive correlation between seed water content and susceptibility to electromagnetic energy. Lower frequencies have broken seed dormancy. Commercial weed-control devices using UHF fields have been developed, patented, and commercialized, but without lasting commercial success. These have been used for selective vegetation control in cotton and in water, but have not achieved great commercial success. They require a great deal of power and can be used preemergence or postemergence. Postemergence use forces plants to conduct current and in effect "boils" plant solutions and ruptures cell walls. Vigneault *et al.* (1990) reviewed what they called electrocution for weed control. They concluded that use of electricity may have a place in high-value, specialty crops such as fine herbs. It may be especially appropriate when the treated area is small, no herbicides are available, and cultivation is undesirable because of the potential for root damage and the risk of soil erosion. Advantages include lack of any chemical residue and no soil disturbance.

V. CULTURAL WEED CONTROL

A. Crop Competition

The techniques of cultural weed control are well known to farmers and weed scientists. In fact, they are employed regularly, but often are not conscious attempts to manage weeds. Planting a crop is a sure way to reduce growth because the crop interferes with the weeds. It is a fundamental method of weed management, but most often cultural weed control just happens rather than occurring as a planned addition to weed management programs. Methods of cultural weed management include conscious use of crop interference, use of cropping pattern, intercropping, and no or minimum tillage.

Weed scientists have investigated the relative competitiveness of crop cultivars. Several crops exhibit genotype differences in competitiveness (Burnside, 1972; Monks and Oliver, 1988). Weed biomass differences up to 45% have been reported among soybean genotypes (Rose et al., 1984). Wild oat competition with wheat was greater than intraspecific competition in wheat. The competitiveness of six wheat cultivars with wild oat was similar for all factors measured (Gonzalez-Ponce, 1988). The most weed-suppressive of 20 winter wheat cultivars reduced weed biomass 82% compared to the least suppressive cultivar (Wicks et al., 1986). With weed interference, the lowest-yielding varieties produced 66 and 54% as much as the highest-yielding varieties of wheat (Ramsel and Wicks, 1988) and rice (Smith, 1974), respectively. Other work showed that short-stemmed cultivars were more affected by taller wild oats because of light competition (Wimschneider and Bacthaler, 1979). The quest to develop integrated weed management systems has encouraged research on the competitiveness of crop cultivars. That cultivars differ in competitive ability was amply demonstrated a number of years ago in soybeans (McWhorter and Hartwig, 1972) (Table 9.12). Recent work in Denmark showed that spring barley varieties vary in weed suppression ability (Christensen, 1995). Weed dry matter in the most suppressive variety was 48% lower than the mean dry matter of all varieties, whereas it was 31% higher in the least suppressive variety.

Table 9.12

**Yield Reduction in Selected Soybean Varieties due to
Johnsongrass or Cocklebur Competition (McWhorter
and Hartwig, 1972)**

| Soybean variety | Yield reduction (%) with weed competition from | |
	Johnsongrass	Cocklebur
Davis	34	56
Lee	41	67
Semmes	23	53
Bragg	24	57
Jackson	30	67
Hardee	23	26

More vigorous, taller, faster-growing cultivars are likely to be better competitors, but too little is known about what makes a cultivar competitive and whether it is a trait that plant breeders can select for and develop. Christensen's (1995) work, however, demonstrated no correlation between varietal grain yields in pure stands and competitiveness, suggesting that breeding to optimize yield and competitive ability may be possible. Research is being done to develop crop cultivars that can be bred or managed for high levels of crop interference via high rates of resource uptake or possible allelopathic (see Chapter 7) interference with weeds (Jordan, 1993).

Alfalfa and other hay crops are smother or cleaning crops. Land is not plowed when these crops are grown, making it hard for annuals to succeed, but perennial weeds do well in perennial crops such as alfalfa. Sudangrass, planted in dense stands, can compete effectively against many, but not all, weeds.

Crops can be favored by knowing and using the effect of row width and crop seeding rate. Kahn *et al.* (1996) showed that spring wheat yields were as great or greater when early seeding or a double seeding rate were used as a substitute for a postemergence herbicide to control foxtail species. Early and middle seeding dates favored the increase of green foxtail over yellow foxtail, whereas late seeding favored yellow over green. Spring wheat competing with foxtail had a higher yield when the seeding rate was 270 kg/ha (twice the normal rate) than when it was 130 or 70 ($\frac{1}{2}$ normal rate) kg/ha unless the seeding was late. Early, high seeding rates increase crop density and biomass early in the season, and this suppresses weed growth. Decreases in weed growth have been observed in narrow (about 8 inch) vs wide (about 30 inch) row spacing in several crops. For example, weed growth was reduced 55% in peanut (Buchanan and Hauser, 1980) and 37% in sorghum (Wiese et al., 1964).

Varying row width is using the principles of plant population biology to achieve competitive interactions that favor the crop. Much work is proceeding in the midwestern United States to devise narrow row production techniques for soybeans. When these are combined with minimal tillage and the right herbicides, yield is maintained or increased, soil erosion is reduced, and excellent weed management is obtained.

Intercropping is a common, small-scale farming system among farmers of the developing world. The main reasons for mixing crops or planting in close sequence are to maximize land use and reduce risk of crop failure. Intercropping maintains soil fertility, reduces erosion, and may reduce insect problems (Altieri et al., 1983). Intercropping also gives greater stability to yield over seasons and provides yield advantages over single-crop agriculture (Altieri, 1984).

It has been claimed (Altieri et al., 1983) that one reason for intercropping is weed suppression, but there is little experimental evidence to support this conclusion (Shaw, 1982). Similarly, there is little evidence that intercropping requires less weed control. It is assumed that intercropping saves labor because weeding is less critical and some operations such as planting a second crop and weeding the first can be combined (Norman, 1973). Intercropping's effectiveness for weed control depends on the species combined, their relative proportions, and plant geometry in the field. No reports recommend no weeding with intercropping, and weeds can often be worse than in sole crops (Moody and Shetty, 1981).

Studies of intercropping do not confirm that any plant grown with a crop will always provide adequate weed control. Intercropping is a common practice in many agricultural systems, and these systems should be studied to develop complementary plants, to control soil erosion, and to prevent or reduce weed growth. It is undoubtedly true that plants that are not crops are classified by most farmers in the developed world as weeds. Other farmers classify noncrop plants in a way that judges their potential use or their effects on soil and crops. Subsistence farmers have a different understanding of the use and value of plants that are neither crop nor weed.

B. Planting Date and Population

The trend in crop production is early planting to optimize yield. Yield is increased because crops have a longer growing season and photosynthesize for more days (Barrett and Witt, 1987). Early planting provides a competitive edge to adapted crop cultivars because of early season establishment of a crop such as corn compared to yellow nutsedge, a warm-season weed (Ghafar and Watson, 1983). The competitive advantage could be due to the weed's light requirement for growth and to shading by the crop that emerged first. Choice of planting date should be considered part of integrated weed management. Planting date of any crop plays a role as illustrated by a 60% reduction in kochia population when proso millet was planted June 1 rather than May 15, although millet yield was not affected (Anderson, 1988). Planting date can also play a role in crop choice. Longspine sandbur emerges in late May and June in Colorado and flowers in late July. The seed, in its bur, reduces the value of hay. Foxtail millet is planted in early June and, if harvested as hay in late August (Lyon and Anderson, 1993), will be contaminated with the burlike seed, if longspine sandbur is present. Oats can also be grown for hay. It is planted in early April and harvested in late June before the longspine sandbur seed develops, and the hay will not be contaminated with the burlike seed.

Sunflower and safflower are grown as oil crops in the U.S. Great Plains states. Safflower is planted in early April and sunflower in early June. Because of its early planting, more than 70% of weed seedlings emerge within 10 weeks of planting safflower. These weeds are easily controlled by tillage or herbicides, and sunflower is planted in a more weed-free field after mid-June (Anderson, 1994). Early planting requires weed control for longer periods. Late planting is usually preceded by tillage that destroys emerged weeds and reduces the population in the crop. Advantages gained by later planting are often outweighed by decreased crop yield over a shorter growing season.

In Minnesota, delaying soybean planting from early May until early June permitted use of preplant tillage to control early germinating weeds (Gunsolus, 1990). This reduced maximum soybean yield potential 10%. When corn planting was delayed from the normal time, in the beginning of May, until after May 25, maximum yield potential was reduced 25% (Gunsolus, 1990). This small set of data illustrates the complexity of agriculture; extrapolations cannot be made between crops, and certainly not between regions. The same study also showed that rotary hoeing for weed control when either crop was young reduced corn plant stand up to 10% but did not affect soybean stand. In Minnesota, a 10% loss in corn stand reduced final yield 2% but did not affect soybean yield.

Khan et al. (1996), in a different kind of study about planting date, reported that crop management practices related to planting date could substitute for herbicide use to control foxtail species in wheat. Spring wheat yields in North Dakota were the same or greater when early seeding or a doubled seeding rate were substituted for postemergence foxtail control with an acceptable herbicide. Yield of spring wheat was greater with a high seeding rate (240 lb/A) than with normal (116 lb/A) or low (62 lb/A) seeding rates for early (late April to mid-May) or mid-season (mid- to late May) seeding, but not for late (early to mid-June) seeding. It is interesting to note how seeding date, in this work, affected certain weeds. Early and middle seeding dates favored the relative increase of green foxtail and the late date favored yellow foxtail. In weed management, as in ecology, no one can do just one thing.

Planting date is often dictated by considerations other than weed management. Similarly, plant population is dictated by agronomic studies that have shown the population that gives the best yield. Populations are also determined by row spacings required by planting, cultivating, and harvesting machines. Increasing crop plant populations can often decrease weed density and growth. Wiese showed more than 30 years ago how row width and seeding rate interacted to reduce competition from weeds in grain sorghum in Texas (Wiese *et al.,* 1964) (Table 9.13). With 25-cm rows, yield loss from weeds was lower with the higher of two seeding rates. This relationship remained true until rows were 102 cm wide.

C. COMPANION CROPPING

Cover crops or living mulches (Akobundu, 1980b) can be used as intercrops or companion plants to suppress weeds (Deat *et al.,* 1977; Shetty

Table 9.13

Effect of Row Width, Seeding Rate, and Cultivation on Yield of Grain Sorghum (Wiese *et al.,* 1964)

Row width (cm)	Seeding rate (kg/ha)	Grain yield (kg/ha)		Yield loss (%)
		Weedy	Hand weeded	
25	5.6	3,326	4,861	31
	11.2	4,188	5,466	23
51	5.6	3,125	5,152	39
	11.2	3,987	4,715	16
76	5.6	3,237	5,365	40
	11.2	3,606	5,029	28
102	5.6	3,058	4,491	32
	11.2	3,203	4,637	31

and Krantz, 1980). Appropriate weed control practices, for many farming systems, must consider the need to maintain soil fertility and prevent erosion, and open row crops are inimical to these needs. Akobundu (1980a) developed integrated low- or no-tillage weed management systems, compatible with more than one crop plant in a field, that reduced herbicide use, fertilizer requirements, and soil erosion. Combinations of a legume or Eugusi melon and sweet potato with corn showed that these companion crops or living mulches maintained corn yield, contributed to nitrogen supply, suppressed weed growth, and reduced soil erosion. Groundnut, centro, and wild winged bean have been used as living mulches with corn. Living mulches incorporate organic mulch, no-tillage, and weed control. Centro and wild winged bean grew so vigorously that a growth retardant had to be applied to bands over corn rows to gain a growth advantage for corn (Akobundu, 1980b). In unweeded no-till plots, corn grain yield was 1.6 T/ha, whereas with conventional tillage it was 2.3 T/ha. Corn yield in unweeded, live mulch plots averaged 2.7 T/ha. Yields were not different and live mulch plants did not reduce yield; they were complementary, not competitive. Further studies (IITA, 1980) verified these results (Table 9.14).

Clover has been grown successfully with corn and has reduced weed growth (Vrabel *et al.,* 1980). Crimson clover and subterranean clover were the most promising cover crops in cucumbers and peppers in Georgia and contributed to effective management of diseases, nematodes, and insects (Phatak, *et al.,* 1991). Sweet corn in a living mulch of white clover had high yields in early years, but lower yields later because a contact herbicide used over the corn row allowed invasion of perennial weeds that were not

Table 9.14

Effect of Weeding Frequency and Ground Cover on Weed Competition and Maize Yield (IITA, 1980)

	Unweeded check[a]	
Ground cover	Weed dry weight (T/ha)	Grain yield (T/ha)
Conventional tillage	1.5 a	1.1 e
No tillage	1.4 a	1.8 bcd
Maize stover	1.3 a	1.6 cde
Maize and groundnut	0.3 c	1.3 de
Maize and wild winged bean	0.1 c	2.1 abc

[a]Values in one column followed by the same letter are not statistically different at the 95% level of probability.

suppressed by white clover (Mohler, 1991). A dead rye mulch decreased weed biomass and did not decrease corn yield (Mohler, 1991). A living mulch of spring-planted rye reduced early season biomass of common lambsquarters 98%, large crabgrass 42%, and common ragweed 90% compared to unmulched controls. Barnes and Putnam (1983) also reported that the age of rye when it was killed with herbicides was important to the subsequent emergence of yellow foxtail and lettuce.

Companion cropping can be a good weed control technique but research is needed to determine how appropriate it may be in specific situations. Limited evidence supports the contention that it can provide weed competition, build soil organic matter, reduce soil erosion, and improve water penetration (Andres and Clement, 1984). In some climates when spring soil moisture is limiting, cover or companion crops can deplete moisture and be detrimental to crops in spite of weed control advantages. Companion crops may also have to be killed before a crop is planted or they become competitors.

In Pennsylvania, crownvetch, a legume, was tried as a living mulch in a no-tillage corn (Cardina and Hartwig, 1980; Hartwig, 1987). Crownvetch is difficult to establish, but once established, it provides soil erosion control and it improves fertility by reducing nutrient loss via erosion and by contributing nitrogen and weed control. Weed control must be supplemented with herbicides that will not kill the crownvetch. The system is amenable to rotation of corn with other crops. Work in Ohio demonstrated use of hairy vetch for weed management (Table 9.15). Unsuppressed hairy vetch

Table 9.15

Corn Grain Yield after Planting in Hairy Vetch at Three Growth Stages (Hoffman *et al.*, 1993)

Weed control treatment	Corn grain yield Kg/ha when planted into hairy vetch growth stage		
	Early bud	Mid-bloom	Late bloom
Untreated	130*a	7,350 b	6,520 b
Rolled with water-filled roller	40*a	7,630 b	7,510 b
Mowed with flail chopper	3,000*a	6,830*b	5,900 b
Glyphosate, 2.8 kg/ha	8,020*a	7,700 a	5,630 b
Weed-free control	9,770 a	8,560 a	5,310 b

*Values are statistically different from the weed-free control in a column; lowercase letters indicate statistical differences across a row.

reduced weed biomass in corn 96% in one year and 58% in another. When corn was planted in late April into hairy vetch in the early bud stage of growth, its yield was reduced up to 76%. Hairy vetch competition was reduced or eliminated when corn was planted into hairy vetch in mid- or late bloom in May or early June. Because of the shortened growing season and competition from hairy vetch, corn planted in May into untreated hairy vetch yielded similarly to corn planted in the no-cover-crop, weed-free check. Use of the contact, nonresidual herbicide glyphosate to kill vetch and eliminate competition with corn was helpful with early and mid-bloom planting but not with late planting, because of the lack of continuing weed control.

In Wisconsin, spring-planted winter rye has been a successful living mulch for weed control in soybean (Ateh and Doll, 1996). A system employing just rye for weed control reduced weed shoot biomass from 60 to 90% over 3 years. Rye worked best for weed control and did not reduce soybean yield when weed density was low. Ground cover from the mulch and soil moisture were adequate for growth, and rye interference with soybean was minimal if rye was killed within 45 days after soybean planting.

Other successful companion crops have been low-growing plants such as cowpea and mungbean in India (Shetty and Rao, 1981). In Colombia (CIAT, 1976), green cover crops were established with hetero (a legume), and when these were intercropped with beans, weed control was similar to that achieved with continuous manual weeding. Seed costs of companion plants and expected competition to the primary crop were offset by the value of companion plant yield, a more permanent soil cover (less erosion), reduced nitrogen fertilizer requirement, and reduced cost of hand weeding. Attempts have also been made to try different cover crops to manage noxious weeds such as cogongrass in India, Malaysia, Nigeria, and Kenya (Vayssierre, 1957).

Another example of a weed used to gain interspecific competition is the use of azolla as a weed control technique in lowland rice. *Azolla pinnata,* a free-floating fern, has been used in Asian rice culture because of its symbiotic relationship with *Azolla anabena,* a nitrogen-fixing blue-green algae. This symbiotic relationship can contribute up to 100 kg of nitrogen/ ha. A second use of *Azolla* is for weed control due to the competitive effect of an *Azolla* blanket over the surface of paddy water.

When *Azolla* is used, some farmers can grow rice without the addition of nitrogen fertilizer. Success of the *Azolla* technique depends on the ability of the farmer to control water supply and on the weed species present. Perennial weeds such as rushes and annuals with strong culms (e.g., barn-yardgrass) are not suppressed and must be controlled in other ways. Many other weeds are controlled well.

Azolla has been successful but cannot be universally recommended because there is an increase in labor just to manage it. Some land must be devoted to supplying a continuing source of inoculum for paddies, and *Azolla* may complicate other pest problems. In fact, *Azolla* may become a weed.

D. CROP ROTATIONS

Crop rotation is done for economic, market, and agronomic reasons. Some weeds associate with certain crops more than with others. Barnyardgrass and junglerice are common in rice. Wild oat is common in irrigated wheat and barley, but almost never occurs in rice. Nightshades are common in potatoes, tomatoes, and beans, and kochia and lambsquarters are frequent in sugarbeets. Dandelions are common in turf, but not as common in row crops, although without management, dandelions can increase in row crops and in pastures and permanent hay crops such as alfalfa.

These associations occur because of similarity in crop and weed phenology (naturally occurring phenomena that recur periodically, e.g., flowering), adaptation to cultural practices (e.g., tillage, mowing, irrigation), similar growth habits (e.g., time to mature or to reach full height), and perhaps of most importance, resistance or adaptation to imposed weed control methods. When one crop is grown for many years (monoculture), some weeds, if they are present in the soil seedbank, will be favored and their populations will increase. Weed–crop associations are not accidental and can be explained. Associations can be changed by changing crop, time of planting, or weed control methods. Wild oats can be reduced in small grain crops by growing corn in the rotation and using herbicides selective in corn plus cultivation to control wild oats when corn is grown. The same herbicides and cultivation cannot be used in small grain crops.

A good rotation includes crops that reduce weeds that are especially troublesome in succeeding crops. Removal is accomplished by competition or through use of different weed control techniques in different crops. In Canada, yellow foxtail populations in flax were highest when flax followed oats, lowest after flax, and intermediate after wheat, corn, and sorghum (Kommedahl and Link, 1958). Sugarbeets grown after beans in Colorado were always more weed-free than sugarbeets grown after sugarbeets, barley, or corn (Dotzenko *et al.,* 1969). Beans are cultivated well and intensive weed control is practiced. The number of weeds was highest where corn preceded sugarbeets, and lowest with beans. Barley was intermediate (Table 9.16).

Table 9.16

Effect of the Preceding Crop on Weed Numbers (Dotzenko et al., 1969)

Preceding rotation	Number of				
	Kochia	Pigweed	Annual grass	Lambsquarters	Total
Barley–beets	32	15	18	18	109
Corn–beets	67	44	48	7	166
Beans–beets	16	7	11	9	44

[a]Weed numbers are those that germinated in a 400-g sample of soil following three years of each sequence.

In many places, barley is planted in spring before soil temperatures are ideal for germination of most weeds. An exception is lambsquarters, which can be a serious weed in barley. Beans, on the other hand, are planted in late spring after tillage has destroyed most summer annual weeds.

Ball and Miller (1990) showed that weed species composition varied with cropping sequence among rotations of corn for 3 years, pinto beans for 3 years, or 2 years of sugar beets followed by 1 year of corn (Figure 9.4). Hairy nightshade seedbank population increased after 3 years of pinto

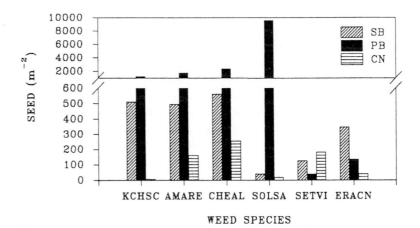

Figure 9.4. Influence of cropping sequence on dominant weed species in the soil seedbank 15 cm deep. SB = 2 years sugarbeets + 1 year corn, PB = 3 years pinto beans, and CN = 3 years corn. KCHSC = kochia, AMARE = redroot pigweed, CHEAL = common lambs-quarters, SOLSA = hairy nightshade, SETVI = green foxtail, and ERACN = stinkgrass. (Ball and Miller 1990, Reproduced with permission of Weed Sci. Soc. of America.)

beans, green foxtail increased after 3 years of corn, and the sugar beet–corn sequence caused an increase in kochia. Ball and Miller attributed the differences to the herbicides used in each cropping sequence. Crop cultivation, land preparation time and method, and time of planting and harvest may also favor one weed and discourage others.

Crop rotation regularly changes the crop in each field, soil preparation practices in one field, subsequent soil tillage, and weed control techniques. All of these affect weed populations, and although crops are not commonly rotated to control weeds, the effect of rotation as a determinant of weed problems must be recognized.

The relative dry weight of weeds in four cropping systems in the Philippines is shown in Table 9.17. Two weeds dominated, but their relative magnitude in the cropping systems, on the same soil, was different. In a rice–sorghum rotation, itchgrass dominated, but with continuous sorghum, itchgrass nearly disappeared and spiny amaranth dominated. Different cropping systems affect weed populations and may favor or deter species. This is observed in vegetable crops where intensive cultivation and weed control are regularly practiced, and weed populations can be reduced (Roberts and Stokes, 1965).

Long-term studies to determine the effect of different cropping sequences on the population dynamics of winter wild oat (Fernandez-Quintanilla *et al.*, 1984) showed that continuous winter cereal cropping (with or without herbicides) increased the winter wild oat soil seedbank from 26 to 80% per year. With spring barley the soil seedbank declined 10% per year. When sunflower was a summer crop *or* a 12 month fallow was included in the rotation to prevent new seed production, the soil seed reserve declined 57 to 80% annually. There was a great reduction in the size of the soil seedbank of winter wild oats if the cropping program was other than continuous winter cereals (Fernandez-Quintanilla, *et al.*, 1984).

Table 9.17

Relative Dry Weight of Weeds in Unweeded Plots in Four Cropping Systems 5 Weeks after Crop Emergence (Pablico and Moody, 1984)

Cropping system	Spiny amaranth (% dry weight)	Itchgrass (% dry weight)
Corn–corn–corn	65	21
Rice–corn	42	48
Rice–sorghum	12	83
Sorghum–sorghum–sorghum	95	3

E. Fertility Manipulation

Manipulation of soil fertility solely to manage weed populations is virtually unknown. However, as is true of most soil manipulations, fertility affects weeds. Walters (1991) suggests that most weeds can be controlled by simple manipulation of soil nutrient levels. His claims are supported by abundant anecdotal evidence, not by planned research. Nevertheless, they should not be dismissed as idle speculation. Farmers fertilize to maximize yield and attain greater assurance of success. They do not fertilize or withhold fertilizer to manipulate weed populations.

Fertilizer is added to improve crop yield, but weeds are often more competitive with crops at higher nutrient levels (DiTomaso, 1995). When weed density is low, added fertilizer, particularly nitrogen, increases crop yield and makes the crop a more vigorous competitor with weeds. But when weed density is high, added nutrients favor weed over crop growth. DiTomaso (1995) has summarized much of the literature on this subject. The subject has also been discussed in Section V,C of Chapter 6, and the crop yield reduction when additional nitrogen fertilizer is added in the presence of weeds is illustrated in the data in Tables 6.8, 6.9, and 6.10.

A good illustration of the potential of fertility manipulation as an instrument for changing plant populations is the Park Grass Experiment at the Rothamstead Agricultural Experiment Station in England. The official title of the experiment is "The park grass experiment on the effect of fertilizers and liming on the botanical composition of permanent grassland and on the yield of hay." The work was started in 1865 by Sir John Lawes, the son of the manor and founder of Rothamstead as an agricultural research center. In many ways, the experiment continues in its original form.

In unlimed plots amended with a complete fertilizer with nitrogen primarily as ammonium sulfate, a pure stand of common velvetgrass has developed. It was selected out of the original mixture solely by fertility manipulation and lack of lime. It has one of the heaviest hay yields of any plot, but is unpalatable. With complete fertilizer and lime, plots have one of the heaviest hay yields and a very diverse flora, including orchardgrass and meadow foxtail. In unlimed plots amended with ammonium sulfate and no phosphorus, the vegetation is completely different from either of the above. If potassium is absent, dandelions are absent because they flourish only with potassium and a pH above 5.6.

In winter wheat, downy brome was least responsive to nitrogen applied during fallow (Anderson, 1991). Nitrogen applied during winter wheat's growing season increased downy brome growth and decreased wheat yield. When crop season rainfall was only 70% of normal (21 vs 62 mm), nitrogen fertilization reduced wheat yield 12–20%.

Competition for nutrients is not independent of competition for light and water. The complexity and opportunity of fertility manipulation is well illustrated in work by Liebman and Robichaux (1990). They demonstrated improved weed control because of differing nitrogen use efficiency of crops and weeds (Table 9.18). With no added nitrogen, total crop seed yield was identical for the long-vined Century or short-vined Alaska pea cultivars. Century's yield was 45% greater than Alaska's under these conditions. Adding nitrogen dramatically increased barley yield and reduced yield of Alaska peas. Barley can compete for the added nitrogen and Alaska cannot, but the latter cultivar does well with no added N. The seed yield of white mustard increased with nitrogen fertilization and it was much more competitive with short-vined Alaska than with long-vined Century peas.

Further evidence of the potential role of soil fertility in weed management is in studies done in Alabama (Hoveland *et al.,* 1976). Soils with low potassium were dominated by buckhorn plantain and curly dock. Soils with low soil phosphorus were dominated by showy crotalaria, morningglory, coffee senna, and sicklepod. Adjusting soil nutrient status can change weed populations and should be regarded as a potential weed management technique.

THINGS TO THINK ABOUT

1. Why is preventing weeds so difficult?
2. Why is eradicating weeds so difficult?
3. What are the advantages and disadvantages of each weed control method?

Table 9.18

Effect of Pea Cultivar and Nitrogen on Seed Yield and Final Above-ground Biomass of White Mustard in a Barley/Pea Intercrop (Liebman, 1989)

	Seed yield			Dry weight
N treatment/pea cultivar	Barley	Pea (g/m^2)	Total	White mustard (g/m^2)
No nitrogen				
Alaska	133(37)	230(63)	363	189
Century	16(5)	334(95)	350	105
90 + 90 kg/ha				
Alaska	262(79)	69(21)	331	1,766
Century	204(33)	406(67)	610	948

4. Why are perennial weeds so hard to control by mechanical methods?
5. What is the principle of carbohydrate starvation?
6. How does timing and type of tillage affect weed presence and weed control?
7. Can mowing really be used as a method of weed control? How?
8. Where could soil solarization be used?
9. How can living mulches and companion cropping be incorporated in modern cropping systems?
10. What role does crop rotation play in weed management?
11. What role can fertility manipulation play in weed management systems?

LITERATURE CITED

Akobundu, I. O. 1980a. Live mulch: A new approach to weed control and crop production in the tropics. Proc. Br. Crop Protect. Conf.—Weeds, pp. 377–380.

Akobundu, I. O. 1980b. Weed science research at the International Institute of Tropical Agriculture and research needs in Africa. *Weed Sci.* **28:**439–445.

Akobundu, I. O. 1982. The status and effectiveness of no-tillage cropping at the smallholder farmer level in the developing countries. Presented at the U.N./Food & Agric. Org. Expert Consultation on Weed Management Strategies for the 1980s in the Less Developed Countries, FAO, Rome, Italy, Sept. 33 pp.

Almeida, F. S., B. N. Rodrigues, and V. F. Oliveira. 1984. Influence of winter crop mulches on weed infestation in maize. Proc. Third Eur. Weed Res. Soc. Symp. on Weed Problems in the Mediterranean Area, pp. 351–358.

Altieri, M. A. 1984. The ecological role of weeds in agroecosystems with special emphasis on crop-weed-insect interactions. Proc. Third Eur. Weed Res. Soc. Symp. on Weed Problems in the Mediterranean Area, pp. 359–364.

Altieri, M. A., D. K. Letourneau, and J. R. Davis. 1983. Developing sustainable agroecosystems, *Bioscience* **33:**45–49.

Anderson, R. L. 1988. Kochia infestation levels in proso millet as affected by planing date. Res. Prog. Rpt. West. Soc. of Weed Sci., pp. 292–293.

Anderson, R. L. 1991. Timing of nitrogen application affects downy brome (*Bromus tectorum*) growth in winter wheat. *Weed Technol.* **5:**582–585.

Anderson, R. L. 1994. Characterizing weed community seedling emergence for a semiarid site in Colorado. *Weed Technol.* **8:**245–249.

Andres, L. A., and S. L. Clement. 1984. Opportunities for reducing chemical inputs for weed control, pp. 129–140 *in* "Organic Farming: Current Technology and Its Role in a Sustainable Agriculture" (D. F. Bexdicek and J. F. Power, Eds.). Am. Soc. Agron. Spec. Pub. No. 46. Am. Soc. Agron., Madison, Wisconsin.

Ascard, J. 1994. Dose–response models for flame weeding in relation to plant size and density. *Weed Res.* **34:**377–385.

Ascard, J. 1995. Effects of flame weeding in weed species at different developmental states. *Weed Res.* **35:**397–411.

Ateh, C. M., and J. D. Doll. 1996. Spring-planted winter rye (*Secale cereale*) as a living mulch to control weeds in soybean (*Glycine max*). *Weed Technol.* **10:**347–353.

Ball, D. A., and S. D. Miller. 1990. Weed seed population response to tillage and herbicide use in three irrigated cropping sequences. *Weed Sci.* **38:**522–527.

Barnes, J., and J. Barnes. 1960. "Weeds and Weed Seeds." Warrex Co. Chicago.

Barnes, J. P., and A. R. Putnam. 1983. Rye residues contribute weed suppression in no-tillage cropping systems. *J. Chem. Ecol.* **9:**1045–1057.

Barrett, M., and W. W. Witt. 1987. Alternative pest management practices, pp. 197–234 *in* "Energy in Plant Nutrition and Pest Control" (Z. R. Helsel, Ed.), Vol. 2, "Energy in World Agriculture." Elsevier, Amsterdam.

Brosten, D., and B. Simmonds. 1989. Crops gone wild. *Agrichemical Age,* May, pp. 6, 7, 26, 28.

Buchanan, G. A., and E. W. Hauser. 1980. Influence of row spacing and competitiveness on yield of peanuts (*Arachis hypogaea*). *Weed Sci.* **28:**401–409.

Buckley, N. G. 1980. No-tillage weed control in the tropics, pp. 12–21 *in* "Weeds and Their Control in the Humid and Subhumid Tropics" (I. O. Akobundu, ed.) Int. Inst. Trop. Agric. Proc. Ser. No. 3, Ibadan, Nigeria.

Buhler, D. D. 1992. Population dynamics and control of annual weeds in corn (*Zea mays*) as influenced by tillage systems. *Weed Sci.* **40:**241–248.

Buhler, D. D., and K. A. Kohler. 1994. Tillage in the dark and emergence of annual weeds. *Proc. North Central Weed Sci. Soc.* **49:**142.

Buhler, D. D., and E. S. Oplinger. 1990. Influence of tillage systems on annual weed densities and control in solid-seeded soybean (*Glycine max*). *Weed Sci.* **38:**158–165.

Buhler, D. D., D. E. Stoltenberg, R. L. Becker, and J. L. Gunsolus. 1994. Perennial weed populations after 14 years of variable tillage and cropping practices. *Weed Sci.* **42:**205–209.

Burnside, O. C. 1972. Tolerance of soybean cultivars to weed competition and herbicides. *Weed Sci.* **20:**294–297.

Burnside, O. C., R. G. Wilson, S. Weisberg, and K. G. Hubbard. 1996. Seed longevity of 41 weed species buried 17 years in eastern and western Nebraska. *Weed Sci.* **44:**74–86.

Callaway, M. B. 1992. A compendium of crop varietal tolerance to weeds. *Am J. Alternative. Agric.* **7:**169–180.

Cardina, J., and N. L. Hartwig. 1980. Suppression of crownvetch for no-tillage corn. *Proc. Northeast Weed Sci. Soc.* **34:**53–58.

CIAT (Centro Internacional de Agricultura Tropical). 1976. Annual Report for 1975. Cali, Colombia.

Christensen, S. 1995. Weed suppression ability of spring barley varieties. *Weed Res.* **35:**241–247.

Crutchfield, D. A., and G. A. Wicks. 1983. Effect of wheat mulch level on weed control in ecofarming corn production. *Weed Sci. Soc. Am. Abstr.,* 1.

DiTomaso, J. 1995. Approaches for improving crop competitiveness through manipulation of fertilization strategies. *Weed Sci.* **43:**491–497.

Donald, W. W. 1990. Primary tillage for foxtail barley (*Hordeum jubatum*) control. *Weed Technol.* **4:**318–321.

Dotzenko, A. D., M. Ozkan, and K. R. Storer. 1969. Influence of crop sequence, nitrogen fertilizer, and herbicides on weed seed populations in sugar beet fields. *Agron. J.* **61:**34–37.

Egley, G. 1983. Weed seed and seedling reductions by soil solarization with transparent polyethylene sheets. *Weed Sci.* **31:**404–409.

Fernandez-Quintanilla, C. L. Navarrete, and C. Torner. 1984. The influence of crop rotation on the population dynamics of *Avena sterilis* (L.) ssp. *ludoviciana* Dur., in Central Spain. Proc third Eur. Weed Res. Society Symp. on Weed Problems of the Mediterranean Area, pp. 9–16.

Fischer, A. J., J. H. Dawson, and A. P. Appleby. 1988. Interference of annual weeds in seedling alfalfa (*Medicago sativa*). *Weed Sci.* **36:**583–588.

Froud-Williams, R. J., R. J. Chancellor and D. S. H. Drennan. 1981. Potential changes in weed flora associated with reduced-cultivation systems for cereal production in temperate regions. *Weed Res.* **21:**99–109.

Ghafar, Z., and A. K. Watson. 1983. Effect of corn (*Zea mays*) seeding date on the growth of yellow nutsedge (*Cyperus esculentus*). *Weed Sci.* **31:**572–575.

Gill, N. T. 1938. The viability of weed seeds at various stages of maturity. *Ann. Appl. Biol.* **25:**447–456.

Gonzalez-Ponce, R. 1988. Competition between *Avena sterilis* ssp. *macrocarpa* Mo. and cultivars of wheat. *Weed Res.* **28:**303–307.

Groves, R. H. 1986. Plant invasions of Australia: an overview, *in* "Ecology of Biological Invasions: an Australian perspective" (R. H. Groves and J. J. Burdin, Eds.), pp. 137–149. Australian Acad. of Sci., Canberra.

Gunsolus, J. L. 1990. Mechanical and cultural weed control in corn and soybeans. *Am. J. Alternative Agric.* **5:**114–119.

Hartmann, K. M., and W. Nezadal. 1990. Photocontrol of weeds without herbicides. *Naturwissenschaften* **77:** 158–163.

Hartwig, N. P. 1987. Crownvetch and no-tillage crop production for soil erosion control. Pennsylvania State Univ. Coop Ext. Serv. 8 pp.

Heinen, E. T., and S. H. Ahmad. 1964. Water hyacinth control on the Nile river. Sudan Publ. Inf. Proc. Center, Dep. Agric. Khartoum. 56 pp.

Hoekstra, B. 1992. Killer lens blasts California's weeds. *New Scientist,* June 13, p. 21.

Hoffman, M. L., E. E. Regnier, and J. Cardina. 1993. Weed and corn (*Zea mays*) responses to a hairy vetch (*Vicia villosa*) cover crop. *Weed Technol.* **7:**594–599.

Holm, L. G., D. L. Plucknett, J. V. Pancho, and J. P. Herberger. 1977. "The World's Worst Weeds: Distribution and Biology." Univ. of Hawaii Press. Honolulu. 609 pp.

Horowitz, M. Weed research in Israel. 1980. *Weed Sci.* **31:**457–460.

Horowitz, M., Y. Regev, and G. Herzlinger. 1983. Solarization for weed control. *Weed Sci.* **31:**170–179.

Hoveland, C. S., G. G. Buchanan, and M. Harris. 1976. Response of weeds to soil phosphorus and potassium. *Weed Sci.* **24:**194–201.

IITA. Annual Report. 1980. Int. Inst. Trop. Agric., Ibadan, Nigeria. 185 pp.

Jordan, N. 1993. Prospects for weed control through crop interference. *Ecological Applications* **3:**84–91.

Khan, M., W. W. Donald, and T. Prato. 1996. Spring wheat (*Triticum aestivum*) management can substitute for diclofop for foxtail (*Setaria* spp.) control. *Weed Sci.* **44:**362–372.

Klein, R. N. 1988. Economics of winter wheat production. Nebraska Ecofarming Conf. **12:**4–47.

Kommedahl, T., and A. J. Link. 1958. The ecological effects of different preceding crops on *Setaria glauca* in flax. *Proc. Minn. Acad. of Sci.* **25–26:**91–94.

Liebl, R. A., D. G. Shilling, and D. A. Worsham. 1984. Suppression of broadleaf weeds by mulch in four no-till cropping systems. Abstr. 70 Div. Pest. Chem. 187th Mtg. Am. Chem. Soc., St. Louis.

Liebman, M. 1989. Effects of nitrogen fertilizer, irrigation, and crop genotype on canopy relations and yields of an intercrop/weed mixture. *Field Crops Res.* **22:**83–100.

Lyon, D. J., and D. L. Anderson. 1993. Crop response to fallow applications of atrazine and clomazone. *Weed Technol.* **7:**949–953.

Mack, R. N. 1981. Invasion of *Bromus tectorum* L. into western North America: An ecological chronicle. *Agro-ecosystems* **7:**145–165.

McCarty, M. K. 1982. Musk thistle (*Carduus thoermeri*) seed production. *Weed Sci.* **30:**441–445.

McWhorter, C. G., and E. E. Hartwig. 1972. Competition of johnsongrass and cocklebur with six soybean varieties. *Weed Sci.* **20:**56–59.

Mohler, C. L. 1991. Effects of tillage and mulch on weed biomass and sweet corn yield. *Weed Technol.* **5:**545–552.

Mohler, C. L. 1993. A model of the effect of tillage on emergence of weed seedlings. *Ecological Applications* **3:**53–73.

Monks, D. W., and L. R. Oliver 1988. Interactions between soybean (*Glycine max*) cultivars and selected weeds. *Weed Sci.* **36:**770–776.

Moody, K., and S. V. R. Shetty. 1981. Weed management in intercropping systems, pp. 229–237 *in* Int. Crops Res. Inst. for the Semi-Arid Tropics. Proc. Int. Workshop on Intercropping. Hyderabad, India.

Musick, G. J., and L. E. Beasley. 1978. Effect of the crop residue management system on pest problems in field corn (*Zea mays* L.) production, pp. 173–186 *in* "Crop Residue Management Systems," Amer. Soc. Agron. Spec. Publ. 31. Am. Soc. Agron., Madison, Wisconsin.

Muzik, T. J. 1970. "Weed Biology and Control," p. 41. McGraw-Hill, New York.

Norman, D. W. 1973. Crop mixtures under indigenous conditions in northern part of Nigeria, pp. 130–144 *in* "Factors of Agricultural Growth in Africa" (I. M. Ofori, Ed.). Inst. Soc. Econ. Res., Univ. Ghana, Ghana.

Norris, R. F., and D. Ayres. 1991. Cutting interval and irrigation timing in alfalfa: Yellow foxtail invasion and economic analysis. *Agron. J.* **83:**552–558.

Pablico, P., and K. Moody. 1984. Effect of different cropping patterns and weeding treatments and their residual effects on weed populations and crop yield. *Philipp. Agric.* **67:**70–81.

Panetta, F. D., and N. D. Mitchell. 1991. Homocline analysis and the prediction of weediness. *Weed Res.* **31:**273–284.

Penfound, W. T., and T. T Earle. 1948. The biology of the waterhyacinth. *Ecol. Monographs* **18:**447–472.

Phatak, S. C., R. L. Bugg, D. R. Sumner, J. D. Gay, K. E. Brunson, and R. B. Chalfant. 1991. Cover crops' effect on weeds, diseases, and insects of vegetables, pp. 153–154 *in* "Proc. Int. Conf. Cover Crops for Clean Water" (W. L. Hargrove, Ed.). Soil and Water Cons. Soc., Ankeny, Iowa.

Phillips, S. H. 1979. No-tillage, past and present, pp. 1–6 *in* "No-tillage Research" (R. E. Phillips, G. W. Thomas, and R. L. Blevins, Eds.). Res. Rpts. and Rev. Univ. Kentucky, Lexington.

Ramsel, R. E., and G. A. Wicks. 1988. Use of winter wheat (*Triticum aestivum*) cultivars and herbicides in aiding weed control in ecofallow corn (*Zea mays*) rotation. *Weed Sci.* **36:**394–398.

Roberts, H. A., and F. G. Stokes. 1965. V. Final observations on an experiment with different primary cultivations. *J. Appl. Ecol.* **2:**307–315.

Rose, S. J., O. C. Burnside, J. E. Specht, and B. A. Swisher. 1984. Competition and allelopathy between soybean and weeds. *Agron J.* **76:**523–528.

Schweizer, E. E., and R. L. Zimdahl. 1984. Weed seed decline in irrigated soil after six years of continuous corn (*Zea mays*) and herbicides. *Weed Sci.* **32:**76–83.

Schweizer, E. E., P. Westra, and D. W. Lybecker. 1994. Controlling weeds in corn (*Zea mays*) rows with an in-row cultivator versus decisions made by a computer model. *Weed Sci.* **42:**593–600.

Scopel, A. L., C. L. Ballare, and S. R. Radosevich. 1994. Photostimulation of seed germination during soil tillage. *New Phytol.* **126:**145–152.

Shaw, W. C. 1982. Integrated weed management systems technology for pest management. *Weed Sci.* **30** (Suppl.):2–12.

Shetty, S. V. R., and B. A. Krantz. 1980. Weed research at ICRISAT. *Weed Sci.* **28:**451–453.
Shetty, S. V. R., and A. N. Rao. 1981. Weed management studies in sorghum/pigeon pea and pearl millet/groundnut intercrop systems. Some observations, pp. 238–248 *in* Int. Crops Res. Inst. for Semi-Arid Tropics. Proc. Int. Workshop on Intercropping. Hyderabad, India.
Shurtleff, W., and A. Aoyagi. 1977. "The Book of Kudzu," pp. 8–12. Autumn Press, Brookline, Massachusetts.
Silveira, H. L., and M. L. V. Borges. 1984. Soil solarization and weed control, pp. 345–349 *in* Proc. 3rd Eur. Weed Res. Soc. Symp. on Weed Problems of the Mediterranean Area.
Slife, F. W. 1981. Environmental control of weeds, pp. 485–491 *in* "Handbook of Pest Management" (D. Pimentel, Ed.). CRC Press, Boca Raton, Florida.
Smith, R. J. 1974. Competition of barnyardgrass with rice cultivars. *Weed Sci.* **22:**423–426.
Smith. E. V., and E. L. Mayton. 1938. Nutgrass eradication studies. II. The eradication of nutgrass *Cyperus rotundus* L. by certain tillage treatments. *J. Am. Soc. Agron.* **30:**18–21.
Suryatna, E. S. 1976. Nutrient uptake, insects, diseases, labor use and productivity characteristics of selected traditional intercropping patterns which together affect their continued use by farmers. Ph.D. Thesis, Univ. Philippines, Los Baños College, Laguna, Philippines. 130 pp.
Tabor, P. 1952. Comments on cogon and torpedo grasses. A challenge to weed workers. *Weeds* **1:**374–375.
Triplett, G. B., Jr. 1982. Tillage and crop productivity, pp. 251–262 *in* "CRC Handbook of Agricultural Productivity," Vol. I (M. Recheigl, Jr., Ed.). CRC Press, Boca Raton, Florida.
U.S. Congress. 1993. Office of Technology Assessment, Harmful non-indigenous species in the United States. OTA-F-365. Washington, D.C. U.S. Govt. Printing Office. 391 pp.
VanGessel, M. J., E. E. Schweizer, D. W. Lybecker, and P. Westra. 1995. Compatibility and efficiency of in-row cultivation for weed management in corn (*Zea mays*). *Weed Technol.* **9:**754–760.
Vayssierre, P. 1957. Weeds in Indo-Malaya. *J. d'Agriculture Tropicale et Botanique Applique* **4:**392–401.
Vietmeyer, N. D. 1975. The beautiful blue devil. *Nat. History Mag.* **84:**64–73.
Vigneault, C., D. L. Benoit, and N. B. McLaughlin. 1990. Energy aspects of weed electrocution. *Rev. Weed Sci.* **5:**15–25.
Vrabel, T. E., P. L. Minotti, and R. D. Sweet. 1980. Seeded legumes as living mulches in sweet corn. *Proc. Northeastern Weed Sci. Soc.* **34:**171–175.
Walters, C. 1991. "Weeds: Control without Poisons." Acres U.S.A., Metairie, Louisiana. 352 pp.
Wesson, G., and P. F. Wareing. 1969. The induction of light sensitivity in weed seeds by burial. *J. Exp. Bot.* **20:**414–425.
Wicks, G. A. 1971. Influence of soil type and depth of planting on downy brome seed. *Weed Sci.* **19:**82–86.
Wicks, G. A., R. E. Ramsel, P. T. Nordquist, J. W. Schmidt, and Challaiah. 1986. Impact of wheat cultivars on establishment and suppression of summer annual weeds. *Agron. J.* **78:**59–62.
Wiese, A. F., J. F. Collier, L. E. Clark, and U. D. Havelka. 1964. Effects of weeds and cultural practices on sorghum yields. *Weeds* **12:**209–211.
Williams, M. C. 1980. Purposefully introduced plants that have become noxious or poisonous weeds. *Weed Sci.* **28:**300–305.
Williamsen, M. H., and K. C. Brown. 1986. The analysis and modeling of British invasions. *Phil. Trans. Royal Soc., London, Series B* **314:**505–522.
Wimschneider, W., and G. Bacthaler. 1979. Untersuchungen uber die Lichtkonkurrenz zwischen *Avena fatua* L. und verschiedenen Sommerweizensorten, pp. 249–256 *in* Proc. Eur. Weed Res. Soc. Symp. Influence of Different Factors on the Development and Control of Weeds, Mainz, FRG.

Chapter 10

Biological Weed Control

FUNDAMENTAL CONCEPTS

- Biological control is the action of parasites, predators, or pathogens to maintain another organism's population at a lower average density than would occur in their absence.
- Most biological control organisms have not escaped to become pests.
- Biological weed control cannot solve all weed problems and is best regarded as a technique to be used in integrated weed management systems.

LEARNING OBJECTIVES

- To know the advantages and disadvantages of biological weed control.
- To understand the importance of specificity in development of biological control strategies.
- To know the different kinds of organisms that have been used for biological control of weeds.
- To know the ways in which biological weed control can be used.
- To appreciate the opportunities for integration of biological and other weed control methods.

I. GENERAL

Plant distribution is determined by edaphic, climatic, and biotic factors. On a given site, soil type and climate can be discussed and studied, but not controlled by humans. The biotic environment can be manipulated. If manipulation is through stable interactions, biological control may be possible.

A. DEFINITION

Biological weed control is defined as the action of parasites, predators, or pathogens in maintaining another organisms's population at a lower average density than would occur in their absence. The term was first used by H. S. Smith (DeBach, 1964). Biological control is usually thought of as intentional introduction of parasites, predators, or pathogens to achieve control, but it is also a natural phenomenon. Scientists can discover the control potential of natural parasites, predators, or pathogens and exploit it to achieve human ends. The aim is to maintain the offending organism's population at a lower average density, not to eradicate it. The objective of biological control is not to eradicate, but to reduce populations to a noneconomic level. It is unlikely that biological control will ever be the solution to every weed problem. Biological control is properly employed as one weed management practice among many. It is likely that biological control of weeds will become more important relative to other control techniques, but it will never be the solution to all weed problems in intensive monocultural agriculture.

B. ADVANTAGES

In its classical or idealized form, biological weed control can be permanent weed management because once an organism is released, it may be self-perpetuating and control will continue without further human intervention (see Table 10.1.) This is true when some fungi are released in an inundative approach to control a weed. The weed need not be observed by people to be controlled, it just happens. This, ideal biocontrol, is not always achieved. Self-perpetuation is an advantage other weed control techniques do not have. There are no chemical environmental residues from biological control other than the organism, which some consider to be a potential pollutant because it is foreign or unnatural in the environment in which it is released. In the classical or idealized version of biological control, this does not happen because extensive research before its release establishes, one hopes, that the organism is environmentally benign. There is no environmental pollution from biocontrol organisms and no environmental or mammalian toxicity. In most cases, initial costs are nonrecurring

Table 10.1

Summary of the Advantages and Disadvantages of Biological Weed Control
(Wapshere *et al.*, 1989)

Advantages	Disadvantages
1. Reasonably permanent.	1. Control is slow.
2. Self-perpetuating.	2. No guarantee of results.
3. No additional inputs required once agent is established successfully.	3. Establishment may fail for many reasons.
4. No harmful side effects.	4. There may be unknown ecological effects. Mutation to an undesirable form is possible.
5. Attack is limited to target weed and a few close relatives.	5. If target is related to a crop, the number of potential biocontrol agents is low.
6. Risks are known and evaluated before release.	6. Some risks may not be known and cannot be evaluated.
7. Control often dependent on host density.	7. Does not work well in short-term cropping cycles. Works best in stable environments.
8. Self-dispersing spread to suitable host habitats.	8. Restriction of spread to area of initial dispersal is impossible.
9. Costs are nonrenewing.	9. Initial investment of time, money, and personnel can be very high.
10. High benefit: cost ratio for successful programs.	10. Eradication is not possible. Must maintain host population at low level to maintain control agent.

and usually, once the organisms are established, no further inputs are needed. Development costs have been much less than those for herbicides (Auld, 1991). Although not all of these advantages accrue to all organisms developed for biological weed control, they are cited commonly to justify research and greater employment of biological control.

C. Disadvantages

There are some situations where biological control is not appropriate. If a plant is a weed in one place and valued in another place, in the same general geographic region, biological control is inappropriate (see Table 10.1). Spread of a biological control organism, once it has been introduced, cannot be controlled. The control organism cannot always distinguish valuable plants from weedy relatives. For example, artichoke thistle (also called cardoon) is a rangeland weed on some California rangeland. It is closely related to cultivated artichoke. Introduction of a biocontrol agent to control the weedy artichoke thistle is discouraged by artichoke growers because the biocontrol agent would lack specificity. There are ornamental species of delphinium related to weedy larkspur. Other weedy species may be related to valuable native plants. The absolute demand for specificity in a biocontrol agent means development is research intensive: It requires a large budget and several years of research.

Biological control is inherently slow and results are not guaranteed. In many crops, weeds must be controlled during a brief, critical period, often within days or weeks, to prevent yield reduction. Biological control does not achieve rapid results. In addition, because eradication is not an appropriate goal for biological control, weeds that should be eradicated on some sites (e.g., larkspur on rangeland) may be better controlled with other techniques. There are some species that are geographically localized minor weeds and development of a biological control for them would be very expensive. Cropland weeds exist in an ecologically unstable habitat that is often a poor environment for successful introduction, survival, and population growth of biocontrol organisms. Cropland weeds also exist in a weed complex, rarely as a single species. Because biocontrol is necessarily directed at a single species, it is often an inappropriate choice for the weed complex found in most crops. Projects are often constrained by the expense of finding a natural enemy in the native habitat. Locating the natural or native habitat is a difficult research task and, even if found, natural enemies may not be abundant.

Because science can never know all possible ramifications of any technological intervention, other cautions are in order. Release of a biological control organism can induce competitive suppression or extinction of native biological control organisms and other organisms. A corollary is that other harmful or beneficial species may increase in abundance. Such events could lead to loss of biological diversity, loss of existing biocontrol, release of species from competitive regulation, disruption of plant community structure, suppression of essential organisms, and disruption of food chains and nutrient cycling (Lockwood, 1993).

In summary, biocontrol is slow, often less effective, and commonly less certain than herbicides or mechanical control. Biocontrol, particularly in disturbed cropping situations, will not control as many different weeds as other techniques will. It will not eradicate weed problems, but neither will most other techniques. Biocontrol is an intervention technique that may, as herbicides do, have unanticipated effects.

D. USE CONSIDERATIONS

Conscious use of biological control of weeds depends on two things. The first is that it is usually, but not always, easier to control an introduced species that, in the process of introduction, was freed of natural predators. The second requirement is that it is best to introduce predators that have been freed of their natural predators during introduction to the weed's area. These requirements presuppose, and successful biological control depends upon, several assumptions:

(a) The weed to be controlled has a native habitat. Redroot pigweed, groundsel, common lambsquarters, or common chickweed are distributed worldwide, and their origin is unknown. If the native habitat is unknown, one cannot go to it to find a predator. Some suggest that many weeds are homeless, having evolved from diverse parentage under various kinds of human-created agricultural pressure (Ghersa *et al.,* 1994). Their homes cannot be known.

(b) An insect or disease will give control, that is, an effective natural enemy can be found. The assumption is questionable because many plants may not have effective natural enemies. If an effective natural enemy is found, whether it is an insect or a disease, it is assumed that it will thrive in the weed's habitat.

(c) The organism has fecundity or the ability to reproduce in the new habitat, and it will occupy all niches the weed (the host) infests.

(d) The weed's genetic composition in its new home is identical to its (now distant) relatives in its old home. In other words, moving has not changed the weed in any significant way. This validity of this assumption has not been determined.

(e) The intended control organism can be reared in captivity so it will not be necessary to import large quantities, which of course may not be possible.

(f) After the organism is released it will search out the weed to be controlled and will be self-dispersing in the right places.

Each of these assumptions is questionable, and organisms proposed for biocontrol of weeds often fail because one or more of these assumptions are false. Mistakes have been made when all of the complexity was not understood. It is easy to make mistakes when a biocontrol agent is introduced. Each introduction creates a new combination of organism and environment. Both must be understood, and often they are not (U.S. Congress, 1993). Scientists have used the concept of the vacant niche to rationalize introductions. The concept holds that some ecological roles (population-regulating organism) are not filled in a place where biological control is desired; thus, the niche is empty and can be filled. Few species fit the narrow ecological vacancy identified by those who wish to control weeds, and it is virtually impossible to predetermine the role a species will play after release (U.S. Congress, 1993).

There are several examples of poor understanding. The mongoose (*Marathi mangus*) was imported to Hawaii to control rats that reduced yield and made sugarcane harvest unpleasant. A mongoose kills rats it meets. However, rats are nocturnal and the mongoose hunts during the day. They never met! The mongoose ate bird eggs, had no natural enemies in Hawaii, and became a pest. Problems can arise when an introduced species moves beyond the area intended. The cactus moth (*Cactoblastis cactorum*) was introduced to the West Indies to control prickly pear cactus, a task it does well. It has moved north to Florida where, it is feared, it may threaten indigenous, nonweedy prickly pear cacti, 16 species of which are rare (U.S. Congress, 1993). The seven-spotted ladybeetle (*Coccinella septempunctata*), an aphid predator, has dispersed throughout much of the United States. It appears to be outcompeting the native nine-spotted ladybeetle (*C. novemnotata*) and has displaced that species in alfalfa. Finally, the US/EPA and the Oregon Department of Environmental Quality funded a large project to eradicate weeds in Devils Lake on the Northern Oregon coast. About 30,000 weed-eating carp (*Ctenopharyngodon idella*), a successful aquatic weed-control agent, were introduced into the lake to control Eurasian watermilfoil. The liquefied fecal waste from the fish created new, unprecedented algal blooms and new weed crops. Six years after the project was initiated, there was no significant reduction in the total amount of aquatic vegetation, but only 4,000 carp still survived. Intensive real estate development in the lake's pristine watershed, clear-cut logging, and recreation proceeded without inhibition and were all major contributors to the lake's pollution and eutrophication (Larson, 1996). We are residents of the world, not its custodians or rulers. We must learn to understand nature's purposes and our role in aiding or defeating them. Biological weed control, similar to other technology, can lead us toward harmony with nature or away from it.

Scientists must determine the place of biological weed control in nature's scheme. Important control questions include the following:

(a) Will the insect or disease organism remain free of its old predators and not be subject to new ones in its new habitat? Will the imported, potential biological control agent find the neighborhood in which it must live to be a congenial one? This may be a reason some potentially good biological control agents are abandoned: They meet too many new enemies in their new home.

(b) The most important criterion and the absolute rule for successful biological control is that if an insect or disease is able to clear all the aforementioned hurdles, it must be specific. Specificity means that it will attack and control one plant (the weed) and no others. This is the acid test for biocontrol agents. It would be a tragedy if a biocontrol agent were released to control a particular weed and it was discovered after the weed's population was reduced that the biocontrol organism had a natural appetite for rosebushes.

Only a very few of the more than 100 organisms released for biocontrol of weeds worldwide have become pests subsequent to their release. Biological control research is difficult and crucial to success. Plants in the weeds host range that are tested to assure specificity include (Strobel, 1991):

1. Those related to the target weed.
2. Those not adequately exposed to the agent for ecological or geographic reasons.
3. Those for which little is known about their natural enemies.
4. Those with secondary chemicals or morphological structures similar to those of the target weed.
5. Those attacked by close relatives of the agent.

About 40% of the successful instances of pest biocontrol have involved an unrelated natural enemy. These were new associations between a host and biocontrol agent, and the host lacked all natural resistance to the new enemy (Pimentel, 1963, 1991). The real risk in biological control is not in finding an introduced species or in assuring that the proposed biocontrol agent clears all the hurdles. The risk is in misunderstanding the nature of host specificity. Not enough is known about how natural enemies find and control weed hosts. Why do they do it?

In addition to the fundamental biological questions, there are questions those who develop to sell must ask (Auld, 1991). These include concern about the size and stability of the market and what competing products there may be. Manufacturers must also be concerned about the ability to patent a product to protect their investment and create a reasonable guarantee of profit. Finally, they must ask what is known about the organism and how much it will cost to develop a biological control agent (Auld, 1991).

Given the advantages and disadvantages of biological weed control, there are, and will continue to be, conflicting interests when biological organisms are used for weed control. A plant which is a pest in one place may be beneficial in another place, or at least it may be liked. The spread of an organism once it has been released cannot be controlled. Future and present values must be considered as well as minority and majority interests, neighboring nations, and direct and indirect effects on other species and the environment (Huffaker, 1964).

A few examples will illustrate the complexity (Huffaker, 1964). Prickly pear is an example of the success of biological control of a weed by an insect. A moth borer, found in Argentina, was released in Australia. The prickly pear area was transformed, as if by magic, from a wilderness of 60 million acres of prickly pear to prosperous agricultural land. No one in Australia objected. In Hawaii, there were vigorous objections to introduction of the same moth borer. Cattlemen objected because the tree cactus was useful as feed and as a source of otherwise unavailable water on some ranges. The program was also opposed on the U.S. mainland because of similar sentiments in Mexico and in parts of the United States.

In California, control of yellowstar thistle involves cattlemen, beekeepers, fruit growers, and seed crop growers. The weed damages grazing land, grain, and seed crops. Cattlemen, those primarily affected, want to get rid of it. However, the thistle is a key plant in maintenance of the bee industry for pollination of fruit and seed crops in California. The fruit and seed crop industry has dominated the debate and biological control for yellowstar thistle has not been introduced.

Wood-boring insects are important for control of mesquite because trees infested with wood borers are easier to burn, a primary control technique (Ueckert and Wright, 1973). Defoliating mesquite with the herbicide 2,4,5-T caused the wood-boring insects to die and resulted in trees that were more difficult to burn. Control techniques can conflict even when each is designed to accomplish the same end.

Table 10.2 is a list of a few weeds for which biological control efforts have been established and a few others that show promise for the future. The list is included to show the scope of current efforts; it could be much longer.

II. METHODS OF APPLICATION

There are four methods of applying biological control agents: one theoretical and three that are used (Wapshere *et al.,* 1989; Turner, 1992).

A. CLASSICAL, INOCULATIVE, OR IMPORTATION

This method has been limited to weeds that are not closely related to crop plants and that belong to sharply defined genera or families that, theoretically, are taxonomically well separated from other families (Wapshere *et al.,* 1989). Classical biological control is the introduction of host-specific, exotic, natural enemies adapted to introduced weeds. The great majority of weeds and nearly all the worst weeds have been introduced.

Table 10.2

Partial List of Present and Potential Biological Control Programs

Weed	Biocontrol agent	Type of agent
Established biocontrol agents		
Musk thistle	*Rhinocyllus conicus*	Beetle
Hydrilla	*Ctenopharyngodon idella*	Fish-grass carp
St. John's wort	*Chrysolina quadrigemina*	Beetle
Prickly pear	*Dactylopius opuntiae*	Cochineal scale
Tansy ragwort	*Longitarsus jacobaeae*	Beetle
Potential weeds for biocontrol		
Russian knapweed	*Subanguina picridis*	Nematode
	Aceria acroptiloni	Mite
Field bindweed	*Aceria malherbe*	Beetle
Leafy spurge	*Aphthona cyparissae*	Beetle
Dodder	*Smicronyx roridus*	Weevil
Orobanche	*Fusarium oxysporum*	Fungus
	Phytomyza orobanchia	Diptera
Puncturevine	*Microlarinus lypriformis*	Beetle

Source: Turner (1992); Julien (1992). Julien (1992) is a world list of 729 releases of exotic agents for control of weeds from the late 19th century to 1992. It records the place(s) of release and known results.

When the weeds were introduced to a new region or country, they were freed of natural enemies that regulated their population effectively. Often they are innocuous species in their native place. This is an ecological approach. The introduced target weeds often occur on undisturbed rangelands or in infrequently disturbed habitats (e.g., a pasture or perennial crop). Classical control works best in habitats with minimal disturbance from man. It is the most used and most successful method.

All steps enumerated earlier are followed. These include weed identification, identification of native habitat, searching for and importing a natural enemy, research on rearing, specificity, etc., and ultimate release. The method is only appropriate with highly specific natural enemies. Arthropods and fungal pests are first choice because they may be specific. Vertebrate animals are usually nonspecific feeders and not suitable for importation (Turner, 1992). The weeds targeted for classical control have almost always been economically important and no other control has been successful, or their range has expanded to areas where it is not economical to control them with available methods (e.g., puncturevine, Russian thistle, diffuse knapweed, and spotted knapweed).

B. Augmentative or Inundative

When large numbers of control agents are raised and released, their abundance is augmented and the area is inundated with them. Releases can be single or repeated throughout a season. The control and target organisms are usually natives. Inundative control employs ecological knowledge but is essentially technological. The method eliminates costly international searches for a weed's native habitat and an organism suitable for import. It augments the inherent phytotoxicity of organisms by abruptly increasing their population. Biological control is made effective in a short time, perhaps even in an annual crop's season. Specificity, however, must be guaranteed. The best agents must be amenable to large-scale captive rearing and have a reproductive method that allows rapid population increase. This requirement alone has inhibited this method. A stable, but easily changed, resting or spore stage is helpful. Organisms used for inundation have been pathogens or nematodes rather than arthropods, which do not satisfy the aforementioned criteria. A *Cochineal* scale is redistributed each year in some areas to control prickly pear, a natural process that has been going on a long time. The conscious use of inundative techniques by man is relatively recent. The natural process is a result of evolution and is reflected in the balance of nature.

C. Conservation

If the number of native parasites, predators, and diseases of native plants could be conserved or protected and thereby increased, they should be more effective and might give control. This theoretical concept rests on the assumption that if the population of organisms that prey on an organism with biological control potential could be reduced, the potential agent could fulfill its potential. It is the same principle involved in importation, but the approach is different. For example, the insect *Aroga websteri* eats foliage of big sagebrush. It has not been exploited for biological control, but presumably could be increased in its natural habitat.

D. Broad-Spectrum

Broad-spectrum control involves artificial manipulation of a natural enemy's population so a weed is controlled. Whole habitats rather than just a target weed have been modified with this technique. Ecological appropriateness and effectiveness and the organism's virulence are not as important because they can be changed by the population or the stocking rate of the control agent. Safety and specificity are less important for the same reason. The best example is use of selectively grazing animals. Fences or shepherds are required and expenses are high, but control is possible.

III. BIOLOGICAL CONTROL AGENTS

Biological control of weeds began after its use to control insects. It began in the United States in Hawaii in 1902, when eight fruit- and flower-feeding insects were introduced from Mexico to control largeleaf lantana, a perennial shrub native to Central America. It is used throughout the world as an ornamental and escaped to become a weed (Goeden, 1988; Huffaker, 1964). Many early biocontrol efforts emphasized insects that bored in roots, stems, or seed. Boring provides avenues for secondary infection by bacteria and fungi, and boring insects are usually host-specific. Early efforts also emphasized agents that destroyed flowers in contrast to those that fed only on foliage. Experience has shown that leaf-eaters may be just as safe and equally effective. Now many organisms other than insects are used for biological control of weeds (Andres, 1966; Geoden *et al.,* 1974; Holloway, 1964). A summary of 73 biological control agents approved for 26 species and several other potential agents is available for weeds in the western United States (Rees *et al.,* 1996).

A. CLASSICAL OR INOCULATIVE BIOLOGICAL CONTROL

1. Insects

Classical biological control has been used for many years. The earliest record of biological weed control was the release of the cochineal insect *Dactylopius ceylonicus* from Brazil to northern India in 1795 to control prickly pear cactus. (Goeden, 1988). Actually, the insect was not identified correctly and was believed to be a species that produced carmine dye (Goeden, 1988). It readily transferred to its natural host plant and was subsequently introduced in southern India from 1836 to 1838, where it successfully controlled prickly pear cactus. Shortly before 1865, the insect was transferred to Sri Lanka and accomplished the same purpose. This was the first successful transfer of a natural enemy between countries for biological weed control (Goeden, 1988).

A classical example of biological control of prickly pear cactus was introduction of *D. opuntiae* in 1951 to Santa Cruz island off the coast of southern California. It is a good example of successful biological control of a native United States weed with introduced insects (Goeden and Ricker, 1980). Over many years the insect has given partial to complete control of prickly pear (Goeden and Ricker, 1980; Geoden *et al.,* 1967).

A second example of weed control by an insect is the use of the French chrysomelid leaf beetle *Chrysolina quadrigemina* for control of St. John's wort. The beetle's success is due to its great specificity and the synchronization of its requirements with St. John's wort's growth. Adult beetles strip the plant at flowering in spring and early summer, and larvae feed in fall and winter (Huffaker and Kennett, 1959). It is widely distributed in the world's temperate zones and has been associated with sheep movement. St. John's wort is susceptible to herbicides, but their cost and the inaccessibility of infested rangeland were problems. It has been used successfully in many of the western United States, and introduced to British Columbia where the beetle has adapted to colder winters (Peschken, 1972).

Biological control has been successful in Washington and Oregon against a poisonous, biennial weed of rangeland, tansy ragwort. Two insect species were imported from Europe (Pemberton and Turner, 1990). A cinnabar moth (*Tyria jacobaeae*) attacks leafy and flowering shoots, and larvae of the ragwort flea beetle (*Longitarsus jacobaeae*) attack the roots. These have reduced the weed to less than 1% of its density before their introduction (Turner, 1992).

Control of puncturevine in California and Colorado is one of the few victories over an annual weed (Turner, 1992). Two weevils, the seed feeder *Microlarinus lareynii* and the stem and crown feeder *Microlarinus lypriformix,* were introduced from Italy beginning in 1961 (Maddox, 1976). The weevils work best where the climate is warm and they do not overwinter well in cold climates (Turner, 1992).

Research has been conducted on several insects for control of leafy spurge. Seven different insects have been released in the U.S. The leafy spurge hawkmoth (*Hyles euphorbiae*) imported from Austria, Hungary, and India eats leaves and flowers during the caterpillar stage (Harris *et al.,* 1985). A root and stem boring beetle (*Obera erythrocephala*) imported from Hungary and Italy was established in Montana and North Dakota (Leininger, 1988). The beetles puncture stems and lay eggs. Larvae bore into roots where they mature and exist on carbohydrate root reserves. Four species of chrysomelid flea beetles, *Apthona nigriscutis, flava, cyparissae,* and *czwalinea,* were imported to the United States from Europe. Adult *Apthona* beetles live up to 3 months and feed on leaves. Adults females lay an average of 250 eggs on stems. Larvae bore into stems and cause extensive damage by feeding on primary and secondary roots and root hairs.

Another chrysomelid beetle was imported from Argentina to Florida in 1965 for the aquatic alligatorweed and successfully controlled it (Coulson, 1977). Alligatorweed was introduced in the United States about 1894 from South America in ship ballast and had infested nearly 70,000 acres in the southern United States by the 1960s. Impressive control has been achieved, but the insect's success is influenced by temperature, rate of water flow, other plants, water nutrition, and plant vigor. The weed's population has been reduced wherever the beetle has been introduced.

A weevil from southern Germany (*Rhinocyllus conicus* Froelich) was introduced to Canada in 1968 and to West Virgina in 1969 for musk thistle control. The adult weevils are dark brown with small yellow spots on their back and are only $\frac{3}{16}$ to $\frac{1}{4}$ inch long. After feeding and mating on thistles, females lay eggs on the bracts of developing flowers in late spring. The larvae hatch, bore into the base of the flower receptacle, and prevent development of some or all seed. It takes a large number of larvae to completely destroy seed production. Because musk thistle is a biennial, a key to its control is prevention of seed production. They produce seed for 7 to 9 weeks and the average plant produces 4,000 seeds. Egg laying is favored by hot, humid weather and late flowers may not be affected. Use of *Rhinocyllus* has been successful, with up to 90% control on some pasture sites where plant competition provided additional stress, but it is not a complete control for musk thistle. This weevil, unfortunately, may be a bad case of biological control and an exception to the statement that biological controls rarely escape to become pests. Because it attacks other thistles, not just musk thistle, it can move to other species, including the endangered Sacramento thistle in New Mexico. Rhinocyllus may have a fatal flaw for a good biological control—it is not host-specific.

A major effort is underway to find biological control insects for melaleuca and Brazilian pepper in Florida. Over 200 insects that feed on melaleuca have been found in Australia, its natural habitat, and are being tested. Each of these weeds is an aggressive non-indigenous plant that has replaced a large natural plant community (U.S. Congress, 1993; Langeland, 1990).

B. Inundative or Augmentative

1. Fungi

An endemic anthracnose disease of Northern jointvetch, a grassy weed in rice and soybeans in the southeastern United States, is increased and the weed is controlled by application of a dry, powdered formulation of the fungus *Colletotrichum gloeosporioides* (Penz) Sacc. f. sp. *aeschymonene* as a mycoherbicide (trade name-COLLEGO™, Ecogen, Inc.). Daniel *et al.* (1973) introduced the concept of mycoherbicide (Wilson, 1969; TeBeest, 1991). It is possible to spray the formulated fungal spores on rice infested with Northern jointvetch (Daniel *et al.*, 1973). After a 4- to 7-day incubation period, Northern jointvetch dies in 5 weeks. The fungus is specific and can be produced in large quantities in artificial cultures, and the cultures are infective in the field. The phenoxyacid herbicide can do the same job in 2 weeks, so the fungus is slower. The fungus must be sprayed annually and is used only when there is a problem. Introduction does not permanently increase its population level.

Bioherbicides are biological control agents applied as chemical herbicides are. The active ingredient is a living microorganism, applied in inundative doses. The organisms is commonly a fungus and its propagules are spores or fragments of mycelium; in this case the bioherbicide is often called a mycoherbicide. A mycoherbicide has been successful for control of stranglevine in citrus orchards after application of live chlamydospores of *Phytophthora palmivora* (Butl.) Butl. It has been marketed as Devine™ (Abbott Laboratories). Live chlamydospores germinate 6 to 10 hours after application on a wet soil surface. The fungus initiates a root infection that kills stranglevine in 2 to 10 weeks, depending on the vine's size and vigor when Devine is applied. Complete control may not be obtained in 1 year, but the fungus persists and is effective for up to 5 years, which may be a disadvantage for those selling the product. Drift to susceptible plants, including cucumber, squash, watermelon, rhododendron, begonia, and snapdragon, is a potential problem. In addition to its persistence and effect on other species, it rapidly loses viability after preparation; it must be treated like fresh milk. The mycoherbicide is made to order for each user. Refrigerated storage and distribution are required and it cannot be stored for use in another year.

Another mycoherbicide, BioMal™, has been registered in Canada for control of common mallow. There are good reasons that none of these products is presently commercially available (Auld, 1995). The most important reason may be that herbicides have been so successful for control of each targeted weed. Equally important is the fact that each product targeted a specific weed and that inevitably made its market small. Other reasons include the fact that it may be difficult to mass-produce the infective agent and formulate it so it can be applied. Application of mycoherbicides normally requires a long dew period that is difficult to obtain in dry climates. Specificity is the essence of success for biological control agents, but it may lead to commercial failure because weeds usually exist in complex communities. Removal of one weed with a specific biocontrol agent creates a situation where others, released from competition, flourish. It is possible, but it has not been easy, to obtain adequate patent protection so developers can recover development costs and make a profit on bioherbicides.

Fusarium oxysporum f. sp. *cannabis* could provide safe, efficient control of marijuana (McCain, 1978). In inoculation studies and in nature, only marijuana has been infected. All marijuana types tested were susceptible, and cultivars grown only for hemp were resistant. Inoculum for field use can be grown efficiently on mixtures of barley straw combined with alfalfa or soybean oil meal. Inoculum spread at 10 kg/ha resulted in 50% mortality of seeded marijuana. Three-quarters of subsequent marijuana plantings died. The fungus causes disease over a wide temperatue range, and once a field is infested, marijuana cannot be grown for many years. There is no known danger from the fungus to man, animals, or other plants.

A potentially more important application of *Fusarium oxysporum* is control of witchweed, one of the worlds worst parasitic weeds. It is considered by many to be the greatest constraint to food production in Africa, particularly in the sub-Saharan region. (Additional information on witchweed can be found in Chapter 3.) *Fusarium* species from West Africa, grown on sorghum straw, have successfully prevented all emergence of witchweed and increased sorghum dry weight as much as 400%. In growth chambers, the fungus inhibited germination and attachment of witchweed to sorghum roots (Ciotola *et al.*, 1995).

The most extensively studied group of plant pathogens is the fungal genus *Colletotrichum*. *C. coccodes* isolated from Eastern black nightshade (Anderson and Walker, 1985) did not kill velvetleaf, but another isolate did (Wymore *et al.*, 1988). Other strains of the fungus kill tomatoes and potatoes, but this strain is harmless to all crops tested. It causes disease on velvetleaf over a wide range of dew periods and temperatures, but is most effective after a 24-hour dew period at 75°F (Wymore *et al.*, 1988).

Several strains of the rust *Puccinia chondrillina* have been tested for control of rush skeletonweed (Lee, 1986) to find one for importation and, subsequently successful, use in Australia and dry Mediterranean areas. A strain of *P. chondrilla* was released successfully in California in 1976 and spread to Oregon in 2 years (Lee, 1986). It illustrates use of an exotic plant pathogen (a rust) for biological control. The rust has controlled skeletonweed successfully and is specific.

Puccinia punctiformis, another rust fungus, is an obligate parasite specific to Canada thistle (Cummins, 1978), and infection can lead to death. Infection reduces flowering and vegetative reproduction (Thomas *et al.*, 1995).

Wilson (1969) described principles for control of weeds with phytopathogens. The first is that host resistance is the primary deterrent to success and may often restrict disease to insignificant levels. Weeds usually have

several, rarely fatal, disease lesions on their foliage. Natural weed populations resist insects and diseases because of climate and soil variability and
the presence of natural enemies. Disease susceptibility is the exception
rather than the rule. Disease epidemics result from importation of new
diseases or more virulent strains rather than mere presence. These principles, while generally true, may fail in specific cases. Weed scientists have
isolated, cultured, and redistributed local pathogens such as the aforementioned anthracnose disease to achieve weed control. Further work in this
area for terrestrial and aquatic weeds (Zettler and Freeman, 1972) offers
great promise. A 1982 review of biological control with plant pathogens
reported 4 projects with bacteria, 42 with fungi, 3 with nematodes, and 6
with viruses (Charudattan and Walker, 1982). The book by TeBeest (1991)
has many more.

Phytopathogenic bacteria have not been considered to have good potential as biological agents because in spite of their known activity, they do
not penetrate plants well. This deficiency has been overcome by combining
bacteria with surfactants or a cultural operation that injures plants such as
mowing. Spray application of *Pseudomonas syringae* in an aqueous buffer
with a surfactant produced severe disease in several members of the Asteraceae, including Canada thistle (Johnson *et al.,* 1996). Spray application
without surfactant failed to produce disease in any plants. *Xanthomonas
campestris* pv. *poannua* controlled several annual bluegrass biotypes in
bermudagrass golf greens when it was sprayed during mowing, but not
when it was applied without mowing. Mowing injured the grass and allowed
the bacteria to enter and cause lethal systemic wilt (Johnson *et al.,* 1996).
This technique may lead to further development of bacterial herbicides.
They are not obligate biological agents, but they do not persist and so
may escape the disadvantage of lack of specificity. They must be applied
annually. They also have an advantage over fungi because a dew period
(wet period) is not required to activate them.

C. Broad-Spectrum

1. Fish

The white amur or grass carp (*Ctenopharyngodon idella* Valenciennes)
is an herbivorous fish native to the Amur River of China and the Soviet
Union. It can consume 3 to 5 pounds per day of aquatic plants (especially
hydrilla) and adults may weigh 70 to 100 lbs. It does not spawn in warm
water, so it is possible to control its population (Van Zon, 1984). The grass
carp breeds only in large rivers or canals with high water volume and

The white amur or grass carp (an herbivorous fish) eats aquatic vegetation.

velocity. It also feeds on grass and other terrestrial vegetation. Subsequent work has discovered a way to ensure production of sterile fish. Federal researchers have tried to cross the white amur and the big-head carp to produce a voracious weed-eating hybrid. There are 240,000 miles of irrigation canals, ditches, and drains in the 17 western United States, and many have aquatic weeds. An advantage of fish is that they may be harvested for food; if the fish are sterile, their population should be controllable and they should not threaten other species.

Resistance to introduction of the grass carp or its hybrids centers around their potential to cause problems similar to those that occurred after introduction of the common carp. These include degradation of water quality due to the carp's bottom feeding, which disturbs sediments and muddies the water, and crowding out of desirable fish species because of the carp's rapid population growth in the absence of natural enemies. A single female grass carp may produce up to a million eggs, and therefore, research has emphasized sterility in released populations.

There is concern when a fish is introduced about whether it will prefer and eat selectively the weeds those who introduce it wish to control. Will the fish consider the same species desirable that a Department of Natural Resources or biological control effort considers undesirable?

2. Aquatic Mammals

The sea manatee (*Trichechus* spp.) eats cattails and waterhyacinth and can weigh more than a ton. It is not discriminatory in its diet and eats many kinds of aquatic vegetation.

It is said that sailors may have seen the sea manatees with their fish tail and thought they were mermaids. If you see one, you may think the sailors had a little too much grog. The animal really looks a bit like former President Grover Cleveland because it is fat, has whiskers, and thick, wrinkled skin.

The sea manatee eats cattails, water hyacinth, and other aquatic vegetation.

Manatees have cleared up to half a mile of canal and banks of a major aquatic waterway in Florida in 3 weeks. In Florida they had no natural enemies other than man until early 1996, when more than 100 of about 2600 remaining manatees died from a poisonous phytoplankton known as red tide. Manatees breath oxygen and often lie just below the surface of the water, where they are hit by propellers of speedboats. They cannot, or will not, reproduce in fresh water.

3. Vertebrates

Sheep and goats graze plants that cattle will not eat. Sheep eat leafy spurge and can be used to remove it selectively. Goats relish shrubby species and eat more than sheep, but they are difficult animals to manage. Goats can be used as a follow-up to mechanical treatments and have killed root sprouts of gambel oak. Goats prefer oak over other plants and do not compete with cattle for forage. There are problems with goats and sheep. Sheep- and goat-proof fencing is expensive, as is herding. They must be removed when they have eaten 90 to 95% of the weedy foliage, or they will compete with cattle. Goats, because they are aggressive grazers that eat many things, can destroy wildlife habitat.

Geese, ducks, and chickens have been used to weed strawberries, raspberries, and some vegetables. They will selectively remove grasses and small broadleaved weeds without crop damage. Chickens and geese selectively control nutsedge in several crops. They are not selective in grass crops. Experiments have shown advantages for geese for weed control during establishment of tree seedlings (Wurtz, 1995). Geese feed almost exclusively on grasses and broadleaved weeds, whereas chickens are omnivorous and eat weed seedlings, seeds, insects, and soil invertebrates (Clark and Gage, 1996). Chickens do not affect weed abundance or crop productivity, but geese are more effective weeders. The problem with geese is they are picky and do not eat all weeds. Species unpalatable to geese, such as curly dock and daisy fleabane, increase in abundance (Clark and Gage, 1996).

IV. INTEGRATION OF TECHNIQUES

Successful and sustainable weed management systems are those that employ combinations of techniques rather than relying on one. Biological control is easy to combine with other methods because, once established, it can be self-perpetuating. To be successful, an integrated system requires a thorough knowledge of the ecology of the weed–crop system. Knowledge of a farmer's production goals and farming system is necessary but not sufficient. When the goal is weed management, thorough ecological understanding is required (Wapshere *et al.,* 1989). Successful weed management means that the weed population will be reduced and maintained at a non-economic level. When control is the only aim, rougher techniques can be employed that require less biological knowledge and management skill.

The fungus *Cochliobolus lunatus* is endemic on barnyardgrass (Scheepens, 1987). It has potential as a biological control agent but does not have sufficient activity alone to kill barnyardgrass. It has been successful when combined with sublethal doses (a dose that will not control the weed) of the herbicide atrazine. Under appropriate conditions the fungus produces leaf necrosis and kills seedlings with less than 2 leaves. Plants with more than 2 leaves recover, although their growth is slowed. It can be used successfully in beans, barley, corn, oats, rye, tomatoes, and wheat. Combination of the fungus with sublethal doses of atrazine enhances control over the fungus or atrazine alone (Table 10.3). This is especially true as the weed gets older.

The success of the Chrysolina beetle for control of St. John's wort has already been mentioned. A successful, integrated system for pastures was developed in Australia (Campbell, 1979) and combined the beetle (biological control agent), a herbicide, and use of plant competition through reseeding in areas where the weed's population had been reduced. On arable land a combination of mechanical and cultural control techniques was integrated with biological control. Land is plowed in summer to expose and dry roots, then cultivated in late summer to continue drying and to prepare for seeding an improved pasture mixture. Adequate fertilization is required to guarantee the seeding's success.

Table 10.3

Effect of *Cochliobolus lunatus* and Atrazine on Barnyardgrass in a Growth Chamber (Scheepens, 1987)

| | % necrosis after 9 days | | |
Treatment	22-day old plants	30-day old plants	47-day old plants
Untreated	0	0	0
C. lunatus	60 ± 21	60 ± 18	15 ± 9
Atrazine at 40 g/A	60 ± 19	60 ± 19	3 ± 3
C. lunatus + Atrazine at 40 g/A	100	100	75 ± 13

On nonarable land, five techniques are combined. In addition to the beetle, heavy grazing by sheep or cattle is used to remove plants that shade the weed. This is followed by spraying with 2,4-D, planting the proper pasture mixture with adequate fertilizer to take full advantage of plant competition, and well-managed, light grazing and additional fertility to maximize the crop's advantage and competitive pressure on the weed. These methods seem so obvious that one is inclined to say, "Of course, that is what whould be done." However, if these or similar methods are tried in other environments and cropping systems, they might fail unless the ecological relationships have been analyzed and understood. Ecological understanding leads to selection of the best combination of techniques to manage the weed population, rather than the best method to obtain a quick kill but no long-term reduction of the weed's population. Chapter 19 has additional examples of integrated weed management.

THINGS TO THINK ABOUT

1. What applications are there for biological control?
2. Why has biological control not been used more widely?
3. What are some good examples of successful biological control of a weed?
4. What is a mycoherbicide and how are they used?
5. Where are vertebrate animals used for biological weed control?
6. How can biological control be integrated with other methods?
7. What are the economic advantages of biological control?
8. Compare and contrast the advantages and disadvantages of biological weed control.

LITERATURE CITED

Anderson, R. N., and H. L. Walker. 1985. *Colletotrichum coccodes:* a pathogen of eastern black nightshade (*Solanum ptycanthum*). *Weed Sci.* **33:**902–905.

Andres, L. A. 1966. The role of biological agents in the control of weeds. Symp. Pest Control by Chemical, Biological, Genetic, and Physical Means, ARS Pub. No. 33-110, pp. 75–82.

Auld, B. A. 1991. Economic aspects of biological weed control with plant pathogens, pp. 262–273. "Microbial Control of Weeds" *in* (D. O. TeBeest, Ed.). Chapman and Hall, New York.

Auld, B. A. 1995. Constraints in the development of bioherbicides. *Weed Technol.* **9:**638–652.

Campbell, M. H. 1979. St. John's Wort. New South Wales Dept. of Agriculture. Agdex 642. New South Wales, Australia. 16 pp.

Charudattan, R. and H. L. Walker (Eds.). 1982. Biological control of weeds with plant pathogens. Wiley, New York. 293 pp.

Ciotola, M., A. K. Watson, and S. G. Hallett. 1995. Discovery of an isolate of *Fusarium oxysporum* with potential to control *Striga hermonthica* in Africa. *Weed Res.* **35:**303–309.

Clark, M. S. and S. H. Gage. 1996. Effects of free-range chickens and geese on insect pests and weeds in an agroecosystem. *American J. Alternative Agric.* **11:**39–47.

Coulson, J. R. 1977. Biological control of alligatorweed, 1959–1972. A review and evaluation. *U.S. Dept. Agric. Tech. Bull* **1547:**98.

Cummins, G. B. 1978. "Rust Fungi on Legumes and Composites in North America," p. 138. Univ. Arizona Press, Tucson.

Daniel, J. T., G. E. Templeton, R. J. Smith, Jr., and W. T. Fox. 1973. Biological control of northern jointvetch in rice with an endemic fungal disease. *Weed Sci.* **21:**303–307.

DeBach, P. 1964. The scope of biological control. pp. 1–20. *In* "Biological Control of Insect Pests and Weeds" (P. DeBach, Ed.) Reinhold, New York.

Ghersa, C. M., M. L. Roush, S. M. Radosevich, and S. M. Cordray. 1994. Coevolution of agroecosystems and weed management. *Bioscience* **44:**85–94.

Goeden, R. D. 1988. A capsule history of biological control of weeds. *Biocontrol News and Information* **9:**55–61.

Goeden, R. D. and D. W. Ricker. 1980. Santa Cruz island—revisited. Sequential photography records the causation, rates of progress, and lasting benefits of successful biological weed control, *in* Proc. V Int. Symp. Biol. Contr. Weeds, Brisbane, Australia, pp. 355–365.

Goeden, R. D., C. A. Fleschner, and D. W. Ricker. 1967. Biological control of prickly pear cacti on Santa Cruz Island, California. *Hilgardia* **38**(16):579–606.

Goeden, R. D., A. Andres, T. E. Freeman, P. Harris, R. L. Peinkowski, and C. R. Walker. 1974. Present status of projects on biological control of weeds with insects and plant pathogens in the United States and Canada. *Weed Sci.* **22:**490–495.

Harris, P., P. H. Dunn, D. Schroeder, and R. Vormos. 1985. Biological control of leafy spurge in North America, pp. 79–82. *in* "Leafy Spurge Monograph No. 3. (A. K. Watson, Ed.). Weed Sci. Soc. America, Champaign. Illinois.

Holloway, J. K. 1964. Projects in biological control of weeds. Chapter 23 *in:* "Biological Control of Insect Pests and Weeds" (P. DeBach, Ed.). Reinhold, New York.

Huffaker, C. B. 1964. Fundamentals of biological weed control. Chapter 22 *in* "Biological Control of Insect Pests and Weeds" (P. DeBach, Ed.). Reinhold, New York.

Huffaker, C. B. and C. E. Kennett. 1959. A ten-year study of vegetation change associated with biological control of Klamath weed. *J. Range Management* **12:**69–82.

Johnson, D. R., D. L. Wyse, and K. J. Jones. 1996. Controlling weeds with phytopathogenic bacteria. *Weed Technol.* **10:**621–624.

Julien, M. H. (Ed.). 1992. "Biological Control of Weeds: A World Catalogue of Agents and Their Target Weeds," 3rd. ed. C. A. B. Int., U.K.

Langeland, K. 1990. Exotic woody plant control. Circular 868. Florida Coop. Ext Serv. Univ. of Florida, Gainesville. 16 pp.

Larson, D. W. 1996. Curing the incurable. *Am. Scientist* **84:**7–9.

Lee, G. A. 1986. Integrated control of rush skeletonweed (*Chondrilla juncea*) in the western U.S. *Weed Sci.* **34**(Suppl.):2–6.

Leininger, W. C. 1988. Non-chemical alternatives for managing selected plant species in the western United States. Colo. State Univ. Ext. Ser. Pub. No. XCM-118. 40 pp.

Lockwood, J. A. 1993. Environmental issues involved in biological control of rangeland grasshoppers (Orthoptera: Acrididae) with exotic agents. *Env. Entomol.* **22:**503–518.

Maddox. D. M. 1976. History of weevils on puncturevine in and near the United States. *Weed Sci.* **24:**414–416.

McCain, A. H. 1978. The feasibility of using Fusarium wilt to control marijuana. *Phytopath. News* **12:**129.

Pemberton, R. W. and C. E. Turner. 1990. Biological control of *Senecio jacobea* in northern California, an enduring success. *Entomophaga* **35:**71–77.

Peschken, D. P. 1972. *Chrysolina quadrigemina* (Coleoptera: Chrysomelidae) introduced from California to British Columbia against the weed *Hypericum perforatum:* comparison of behavior, physiology, and colour in association with post-colonization adaptation. *Can. Entomol.* **104:**1689–1698.

Pimentel, D. 1963. Introducing parasites and predators to control native pests. *Can. J. Ent.* **95:**785–792.

Pimentel, D. 1991. Diversification of biological control strategies in agriculture. *Crop Prot.* **10:**243–253.

Rees, N. E., P. C. Quimby, Jr., G. L. Piper, E. M. Coombs, C. E. Turner, N. R. Spencer, and L. V. Knutson (Eds.). 1996. "Biological Control of Weeds in the West." Montana State Univ., Bozeman, Montana.

Scheepens, P. C. 1987. Joint action of *Cochliobolus lunatus* and atrazine on *Echinochloa crusgalli* (L.) Beauv. *Weed Res.* **27:**43–47.

Strobel, G. 1991. Biological control of weeds. *Sci. Am.* July:72–78.

TeBeest, D. O. (Ed.). 1991. "Microbial Control of Weeds." Chapman and Hall, New York. 284 pp.

Thomas, R. F., T. J. Tworkoski, R. C. French, and G. R. Leather. 1995. *Puccinia punctiformis* affects growth and reproduction of Canada thistle (*Cirsium arvense*). *Weed Technol.* **8:**488–493.

Turner, C. E. 1992. "Beyond Pesticides: Biological Approaches to Weed Management in California," pp. 32–67. Div. Agric. and Nat. Res. Univ. Of Calif. Albany, California.

U.S. Congress, Office of Technology Assessment. 1993. Harmful Non-indigenous species in the United States, OTA-F-565. Washington, D.C.: U.S. Govt. Printing Office.

Ueckert, D. N. and H. A. Wright. 1973. Wood boring insect infestations in relation to mesquite control practices. *J. Range Management* **27:**383–386.

Van Zon, J. C. J. 1984. Economic weed control with grass carp. *Tropical Pest Man.* **30:**179–185.

Wapshere, A. J., E. S. Delfosse, and J. M. Cullen. 1989. Recent developments in biological control of weeds. *Crop Prot.* **8:**227–250.

Wilson, C. L. 1969. Use of plant pathogens in weed control. *Ann. Rev. Phytopath.* **7:**411–434.

Zettler, F. W. and T. E. Freeman. 1972. Plant pathogens as biocontrols of aquatic weeds. *Ann. Rev. Phytopath.* **10:**455–470.

Wymore, L. A., A. K. Watson, and A. R. Gotlieb. 1987. Interaction between *Colletotrichum coccodes* and thidiazuron for control of velvetleaf (*Abutilon theophrasti L.*) *Weed Sci.* **35:**377–382.

Wurtz, T. L. 1995. Domestic geese: Biological weed control in an agricultural setting. *Ecological Applications* **5:**570–578.

Chapter 11

Introduction to Chemical Weed Control

FUNDAMENTAL CONCEPTS

- Herbicides created a major change in the way agriculture is practiced by substituting chemical energy for human and animal energy.
- Herbicides have several advantages and disadvantages that must be considered prior to use.
- Herbicides can be classified in several useful ways, but no way integrates the entire subject.
- Classification based on chemical structure and mode of action is common.

LEARNING OBJECTIVES

- To understand the history of chemical weed control.
- To know and understand the advantages and disadvantages of herbicides.
- To understand the different ways of classifying herbicides and the use of each classification system.

Versatility in vegetation management has been extended through the use of selective agricultural chemicals. Herbicide comes from the Latin *herba,* or plant, and *caedere,* to kill. Therefore, herbicides are chemicals that kill plants. A better definition is that a herbicide is a chemical that disrupts the physiology of a plant over a long enough period to kill it or severely reduce its growth. Pesticides are chemicals used to control pests. Herbicides are pesticides used to control weeds. They are different from other pesticides because their sphere of influence extends beyond their ability to kill or control weeds. Herbicides change the chemical environment of plants, which can be more readily manipulated than the climatic, edaphic, or biotic environment.

Herbicides reduce or eliminate labor and machine requirements and modify crop production techniques. Herbicides, when used appropriately, are production tools that increase farm efficiency, reduce horsepower, and perhaps reduce energy requirements. Herbicides do not, of course, eliminate energy requirements because they are petroleum-based.

Understanding the nature, properties, effects, and uses of herbicides is essential if one is to be conversant with modern weed management. Weed management is not accomplished exclusively by herbicides, but they dominate in the developed world. Whether one likes them or deplores them, they cannot be ignored. To ignore them is to be unaware of the opportunities and problems of modern weed management. Ignoring or dismissing herbicides may lead to an inability to solve weed problems in many agricultural systems and may delay development of better weed management systems.

I. HISTORY OF CHEMICAL WEED CONTROL

A. THE BLOOD, SWEAT, AND TEARS ERA

Agriculture can be thought of as having three eras. The first might be characterized as the blood, sweat, and tears era. Famine and fatigue were common and inadequate food supplies occurred all too frequently. Most people were farmers, and many farms were small and operated at a subsistence lvel. In the words of the British philosopher Thomas Hobbes (1588–1679), life

> wherein men live without other security, than their own strength, and their own invention shall furnish them. . . . In such conditions there is . . . no knowledge of the face of the earth; no account of time; no arts; no letters, no society; and which is worst of all, continual fear and danger of violent death; and the life of man, solitary, poor, nasty, brutish, and short.

B. THE MECHANICAL ERA

The mechanical era of agriculture began with the invention of labor-saving machines. In 1793, Eli Whitney invented the first workable cotton gin. In 1834, Cyrus McCormick invented the reaper and began manufacture in 1840. John Deere perfected the steel moldboard plow in 1837. In 1830, 4 farmers in the United States supported 5 nonfarmers. In 1910 a farmer fed himself and 6 others. By 1930, each farmer was able to support 10 nonfarmers. One farmer supported 40 nonfarmers in 1965. Much of this increase in a farmer's productivity was due to agricultural mechanization and improved technology.

C. THE CHEMICAL ERA

The third era of agriculture, the chemical era, boosted production again. This occurred primarily after 1945, when fertilizers and pesticides were developed and became widely available. In 1992, about 1% of U.S. citizens were farmers (about 2.8 million) and each farmer fed 128 others (94.3 Americans and 33.7 people in other countries) (Krebs, 1992). These changes are not unique to American agriculture. In 1938, Britain employed a million people to produce a third of the food needed for a nation of 48 million. In 1988 only 450,000 British farmers and farm workers produced three-quarters of the food for 58 million people (Malcolm, 1993). Production per British agricultural worker increased at about twice the rate of increase for the rest of the economy (Malcolm, 1993). Less than 3% of the population of Germany works on farms. Increases in crop production and labor productivity with each new agricultural era were caused by extensive farm mechanization, the use of agricultural chemicals, increased education of farmers, improved crop varieties, and improved farming practices. Developed country agriculture is now in the era of extensive and intensive use of chemical fertilizers and pesticides.

The chemical era of agriculture developed rapidly after 1945, but it did not begin then. In 1000 B.C. the Greek poet Homer wrote of pest-averting sulfur. Theophrastus, regarded as the father of modern botany (372?–287? B.C.), reported that trees, especially young trees, could be killed by pouring oil, presumably olive oil, over their roots. The Greek philosopher Democritus (460?–370? B.C.) suggested that forests could be cleared by sprinkling tree roots with the juice of hemlock in which lupine flowers had been soaked. In the first century B.C., the Roman philosopher Cato advocated the use of amurca, the watery residue left after the oil is drained from crushed olives, for weed control (Smith and Secoy, 1975).

History tells us of the sack of Carthage by the Romans in 146 B.C. and that salt was used on the fields to prevent crop growth. Later, salt was used as a herbicide in England. These examples show that herbicidal chemicals have been used in agriculture for a long time, but their use was sporadic, frequently ineffective, and lacked any scientific base (Smith and Secoy, 1975, 1976).

In 1755, mercurous chloride ($HgCl_2$) was used as a fungicide and seed treatment. In 1763, nicotine was used for aphid control. As early as 1803, copper sulfate was used as a foliar spray for diseases. Copper sulfate (blue vitriol) was first used for weed control in 1821. In 1855, sulfuric acid was used in Germany for weed control in cereals and onions. In 1868, Paris green (copper acetoarsenite) was used for control of the Colorado potato bettle (*Leptinotarsa decemlineata*). The U.S. Army Corps of Engineers used sodium arsenite in 1902 to control waterhyacinth in Louisiana.

Bordeaux mixture, a combination of copper sulfate, lime, and water, has been applied to grapevines for the control of downy mildew. Someone in Europe in the late 19th century noted that it turned yellow charlock leaves black. That led Bonnet, in France in 1896, to show that a solution of copper sulfate would selectively kill yellow charlock plants growing with cereals. In 1911, Rabaté demonstrated that dilute sulfuric acid could be used for the same purpose. The discovery that salts of heavy metals might be used for selective control led, in the early part of the 20th century, to research by the Frenchmen Bonnett, Martin, and Duclos, and the German Schultz (cited in Crafts and Robbins, 1962). Nearly concurrently, in the United States, Bolley (1908) studied iron sulfate, copper sulfate, copper nitrate, and sodium arsenite for selective control of broadleaved weeds in cereal grains. Bolley, a plant pathologist who worked in North Dakota, is widely acknowledged as the first to report on selective use of salts of heavy metals as herbicides in cereals. The action was caustic or burning with little, if any, translocation. Succeeding work in Europe observed the selective herbicidal effects of metallic salt solutions or acids in cereal crops (Zimdahl, 1995). The important early workers were Rabaté in France (1911, 1934), Morettini in Italy (1915), and Korsmo in Norway (1932).

Use of the inorganic herbicides developed rapidly in Europe and England, but not in the United States. In fact, weed control in cereal grains is still more widespread in Europe and England than in the United States. Some of the reasons for slow development in the United States include lack of adequate equipment and frequent failure to obtain weed control because the heavy metal salts were dependent on foliar uptake that did not readily occur in the low humidity of the primary U.S. grain-growing areas. The heavy metal salts worked well only with adequate rainfall and high relative humidity. There were other agronomic practices such as increased use of fertilizer, improved tillage, and new varieties that increased U.S. crop yield without weed control. U.S. farmers could always move on to the endless frontier and were not as interested, as they would be later, in yield enhancing technology.

Carbon bisulfide was first used in agriculture in 1854 as an insecticide in France. It was applied as a soil fumigant in Colorado to control *Phylloxera,* a root-borne disease of grapes. In 1906, it was introduced as a soil fumigant for control of Canada thistle and field bindweed. It smells like rotten eggs and may have reached its peak usage in Idaho in 1936 when over 300,000 gallons were used.

Petroleum oils were introduced for weed control along irrigation ditches and in carrots in 1914. They are still used in some areas for weed control. Field bindweed was controlled successfully in France in 1923 with sodium chlorate, which is now used mainly as a soil sterilant in combination with organic herbicides. Arsenic trichloride was introduced as a product called KMG (kill morningglory) in the 1920s. Sulfuric acid was used for weed control in Britain in the 1930s. It was and still is a very good herbicide, but is very corrosive to equipment and harmful to people.

The first synthetic organic chemical for selective weed control in cereals was 2-methyl-4,6-dinitrophenol (DN or dinitro), introduced in France in 1932 (cited in King, 1966, p. 285). It was used for many years for selective control of some broadleaved weeds and grasses in large seeded crops such as beans. Dithiocarbamates were patented as fungicides in 1934. In 1940, ammonium sulfamate was introduced for control of woody plants.

Future historians of weed science will note 1941 as an important year in herbicide history. Pokorny (1941) first synthesized (2,4-dichlorophenoxy) acetic acid (2,4-D). It had no activity as a fungicide or insecticide. Accounts vary about when the first work on growth-regulator herbicides was done (Akamine, 1948). Zimmerman and Hitchcock (1942) of the Boyce-Thompson Institute (formerly in Yonkers, NY, now at Cornell University, Ithaca, NY) first described the substituted phenoxy acids (2,4-D is one) as growth regulators, but did not report herbicidal activity. They also worked with other compounds that eventually became herbicides. They were the first to demonstrate that these molecules had physiological activity in cell elongation, morphogenesis, root development, and parthenocarpy (King, 1966). A Chicago carnation grower's question, "What is the effect of illuminating gas[1] on carnations?" led to the work and eventual discovery of plant growth regulating substances by Boyce-Thompson scientists (King, 1966).

E. J. Kraus was Head of the University of Chicago Botany Department and had worked with plant growth regulation for several years. He had supervised the doctoral programs of J. W. Mitchell and C. L. Hamner who in the early 1940s were working as plant physiologists with the U.S. Department of Agriculture Plant Industry Station at Beltsville, MD. Kraus thought these new, potential plant growth regulators which often distorted

[1]Acetylene gas.

plant growth when used at higher than growth-regulating doses and even killed plants, might be used beneficially. He thought they could be used to selectively kill plants. He saw potential use as chemical weed killers or herbicides and was the first to advocate purposeful application in toxic doses for weed control. Because of World War II, much of this work was done under contract from the U.S. Army related to its potential for biological warfare against the enemy's crops (Peterson, 1967). Similar work for similar reasons was done in Great Britain (Kirby, 1980).

Hamner and Tukey (1944a and b) reported the first field trials with 2,4-D for successful selective control of broadleaved weeds. They also worked with 2,4,5-T as a brush killer. At nearly the same time, Slade *et al.* (1945), working in England, discovered that naphthaleneacetic acid at 25 lbs/acre would selectively remove charlock from oats with little injury to oats. They (Slade *et al.,* 1945) also discovered the broadleaved herbicidal properties of the sodium salt of MCPA (later called Methoxone; King, 1966), a compound closely related to 2,4-D. Slade *et al.* (1945) confirmed the selective activity of 2,4-D. Marth and Mitchell (1944), former students of E. J. Kraus, first reported the differential use of 2,4-D for killing dandelions and other broadleaved weeds selectively in Kentucky bluegrass turf. Marth and Mitchell (1944) attribute the quest for selective activity of these compounds to Kraus. These discoveries were the beginning of modern chemical weed control. All previous herbicides were just a prologue to the rapid development that occurred following discovery of the selective activity of the phenoxyacetic acid herbicides. The first U.S. patent (No. 2,390,941) for 2,4-D as a herbicide was obtained by F. D. Jones of the American Chemical Paint Co. in 1945 (King, 1966). There had been an earlier patent (No. 2,322,761) in 1943 of 2,4-D as a growth-regulating substance (King, 1966). It is interesting to note that Jones patented only its activity and made no claim about selective action (King, 1966).

The effectiveness of monuron for control of annual and perennial grasses was reported by Bucha and Todd (1951). This was the first of many new selective chemical groups with herbicidal activity. The first triazine herbicide appeared in 1956 and the first acylanilide in 1953 (Zimdahl, 1995), followed by CDAA, the first alphachloroacetamide, in 1956 (Hamm, 1974).

The great era of herbicide development came at a time when world agriculture was involved in a revolution of labor reduction, increased mechanization, and new methods to improve crop quality and produce higher yields at reduced cost. Herbicide development built on and contributed to agriculture's change. Agriculture was ready for improved methods of selective weed control. It is true that no weed control method has ever been abandoned; instead, new ones have been added as the relative importance of methods changed. The need for cultivation, hoeing, etc., has not disap-

Table 11.1

The Evolution of Weed Control Methods in the United States
(Alder et al., 1977)

	% Control by year in U.S.			
Year	Human energy	Animal energy	Mechanical energy (tractor)	Chemical energy
1920	40	60		
1947	20	10	70	
1975	5	TR[a]	40	55
1990	<1	TR	24	75

[a]TR = trace.

peared. These methods persist in small-scale agriculture (e.g., I hoe my garden). They have become less important in development world agriculture because of the rising costs of labor and narrower profit margins (Table 11.1).

Rapid development of herbicides occurred after WW II. Today, there are over 180 different selective herbicides in use in the world and several experimental herbicides in some stage of progress toward marketability (Hopkins, 1994). If proprietary labels are considered, there may be more than 1,000 chemical and biological compounds used for pest control in the world (Hopkins, 1994). More than 1 billion pounds of pesticides are sold in the United States each year, and over 50% of that amount is herbicides (Table 11.2). The global herbicide market was estimated to be $13.5 billion from 1990 to 1993, and one-third ($4.5 billion) was the United States market alone. Japan was the next largest with $1.5 billion in sales. When the

Table 11.2

World Sales of Crop Protection Products, 1960 to 1990, with 2000 Estimated in Billions of Dollars (Hopkins, 1994)

Pesticide	1960	1970	1980	1990	2000
Herbicides	160	918	4,756	12,600	16,560
Insecticides	288	945	3,944	7,840	9,360
Fungicides	320	702	2,204	5,600	7,560
Other	32	135	696	1,960	2,520
TOTAL	800	2,700	11,600	28,000	36,000

European market is considered as a whole, it is second largest, with France ($1.250 billion) the largest single country (Hopkins, 1994).

In 1990, for the world, about 45% of total pesticide sales volume was herbicides, insecticides were 28%, and fungicides approximately 20% of total sale volume (Hopkins, 1994). Over 85% of herbicides are used in agriculture. The large, worldwide market is becoming increasingly concentrated in the hands of a few multinational corporations. About 90% of the international pesticide market is controlled by about 12 companies and, at this writing, about half of the market is shared by five companies, four of which are American. Nearly half the companies in pesticide discovery in 1994 were Japanese (Hopkins, 1994).

II. ADVANTAGES OF HERBICIDES

Any successful technology should create gains in value that are reproducible. Table 11.3 shows energy relationships for weed control in corn in Minnesota (Nalewaja, 1984), and Table 11.4 shows similar data for cotton in Georgia (Dowler and Hauser, 1975). The weed density in corn was low, but it was high in cotton. In both cases the cost–benefit analysis favored herbicides over other methods. Hand labor gave the greatest energy output/input ratio. The data do not consider the energy to house and feed workers or the fact that such work is seasonal (Barrett and Witt, 1987). Soil was plowed, disked, and prepared for planting corn or cotton in the conventional manner and all other cultural practices were uniform. With no weed control,

Table 11.3

Energy Relationships in Weed Control in Six Experiments on Corn in Minnesota (Nalewaja, 1984)

Method of weed control	Energy for weed control (MJ/ha)	Corn yield kg/ha MJ/ha	Net profit from weed control $/ha MJ/ha	Energy output/input ratio	Man-hour/ha
None	0	3,387 56,528	— — —	0	
Cultivation	579	5,080 84,387	151 27,550	48/1	1.41
Herbicide	391	5,645 93,763	194 37,113	95/1	0.12
Hand labor	337	5,770 —	163 39,251	116/1	148.15

[a]The land was plowed, disked, and prepared for planting of corn in the conventional manner. Calculations are based on an average of 2.5 cultivations using 2.83 liter gasoline/ha and spraying using 0.84 liter gasoline/ha. Herbicide was 3.36 kg/ha of atrazine.

Table 11.4

Energy and Cost for Weed Control in Cotton (Dowler and Hauser, 1975; Barrett and Witt, 1987)

Weed control method	Energy for weed control (MJ/ha)	Cotton yield			Ratio of energy in crop to weed control input	Values $/ha		
		Lint kg/ha	Seed kg/ha	Energy MJ/ha		Cost of Weed control	Income from lint and seed	Income above cost of weeding
Four herbicides, no cultivation	2,093	619	856	28,310	13/1	64.47	901.11	836.64
Three herbicides, 2 cultivations	1,898	545	754	24,935	13/1	56.96	860.91	803.95
No herbicides, 5 cultivations	1,220	177	244	8,082	7/1	24.69	257.43	232.74
No herbicides, 5 cultivations, 185 man-hour/ha hand hoeing	1,641	592	819	27,085	17/1	645.06	862.02	216.97

Note. Adapted by Barrett and Witt (1987) from Nalewaja (1984) and Dowler and Hauser (1975). Hoeing cost was estimated to be $91 for a 40-hour week.

there was, of course, no profit due to weed control and a low yield and crop value. Cultivation and herbicides were not very different but hand labor produced a net loss because of its high cost and poor weed control.

Herbicide energy efficiency is reinforced by data from the same study showing the energy relationships for methods of weed control. Herbicides consume more energy than hand labor but less than cultivation. Herbicides compare favorably to other methods in net energy profit due to weed control because yield was nearly as high as that achieved with hand labor (Table 11.5).

The total energy required for herbicides for corn, a crop that requires a great deal of energy, is relatively small (Table 11.6) (Primentel and Pimentel, 1979). About 3% of the total energy input for the U.S. corn production system is directly related to herbicides. The major energy consumers for corn production are nitrogen fertilizer, diesel fuel, and irrigation. It is true that the U.S. corn production system has one of the lowest energy efficiencies among the world's crop production systems (Table 11.7). the data show that U.S. energy efficiency is low, but yield is high.

An important criterion for a grower is profit or return on investment in technology. Becker (1983, cited in Barrett and Witt, 1987) attributed corn and soybean yield increases after herbicide use to improve weed control and earlier planting when herbicides were available. Combining these two factors reduced cost of production and thus increased profit about 10% (Becker, 1983) (Table 11.8). Abernathy (1981) calculated the additional land required to maintain production of seven major U.S. crops without herbicides. He used estimates from several sources to determine likely losses and their value. All aspects of loss were considered, including addi-

Table 11.5

Weed Control Energy Relationships for Six Corn Experiments in Minnesota (Nalewaja, 1974)

Method of weed control	Energy input for weed control (kcal/A)	Yield of corn/A		Net profit due to weed control (kcal/A)
		Bushels	Kilocalories	
None	0	54	5,443,200	—
Cultivation	56,005	81	8,164,800	2,665,595
Herbicide	37,920	90	9,072,000	3,590,880
Hand labor	32,655	92	9,273,600	3,797,745

Note. The land was plowed, disked, and prepared for planting of corn in the conventional manner.

Table 11.6

**Average Energy Input to U.S. Corn Production
System in 1975 (Pimentel and Pimentel, 1979)**

Input	% of total (kcal/ha)
Labor	0.09
Machinery	8.9
Diesel fuel	19.6
Nitrogen	28.8
Phosphorus	3.3
Potassium	2.0
Lime	0.5
Seed	8.0
Irrigation	11.9
Insecticide	1.3
Herbicide	3.1
Drying	6.5
Electricity	5.8
Transportation	0.5
kcal output/kcal input	2.93

tional cultivations required to control weeds when herbicides are not used. Abernathy (1981) proposed a net loss greater than $23 billion and the need for an additional 28.22 million acres to maintain production. Pimentel *et al.* (1978) also estimated costs and losses with herbicide use and with alternative

Table 11.7

**Energy and Yield Comparisons of Corn Production Systems (Pimentel and
Pimentel, 1979)**

Corn production system	Total kcal output/input	Yield kg/ha	T/ha
United States	2.93	5,394	2.4
Philippines w/animal power	5.06	941	0.42
Mexico w/oxen	4.34	941	0.42
w/manpower	10.74	1,944	0.87
Nigeria w/human labor	6.41	1,004	0.45

Table 11.8

Cost–Benefit Assessment of Herbicide Use in Corn and Soybean (Barrett and Witt, 1987)

	Corn (cost in $/ha)		Soybean (cost in $/ha)	
	Cultural weed control	Herbicide	Cultural weed control	Herbicide
Yield (kg/ha)	7,212	8,179	2,554	2,084
Herbicide + application cost	0	43.51	0	50.32
Savings in tillage cost	0	9.70	0	7.68
Total herbicide cost and tillage savings	0	33.80	0	44.64
Total crop production				
Costs	799.75	833.59	587.65	632.30
Cost per kg produced	0.11	0.10	0.23	0.21

Data also available in Becker 1983.

methods of weed control, but reached a different conclusion. The primary reason for the difference is that Pimentel *et al.* (1978) assumed that with careful management, little additional crop loss (only $341 million) would occur by switching from herbicides to alternative weed management techniques. More intensive weed management is proposed frequently as a necessary part of alternative (i.e., nonherbicidal) weed management.

No one knows who is correct in this ongoing debate. Clements *et al.* (1995) provide a clue. They confirm the proposition put forth earlier (see Table 11.6) that energy for weed management represents a small proportion of on-farm energy use for food production. Clements *et al.* (1995) propose that a "large portion of energy allocated to weed control could be conserved in alternative weed management systems by elimination or reduction strategies for tillage and/or herbicide use." They showed that potential energy savings from reduction or elimination of tillage was greater than for elimination of herbicides. It was also suggested that there would be a potentially high energy requirement for tillage when herbicides are eliminated, particularly when numerous inter-row cultivations are required for weed control. Most alternative methods of weed management are more energy efficient than methods based on herbicides (Clements *et al.*, 1995). Energy savings are being achieved by alternative herbicide use techniques such as reducing the total area of application, using band application instead of broadcast spraying, and choosing herbicides that require less energy to produce (e.g., trifluralin or atrazine as opposed to paraquat or bentazon; Clements *et al.*, 1995).

It is often argued, by those in agriculture, that the purpose of agriculture is to produce food, not energy. Others argue that the U.S. system is so dependent on petroleum energy sources that it is not sustainable. In any case, many believe that modern weed control with herbicides is essential to maintain the present U.S. agricultural system and is justified because herbicides represent only a small part of the total energy input.

Herbicides are not only beneficial and profitable where labor is scarce or expensive. They may be advantageous where labor is plentiful and cheap. Herbicides control weeds in crop rows where cultivation is not possible. They can be used in places where other methods do not work. Preemergence herbicides provide early season weed control when competition results in the greatest yield reduction and when other methods are less efficient or impossible to use (e.g., it is impossible to mechanically cultivate when soil is wet).

Cultivation can injure crop roots and foliage. Selective herbicides reduce the need for tillage and control of weeds in crop rows where tillage is not effective. Herbicides reduce the destruction of soil structure by decreasing the need for tillage and the number of trips over the field with heavy equipment.

Erosion in orchards and in other perennial crops can be prevented by maintenance of a sod cover with selective herbicides. Tillage to eliminate weeds is not required, or not required as often, when herbicides are used. Many perennial species cannot be controlled effectively with hand labor and herbicides are often the only reasonable option.

Herbicides save labor and energy by reducing the need for hand labor and mechanical tillage. They can reduce fertilizer and irrigation requirements by eliminating competing weeds. They reduce harvest costs by eliminating interfering weeds and can reduce grain drying costs because green, weedly plant material is absent. Other methods of weed control will, of course, also accomplish these things, but not as efficiently and often not as cheaply.

III. DISADVANTAGES OF HERBICIDES

It is often suggested that herbicides reduce crop production costs. Many disagree and suggest herbicides are a net cost because they are expensive and the equipment for applying them is an added cost. The argument cannot be concluded in general terms; it must be decided for each specific crop production system.

One of the major concerns about herbicides is their undeniable mammalian toxicity. All have some degree of toxicity to humans and other plant and animal species. Some are no more toxic in terms of their LD_{50} than many chemicals (e.g., aspirin, mothballs, gasoline, and table salt) that we use commonly. All of these common chemicals are toxic. Many worry about herbicides because, in their opinion, they have no choice about exposure to toxicity.

Some herbicides persist in the environment. None persist forever, but all have a measurable environmental life. In some cases, but not all, a herbicide can carry over from one crop season to the next. This restricts rotational possibilities and may injure succeeding crops. Therefore, herbicides can be hazards to plants planted after they are used: plants that are not targets, but are affected by drift or inappropriate application.

Herbicides permit selective weed control in orchards. Proper herbicide selection maintains plant cover and reduces or eliminates the need for tillage that encourages soil erosion. Their use in many crops eliminates all plants except the crop and may increase soil erosion by reducing vegetative cover.

For a long time, it was assumed that because weeds have relatively long life cycles and the same herbicide is not used repetitively on the same land, weeds would not develop resistance to herbicides as insects developed resistance to insecticides. That is not true, and more than 100 cases of herbicide resistance have now been reported in one or more of 15 herbicide chemical families (Holt and LeBaron, 1990; LeBaron and McFarland, 1990).

Herbicides are often inconsistent in their control because they are affected by environmental conditions, and results of these interactions are not always predictable.

American agriculture is characterized by monoculture—large land areas devoted to a single crop. This is ideal for use of herbicides, and many have criticized herbicides because they encourage monoculture and discourage diversity. Unquestioned expansion of herbicide technology into developing countries is not always wise. There is strength in the agricultural diveristy of these countries, and it should not be inhibited or reduced by extensive use of herbicides for weed control, especially where their consequences have not been thoroughly examined.

Precision is required when herbicides are used. They cannot be used casually; one must think carefully about what herbicide to use, when to use it, how much to use, and how surplus chemical and empty containers will be disposed of.

Finally, because herbicides are so good at what they do, they may actually create problems after their use. Herbicides control certain weeds while leaving a crop unscathed. Natural plant communities are usually a polyculture (although this is not an absolutely true generalization). Diversity is the rule. When all plants are eliminated save the crop, other plants (weeds)

Herbicides can control weeds in crop rows where most mechanical methods are ineffective.

will move into the environment created, and they may be more difficult to control than the ones just eliminated.

For example, Florida pusley was a common weed in peanut production before herbicides were introduced (Johnson and Mullinix, 1995). When herbicides became integral to peanut production, Florida pusley was controlled, but the previously minor weeds, Florida beggarweed, Texas panicum, and yellow nutsedge, increased. When herbicides were discontinued, even after several years of use, Florida pusley again became the dominant weed (Johnson and Mullinix, 1995). A second example of replacement is from a rice–corn–soybean rotation in Peru. The weeds prior to herbicide use were 60% grass, 25% sedges, and 15% broadleaved. The grasses were large crabgrass and goosegrass. After 6 years, the weeds were 80% grass, 13% broadleaved, and 7% species of dayflower; 85% of the grass was itchgrass (Mt. Pleasant and McCollum, 1987). A disadvantage of selective herbicides is their ability to control some weeds and create open niches in which other weeds succeed.

Herbicides, like any technology, have advantages and disadvantages that must be weighed carefully to consider intended and unintended consequences prior to use.

IV. CLASSIFICATION OF HERBICIDES

An adequate classification system should be more than an index and should, so far as possible, integrate the objects being classified. Although there are several methods of herbicide classification, no one method is completely adequate. A problem for development of an adequate system of classification is the great diversity in chemistry and mode of action among herbicides. Indeed, not many years ago, it was possible to classify herbicides on the basis of chemical structure by including one small group with several different structures. That is no longer possible because diversity of structures and mode of action has increased. In spite of the inadequacy of most systems of classification, all are used because each has some utility. To become familiar with chemical weed control, one must understand some of the jargon, and much of it is found in the language surrounding herbicides and their classification. The objectives of this section are to understand why herbicides are grouped as they are and to enable use of the several systems for classification to discuss herbicides. Understanding systems of classification will permit explanation of field observations in terms users will understand.

A. CROP OF USE

One often hears that a particular herbicide is a corn herbicide or a turf herbicide. This is useful information because it immediately reveals the crop or site where the herbicide is used. Frequently, such statements represent only the narrow geographic or crop perspective of the speaker, and therefore crop of use is not a complete classification system. To illustrate its inappropriateness, one need only consider herbicides used to control wild oats in small grain crops. A partial list includes six herbicides: diclofop, difenzoquat, fenoxaprop, flamprop-methyl, imazamethabenz, and triallate. These herbicides, from six different chemical families, have five different modes of action and are applied at two different times relative to plant growth. Similarly, describing 2,4-D as a turf herbicide is accurate but not reflective of its many other uses. Classification by crop is essential knowledge but includes such a diversity of other factors that it is impossible to integrate the subject. Table 11.9 shows different crops or sites in which herbicides are used and the range of chemical groups used on each. If one does not know the crops in which a particular herbicide can be used or conversely, what crops it cannot be used in, one is not conversant with modern weed management.

Table 11.9

Partial Classification of Herbicides Based on Crop of Use

General group	Alfalfa	Aquatic weeds	Soybean	Corn	Cotton	Industrial sites	Orchards	Peanuts	Potatoes	Perennial weeds	Small grains	Sorghum	Turf or ornamental	Woody plants and brush
Chloroacetamides														
Acetochlor			X	X										
Alachlor			X	X				X	X			X		
Dimethenamid			X	X										
Metolachlor				X					X					
Propachlor				X								X		
Foliar contact and systemic														
Arsenicals	X				X		X						X	
Bromoxynil				X							X	X		
Difenzoquat											X			
Diquat		X												
Glufosinate						X	X			X			X	
Glyphosate					X	X	X			X			X	
Paraquat	X		X	X		X			X			X	X	
Dinitroanilines—toludines														
Benefin	X													
Ethalfluralin			X					X						
Oryzalin							X							
Pendimethalin			X	X	X				X			X		
Trifluralin	X		X		X						X	X		
Growth regulators														
Clopyralid										X				
Dicamba				X		X				X	X	X	X	X
Phenoxy Acids		X		X		X				X	X	X	X	X
Picloram						X					X	X		X

(*continues*)

Table 11.9 (*continued*)

General group	Crop or site													
	Alfalfa	Aquatic weeds	Soybean	Corn	Cotton	Industrial sites	Orchards	Peanuts	Potatoes	Perennial weeds	Small grains	Sorghum	Turf or ornamental	Woody plant and brush
Imidazolinones														
Imazamethabenz											X			
Imazapyr						X								
Imazaquin			X											
Imazethapyr			X					X						
Trazolinone														
Sulfentrazone			X											
p-Nitro-substituted diphenyl ethers														
Acifluorfen			X					X			Rice			
Bifenox			X								Rice			
Oxyfluorfen				X	X								X	
Lactofen			X		X									
Soil sterilants														
Sodium metaborate						X				X				
Sodium chlorate						X								X
Sulfonylureas														
Bensulfuron											Rice			
Chlorsulfuron											X			
Metsulfuron											X			
Nicosulfuron				X										
Primisulfuron				X										
Sulfometuron						X								
Thifensulfuron											X			
Carbamothioates—thiolcarbamates														
Butylate				X										
EPTC	X			X	X				X					
Triallate											X			

288

Triazines
  Ametryn
  Atrazine
  Cyanazine
  Hexazinone
  Prometon
  Simazine
Triazinones
  Metribuzin
N-Phenylphthalimides
  Flumiclorac
Triazolopyrimidine
  Flumetsulam
Uracils
  Bromacil
  Terbacil
Ureas
  Diuron
  Linuron
  Siduron
  Tebuthiuron
Other
  Amitrol
  Sethoxydim
  DCPA
  Oxadiazon

B. Observed Effect

A second system of classification is based on effects observed after emergence. Some herbicides, including sulfuric acid, bipyridiliums, dinitrophenols, and petroleum oils, having a burning effect. This describes what one sees, but not how the herbicide actually works. Other herbicides cause chlorosis (amitrole, clomazone) or gradual chlorosis, which is characteristic of photosynthetic inhibitors. Still other herbicides cause what is called a hormonal effect or an obvious growth abnormality. Growth abnormalities are so imprecisely defined, and so many herbicides affect growth, that the category merely serves to distinguish these effects from chlorosis or burning but does not describe what happens. Therein is the problem with effects as a system of classification.

C. Site of Uptake

A third, frequently used system of classification is based on site of uptake and distinguishes between foliar and soil-applied herbicides. A herbicide that acts after contact with plant foliage falls in the foliar-active group. Other herbicides can be foilar-active *and* soil-active, with the distinction often based on rate of application. The diphenyl ether herbicides, most of the phenoxy acids, the arsenicals, selective oils, and bipyridilium herbicides act primarily via foliage. Phenoxy acid herbicides and arsenicals translocate readily, whereas bipyridiliums and selective oils do not. The sulfonylureas, imidazoliones, triazines, chloroacetamides, carbamothioates, and dinitroanilines are taken up by roots. An adequate classification cannot be created for any group as large as the herbicides by dividing the group in two.

D. Contact vs Systemic Activity

Many herbicides are defined by noting they have contact as opposed to systemic activity. Translocation from point of application to site of action is synonymous with systemic activity. Some herbicides move only upward or acropetally in plants; others move acropetally and basipetally, or down. This system, like many others, is useful because it reveals how a herbicide is likely to behave, but it does not tell us how it does what it does, nor does it mesh well with any other category.

E. SELECTIVITY

Knowledge of selectivity is essential for wise use of any herbicide because it reveals the plants affected and unaffected. The first herbicides, iron and copper salts and dilute sulfuric acid, were selective because of differential wetting. Droplets of water solutions or suspensions of these herbicides bounced off upright cereal leaves and stayed on broad leaves. Selective herbicides kill or stunt weeds in a crop without harming the crop beyond the point of economic recovery. Nonselective herbicides kill all plants when applied at the right rate. No herbicide belongs rigidly to either group because selectivity is a function of rate. Selectivity is also based on many other factors, including the following:

(a) Plant age and stage of growth
(b) Plant morphology
(c) Absorption
(d) Translocation
(e) Type of treatment (e.g., broadcast vs band or specific application)
(f) Time and method of application
(g) Herbicide formulation
(h) Environmental conditions

Because selectivity is a function of the combined action of these variables, it is not a precise system of classification. It is essential knowledge, but it does not integrate the subject.

F. TIME OF APPLICATION

Almost all herbicides must be applied at a particular time to maximize control and selectivity. Therefore, knowledge of when to apply to obtain a desired goal is essential to wise use. Unfortunately, some herbicides can be applied successfully at different times, and this system, like preceding systems, does not integrate the subject even though it is essential knowledge for wise use. There are three times when herbicides are applied, and each can be specified relative to weed or crop. The first is prior to planting or preplanting. Sometimes application is immediately before planting or as early as several weeks prior to planting. Often preplanting applications include soil incorporation or mixing into soil. Incorporation can be combined with any time of application. It is not linked with a particular time. It is related to the herbicide and control goal. The second application time is preemergence to crop, weed, or both. It is after planting, but prior to emergence of the crop or weed. Postemergence applications are after the crop, weed, or both have emerged. Postemergence herbicides are often applied to foliage, but can be applied to soil. The exact time for postemergence application varies with the crop, the herbicide, and the weed.

G. Chemical Structure

There is no simple relationship between a herbicide's chemical structure—its chemical family—and its biochemical behavior. Classification based on structural formulas has been used, but with the ever-increasing number of new structures it cannot integrate the subject. The structural formula, especially when presented in only two dimensions, is a code that bears little relationship to three-dimensional shape, physical properties, electronic disposition, and steric factors, which determine biological behavior. To illustrate, there are two chemicals with nearly identical structures, a similar mode of action, but quite different outcomes. One is testosterone, the male hormone, which differs from progesterone, the female hormone, by two carbon and two hydrogen atoms. These structures illustrate the lack of desirability of classifying herbicides based only on structure. Table 11.10 further illustrates why structure is not a perfect classification system. The four structural groups shown illustrate at least two different modes of action.

Table 11.10

Examples of Why Chemical Structure Is Not a Perfect Classification System

1. Carbamates	
Barban, IPC, CIPC (no longer used in the United States)	Cell division inhibitors
Desmedipham and phenmedipham	Inhibit photosynthesis
2. Substituted nitriles	
Ioxynil and bromoxynil	Uncouple oxidative phosphorylation
Dichlobenil	Interferes with cell wall synthesis
3. Aryl aliphatic acids	
Naptalam	Interferes with IAA transport, anti-geotropic agent
Most other aliphatic acids	Mimic IAA action, hormone inhibitors
4. Phenyl pyridazinones	
Pyrazon	Photosynthetic inhibitor
Norflurazon	Interferes with carotenoid synthesis

H. MODE OF ACTION

Why do herbicides affect growth of, or kill, some plants and not others? What is their mode of action? Knowing a herbicide's mode of action may not lead directly to better weed control, but it gives a better knowledge base from which to derive conclusions based on field observations. This text does not emphasize detailed knowledge of chemistry or biochemistry. Knowledge of the difference between photosynthesis and respiration is assumed. Knowledge of the details of the light reactions of photosynthesis or the tricarboxylic acid cycle is not required. Determination of mode of action is a complex study of chemistry, biochemistry, and plant physiology. Mode of action is defined as the entire chain of events from first contact to final effect. It is distinguished from mechanism of action, defined as the ultimate biochemical or biophysical event or events that express the herbicide's effect. Both will be referred to in the next chapter.

Mode of action is another system of classification. However, if one knows only mode of action and nothing about the other, albeit incomplete, systems of classification, knowledge is incomplete.

The discussion in Chapter 12 will use the several systems of classification mentioned and not depend solely on one. The primary system of classification will be based on chemical structure with some essential discussion of mode of action.

THINGS TO THINK ABOUT

1. When did chemical weed control begin. How long is its history?
2. How did herbicides change the practice of agriculture?
3. Do herbicides affect energy use in American agriculture?
4. What do herbicides do that other weed control techniques cannot do?
5. What are the advantages and disadvantages of herbicides?
6. What are the attributes of a good classification system?
7. Why are there so many ways to classify herbicides?
8. What is wrong with classifying herbicides based on crop of use or time of application?
9. Can all herbicides be classed as contact or systemic?
10. What is the most important determinant of selectivity?
11. What are the problems with classifying only by chemical structure or mode of action?

LITERATURE CITED

Abernathy, J. R. 1981. Estimated crop losses due to weeds with nonchemical management, pp. 159–167 *in* "Handbook of Pest Management in Agriculture," Vol 1 (D. Pimentel, Ed.). CRC Press, Boca Raton, Florida.

Akamine, E. K. 1948. Plant growth regulators as selective herbicides. *Hawaii Agric. Exp. Stn. Circ.* **26:**1–43.

Alder, E. F., W. L. Wright, and G. C. Klingman. 1977. Development of the American herbicide industry, Chapter 3 *in* "Pesticide Chemistry in the 20th Century" (J. R. Plimmer, Ed.), Amer. Chem. Soc. Symp. Ser. 37. ACS, Washington, D.C.

Barrett, M., and W. W. Witt. 1987. Alternative pest management practices, pp. 195—234 *in* "Energy in Plant Nutrition and Pest Control," Vol. 2 (Z. R. Helsel, Ed.), Energy in World Agriculture. Elsevier, New York.

Becker, R. 1983. Selling the benefits of weed control. Proc. 35th Ann. Fertilizer and Agric. Chem. Dealers Conf., Bul. CE-184-4e. Coop. Ext. Serv., Iowa St. Univ., Ames, Iowa.

Bolley, H. L. 1980. Weeds and methods of eradication and weed control by means of chemical sprays. *N. Dak. Agric. Coll. Exp. Stn. Bul.* **80,** 511–574.

Bucha, H. C. and C. W. Todd. 1951 3-(*p*-chlorophenyl)-1,1-dimethylurea—A new herbicide. *Science* **114,** 493–494.

Clements, D. R., S. F. Wiese, R. Brown, D. P. Stonehouse, D. J. Hume, and C. J. Swanton. 1995. Energy analysis of tillage and herbicide inputs in alternative weed management systems. *Agric., Ecosystems and Environ.* **52:**119–128.

Crafts and Robbins 1962. "Weed Control." McGraw-Hill, New York p. 173.

Dowler, C. C. and E. W. Hauser. 1975. Weed control systems in cotton in Tifton loamy sand soil. *Weed Sci.* **23:**40–42.

Hamm, P. C. 1974. Discovery, development, and current status of the chloroacetamide herbicides. *Weed Sci.* **22:**541–545.

Hamner, C. L., and H. B. Tukey. 1944a. The herbicidal action of 2,4-dichlorophenoxyacetic and 2,4,5-trichlorophenoxyacetic acid on bindweed. *Science* **100,** 154–155.

Hamner, C. L. and H. B. Tukey. 1944b. Selective herbicidal action of midsummer and fall applications of 2,4-dichlorophenoxyacetic acid. *Bot. Gaz.* **106:**232–245.

Hobbes, T. 1651. "Leviathan," Part 1, Chapter 13.

Holt, J. S., and H. M. LeBaron. 1990. Significance and distribution of herbicide resistance. *Weed Technol. No* **4:**141–149.

Hopkins, W. L. 1994. Global Herbicide Directory, 1st Ed. Ag Chem Information Services. Indianapolis, IN.

Johnson, W. C. III, and B. C. Mullinix, Jr. 1995. Weed management in peanut using stale seedbed techniques. *Weed Sci.* **43:**293–297.

King, L. J. 1966. "Weeds of the World: Biology and Control." Interscience, New York.

Kirby, C. 1980. The hormone weedkillers: A short history of their discovery and development. *Brit. Crop. Prot. Council Pub. Croyden,* UK. 55 pp.

Korsmo, E. 1932. Undersok elser. 1916–1923. Over ugressets skadevirkninger og dets bekjempelse. I. Aker brucket. Johnson and Nielsens Boktrykkeri, Oslo, Norway.

Krebs, A. V. 1992. "The Corporate Reapers: The Book of Agribusiness," p. 160, Essential Books, Washington, D.C.

LeBaron, H. M., and J. McFarland. 1980. Herbicide resistance in weeds and crops. An overview and prognosis, *ACS Symp. Ser.* **421:**331–352. Amer. Chem. Soc., Washington, D.C.

Malcolm, J. 1993. The farmer's need for agrochemicals, pp. 3–9 *in* "Agriculture and the Environment" (J. Gareth Jones, ed.). Ellis Horwood, London.

Marth, P. C., and J. W. Mitchell. 1944. 2,4-Dichlorophenoxyacetic acid as a differential herbicide. *Botan. Gaz.* **106:**224–232.

Morettini, A. 1915. L'impegio dell'acido sulfurico per combattere le erbe infeste nel frumento. *Staz. Sper. Agr. Ital.* **48:**693–716.

Mt. Pleasant, J., and R. McCollum. 1987. Effect of weed control practices on weed population dynamics in intensive-managed continuous cropping system in the Peruvian Amazon. *Agron. Abstrs.*, p. 41.

Nalewaja, J. D. 1974. Energy requirements for various weed control practices. *Proc. N. Central Weed Control Conf.* **29:**19–23.

Peterson, G. E. 1967. The discovery and development of 2,4-D. *Agric. History* **41:**243–253.

Pimentel, D., and M. Pimentel. 1979. "Food, Energy, and Society," pp. 65–69. E. Arnold, London.

Pimentel, D., J. Krummel, D. Gallahah, J. Hough, A. Merrill, I. Schreiner, P. Vittum, F. Koziol. E. Back, D. Yen, and S. Fiancé. 1978. Benefits and costs of pesticides in U.S. food production. *Biosci.* **28:**772–783.

Pokorny, R. 1941. Some chlorophenoxyacetic acids. *J. Am. Chem. Soc.* **63:**1768.

Rabaté, E. 1911. Destruction des revenelles par l'acid sulfurique. *J d'agr. Prat (N.S.21)* **75:**497–509.

Rabaté, E. 1934. "La destruction des mauvaises herbes," 3rd ed. Paris.

Slade, R. E., W. G. Templeman, and W. A. Sexton. 1945. Plant growth substances as selective weed killers. *Nature (London)* **155:**497–498.

Smith, A. E., and D. M. Secoy. 1975. Forerunners of pesticides in classical Greece and Rome. *J. Agric. Food Chem.* **23:** 1050–1055.

Smith, A. E., and D. M. Secoy. 1976. Early chemical control of weeds in Europe. *Weed Sci.* **24:**594–597.

Zimdahl, R. L. 1995. Introduction, pp. 1–18 *in* A. E. Smith, ed. "Handbook of Weed Management Systems." Dekker, New York.

Zimmerman, P. W., and A. E. Hitchcock. 1942. Substituted phenoxy and benzoic acid growth substances and the relation of structure to physiologically activity. *Contrib. Boyce Thompson Inst.* **12:**321–343.

Chapter 12

Properties and Uses of Herbicides

FUNDAMENTAL CONCEPTS

- There are many ways to classify herbicides and, with present knowledge, no single way integrates the subject.
- There are eight major mechanisms of herbicide action.
- Research on herbicide action is a rapidly progressing field.

LEARNING OBJECTIVES

- The integration of herbicide mechanism of action and chemical structure in a classification system.
- To be aware of the great diversity of chemical structures and mechanisms of herbicide action.
- To be able to use and expand the classification system with new herbicides.

I. INTRODUCTION

The classification scheme used herein is based on mechanism of action and secondarily on consideration of chemical structure. Where appropriate, time of use, observed effect, site of uptake, and selectivity are mentioned. Details of chemical structure and biochemical action are included when essential to understanding, but these are not emphasized.

The glossary at the end of the text defines many terms. An herbicide's mechanism of action is the precise biochemical (e.g., inhibition of a specific enzyme) or biophysical lesion (e.g., inhibition of electron flow, binding to a protein, or disrupting cell division) that creates the herbicide's final effect. This text uses mechanism of action rather than the frequently encountered and often confusing term mode of action.

It used to be common for weed scientists to speak of herbicides as members of a structurally related chemical family. The family defined performance characteristics and mechanism of action. This is still possible for some families (e.g., triazines), but herbicide chemistry is now so diverse that such generalizations are no longer as meaningful. Herbicides usually have one primary mechanism of action and often several secondary actions. For example, diphenyl ethers inhibit protoporphyrinogin oxidase (PPO or Protox inhibition) and secondarily affect photosynthetic electron transport, carotenoid synthesis, and ATP synthesis. Bentazon inhibits photosynthetic electron transport at two sites. Even though there is a diversity of sites of action among herbicides, it is important to understand that the precise site of action is known for most herbicides.

Without very detailed information on the relationship between structure and activity, it is not possible to predict the mechanism of action from examination of an herbicide's chemical structure. Structure and activity are related, but the relationship is often not clear except to manufacturers who regard such information as proprietary and important to future herbicide development. No herbicides developed specifically to target a new site of action have subsequently achieved commercial success. The mechanism of action of most herbicides has been discovered during development or after patenting. Study of herbicide mechanism of action is advancing rapidly, as is discovery of activity and selectivity of new chemical structures. It is not possible or wise in a book of this kind to describe all herbicides or all mechanisms of action. This chapter will not even attempt to describe structure–activity relationships (SAR) or quantitative structure activity relationships (QSAR). Both are active research areas but are beyond the intent of this book. This chapter provides a description of most major herbicide activity and structural groups. It is not intended to be a complete description of herbicide action or of all herbicides. Books that do these things are listed in the chapter's references. This chapter is designed to acquaint students with the diversity of herbicide's mechanisms of action and the classification of some important herbicides.

In this text herbicides have been divided into eight mechanism of action groups in accordance with the classification scheme used by Devine *et al.* (1993):

1. Inhibitors of photosynthesis	Section II
2. Inhibitors of pigment production	Section III
3. Lipid biosynthesis inhibitors	Section IV

II. INHIBITORS OF PHOTOSYNTHESIS

A fundamental feature of photosynthesis is conversion of light energy to chemical energy: a process on which all life depends. Light quanta falling on green leaves energize electrons in chlorophyll. The energy is converted to chemical energy by reducing (adding an electron) to an acceptor in the plant.

The two photosynthetic light reactions are coupled by the photosynthetic electron transport chain where photophosphorylation (production of ATP) occurs. Light reaction I produces reduced nicotine adenine dinucleotide phosphate (NADP). Herbicides that act in relation to light reaction I divert electrons away from photosystem I *and* generate toxic molecular species. Light reaction II begins with removal of electrons from water and production of oxygen (the Hill reaction) and is where many herbicides exert their effect. Most of the large and diverse group of chemical structures that inhibit light reaction II block electron transport by binding to adjacent sites on the D-1 quinone protein of the photosynthetic system that functions in the electron transport chain between the primary electron acceptor from chlorophyll a and plastoquinone.

These herbicides cause gradual chlorosis in plants. The chemical groups include the phenylureas, triazines, uracils, acylanilides, pyridazinones, biscarbamates, hydroxybenzonitriles, and a few other groups represented by single herbicides. More herbicide chemical groups act on photosynthesis than on any other physiological process. A summary of information on some of these herbicides is in Table 12.1.

A. PHOTOSYSTEM I ELECTRON ACCEPTORS

In some systems of classification, these and other herbicides (see Section III,B) are called photodynamic, a name that has not achieved universal acceptance. To call a herbicide photodynamic is to identify how it acts, but only in a general way. The bipyridiliums are photodynamic because they cause photo-oxidative stress by diversion of photosynthetic energy from photosystem I. The specific action of these herbicides is reduction of molecular oxygen to a toxic superoxide radical. Rapid bleaching of photosynthetic

Table 12.1

Some Herbicides That Inhibit Photosynthesis[a]

Herbicide name		
Common	Trade	Applications
Phenylureas		
Diuron	Karmex	Alfalfa, cotton, sugarcane, pineapple, grapes, tree fruits
Fluometuron	Cotoran	Cotton, sugarcane
Linuron	Lorox	Soybean, carrot, corn, sorghum, potato
Siduron	Tupersan	To control annual grasses in turf
Tebuthiuron	Spike	Non-cropland
Triazines		
Chlorotriazines		
Atrazine	Aatrex	Corn, sorghum, sugarcane, conifers
Cyanazine	Bladex	Corn, cotton
Hexazinone	Velpar	Alfalfa, pineapple, conifers
Simazine	Princep	Strawberries, tree and citrus fruits
Methoxytriazines		
Prometon	Pramitol	Non-cropland
Thiomethyl triazines		
Ametryn	Evik	Banana, corn, pineapple, sugarcane
Prometryn	Caparol	Cotton, celery
Triazinones		
Metribuzin	Sencor/Lexone	Potato, soybean, sugarcane, alfalfa, asparagus, tomato
Uracils		
Bromacil	Hyvar	Citrus, pineapple, brush on non-cropland
Terbacil	Sinbar	Alfalfa, mint, sugarcane

[a]Current herbicide label directions must be consulted for complete use information.

tissue occurs. Affected plants initially appear water-soaked but rapidly (several hours to a few days) become necrotic and die. There is severe disruption of cell membranes directly attributable to production of toxic oxygen species.

The bipyridiliums are almost completely dissociated in solution. Their action is due to the positive bypridinium ion, reduced by drawing an electron from photosystem I (the primary site of action) to form a relatively stable free radical that continues to react and produces hydrogen peroxide, a superoxide radical O^{2-}, a hydroxyl radical (OH^-), and singlet oxygen $^1O_2$, each of which is potentially toxic to cell membranes, where damage occurs.

Diquat and paraquat were discovered by Imperial Chemical Industries of England, and paraquat (Figure 12.1) was released in 1958. These are cations in solution and the active form is a cation. They act only when absorbed by foliage and have almost no soil activity because of complete adsorption. Both herbicides act quickly; effects are normally seen within several hours and certainly within a few days. Translocation is poor and complete foliar coverage is essential to good weed control. They kill a wide range of annual plants and will desiccate shoots of perennials, but are not translocated to roots: one reason they do not provide permanent control of perennials. The addition of a wetting agent improves control of many species because it aids foliar dispersal and cuticular penetration.

Paraquat is more active on grasses and diquat on broadleaves. Paraquat has found extensive use in chemical fallow and diquat is preferred for submerged aquatic species. Both are toxic to humans from skin contact or ingestion. These are nonselective herbicides that kill or affect almost any plant foliage they contact. Because of poor translocation, foliage not contacted directly is not affected. They can be used as preharvest desiccants to speed drying of some crops.

B. PHOTOSYSTEM II ELECTRON TRANSPORT INHIBITORS

1. Phenylureas

Phenylurea herbicides (Table 12.1) take their name from the organic compound urea (Figure 12.2), the core of all urea herbicides (see diuron structure, Figure 12.2). They are broad-spectrum herbicides applied to soil. Most are nonvolatile, noncorrosive compounds, with low mammalian toxicity, absorbed by plant roots from soil and translocated to shoots in the apoplast. Leaching is variable. At doses of less than 2 kg/ha, ureas are selective and control seedling growth of weeds in some crops. At higher doses they are soil sterilants. Soil persistence ranges from 2 to 6 months, but they usually do not affect succeeding crops. Ureas are most effective on young germinating weeds following uptake from soil. Because they are

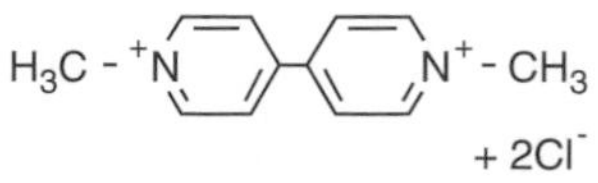

$$H_3C - {^+}N \diagup\!\!\!=\!\!\!\diagdown N^+ - CH_3$$
$$+ 2Cl^-$$

Figure 12.1 Structure of paraquat.

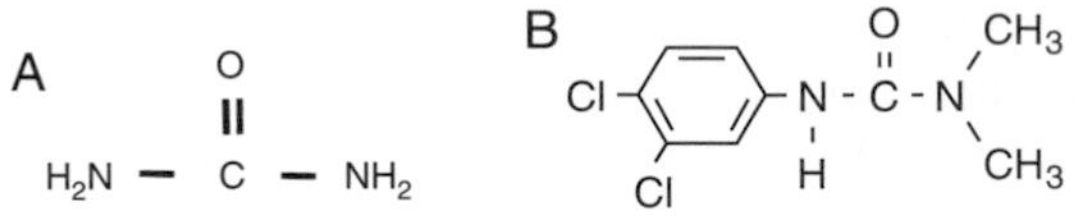

Figure 12.2   Structure of urea (A) and diuron (B).

photosynthetic inhibitors, death occurs after emergence, as photosynthesis begins. Weeds germinating over time are controlled because of their soil persistence. Photosynthetic inhibition is the primary mechanism of action, but they cause chlorosis and necrosis and can have a burning effect at higher doses.

2. *s*-Triazines

A large number of herbicides is based on the generalized symmetrical diamine triazine (Figure 12.3). They all followed the discovery and release of simazine by Geigy Chemical Co. (Switzerland) in 1956. Most triazines are inhibitors of photosynthesis following root uptake, but some are absorbed by foliage. Those with higher water solubility also have foliar activity. Translocation is apoplastic after root uptake. The primary mechanism of action is inhibition of photosynthesis. There are secondary mechanisms because some seedlings fail to emerge and become photosynthetic and there is activity in the dark. All have relatively low mammalian toxicity. Residues of many persist in soil. This is an advantage where long-term, nonselective weed control is desired. Others (e.g., cyanazine and ametryn) have short soil lives. Table 12.1 includes several triazines.

a. Chlorotriazines

Chlorotriazines are selective in corn. Simazine was developed for use in corn but was largely replaced by atrazine. Both can be used as soil sterilants at doses over 20 kg/ha. The usual selective crop rates are between 1 and 4 kg/ha. Atrazine (Figure 12.3) is more water-soluble than simazine (33 vs

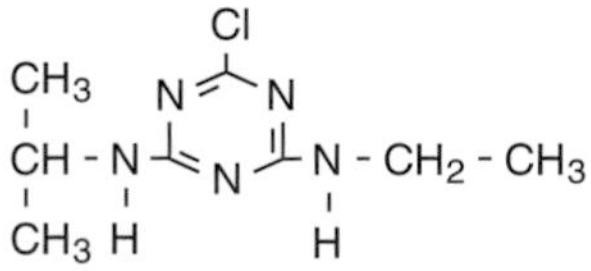

Figure 12.3   Structure of atrazine.

6.2 ppm) and is therefore less dependent on, but not completely indepen-dent of, rainfall or irrigation for activity. Simazine's low water solubility and high soil adsorption permit use in orchards where deep-rooted trees do not come in contact with it because it does not leach.

Simazine is used for weed control in corn, turf grown for sod, ornamen-tals, nursery plantings, Christmas trees, asparagus, and artichokes. Weeds in alfalfa have been controlled with simazine, which is not very mobile in soil, but not with the more mobile atrazine. Because of its persistence and the problem of soil residues affecting a succeeding crop, some uses of atrazine were supplanted by cyanazine, which is selective in corn, grain sorghum, and cotton, but has shorter soil persistence and a slightly different weed-control spectrum. Often, cyanazine and atrazine have been combined to broaden the weed-control spectrum and shorten total soil life by reducing the amount of atrazine. Cyanazine use will end in the United States in 1999 because it is suspected of being a human carcinogen.

b. Methoxytriazines

Methoxytriazines (OCH_3) are usually more water-soluble than their chloro analogs and have more foliar activity. They are used mainly for industrial and non-cropland weed control. All are more toxic to corn than chloro derivatives. Prometon is used for soil sterilization on non-cropland at 10 to 60 kg/ha.

c. Methylthiotriazines

As a result of higher vapor pressure and higher phytotoxic activity via foliar application, the methylthiotriazines (SCH_3) show high variability in selectivity. Prometryn is selective in cotton and celery, whereas ametryn is selective in pineapple, sugarcane, banana, and plantain. Ametryn can be used as a post-directed spray in corn and has been used in combination with other herbicides for chemical fallow because of its burning, contact activity. Volatility is not a problem but persistence can be. Persistence of the methylthiotriazines is usually shorter than that of the chlorotriazines. Part of their selectivity is related to greater soil adsorption. Banana plants, for example, are sensitive to atrazine because of its mobility, but not sensi-tive to ametryn, which is far less mobile and does not reach the banana root zone.

3. as-Triazinones

Triazinone or asymetrical (*as*) triazine herbicides were first developed in 1971. Metribuzin (Figure 12.4) is selective in potato and soybean and

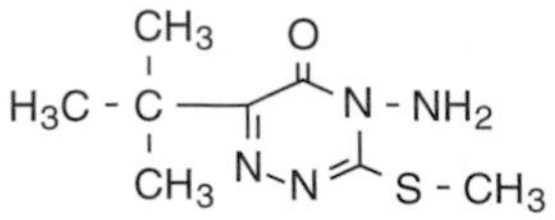

Figure 12.4   Structure of the *as*-triazin metribuzin.

has a shorter soil life than most symmetrical triazines. It is used as a dormant spray for weed control in alfalfa. Ethiozin is 3-ethylthio rather then 3-methylthio and thus closely related to metribuzin. It has shown promise for selective control of annual grasses (particularly bromes) in wheat.

4. Uracils

The basic uracil structure (Figure 12.5) is the core structure of uracil, thymine, cytosine, and guanine, the building blocks of nucleic acids. Only two herbicides, bromacil and terbacil, are based on the uracil structure, an asymmetrical ring with two nitrogens. Bromacil (Figure 12.5) is nonvolatile and moderately leachable. Its phytotoxic residues may persist up to 1 year in soil, and it can be used as a soil sterilant.It is excellent for control of perennial weeds at 12 to 24 kg/ha. It is selective only in some perennial crops such as pineapple and in lemon, orange, and grapefruit orchards. Bromacil is toxic to a wide range of grass and broadleaved species. Uracils are absorbed primarily by roots. Terbacil is used selectively in peppermint, spearmint, sugarcane, small fruits, deciduous tree fruits, and as a postemergence, dormant spray in alfalfa. Foliar chlorosis is a normal symptom, but general root and shoot inhibition and leaf necrosis are observed.

5. Hydroxybenzonitriles

Hydroxybenzonitriles were introduced in the early 1960s in the United States and the United Kingdom. They are contact herbicides, selective in

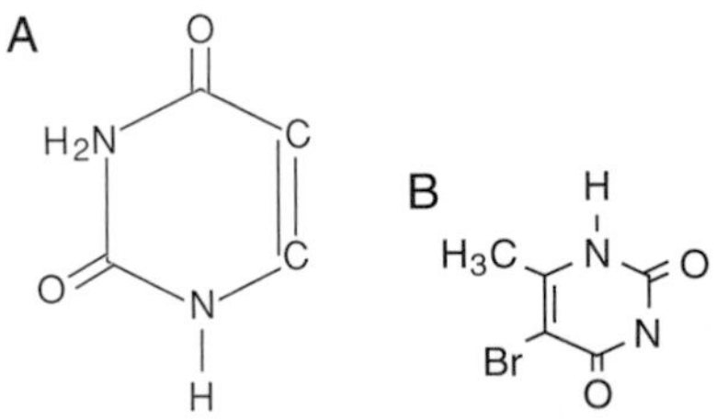

Figure 12.5   Basic uracil structure (A) and structure of bromacil (B).

grasses, with limited translocation in shoots of some species. Because of soil sorption, they have no soil activity, and they were developed for control of broadleaved weeds not controlled by other herbicides in cereal crops. They interfere with photosynthesis (Gunsolus and Curran, 1991). They are nonmobile photosynthetic inhibitors because their site of action is the D-1 quinone protein of the photosynthetic electron transport system. At doses of 0.21 to 0.56 kg ae/ha, bromoxynil (Figure 12.6) kills a wide range of annual weeds such as chickweed, mayweeds, and members of the Polygonaceae without injury to wheat, barley, or oats. They are only effective on seedling weeds. Bromoxynil and ioxynil are often marketed in combination with a phenoxyacid herbicide to broaden the weed control spectrum and reduce cost. Benzonitrile herbicides are moderately toxic to mammals.

A third benzonitrile, dichlobenil, is structurally similar to ioxynil and bromoxynil, but has little foliar activity, is volatile, primarily soil-active, and does not inhibit photosynthesis. It may inhibit cell-wall synthesis through action on cellulose synthesis.

6. Acylanilides

Propanil is used to control some annual broadleaved weeds and grasses in wheat and barley and, at higher rates, several annual grasses and barnyardgrass in rice. It is foliarly applied, whereas its chemical relatives, the substituted acetamides, are soil-applied. It is a photosynthetic inhibitor but also inhibits root and coleoptile growth when applied to those tissues. Propanil causes cessation of growth and gradual leaf necrosis with little or no residual soil activity. It also has been reported to inhibit anthocyanin, RNA, and protein synthesis, but these are secondary effects. It breaks down rapidly in soil.

7. Pyridazinones

Pyrazon (Figure 12.7) enters weeds by foliar and root uptake and can be used pre- or postemergence, but is no longer used in the United States. A variety of broadleaved weeds are susceptible. The most important selec-

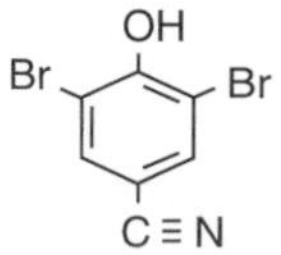

Figure 12.6 Structure of bromoxynil.

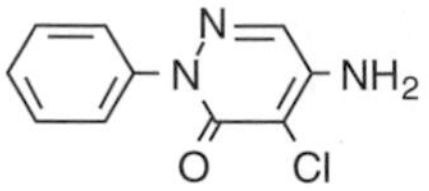

Figure 12.7   Structure of pyrazon.

tive use is in sugar and red beets. It has low mammalian toxicity and persists from 4 to 8 weeks in soil, an advantage for use in beets.

Pyridate, a phenyl pyridazine includes an S (sulfur)-octyl (eight-carbon chain) carbamothioate group. It is a rapidly absorbed, postemergence, contact (foliar) herbicide that controls broadleaved weeds and some grasses in peanut and corn. It has no soil persistence and does not leach. Susceptible plants turn yellow and necrotic and a rapidly formed metabolite inhibits electron transport in photosystem II.

8. Bis-carbamates

Two closely related bis-carbamates inhibit photosynthesis. Phenmedipham and desmedipham are used for postemergence weed control, especially of broadleaved weeds, in sugar and red beets.

9. Others

The benzothiadiazine bentazon is the only herbicide in its chemical group. It inhibits photosynthesis and is used for selective control of broadleaved weeds in leguminous crops such as soybean, dry bean, pea, and peanut. It can be used for postemergence control of foliage of Canada thistle and yellow nutsedge in some crops. It is not as effective for control of Canada thistle as some other herbicides.

III. INHIBITORS OF PIGMENT PRODUCTION

A. CAROTENOID BIOSYNTHESIS INHIBITORS

Carotenoids are essential to plant survival because they protect individual pigment–protein complexes, especially chlorophyll, and ultimately the chloroplast, against photo-oxidation. With high light intensity or under stressed conditions, chlorophyll molecules receive more light energy than they can transfer effectively into electron transport. The excess energy can be dissipated in several ways, including production of singlet oxygen that is destructive of tissue integrity. Carotenoids protect against this by quenching excited chlorophyll molecules and by quenching singlet oxygen. Destruction of carotenoids or their biosynthesis leads to loss of the protective role (Young, 1991). A few herbicides from different chemical groups inhibit this biosynthetic pathway, but they do not all act in the same way.

1. Amitrole

Amitrole (Figure 12.8), a unique five-membered heterocyclic ring, is structurally unlike any other herbicide. It is usually sprayed directly on foliage at 2 to 10 kg a.i./ha and is active on many plants, but does not have sufficient selectivity to be used in crops. It produces chlorotic, white foliage because of its interference with carotenoid production. The intensity of the effect and extent of recovery depend largely on dose. Chlorosis results, in part, from failure of the chloroplasts to develop from the proplastid stage. Amitrole is readily inactivated in soil, with 4 kg/ha normally dissipated in about 7 days. It is very good for control of poison ivy and poison oak.

Activity of amitrole has been enhanced by mixture with ammonium thiocyanate. A 1:1 mixture is 2 to 4 times as effective as amitrole alone against some weed species. This synergistic mixture kills foliage more slowly than amitrole alone, probably because of protection of foliage against rapid contact action and a longer time for absorption and translocation of amitrole.

2. Norflurazon and Fluridone

Norflurazon and fluridone inhibit the phytoene desaturase enzyme system. Norflurazon's structure is similar to that of the phenylpyridazinones (see Section II,B,7) that inhibit photosynthesis. It controls grasses, sedges, rushes, and broadleaved weeds preemergence in cotton, cranberries, citrus, nut, and some fruit crops.

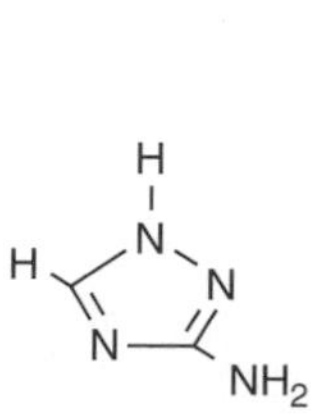

Figure 12.8 Structure of amitrole.

Fluridone, a unique structure (Figure 12.9), controls submerged and emerged aquatic weeds. It has little effect on algae and gives partial control of cattails. Cotton is tolerant, but fluridone is used only as an aquatic herbicide.

3. Isoxazolidinones

In 1986 clomazone was released for selective control of several weeds in soybean. It used preemergence and has no or limited postemergence activity. It rapidly turns plants white, and if more than 75% of the plant is affected, the plant dies. There have been some important instances of drift from clomazone, made readily apparent because of its bleaching symptoms.

B. Protox Inhibitors

Herbicides in this group are often also called photodynamic (see Section II,A). They act independently of the photosynthetic mechanism but require light for activity. These herbicides inhibit the enzyme protoporphyrinogen oxidase, PPO or protox, the last step in the porphyrin pathway that produces chlorophyll. In light, protox inhibitors cause accumulation of large amounts of the phytotoxic molecule protoporphyrin or proto. Proto accumulation quickly damages the plasmalemma and tonoplast and interferes with lipid production.

1. *p*-Nitrodiphenyl Ethers

The ether or p-nitro-substituted diphenyl ethers were introduced in the 1980s for postemergence broadleaved weed control in broadleaved crops. All require light for their action but are not dependent on photosynthesis; light is required to produce a substrate for their action (Duke *et al.,* 1991). They are sometimes called photobleaching herbicides because a primary symptom is bleaching of plant foliage. All are active on broadleaved weeds

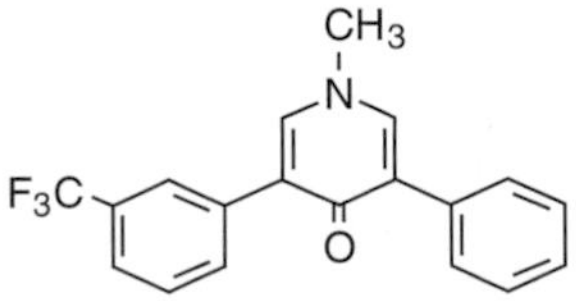

Figure 12.9 Structure of fluridone.

and selective in broadleaved crops, including soybean, peanut, bean, and cotton, when applied after emergence of crop and weeds. They can be used as post-directed sprays in grass crops. Others are used for postemergence weed control in corn and rice (see Table 12.2).

Acifluorfen (Figure 12.10) is used for postemergence control of broadleaved weeds in soybean, peanut, and rice. Complete foliar coverage ensures good activity. Bifenox can be applied postemergence to foliage or preemergence to soil and controls a range of broadleaved weeds and some annual grass weeds in soybean, corn, sorghum, and small grain crops. Its soil life is 6 to 8 weeks.

Oxyfluorfen controls broadleaved weeds and annual grasses pre- or postemergence in soybean, peanut, rice, corn and several tree crops. The soil life, as with most herbicides in this group, is less than 2 months. It and its chemical relatives do not leach in soil. Fomesafen and lactofen are selective in soybean; both are most active postemergence and are rapidly absorbed through leaves.

2. N-Phenylthalamides

There are two currently active herbicides in this group, but much development is occurring. One of the new herbicides is still a numbered compound at this writing, and the other is flumiclorac. Flumiclorac is a fast-acting, contact, postemergence herbicide. It controls several annual broadleaved weeds in soybean and corn that have not always been controlled well with other herbicides. Soil degradation is rapid, with complete disap-

Table 12.2

p-Nitrodiphenylethers

Herbicide name		
Common	Trade	Applications
Acifluorfen	Blazer/Scepter/Tackle	Controls several annual broadleaved weeds and some grasses in peanut and soybean.
Bifenox	Modown	Used in combination with phenoxy acid or grass herbicides in cereals. Also used in rice, sunflower, and soybean in combination.
Fomesafen	Reflex/Tornado	Soybean.
Lactofen	Cobra	Cotton, soybean.
Oxyfluorfen	Goal	Cotton, corn, soybean, and several vegetable crops, fruit and nut trees.

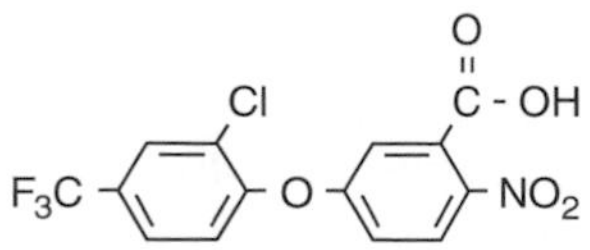

Figure 12.10   Structure of acifluorfen.

pearance in 12 hours to 4 days in a loamy sand soil at pH 7. The herbicides and its metabolites do not leach below 3 inches.

3. Triazolinone

This is another small group that will likely have more members with time. The one available herbicides is sulfentrazone, which controls annual broadleaved weeds, some annual grasses, and *Cyperus* (nutsedge) spp. in soybean. It is absorbed by roots and foliage and is applied preemergence or preplanting.

4. Oxadiazoles

Once again, there is only one registered herbicide is this group, but it has been available for some time and no other candidates are being developed. Oxadiazon is of interest because of its ability to control several annual grasses and annual broadleaved species in ornamentals. It is strongly adsorbed by soil colloids, rarely leaches, persists in soil, and acts mainly preemergence.

IV. LIPID BIOSYNTHESIS INHIBITORS

Salt and acid herbicides used prior to World War II were contact chemicals that destroyed plant structure by acting on membranes. Their exact mechanism of action has never been determined. Many presently available contact herbicides act by modifying membrane structure (Ashton and Crafts, 1981) through effects on lipid biosynthesis or production of toxic radicals. Lipids include fatty acids, neutral fats, and steroids. Plant surfaces are covered with and composed of a complex mixture of lipids, often in crystalline form. These are generally referred to as plant waxes. They form the cuticle or noncellular outer skin of plants and are integral components of intracellular plant membranes. Five herbicide families inhibit lipid biosynthesis: thiocarbamates (often called thiol carbamates or carbamothioates), chloroacetamides, cyclohexanediones, aryloxyphenoxypropionic acids, and substituted pyridazinones (Gronwald, 1991b). Thiocarbamates and chloroacetamides inhibit growth of emerging seedlings when applied to soil prior to weed emergence. Other substituted pyridazinones have been discussed previously (see Sections II,B,7 and III,A,2). Each herbicide in the group also may secondarily inhibit photosynthesis and carotenoid biosynthesis. Rate of application and plant species determine the dominant mechanism of action. There are three primary mechanisms of action that result in inhibition of lipid biosynthesis: Inhibition of acetyl-CoA carboxylase (ACCase, Section IV,A), inhibition of fatty acid elongation (Section IV,B), and inhibition of other targets (Section IV,C).

A. ACCASE INHIBITORS

Aryloxyphenoxypropionic acids and cyclohexanediones are commonly called "fops" and "dims," respectively. Each is used for postemergence selective control of annual and perennial grasses in some dicotyledonous crops and in some cereal crops. They are often referred to as graminicides (Gronwald, 1991b). Herbicides in these groups are foliar applied, readily absorbed, and translocated to meristems where they are toxic to grasses (Gronwald, 1991b) and have similar selectivity. Observed symptoms in susceptible plants are also similar because both inhibit activity of the enzyme acetyl-CoA carboxylase (ACCase), a key lipid biosynthetic enzyme. Tables 12.3 and 12.4 list representatives of these two chemically different but mechanistically similar families.

Table 12.3

Aryloxyphenoxypropionate Herbicides (the fops)[a]

Herbicide name		
Common	Trade	Applications
Diclofop-methyl	Hoelon/Hoe-Grass	Wheat, barley, lentils, flax, and sugarbeet
Fenoxaprop-ethyl	Acclaim/Whip/and several others	Soybean, wheat, and turf
Fluazifop-butyl	Fusilade	Cotton, soybeans, and several horticultural crops
Haloxyfop-methyl	Galant/Verdict, and others	Soybean, sunflower, rape, potato, bean, flax, and peanut
Quizalofop-ethyl	Assure	Soybean

[a]In all cases only annual and some perennial grasses are controlled.

Table 12.4

Cyclohexanedione Herbicides (the dims)

Herbicide name		
Common	Trade	Applications[a]
Clethodim	Select	Cotton and soybeans
Cycloxydim	Focus/Laser	Cotton, soybeans, and sugarbeets
Sethoxydim	Poast and several others	Soybean, peanut, alfalfa, sugarbeet, sunflower, and cotton
Tralkoxydim	Achieve	Cereal crops; compatible with several herbicides active against dicot weeds, but incompatible with sulfonylureas

[a]In all cases, only annual and some perennial grasses are controlled.

Aryloxyphenoxypropionates (fops) are foliar graminicides that selectively remove annual grasses from grass crops such as wheat and barley. They are used for selective grass control in many broadleaved crops.

Diclofop (Figure 12.11) is a phenoxyphenoxy derivative, as are all "fops." Its selectivity is due to differential rates of metabolism to inactive products in susceptible and tolerant species. Control of wild oats and other grasses is growth stage dependent, with the best control when grasses have two to four leaves.

Fluazifop, another postemergence grass herbicide, is selective in broadleaved crops. It, and other members of this group, control young (three to six leaves), actively growing grasses best. Quizalofop, haloxyfop, and fenoxaprop control a broad range of annual and perennial grasses, but no broadleaved weeds. All are selective after postemergence, foliar application in many important broadleaved crops.

Among the cyclohexanediones (dims), sethoxydim (Figure 12.12) and clethodim, applied postemergence, selectively control nearly all annual and perennial grasses in all broadleaved crops. Sethoxydim is also selective in many ornamental trees, shrubs, flowers, and ground covers. These are

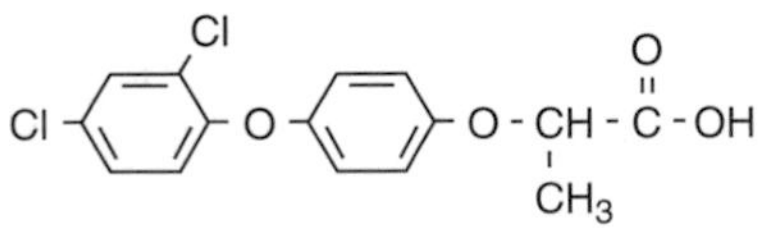

Figure 12.11   Structure of diclofop.

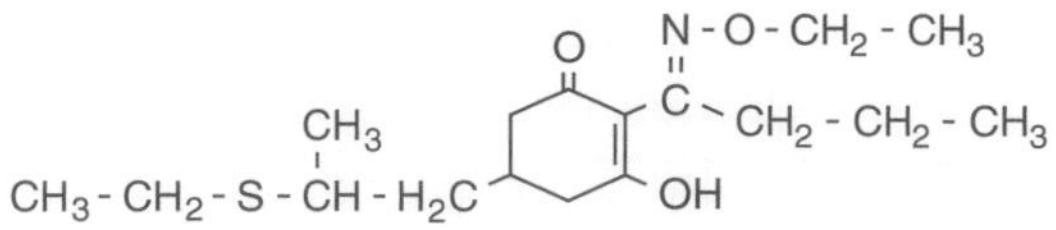

Figure 12.12   Structure of sethoxydim.

unique structurally because they are based on a hexane rather than a benzene ring. Rate of application varies with the grass species to be controlled; higher rates are needed for larger plants and perennials. Combination with cultivation often improves control of perennials.

B. FATTY ACID ELONGATION INHIBITORS

Either thiocarbamate or carbamothioate is a correct name for this family. Carbamothioates is more technically accurate, but thiocarbamate or thiolcarbamate is used more frequently in weed science. These herbicides inhibit fatty acid elongation and thereby production of lipids, but that alone may not be sufficient to kill plants. Cuticles protect plants against water loss, injury from wind, physical abrasion, frost, radiation, pathogens, and chemical entry. Loss of one or more of these functions because of the inability to synthesize cuticular lipids may lead to death. Interference with the integrity of internal plant membranes also leads to death. The primary mechanism of action of these herbicides is inhibition of fatty acid elongation, although present evidence is inconclusive. Secondarily, they may play a role in gibberellin biosynthesis (Wilkinson, 1983, 1986).

The primary visual symptoms of thiocarbamate injury are shoot inhibition (aberrant morphology), abnormal growth and emergence of leaves in grasses, often seen as the leaf's inability to emerge or unfurl, and formation of leaf loops as leaves fail to emerge properly. Many thiocarbamates are volatile and must be incorporated to prevent loss. Soil persistence is short and carryover problems have not occurred. In general, they are much more effective against annual grass weeds than against annual broadleaved weeds. Table 12.5 summarizes information about these herbicides.

Diallate and triallate differ by one chlorine atom and are soil-applied specifically for control of wild oats in small grains, lentil, pea, and sugarbeet. Triallate is more selective in small grains and has been used widely. Neither is effective postemergence and both must be incorporated. Incorporation throughout the top 4 to 5 inches (10 to 13 cm) of soil often leads to crop injury, whereas incorporation in the top 1 to 3 inches (2.5 to 7.5 cm) does not. This selectivity is because of the different growth habits of the

Table 12.5
Thiocarbamate Herbicides

Herbicide name		
Common	Trade	Applications
Butylate	Sutan	Preplant incorporated application controls several annual grasses, yellow and purple nutsedge, and a few broadleaved species in corn
Cycloate	Ro-Neet	Preplant incorporated application controls annual grasses and some annual broadleaved species in sugarbeet, table beet, and spinach
Diallate	Avadex	Pre- or postplant incorporated application controls wild oats in barley, flax, corn, lentil, pea, potato, forage legumes, and soybean
EPTC	Eptam	Preplant incorporated application controls weeds in several crops, including alfalfa, bean, flax, potato, sugarbeet, sunflower, citrus, pea, walnut, almond, and tomato
	Eradicane	EPTC formulated with a safener for preplant incorporated weed control in corn
Triallate	Avadex B-W	Pre- or postplant incorporated control of wild oats in spring wheat, barley, lentil, and pea

mesocotyl of small grain and wild oats. In small grains, the mesocotyl, the primary area for uptake of these herbicides, and the apical meristem remain near the seed as the seedling emerges. In wild oats, these regions are pushed toward the surface and more herbicide is absorbed.

EPTC was released in 1954. In the 1980s it was discovered that after use, its activity could completely disappear within a matter of days. This enhanced degradation in soil occurs after repeated use. Microorganisms adapt to EPTC, and other thiocarbamates, and are able to degrade them quickly. The problem can be avoided by identifying soils where enhanced degradation is likely, by rotating crops, using different herbicides, using diverse weed management techniques, or, if the same crop must be grown, by rotating herbicides.

EPTC is used in corn to control grassy weeds, not controlled well by other herbicides. If EPTC is used alone, it injures corn, but a safener that permits higher rates and crop tolerance is available. The mechanism of action is enhanced metabolism. EPTC is not an active herbicide until it is converted in plants to EPTC sulfoxide, which interferes with vital plant processes. Resistant plants detoxify sulfoxides by converting them to a glutathione derivative via conjugation. The safener increases levels of the necessary enzyme and of glutathione. It is also possible that there is a direct competition between the safener and EPTC for sites of action. The safener has expanded the selectivity range of a common herbicide. EPTC, with a safener, is used in corn under the trade name Eradicane.

Two other thiocarbamate herbicides illustrate the diversity of uses found in this group. Cycloate is soil incorporated for control of weeds preemergence in sugarbeet and spinach. EPTC can be used in sugarbeet, but cycloate controls as many weeds with less crop injury. Butylate is the only thiocarbamate herbicide selective in corn without a safener. It is primarily effective on grasses and some broadleaved weeds and has a relatively short (about 2 weeks) soil persistence. The thiocarbamate group had been known for many years prior to the discovery of butylate's selectivity in corn.

C. OTHER TARGETS FOR INHIBITION OF LIPID BIOSYNTHESIS

The chloroacetamide herbicides inhibit shoot growth of emerging seedlings and produce abnormal seedlings that may not emerge from soil. There is contradictory evidence of their effect on *de novo* fatty acid biosynthesis and thus on membranes. The primary mechanism of action has not been determined and their classification could change.

Chloroacetamides, a small chemical group, are important because of their widespread use in several major crops (Table 12.6). They are relatively water-soluble, soil-applied, readily degraded herbicides that are not hazards to succeeding crops. They affect germinating seedlings. Alachlor (Figure 12.13) is used preemergence to control many broadleaved and annual grass weeds. It is structurally similar to metolachlor, and they are used in the same crops. Metolachlor has longer soil persistence and a slightly different weed control spectrum.

Propachlor has been combined with triazines in corn to improve grass control. It was one of the first members of this group released. The range of activity in the group is illustrated by comparing butachlor, which controls many annual grasses and some broadleaved weeds selectively in rice, with the other herbicides in Table 12.6.

The chemical nature of chloroacetamides continues to evolve as shown by development of dimethenamid, wherein a five-membered thiophene (sulfur-containing) ring replaces the six-membered benzene ring common to the other chloroacetamides. Crop selectivity has not changed and the weed control spectrum is similar.

The chloroacetamides have been complemented in recent years by the development of safeners. Safeners, also called antidotes or protectants, were developed to broaden the range of crop selectivity for particular

Table 12.6

Alphachloroacetamide Herbicides

Herbicide name		
Common	Trade	Applications
Acetochlor	Harness	Controls most annual grasses, yellow nutsedge, and certain small-seeded broadleaved weeds in corn and soybean
Alachlor	Lasso/Micro-tech	Control of many annual grasses, yellow nutsedge, and certain broadleaved weeds in soybean, corn, dry bean, peanut and grain sorghum
Butachlor	Machete	Controls many annual grasses, some broadleaved weeds, and many aquatic species in transplant, dry, and wet seeded rice; not marketed in the United States.
Dimethenamid	Frontier	Controls yellow nutsedge, many annual grasses, and some broadleaved weeds in corn and soybean
Metolachlor	Dual	Controls many annual grasses, yellow nutsedge, and some broadleaved weeds in soybean, cotton, corn, potato, peanut, safflower, sorghum, and in nursery and landscape plantings
Propachlor	Ramrod	Controls many annual grasses and some broadleaved weeds in corn and grain sorghum

herbicides. The compound flurazole, sold under the trade name Screen by Monsanto Chemical Co., when applied to sorghum seed, makes it possible to use acetochlor and alachlor selectively in sorghum. Similarly, oxabetrinil (Concep II) and fluxofenim (Concep III), discovered by Ciba Agrochemical, are seed treatments to protect sorghum seed from injury by chloroacetamide herbicides. None of these safeners is a herbicide.

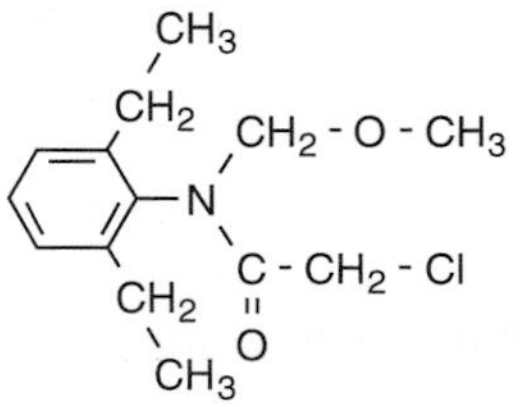

Figure 12.13 Structure of alachlor.

V. AMINO ACID BIOSYNTHESIS INHIBITORS

When a cell divides, the information necessary to form new cells is carried in genes by DNA and is subsequently expressed in structural and enzymatic proteins. Information in plant cells flows from nucleic acids to proteins, but not in the opposite direction. Any disruption of this information flow leads to growth inhibition. Protein synthesis is necessarily preceded by amino acid synthesis. Three sites for amino acid biosynthesis that include three different enzyme systems are important sites of herbicide action (Duke, 1990):

(a) Inhibition of aromatic amino acid synthesis, specifically inhibition of 5-enolpyruvylshikimate 3-phosphate synthase = EPSPS

(b) Inhibitors of branched chain amino acid synthesis, specifically inhibition of acetolactate synthase = acetohydroxyacid synthase = ALS

(c) Inhibition of glutamine synthetase = GS

Plants synthesize all essential amino acids and, in theory, blocking biosynthesis of any one will kill the plant. The three enzymes just listed are firmly established as primary sites of action of three herbicide families and some other herbicides.

A. AROMATIC AMINO ACIDS

Glyphosate was released by Monsanto Chemical Co. in 1971. It is now sold under several trade names by Monsanto and other companies. The structure of the amino acid glycine is underlined in Figure 12.14; glyphosate, the N-phosphonomethyl derivative of glycine, is a nonselective, foliar herbicide with limited to no soil activity because of rapid and nearly complete adsorption. It controls perennial grasses well and has an advantage over paraquat, because glyphosate translocates. It is the only commercial herbicide that inhibits EPSP synthase. The enzyme is common in the synthetic pathways leading to the aromatic amino acids phenylalanine, tyrosine, and tryptophan. These amino acids are essential in plants as precursors for cell

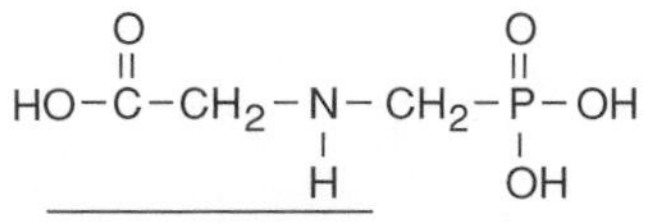

$$HO-\overset{\overset{\displaystyle O}{\|}}{C}-CH_2-\underset{\underset{\displaystyle H}{|}}{N}-CH_2-\overset{\overset{\displaystyle O}{\|}}{\underset{\underset{\displaystyle OH}{|}}{P}}-OH$$

Figure 12.14 Structure of glyphosate with glycine underlined.

wall formation, defense against pathogens and insects, and production of hormones (Duke, 1990). The enzyme is not found in animals and glyphosate has very low mammalian toxicity. Secondarily glyphosate affects respiration, photosynthesis, and protein synthesis. It is active only postemergence because it is completely and rapidly adsorbed on soil colloids. Its nonselectivity means that it will affect, if not kill, almost any green plant it contacts. Low application volume is more effective than high volume and small plants are more readily controlled than large ones. Paraquat, a photosynthetic inhibitor (see Section II,A), acts quickly (1–2 days) on most plants. Glyphosate activity usually cannot be perceived for 7 to 10 days after application. One glyphosate formulation is used as an aquatic herbicide. Transgenic soybean and canola resistant to glyphosate have been created and marketed. There is one report from Australia[1] of annual ryegrass resistance to glyphosate, but resistance has not been noted anywhere else.

B. Branched-Chain Amino Acids

1. Sulfonylureas

In the 1980s the sulfonylureas were introduced by Du Pont Co. The core structure for the sulfonylureas (Figure 12.15 shows the structure of chlorsulfuron) combines the urea and triazine groups, but the primary mechanism of action is inhibition of amino acid synthesis, not photosynthesis. Secondarily, they inhibit photosynthesis, respiration, and protein synthesis. Plant symptoms include chlorosis, necrosis, terminal bud death, and vein discoloration. The site of action for the sulfonylureas catalyzes the first step in the biosynthesis of the three branched-chain aliphatic amino acids valine, leucine, and isoleucine. A secondary effect of their action is cessation of plant growth (stunting) due to cessation of cell division and slow plant death. Tolerance is related to a plant's ability to detoxify the herbicide.

Table 12.7 shows the range of selectivity of sulfonylureas. A notable attribute of these herbicides is that they are active at rates in the range of 8 to 80 g/ha. This is as significant a reduction in the quantity of herbicide

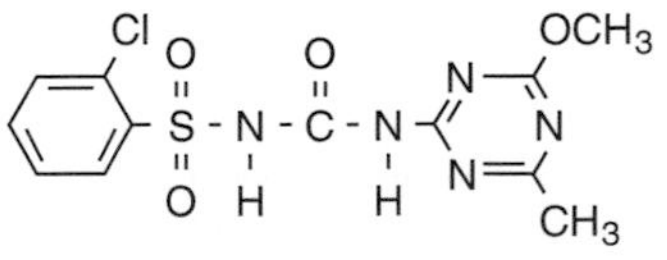

Figure 12.15   Structure of chlorsulfuron.

[1]Personal communication, 1996, M. D. Devine, Univ. of Saskatchewan.

Table 12.7

Sulfonylurea Herbicides

Herbicide		Primary use
Common name	Trade name	
Bensulfuron	Londax	Rice
Chlorimuron	Classic	Soybeans
Chlorsulfuron	Glean/Telar	Cereals, noncrop
Ethametsulfuron	Muster	Canola (rapeseed)
Metsulfuron	Ally/Escort	Cereals, noncrop
Nicosulfuron	Accent	Corn
Oxasulfuron	Expert	Soybean
Primisulfuron	Beacon	Corn
Prosulfuron	Peak	Cereals
Rimsulfuron	Matrix/Basis/Titus	Corn
Sulfometuron	Oust	Noncrop
Triasulfuron	Amber	Cereals
Tribenuron	Express	Cereals
Triflusulfuron	Upbeet/Debut	Sugarbeets
Thifensulfuron	Pinnacle/Harmony	Cereals, corn, soybeans

required as that which occurred when 2,4-D was introduced and replaced heavy metal salts. However, older herbicides in this group tend to be persistent (newer ones, thifensulfuron, triflusulfuron, and nicosulfuron, are not) and very active on several crops. Wheat is not affected by chlorsulfuron until soil concentrations approach 100 ppb. Lentil and sugarbeet, on the other hand, are affected by soil concentrations of 0.1 ppb. This thousandfold range in activity is unprecedented in herbicide chemistry. Great care is required to use these herbicides so that their activity and weed control potential is exploited but untoward environmental problems are avoided. Several species have developed resistance to these herbicides, some in as little as 3 years, after annual use.

2. Imidazolinones

Also developed in the 1980s, the imidazolinones are active at low rates. Their mode of action is the same as the sulfonylureas, but their activity is lower. Imazapyr is not selective in crops, does not leach vertically or laterally, and can be used for weed control in forests. Imazamethabenz is a selective, postemergence herbicide for control of wild oats and some broad-leaved weeds in wheat, barley, and sunflowers. Imazethapyr (Figure 12.16) is used for pre- or postemergence control of broadleaved weeds in soybean. It has a long soil persistence, and small grains and rice can be planted within 4 months of its use, but corn, bean, and sorghum cannot be planted for 1 year. Imazethapyr is selective in leguminous crops such as soybean, bean, pea, peanut, and alfalfa. It controls a wide variety of annual grass and broadleaved species, in most cases with postemergence use. Its half-life is about 4 months, and one must be cautious about rotational crops. Table 12.8 summarizes uses for four imidazolinones. There are problems of persistence, effects on rotational crops, and rapid development of weed resistance.

3. Triazolopyrimidines

At this writing, only one triazolopyrimidine, flumetsulam, has been commercialized in the United States. It is used only in combination with trifluralin in soybean, metolachlor in soybean and corn, and clopyralid in corn to control a range of broadleaved weeds. Soil life is short, so rotational crop injury is not a problem.

4. Other

Pyrithiobac is a benzoate and the only herbicide in its chemical group. When used pre- or postemergence it controls several broadleaved and grass weeds in cotton. Although chemically different from other ALS inhibitors, it acts in the same way.

C. INHIBITORS OF GLUTAMINE SYNTHETASE

Glutamine synthetase (GS) is essential for assimilation of organic nitrogen as ammonia (Duke, 1990). Its lack leads to very high ammonia levels.

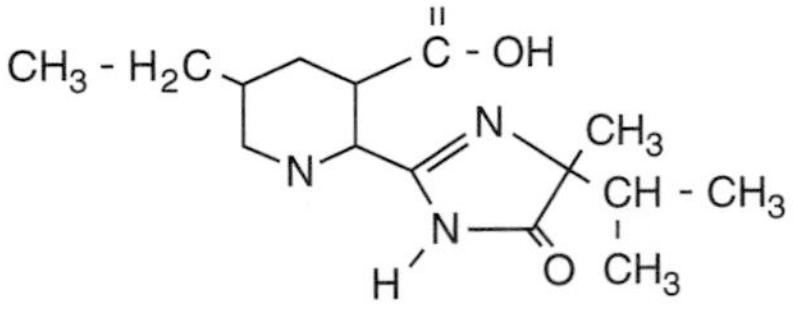

Figure 12.16 Structure of imazethapyr, an imidazolinone herbicide.

Table 12.8

Imidazolinone Herbicides

Herbicide		
Common name	Trade name	Primary use
Imazamethabenz	Assert	Wild oat control in small grains and sunflowers
Imazapyr	Arsenal	Noncrop and forests
Imazaquin	Scepter	Soybeans
Imazathapyr	Pursuit	Legume crops

Two commercial herbicides, glufosinate (phosphinothricin) (Figure 12.17) and bialaphos, inhibit GS. The former is available in the United States for complete weed control in noncrop areas and as a directed spray in field- and container-grown nursery stock. It is rapidly degraded in soil with a half-life of 7 days. Even though it is not adsorbed tightly, it does not leach because it is degraded quickly. Bialaphos, a fermentation product of *Streptomyces* spp., has a limited market in Japan. It is a proherbicide and the only product now in use derived directly from a natural product. It is nearly inactive on the GS system alone, but is rapidly converted to phosphinothricin by susceptible weeds. Glufosinate is nearly nonselective, and bialaphos selectivity seems to be based on whether a plant can convert it to the toxic form. Glufosinate has been made selective in corn because a gene coding for phosphinothricin acetyl transferase activity was isolated from the soil bacterium *Streptomyces hygroscopicus* and cloned into corn. The acetyl transferase enzyme converts glufosinate to its nonphytotoxic acetylated metabolite enabling crops to achieve resistance by rapidly metabolizing glufosinate.

VI. CELL-DIVISION INHIBITORS

All herbicides that inhibit cell division are effective on seedlings. Many herbicides affect mitosis and directly or indirectly affect microtubules (Vaughn

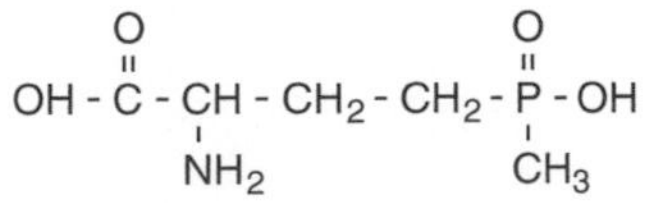

$$OH - \overset{\overset{O}{\|}}{C} - \underset{\underset{NH_2}{|}}{CH} - CH_2 - CH_2 - \overset{\overset{O}{\|}}{\underset{\underset{CH_3}{|}}{P}} - OH$$

Figure 12.17   Structure of glufosinate.

and Lehnen, 1991). In general, they act preemergence and are absorbed by roots and shoots from soil. Many herbicides inhibit cell division as a secondary mechanism of action. The precise mitotic site of action of these herbicides is not known, but it is related to disruption of polar movement and microtubule formation and function during mitosis (Vaughn and Lehnen, 1991).

A. CARBAMATES

Carbamates (Figure 12.18) are important insecticides but less important as herbicides. In 1945, the first tests of arylcarbamate esters as herbicides were reported, and propham and chlorpropham were the first selective solely soil-applied herbicides. They control some annual grasses and, with few exceptions, are ineffective on established weeds. They are effective on dodder, an important shoot parasite in many areas. Each is absorbed by roots and emerging coleoptiles of susceptible species, not by emerged foliage. Both have a short soil life, do not leach, are volatile, and may be lost under hot, windy conditions. Their greatest success has been when used under the cool, moist conditions found in grass-seed crops of the Pacific Northwest. These early carbamates also inhibited mitosis. Their action is somehow related to disruption of normal polar movement of chromosomes during anaphase of cell division. The thiocarbamates do not interfere with mitosis (see Section IV,B).

B. DINITROANILINES

The dinitroanilines or toluidines at one time held 8 to 10% of total U.S. herbicide sales. They are based on *para*-toluidine (Figure 12.19A). Examination of the structure of 2,4,6-trinitrotoluene (Figure 12.19B) illustrates the structural but not functional relationship between it, an explosive (TNT), and trifluralin, a herbicide (Figure 12.19C). These herbicides bind to tubulin, the protein from which the microtubules required for cell division and wall formation are composed (Duke, 1990). The binding inhibits tubulin polymerization, a process essential to cell division.

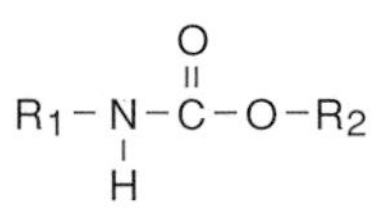

$$R_1 - N - \overset{\overset{\displaystyle O}{\|}}{C} - O - R_2$$
$$\underset{H}{|}$$

Figure 12.18   General carbamate structure.

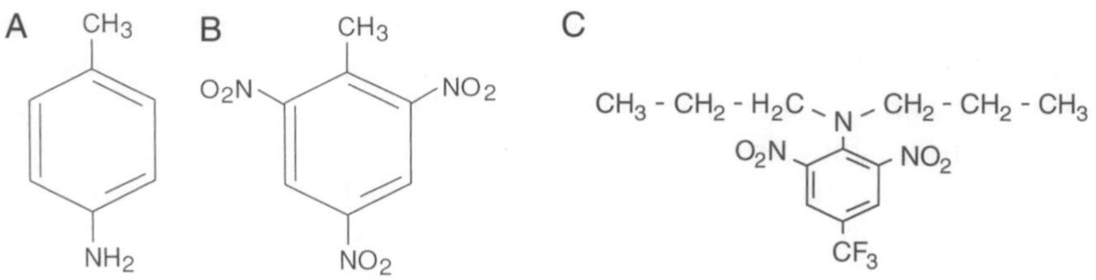

Figure 12.19 Structure of *para*-toludine (A), trinitrotoluene (B), and trifluralin (C).

This large group of herbicides is used to control grasses and some small-seeded broadleaved weeds in cotton, soybean, dry bean, potato, canola (rapeseed), and many horticultural crops (Table 12.9). The compounds all yield a yellow, liquid formulation and have low water solubility, little leaching, and generally preemergence activity. Soil persistence varies and soil residue problems have occurred. Dinitroanilines are usually applied prior

Table 12.9

Dinitroaniline or Toluidine Herbicides

Herbicide name		Application
Common	Trade	
Benefin	Balan	Controls grasses and several annual broadleaved species in lettuce, alfalfa, tobacco, and established turfgrasses
Ethalfluralin	Sonalan	Applied preplant and incorporated, it controls most annual grasses and many annual broadleaved weeds in cotton, soybean, peanut, edible bean, pea, and sunflower
Isopropalin	Paarlan	Controls grasses and several annual broadleaved species in transplant tobacco
Oryzalin	Surflan	Several annual grass and broadleaved species in soybean, tree fruits and nuts, and some ornamentals
Pendimethalin	Prowl	Weed control in corn, soybean, cotton, barley, rice, sunflower, potato, pea, onion
Prodiamine	Barricade	Controls many annual grasses and some broadleaved species in established turf and ornamentals
Trifluralin	Treflan and other names	Controls most grasses and many broadleaved weeds in a wide range of agronomic and horticultural crops

to planting with incorporation, but some can be used post-planting but are effective only when applied prior to seed germination. They are, and are often classified as, root growth inhibitors. They create stunted plants that do not emerge fully from soil. Affected plants have short, thick lateral roots with a swollen root tip. Grass shoots are short, thick, and commonly red or purple. Broadleaved plants have swollen, cracked hypocotyls.

Volatility varies among the dinitroanilines and is not a problem if recognized and controlled, usually by soil incorporation, after application. Incorporation is essential for most dinitroanilines (not for oryzalin or pendimethalin) to prevent loss by volatilization and photodecomposition. They are poorly translocated in plants and not leached in soil. Good incorporation places them in the weed seed germination zone, which enhances their effectiveness.

C. OTHER

Pronamide is a substituted amide that selectively controls many annual grass and broadleaved weeds in several vegetables and small fruits (e.g., raspberry, blackberry), in apple, cherry, and other fruits, and in woody ornamentals. It inhibits mitosis by binding to tubulin and preventing microtubule formation. Dithiopyr is a pyridine, although not a picolinic acid. It is used in rice and for weed control in turf. Dithiopyr does not bind to tubulin as pronamide and other herbicides do. It inhibits mitosis in late pro-metaphase by binding to another protein that is likely associated with tubulin.

VII. AUXIN-LIKE HERBICIDES

There are at least five classes of hormones that affect plant growth: auxins, cytokinins, gibberellins, ethylene, and abscisic acid. Plant hormones are chemicals that, after production in one place, act, in very low concentration, at another place. Auxins stimulate plant growth, particularly growth of excised coleoptile tissue. The name auxin generally refers to indoleacetic acid, but there are other active molecules. Gibberellins have varied effects on plant growth that differ between organs and between plants. They change dwarf to tall plants, affect cell division, induce fruit development, affect bud and seed dormancy, and can substitute for cold or light treatments required to induce sprouting or germination. There are no known herbicides whose primary mechanism of action is interference with gibberellin synthesis or action, but some thiocarbamates may interfere with gibberellin biosynthesis as a secondary action.

Ethylene is a plant hormone involved in many aspects of growth. There are no herbicides whose primary mechanism of action is interference with ethylene action, although some non-herbicidal compounds have been developed to stimulate fruit ripening and stem growth of flowers. Auxin-like herbicides often increase ethylene production that is linked to development of injury symptoms.

There are no herbicides based on cytokinin's or abscisic acid's structure, and there are no known herbicides that interfere with their action.

It is difficult to assign a specific physiological role to a compound within one of the five major hormone groups because they interact with each other and with other factors that influence plant growth. In a similar way, we do not know precisely how all herbicides mimic auxin action, but we know enough about them to use them intelligently. In this text and in most classifications, herbicides that interfere with plant growth are the phenoxyacetic acid or aryloxyalkanoic acids, benzoic or arylcarboxylic acids, and pyridine or picolinic acids. These growth regulator or hormone herbicides act at one or two specific auxin-binding proteins in the plasma membrane. They disrupt hormone balance and also affect protein synthesis to yield a range of growth abnormalities.

A. MIMICS OF THE ACTION OF INDOLEACETIC ACID

1. Phenoxy Acids

The chloro-substituted phenoxyacetic acids 2,4-D and MCPA were developed in 1942 in the United States and United Kingdom, respectively. When they were introduced widely after WW II, they revolutionized weed control because of their ability to kill many annual and perennial broadleaved weeds without harming cereals and other grass crops. Grass tolerance is related to different morphology but more importantly to rapid, irreversible metabolism to nontoxic molecules. Metabolism occurs in susceptible dicotyledons, but reversible, conjugated products are formed. These herbicides were accepted readily by growers because they were inexpensive and easy to apply in low water volumes. The inorganic salt herbicides that preceded them were not expensive, but large amounts had to be applied, in large volumes of water, and cost could be high, even though poor weed control was common. The phenoxyacetic acids are absorbed by roots and shoots and readily translocated in plants. They have low mammalian toxicity, are nonstaining and nonflammable, and do not have long environmental persistence.

Auxin mimics are effective herbicides because of maintenance of high tissue concentrations. They affect proteins in the plasma membrane, interfere with RNA production, and change properties and integrity of the plasma membrane. The rate of protein synthesis and RNA concentration increase as persistent auxin-like materials prevent normal and necessary fluctuation in auxin levels required for proper plant growth. Sugars and amino acids in reserve pools are mobilized by the action of auxin mimics. This is followed by, or occurs concurrently with, increased protein and RNA synthesis and degradation and depolymerization of cell walls. There are chemical structural requirements that must be satisfied for a herbicide to interfere with auxin activity. These include a negative charge on a carboxyl group in a particular orientation with respect to the ring and a partial positive charge associated with the ring that is a variable distance from the negative charge. These spatial and charge requirements enable herbicide molecules to interact precisely with the surface of receptor proteins.

Growth regulator herbicides are not metabolically stable in plants and are metabolized to a variety of different products. They are not resistant to metabolism, but plants cannot control their concentration as they can control concentration of natural plant hormones. This is an important reason for their activity. Physically, their action blocks the plant's vascular system because of excessive cell division and excessive growth with consequent crushing of the vascular transport system. External symptoms include epinastic (twisting and bending) responses, stem swelling and splitting, brittleness, short (often swollen) roots, adventitious root formation, and deformed leaves. All or a few of these symptoms may appear in particular plants and activity is often due to two or more actions at the same time.

Use of these translocated, growth-regulating herbicides offers significant advantages, but they have limitations. Advantages include the need for only small quantities and foliar application that can kill roots deep in soil because of phloem translocation. Low doses keep residual problems to a minimum. However, limitations are just as real and important. Only roots attached to living shoots in the right growth stage are killed. A uniform stage of growth is often required and very difficult to achieve with a variable plant population whose individuals emerge over time and grow at different rates. Residual effects can be important if soil remains dry after application.

2,4-D (Figure 12.20) is a white, crystalline solid, slightly soluble in water. Soon after it and its many relatives were developed, it became obvious that substitution for hydrogen in the carboxyl group affected activity. Therefore, a great deal of work was done on formulation to develop the now-dominant ester and amine formulations (Figure 12.20).

These forms are important because of their differing ability to penetrate plants and differences in volatility. In general, phytotoxicity of esters is

Basic phenoxyacetic acid = 2,4-D

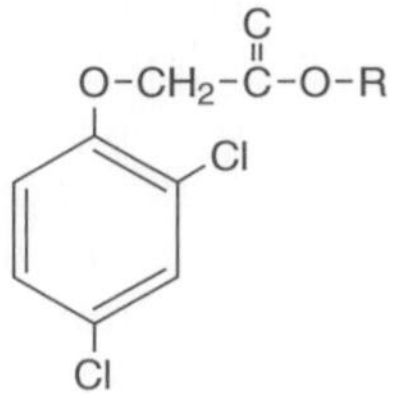

Salts	Amines	Esters
Ammonium NH_4^+	dimethylamine $-N{<}^{CH_3}_{CH_3}$	ethylester $-CH_2CH_3$
Sodium Na		
	isopropylamine $-NH_2-CH{<}^{CH_3}_{CH_3}$	isopropylester $-CH{<}^{CH_3}_{CH_3}$
	triethanolamine $NH-(CH_2-CH_2OH)_3$	butoxyethylester $-CH_2-CH_2-O-(CH_2)_3-CH_3$
		propylene glycol butyletherester $(-CH_2(CH_3)-CH_2-O)_x-(CH_2)_3-CH_3$

Figure 12.20 Salt, amine, and ester forms of phenoxy acids.

higher on an acid equivalent basis than it is for amine or salt forms. Technically, amines are also salts but have been distinguished because of their different chemical properties. Amines are, in general, soluble in water and used in aqueous concentrate formulations. Esters are oil-soluble but may be applied as water emulsions with a suitable emulsifying agent. They are more toxic to plants because they are more readily absorbed by plant cuticle and cell membranes. The methyl, ethyl, and isopropyl esters are no longer commercially available because of high volatility. The butoxyethyl ester and propylene glycol butylether ester have low volatility and thereby reduce, but do not eliminate, volatile movement to adjacent crops.

Symptoms often appear within hours of application, and usually within a day. The most obvious symptom is an epinastic response resulting from differential growth of petioles and elongating stems. Leaf and stem thickening leading to increased brittleness often appear quickly. Color changes, cessation of growth, and sublethal responses occur. Plants often produce tumor-like proliferations and excessive adventitious roots. The effective amount varies with each weedy species, its stage of growth at application, and the formulation applied. As plants mature, they can still be controlled by growth regulator herbicides, but more is required.

MCPA, developed in England, differs from 2,4-D by the substitution of a methyl (CH_3) group for chlorine at the 2 position of the ring. Uses are similar and performance is nearly identical. MCPA is more selective than 2,4-D in oats, but less 2,4-D is required to control many annual weeds. MCPA persists 2 to 3 months in soil, whereas 2,4-D persists about 1 month. Formulations are the same for each compound. MCPA is used more in the United Kingdom and in Europe than 2,4-D. MCPA is used in peas and flax in the United States because they are more susceptible to injury from 2,4-D.

The herbicide 2,4,5-T is no longer available in the United States. It is more effective against woody plants than 2,4-D. It was developed for brushy rangeland weeds and tree control. It can be formulated as an amine and ester just as 2,4-D or MCPA can, but it is more persistent than either. For several years, it was marketed in combination with 2,4-D for broad-spectrum control of broadleaved weeds. It has low mammalian toxicity but longer soil persistence than other phenoxy acids. It is no longer registered because of a dioxin contaminant found during its use for defoliation during the Vietnam War.

It does not take long after activity is found with an acetic acid derivative for the question of activity of the propionic (three-carbon), butyric (four-carbon), pentyl (five-carbon), or longer chain to be asked. Very early in the development of these compounds, it was found that a chain with an even number of carbons had herbicidal activity but a chain with an odd number did not. This is because the even-number carbon chain is broken down through beta oxidation (cleavage of two carbon units) to produce 2,4-D, MCPA, or the appropriate analog with a 2-carbon chain. A chain with 3, 5, 7, etc., carbons will also be broken down by beta-oxidation, but the final product is an alcohol that has no herbicidal activity. Thus, it is only the even-numbered carbon chains that are of interest as herbicides. However, as is true for many generalizations about herbicides, this one is wrong. Straight chains follow the rule, but iso- or branched chains do not.

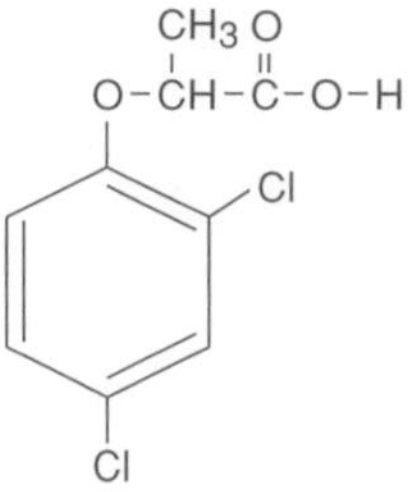

$$\text{O-CH-C-O-H}$$

Figure 12.21   Structure of 2-(2,4-dichlorophenoxy)propanoic acid.

The alphaphenoxypropionic acids are widely used in Europe for weed control in small grains. Their structure (compare Figures 12.20 and 12.21) has three carbons in a branched chain that acts like a two-carbon chain. These compounds are dichloroprop (the analog of 2,4-D, Figure 12.21) or mecoprop (the analog of MCPA). Mecoprop was introduced in Europe as a complement to MCPA because of its ability to control catchweed bedstraw and common chickweed. Previously, these weeds could only be controlled by sulfuric acid or the substituted phenols. Dichloroprop is effective against weeds in the Polygonaceae.

Another interesting part of the history of phenoxy acid herbicides is the phenoxybutyrics. MCPB and 2,4-DB [or 4-(2,4-DB)] were used widely. 2,4-DB is selective postemergence to the crop for annual broadleaved weed control in peanut, soybean, and seedling forage legumes. Plants, through their enzyme composition, determine selectivity of 2,4-DB. Young alfalfa is less susceptible than older alfalfa, because older plants have a more efficient and widespread beta oxidation system and are able to break down 2,4-DB to 2,4-D, which is immediately toxic to alfalfa.

2. Arylaliphatic or Benzoic Acids

Figure 12.22 shows the structure of the benzoic acid aspirin and the structure of the herbicide dicamba, also a benzoic acid. These structures

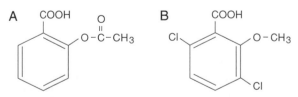

Figure 12.22   Structure of aspirin (A) and dicamba (B).

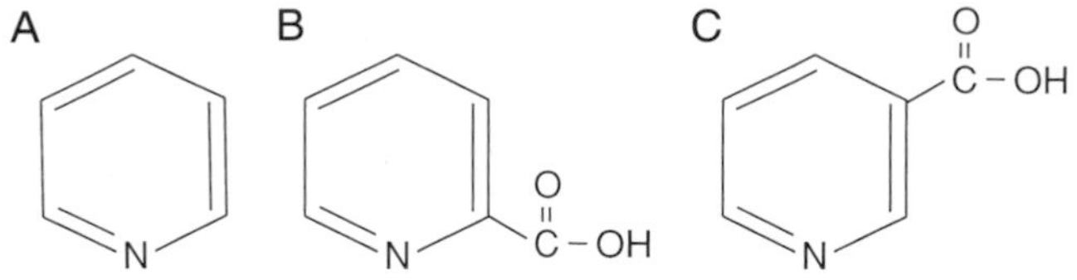

Figure 12.23   Structure of pyridine (A), picolinic acid (B), and nicotinic acid (C).

illustrate that herbicide chemistry is not strange or unique and that herbicides are related to other common chemicals. Dicamba is a growth regulator, with a weed control spectrum similar to that of 2,4-D, but it is more effective on many weeds at lower rates and more effective on perennial weeds, which 2,4-D does not control well. It has more foliar activity than 2,4-D and the other phenoxy herbicides, and is often used in combination with one or more of them for weed control in small grains and turf. It does not control mustards well, but is very effective on weeds in the Polygonaceae, which the phenoxy acids are weak on. This is part of the rationale for combinations. It is approved for use in cereals and corn and persists in soil longer than phenoxy acids do.

3. Picolinic Acids

The final group of growth regulator herbicides is based on the pyridine ring (Figure 12.23A). By adding a carboxyl group, picolinic acid (Figure 12.23B) is created. Moving the carboxyl group creates nicotinic acid (Figure 12.23C), the basis of the essential B vitamin niacin.

The development of the picolinic acid herbicides is an interesting tale. Scientists at Dow Chemical Co. were working with a pyridine-based structure to inhibit nitrification; the conversion of ammonia in soil to nitrate, the available form for plants. The general process is

$$NH_4^+ \quad \underline{\text{Nitrosomonas}} \quad NO_2^- \quad \underline{\text{Nitrobacter}} \quad NO_3^-.$$

↑		↑
Common form In fertilizer	Responsible microorganisms	Readily available to plants and leachable

Nitrification occurs readily in many soils and is desirable. Nitrate ions are readily available to plants but leachable. Therefore, one wants to have enough nitrate available so plants grow well, but not so much that it leaches and might lead to groundwater contamination. Therefore, if nitrification can be slowed but not stopped, leaching will be reduced and plant availability maintained. The scientists were working with the structure shown below and applying it in combination with ammonium fertilizers. The presence of the ammonia fertilizer made it possible for microorganisms to aminate the 4 position, and carboxylation of the trichloromethyl group was also carried out by soil microorganisms

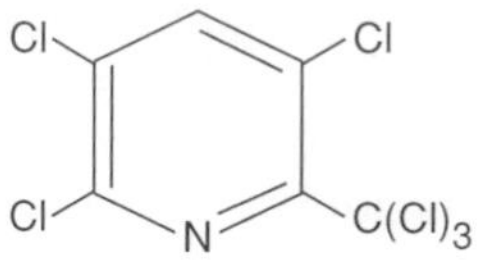

2,3,5-trichloro-6-(trichloromethyl)pyridine.

to yield picloram. The scientists saw plants dying where they were not supposed to, and by studying their work they discovered a herbicide when they had been looking for an inhibitor of nitrification.

Picloram (Figure 12.24) gives excellent control of woody plants and many annual and perennial broadleaved species. It is not effective on grasses, nor is it particularly effective on members of the Brassicaceae. It is chemically similar to, but not directly related to, other growth regulator herbicides. Picolinic acids produce epinastic and other effects typical of growth regulators and are active after uptake through foliage and through roots. Picloram is effective on many perennial broadleaved plants, including field bindweed and Canada thistle. Picolinic acids are translocated in plants after pre- or postemergence application. Doses as low as 0.25 kg/ha are

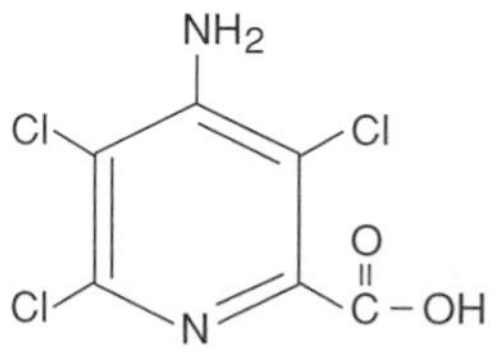

Figure 12.24 Structure of picloram.

effective. Grasses, even as seedlings, are relatively resistant. Picloram is very persistent and lasts for several months up to 1 year or longer to affect succeeding crops. It is water soluble, not highly adsorbed, and therefore susceptible to leaching. These characteristics are undesirable, although its high activity is desirable for control of perennial weeds.

Clopyralid is less persistent and leachable and effective for control of broadleaved species. It is selective in corn and sorghum and not effective on grasses or mustards. It is especially effective on Polygonaceae and Asteraceae in field crops and turf. Clopyralid can be used in sugarbeet, a crop picloram kills. A primary advantage is its high activity against Canada thistle.

Triclopyr is effective on woody plants and broadleaved weeds and has been used for control of ash, oak, and other root-sprouting species. Most grasses are tolerant and while it is not used in many crops (it is used in rice), it is used as a turf herbicide.

4. Other

Quinclorac is a quinolinecarboxylic acid derivative and the only available herbicide from this group. It is used for control of some annual grasses in rice. Given that use, it seems anomalous, but it also has good activity on some annual and perennial broadleaved weeds, including field bindweed.

B. INHIBITORS OF IAA TRANSPORT

Naptalam is a selective, preemergence herbicide for control of a wide range of annual broadleaved weeds and grasses in broadleaved crops, including soybean, peanut, cucumber, melon, and established woody ornamentals. It has a unique antigeotropic property. Because microbial breakdown is slow it provides weed control for 3 to 8 weeks. Technically, it is a phthalmic or benzoic acid, but it can also be regarded as a substituted amide. It is no longer widely used but is interesting because of its ability to interfere with auxin transport.

This ability is shared with compounds known as morphactins. These materials have specific antigeotropic activity and prevent the normal downward movement of roots in soil and of shoots toward light. Their herbicidal activity is minimal to nonexistent, but they have been used to promote activity of growth-regulator herbicides.

VIII. INHIBITORS OF RESPIRATION

Plants obtain energy by transforming the electromagnetic energy of the sun into stored chemical energy of carbohydrate molecules through photosynthesis. They must transform that energy to a form suitable for driving life processes. That transformation, called respiration, is analogous to the conversion of fossil fuel to electric power. Respiration is the removal of reducing power from carbohydrates, fats, or proteins and its transfer to oxygen with the concomitant trapping of released energy in ATP.

A. UNCOUPLERS OF OXIDATIVE PHOSPHORYLATION

Herbicides interfere with respiration by uncoupling oxidative phosphory-lation. Uncoupling is like braking while continuing to press a car's accelera-tor. Energy is released as electrons pass down the electron transport chain to oxygen and is trapped by converting ADP to ATP (oxidative phosphory-lation). If you uncouple and keep the accelerator down, the motor will race and overheat (Corbett, 1974).

1. Arsenites

Arsenic has been known as a biological poison for many years. Arsenic-based insecticides were used in orchards in the late 1800s. Arsenic trioxide, an insoluble soil sterilant, was used at 400 to 800 kg/ha, but is no longer registered for United States use. Its residues remained for many years and weed control could be effective up to 5 years. Livestock were attracted to, and could be poisoned by, plants sprayed with arsenic trioxide because of the release of aromatic compounds.

The acid arsenicals such as sodium arsenite were more effective because they were translocated in plants. They are nonspecific inhibitors of sulfur-containing enzymes and also precipitate proteins and disrupt membranes. Sodium arsenite was used as a preharvest desiccant in cotton. Inorganic arsenicals are poisonous to mammals and are generally regarded as nonse-lective, foliar-contact herbicides with soil sterilant activity. They persist in soil and arsenic is no longer used except in combinations, for soil steril-ization.

2. Metallo-organics

The organic arsenicals or metallo-organics interfere with general plant growth and may affect cell division. These chemicals, based on arsonic or arsenic acids (Figure 12.25), are strong acids that decompose carbonates. They have postemergence contact activity on plant foliage, are rapidly adsorbed by soil, and do not have soil activity. They are most effective at high temperatures, but rapidly lose selectivity above about 27°C. Organic arsenicals are more phytotoxic than inorganic arsenic herbicides. The most

Figure 12.25   Structures of arsonic and arsenic acid.

toxic form of arsenic to mammals is the As^{3+} state, the form in inorganic arsenic compounds. Organics have arsenic in the 5+ state, and because it is not normally reduced to As^{3+}, organic arsenicals are less toxic to mammals.

The principal organic arsenicals, first released in the United States in the 1950s, are monosodium methane arsenate (MSMA) and disodium methane arsenate (DSMA), both derivatives of arsonic acid. Cacodylic acid, an arsinic acid, is less selective than arsonic acids. Organic arsenicals are water soluble, rapidly sorbed by soil, and do not leach except in sandy soils, and then not beyond 20 cm. They are much more toxic to annual than to perennial grasses. DSMA has been used for selective weed control in turf and cotton. Arsenicals have been used in forest weed control. They do not persist in soil because they are rapidly and completely absorbed by soil colloids. They may affect mitosis.

3. Phenols

The first synthetic organic herbicide that achieved success in the field was 2-methyl-4,6-dinitrophenol, released in 1932. Several other substituted phenols followed, but all are now only of historical interest. They are intensely yellow staining compounds, toxic to mammals, and poisonous to humans by ingestion, inhalation, or skin absorption. They were used for selective broadleaved weed control in cereals, and their activity increased directly with temperature.

DNBP [(2-(1-methylpropyl)-4,6-dinitrophenol] was used in several salt forms as a selective broadleaved weed herbicide in pea, legumes, corn, and flax. It was used, with less success, in small grains and for preemergence weed control in bean and cotton. Its selectivity depended on selective retention and uptake by foliage, and good coverage was essential. The phenol derivatives have short soil persistence. They also, secondarily, inhibit other plant processes, including photosynthesis and lipid, RNA, and protein synthesis. Pentachlorophenol was a widely used wood preservative.

IX. UNKNOWN MECHANISM OF ACTION

All present systems of herbicide classification have imperfections because the mechanism of action is unknown for some herbicides. Some of these are older herbicides, described as having a nonspecific action. An unknown mode of action may mean:

(a) It is truly unknown
(b) The herbicide has not been studied completely
(c) The herbicide is too knew to know—it is being studied

The mechanism of action is unknown for most inorganic herbicides. Available studies are old and were done with far less sophisticated analytical techniques and less knowledge than modern studies. Most of these herbicides have been used for many years and their use has declined as organic herbicides have been discovered to provide better weed control with much lower rates. Some are still used in mixtures with organic herbicides for soil sterilization.

Chemical weed control began with inorganic herbicides. Ammonium sulfamate ($NH_2SO_3NH_4$) was patented in 1942. It is a water-soluble contact herbicide used for brush and weed control in industrial and residential areas. It is nonstaining and has low mammalian toxicity. Rates of 100 to 200 kg/ha applied in 400 liters of water are required for effective brush control. Rates of 60 kg/ha in 400 liters of water control poison ivy. These rates illustrate the great change that occurred when the phenoxy acid herbicides were introduced and rates dropped from over a hundred to a few kilos per hectare.

Sodium tetraborate ($Na_2B_4O_7$) and sodium metaborate ($Na_2B_2O_4$) are nonselective, taken up by roots, and have an unknown mechanism of action. Boron accumulates in reproductive structures after translocation from roots. Boron compounds are used for long-term, nonselective weed control in industrial and power-line areas in combination with triazine and urea herbicides.

Sodium chlorate ($NaClO_3$) is a nonselective soil sterilant used on noncrop land or in combination with triazines or ureas for soil sterilization. It leaches, has foliar contact activity, and, in the past, was used widely along railroads. It is flammable when dried on foliage and many railroad fires occurred when sparks from coal-fired engines landed on sprayed plants. Sodium chloride (table salt) is an example of a herbicide that desiccates and disrupts a plant's osmotic balance. It has been used for nonselective weed control for centuries.

Sulfuric, phosphoric, and hydrochloric acid all have burning, contact activity, but because of high toxicity to users, ability to corrode equipment, and the availability of safer alternatives, they are no longer used.

Among the metallic salts, copper sulfate is one of the few still used as a herbicide. Its toxicity is due to a nonspecific affinity for various groups in cells leading to nonspecific denaturation of protein and enzymes. It is used as an algicide.

There are several organic herbicides with presently unknown mechanisms of action. DCPA is a dibenzoic or terephthalic acid. It is actually the dimethyl ester of terephthalic acid and another illustration of the ubiquity of herbicide chemistry. Polymerization of ethylene glycol terephthalic acid yields Dacron, a textile. Dacthal is a turf herbicide and 12 to 17 kg ai/ha give excellent preemergence control of crabgrass and other seedling annual grasses and some seedling broadleaved weeds. It is active only on germinating seedlings and inhibits growth of root and stem meristems. It is also used in horticultural crops and nurseries. It is selective in soybean and corn, but it has been replaced by other herbicides. It is used in vegetable crops.

Dacthal does not interfere with seed germination, only with seedling growth. Therefore, to be effective, it must be applied before plants emerge; it is exclusively a preemergence herbicide and its mechanism of action is unknown, but it has been implicated as an inhibitor of cell division and protein synthesis.

Bromoxynil and terbacil inhibit photosystem II. Dichlobenil is chemically related but has little to no foliar activity, is volatile, primarily soil-active, highly adsorbed, with little leaching, and a soil life of 2 to 6 months—almost the exact opposite of the other substituted benzonitriles (see Section II,B,5). It may exert its toxic action by inhibition of cellulose synthesis (Corbett *et al.,* 1984; Duke, 1990). Cellulose is unique to plants and interference with its synthesis offers avenues for herbicide development. Dichlobenil is effective on a wide range of annual and perennial weeds and is particularly effective preemergence on germinating seedlings. Primary uses are in ornamentals, turf, cranberries, and as an aquatic herbicide.

Difenzoquat, a pyrazolium salt, is used for the selective control of wild oats in spring cereal grain crops. Its primary mechanism of action is unknown, but it inhibits photosynthesis, ATP production, potassium absorption, and phosphorus incorporation into phospholipids and DNA. It is a postemergence herbicide that, like other bipyridiliums, has only contact foliar activity and no soil activity. It has fungicidal properties and controls powdery mildew (*Erisiphe graminis* f.sp. *Hordii*) in barley. In contrast to diquat and paraquat, it does not cause rapid burning and desiccation of plant foliage.

Pelargonic or nonanoic acid was introduced in 1995 as a contact, nonselective, broad-spectrum foliar herbicide. It is a naturally occurring, nine-carbon fatty acid found in several plants and animals. Translocation does not occur, so it is not effective against perennial weeds. There is no soil residual activity. The bipyridyliums are fast acting, but pelargonic acid is faster. Rate of kill is related to temperature, but even in cool conditions plants begin to exhibit damage 15 to 60 minutes after application and die within 1 to 3 hours. Foliage darkens and begins to look water-soaked followed by rapid wilting. The mechanism of action is unknown, but it causes rapid cell death, bleaching of chloroplasts, and general ion leakage. The primary effect may be a sudden drop in intracellular pH that causes rapid membrane deterioration.

The exact molecular site or mechanism of action of many herbicides is unknown. Research is advancing rapidly and classification will become more ordered with time. Any classification of a group as complex as herbicides creates problems because of disagreement about the best way to group them, and about their primary and secondary mechanisms of action. Classification systems are also debatable because of the insufficiency of knowledge about mechanisms of herbicide action. This chapter has integrated the presently available herbicides, but is not a catalog of all of them.

X. SUMMARY

There are many herbicides and a great deal on information about each, its chemical family, and mechanism of action. The amount of information, even in a brief chapter, can be overwhelming. Table 12.10 lists the herbicide families included in this chapter and combines their primary mechanism of action with the major plant function modified or irreversibly changed by the herbicide's activity. It is included to assist organization of the abundant information in this chapter.

THINGS TO THINK ABOUT

1. What is the primary mechanism of action of the herbicides in _____ chemical group?
2. Does chemical structure always predict a herbicide's mechanism of action? Is it a reasonably good predictor?
3. What are the major plant processes affected by herbicides?
4. Are herbicides always unique chemical molecules that are unrelated to other common chemicals?
5. What are the major mechanisms of herbicide action?
6. Is it likely that the next edition of this book will have a modified system of herbicide classification? Why?
7. Why do herbicides have so many different mechanisms of action?

Table 12.10

Summary of Herbicide Mechanisms of Action (Adapted from Corbett *et al.*, 1984)

Herbicide family	Primary action	Major function modified or disrupted
Aryloxyphenoxypropionates		
Cyclohexanediones	Lipid synthesis	
Carbamothioates		Structural organization
Chloroacetamides		
Dichlobenil	Cellulose synthesis	
Oxadiazoles		
p-Nitro-substituted- diphenyl ethers	Photo-oxidation	
Bipyridylliums	Photosynthetic electron transport diverted from light reaction I	
Ureas, triazines, triazinones, uracils, acylanilides, pyridazinones, bis-carbamates, benzonitriles	Photosynthetic electron transport (Hill reaction)	
		Energy supply
Metallorganic arsenicals, dinitrophenols	Oxidative phosphorylation uncoupled	
		DEA
		Chlorophyll destruction
Isoxazolidones, amitrole	Carotenoid synthesis	
		Low carotenoid level
Imidazolinones, sulfonylureas, glyphosate	Amino acid synthesis	
Glufosinate	Glutamine synthesis	
Carbamates, dinitroanilines	Cell division inhibited → Growth and reproduction → Grow
Phenoxyacids, benzoic acids, picolinic acids	Mimic IAA action → Growth
Naptalam	IAA transport	

8. Why aren't there more herbicides that inhibit specific plant biosynthetic processes?
9. Why don't we have a complete understanding of the precise mechanism of action of all herbicides?

LITERATURE SOURCES

NOTE: These sources have not been cited frequently in the text, but have been used extensively.

Anonymous. 1994. "Herbicide Handbook," 7th ed. Weed Sci. Soc. Am., Champaign, Illinois. 352 pp.

Ashton, F. M., and A. S. Crafts. 1981. "Mode of Action of Herbicides," 2nd ed. Wiley Interscience, New York, 525 pp.

Ashton, F. M., and T. J. Monaco. 1991. "Weed Science: Principles and Applications," 3rd ed. Wiley, New York.

Baker, N. R., and M. P. Percival (Eds.). 1991. "Herbicides: Topics in Photosynthesis," Vol. 10. Elsevier, Amsterdam. 382 pp.

Beyer, E. M., M. J. Duffy, J. V. Hay, and D. D. Schueter. 1982. Sulfonylurea herbicides, pp. 118–183 *in* Herbicides: Chemistry, Degradation, and Mode of Action," Vol. III (P. Kearney and D. D. Kaufmann, Eds.). Dekker, New York.

Boger, P., and G. Sandmann (Eds.). 1993. "1989 Target Assays for Modern Herbicides and Related Phytotoxic Compounds." CRC Press, Boca Raton, Florida. 299 pp.

California Weed Conference. 1989. "Principles of Weed Control in California," 2nd ed. Thompson Publications. Fresno, California.

Caseley, J. C., G. W. Cussans, and R. K. Atkin. 1991. "Herbicide Resistance in Weeds and Crops." Butterworth Heinemann, Oxford, UK.

Cobb, A. 1992. "Herbicides and Plant Physiology." Chapman & Hall, New York. 176 pp.

Corbett, J. R. 1974. "The Biochemical Mode of Action of Pesticides." Academic Press, London. 330 pp.

Corbett, J. R., K. Wright, and A. C. Baillie. 1984. "The Biochemical Mode of Action of Pesticides," 2nd ed. Academic Press, London. 382 pp.

DePrado, R., J. Jorrin, and L. Garcia-Torres (Eds.). 1997. "Weed and Crop Resistance to Herbicides." Kluwer Academic, Dordrecht, The Netherlands. 340 pp.

Devine, M., S. O. Duke, and C. Fedtke. 1993. "Physiology of Herbicide Action." PTR Prentice Hall, Englewood Cliffs, New Jersey. 441 pp.

Dodge, A. D. (Ed.). 1989. "Herbicides and Plant Metabolism." Cambridge University Press.

Duke, S. O. (Ed.). 1985. "Weed Physiology," Vol. I and II. "Herbicide Physiology." CRC Press, Boca Raton, Florida.

Duke, S. O. 1990. Overview of herbicide mechanisms of action. *Env. Health Perspectives* **87:**263–271.

Duke, S. O. (Ed.). 1996. "Herbicide-Resistant Crops. Agricultural, Environmental, Economic, Regulatory and Technical Aspects." CRC Press, Boca Raton, Florida. 420 pp.

Duke, S. O., J. Lydon, J. M. Becerril, T. D. Sherman, L. P. Lettnen, Jr. and H. Matsumoto. 1991. Protoporphyrinogen oxidase-inhibiting herbicides. *Weed Sci.* **39:**465–473.

Fedtke, C. 1982. "Biochemistry and Physiology of Herbicide Action." Springer-Verlag, Berlin. 202 pp.

Frehse, H. (Ed.) 1991. "Herbicide Chemistry. Advances in International Research, Development and Legislation." VCH, Weinheim. 666 pp.

Gronwald, J. W. 1991a. Inhibition of carotenoid biosynthesis, pp. 131–171 *in* "Herbicides," Vol. 10. "Topics in Photosynthesis" (N. R. Baker and M. P. Percival, Eds.). Elsevier, Amsterdam.

Gronwald, J. W. 1991b. Lipid biosynthesis inhibitors. *Weed Sci.* **39:**435–449.

Gunsolus, J. L., and W. S. Curran. 1991. Herbicide mode of action and injury symptoms. N. Central Reg. Ext. Pub. 377. U of Minnesota, St. Paul, Minnesota. 15 pp.

Hassall, K. A. 1990. "The Biochemistry and Uses of Pesticides: Structure, Metabolism, Mode of Action, and Uses in Crop Protection." VCH, New York. 536 pp.

Holt, J. S., S. B. Powles, and J. A. M. Holtum. 1993. Mechanisms and agronomic aspects of herbicide resistance. Ann. Rev. Plant Physiol. Plant Mol. Biol. **44:**203–229.

Kearney, P. C., and D. D. Kaufman (Eds.). 1976. "Herbicides: Chemistry, Degradation, and Mode of Action," Vols. 1 and 2. Dekker, New York.

Kearney, P. C., and D. D. Kaufman (Eds.). 1988. "Herbicides: Chemistry, Degradation, and Mode of Action," Vol. 3. Dekker, New York.

Kirkwood, R. C. 1991. "Target Sites for Herbicide Action." Plenum, New York.

Mets, L., and A. Thiel. 1989. Biochemistry and genetic control of the photosystem II herbicide target site, pp. 1–24 *in* "Target Sites of Herbicide Action" (P. Bogor and G. Sandmann, Eds.). CRC Press, Boca Raton, Florida.

Moreland, D. E., J. B. St. John, and F. D. Hess (Eds.). 1982. "Biochemical Responses Induced by Herbicides." Amer. Chem. Soc. Symp. Series 181. 274 pp.

Powles, S. B., and J. A. M. Holtum. 1994. "Herbicide Resistance in Plants: Biology and Biochemistry." Lewis, Boca Raton, Florida. 360 pp.

Shaner, D. L., and S. L. O'Connor (Eds.). 1991. "The Imidazolinone Herbicides." CRC Press, Boca Raton, Florida. 290 pp.

Vaughn, K. C., and L. P. Lehnen, Jr. 1991. Mitotic disruptor herbicides. *Weed Sci.* **39:**450–457.

Weed Science Society of America. 1994. "Herbicide Handbook," 7th ed. Champaign, Illinois. 301 pp.

Wilkinson, R. E. 1983. Gibberellin precursor biosynthesis inhibition by EPTC and reversal by R-25788. *Pestic. Biochem. Physiol.* **19:**321–329.

Wilkinson, R. E. 1986. Diallate inhibition of gibberellin biosynthesis in sorghum coleoptiles. *Pestic. Biochem. Physiol.* **25:**93–97.

Young, A. J. 1991. Inhibition of carotenoid biosynthesis, pp. 131–171 *in* "Herbicides," Vol. 10, "Topics in Photosynthesis." (N. R. Baker and M. P. Percival, Eds.). Elsevier, Amsterdam.

Chapter 13

Herbicides and Plants

FUNDAMENTAL CONCEPTS

- There are several environmental, chemical, and physiological factors that affect a herbicide's activity and selectivity.
- The most important determinants of herbicide selectivity are the rate and amount absorbed, translocated, and metabolized by two species.
- Several plant and environmental factors interact to determine selectivity.

LEARNING OBJECTIVES

- To understand the difference between herbicide drift and volatility and the importance of each.
- To know techniques to control drift and volatility.
- To know the external factors that influence spray retention and herbicide absorption.
- To know the effect of moisture, temperature, and light on herbicide action.
- To know the relative advantages and disadvantages of foliar- and soil-applied herbicides.
- To understand the difference between shoot and root absorption of herbicides.
- To understand the role of absorption, translocation, and metabolism as determinants of selectivity.

I. FACTORS AFFECTING HERBICIDE PERFORMANCE

This discussion of factors affecting herbicide performance in plants assumes that users have an applicator appropriate to the task and that it has been calibrated to apply the correct volume and the proper amount of active ingredient per acre. The discussion also assumes that the correct herbicide has been selected and that it will be applied at the right time. If these things are not ensured, they will affect herbicide performance. However, because such human errors and their results are not predictable, the discussion herein assumes that human error has been avoided.

This chapter discusses factors affecting performance from the time a herbicide molecule leaves the applicator (usually this means the nozzle tip) until it hits a target and acts.

II. GENERAL

A. REACHING THE TARGET PLANT

1. Drift

Spray drift is movement of airborne liquid spray particles. It is often unseen and may be unavoidable. It can be minimized. Drift increases with wind speed and the height above the ground at which drops are released and decreases as spray droplet size increases. Ideally, uniform drops between 500 microns (moderate rain) and 1 mm (1000 microns = heavy rain) in diameter are desired. Drops of this size minimize, but do not eliminate, drift, especially if spraying is done when wind speed is less than 5 mph. It is not uncommon, especially in arid environments, for water to evaporate within 200 to 300 ft of the point of delivery, so only the herbicide and associated organic solvents remain to drift. Table 13.1 shows spray droplet size, droplet lifetime, and the potential effect on drift (Brooks, 1947; Hartley and Graham-Bryce, 1980). For comparison, a number 2 pencil lead is about 2,000 microns in diameter; a paper clip is 850, a toothbrush bristle is 300, and a human hair is about 100 microns in diameter.

Nozzle tips give pattern to sprays and break up the liquid stream into small particles. Hydraulic nozzles produce a range of droplet sizes rather than just one. Droplet size is a function of orifice size, operating pressure, and surface tension of the spray solution. Smaller nozzle orifices, higher pressures, and lower surface tensions produce more small drops. All hydraulic nozzles produce a normal (Figure 13.1) distribution of spray drop sizes. As size decreases and pressure increases, a greater percentage of small droplets is produced.

Table 13.1

Effect of Spray Droplet Size on Evaporation and Drift

Droplet diameter (microns)	Type of droplet	Precipitation (in./hr)	Drops (no./in.^{2})	Evaporating water		Time to fall 10 ft. (sec)	Distance traveled while falling 10 ft in a 3 mph wind
				Drop life (sec)	Lifetime fall distance (in.)		
5	Dry fog	0.04	9,220,000	0.04	<1	3,960	3 miles
20	Wet fog	—	144,000	0.7	<1	—	—
100	Misty rain	0.4	1,150	16	96	10	409 feet
200	Light rain	—	144	65	1,512	—	—
500	Mod. rain	3.9	9	400	>1,500	1.5	7 feet
1,000	Heavy rain	39	1	1,620	≥15,000	1.0	4.7 feet

Source: Adapted from Bode, L. E. and R. E. Wolf. Techniques for applying postemergence herbicides. Univ. Illinois, Urbana, IL. 5 pp. Undated.

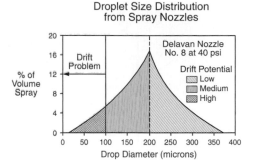

Figure 13.1   Normal distribution of spray drop size from a hydraulic nozzle.

The influence of drop size on drift is illustrated in Table 13.1. Small droplets will drift a long way in a light breeze. Large drops decrease drift problems. Spraying in strong wind should be avoided, but it is difficult when large areas must be sprayed with herbicides that require application at particular growth stages or before crop emergence. Farmers and other applicators must apply herbicides at the proper time. However, if other considerations take precedence over drift avoidance, problems may ensue when the applicator's or a neighbor's crops are injured or our environment is contaminated by improper application.

As discussed in Chapter 15, boom height is normally fixed and not adjusted. However, as illustrated (Table 13.1), release height influences drift potential simply by allowing drops to remain suspended longer.

Because drift is an inevitable problem, several techniques have been developed to control it. The first, and simplest, is to reduce spray pressure and create fewer small drops. Increasing drop size from 20 to 200 microns decreases coverage 200 times and increases drop lifetime from 0.7 to 65 seconds (Table 13.1). Small drops attain a horizontal trajectory quickly and water evaporates before the drop contacts plants. After water evaporation, pesticides can become airborne aerosols that fall out with rain or sprinkler irrigation, but one cannot be sure where they will fall. Droplets larger than 150 microns resist evaporation long enough to reach the target.

Low-volume applicators that use a rotary atomizer reduce water requirements and equipment weight and are known as ULV (ultralow volume) or CDA (controlled drop) applicators. Drop size is controlled usually between 150 and 300 microns and total volume can be as low as 0.5 gal/A. CDA applicators have been available for several years, but have not achieved a high level of commercial acceptance because of high cost, frequent perfor-

mance failures, and the widespread acceptance and availability of hydraulic nozzles.

Nozzles that incorporate air or facilitate use of a foaming adjuvant are
available. They produce coarse droplets, but still up to 5% loss is possible
within 1,000 ft of the point of application. Foam adjuvants increase spray
volume 2 to 3 times and may also act as wetting agents and increase
phytotoxicity. Water-soluble thickening agents (thixotropic agents) increase
average droplet size. These water-imbibing polymers create a particulated
gel spray. The smallest droplet size is predetermined by the polymer's
particle size. There is usually no phytotoxic benefit, but drift is reduced.
Chapter 16 discusses use of invert emulsions to increase spray solution
viscosity and reduce drift.

A recirculating sprayer was developed (McWhorter, 1970) to apply herbicides. The hypothesis was that environmental contamination could be reduced without loss of weed control if spray that did not strike a target
plant was captured, recirculated, and reused. This was done by spraying
horizontally above crops. The system was only successful with a foliar-
applied, postemergence herbicide. The system eliminated vertical spray
movement during spraying. The sprayer successfully applied glyphosate to
control weeds above the canopy of cotton and soybeans in the southern
United States, but was never a commercial success.

Wax bars impregnated with 2,4-D (McWhorter, 1966) have been used
to control weeds in crops and in turf. They were not satisfactory because
2,4-D was hard to impregnate uniformly, bars tended to self-destruct, and
wax melting was not uniform, so herbicide application was not uniform.

The development of glyphosate led to renewed interest in the question,
why spray? Subsequently, wiper technology was developed (Derting, 1987),
first with shag carpet on a roller and eventually with rope wicks. The
ropes or carpet, saturated with glyphosate or a similar herbicide, acted as
conducting wicks. When the ropes were moved horizontally above the crop,
weeds, growing above the crop, contacted the herbicide in the rope and,
through control of solution concentration, sufficient herbicide was applied
to kill weeds without affecting crops. This is an excellent way to eliminate drift.

Drift is not just a problem of historical interest. In the late 1980s, clomazone drift affected many nontarget plants after application to soybeans
in the midwestern United States. More recently, the U.S. Environmental
Protection Agency has expressed concern about drift from sulfonylurea
herbicides that can damage flowers, seeds, and fruits of grapes, alfalfa,
cherries, and asparagus. The state of Washington has prohibited use of
sulfonylurea herbicides within 14 miles of nontarget crops because of drift

potential and requires 24 hours notification of intent to spray (Anonymous, 1996).

2. Volatility

Volatility measures the tendency of a chemical to vaporize, which permits movement as a gas. Drift is movement as a liquid. Volatility is related to a herbicide's vapor pressure and ambient temperature. Volatilized herbicides may cause damage in another place or reduce effectiveness at the point of application. The most common example of volatilization is esters of phenoxyacetic acids (e.g., 2,4-D). Figure 13.2 shows the germination of pea seeds after exposure to volatile 2,4-D (Mullison and Hummer, 1949). Because of the experiment's design, only volatility could have caused the observed effects. Methyl and ethyl (one- and two-carbon) esters are more volatile than the five-carbon amyl ester, and all esters are more volatile than amine or sodium salts. Because of the high volatility of methyl, ethyl, and other short-chain esters of 2,4-D, it is no longer possible to purchase and use them. Volatility is not limited to phenoxyacetic acids. Many carbamothioates and dinitroanilines are volatile and can be lost from the area of application if not incorporated in soil. Volatility problems are not going to go away, and intelligent herbicide use demands continuing attention to the risk.

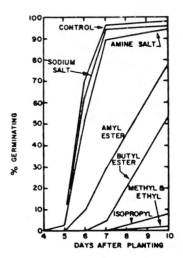

Figure 13.2   Percent germination of pea seeds after exposure to 2,4-D formulations (Mullison and Hummer, 1949).

III. FOLIAR-ACTIVE HERBICIDES

A. SPRAY RETENTION

If herbicides do not drift or volatilize, the next factor that affects performance is retention on plant surfaces. A foliar herbicide must remain on leaves long enough for absorption to occur. Plants differ in their ability to retain water on leaf surfaces (Blackman *et al.*, 1958) (Table 13.2). Barley has upright leaves disposed nearly perpendicular to the soil surface, and sprays run off easily. Peas have a waxy cuticle that makes it difficult for liquids to remain on the leaf surface. Flax is an upright plant with small vertical leaves, but sunflowers and white mustard are large plants with broad leaves. Broadleaved species with large, flat leaves disposed parallel to the ground retain liquid droplets more easily than grass leaves.

The use of propanil for selective control of green foxtail in hard red spring and durum wheat (Eberlein and Behrens, 1984) illustrates the influence of spray retention on herbicide activity (Table 13.3). The data show that wheat has a slightly higher concentration in terms of milligrams of propanil retained per plant, but green foxtail absorbs and retains more. Propanil is used in rice and is selective because of rapid metabolism rather than differential retention (Yih *et al.*, 1968a and b).

1. Leaf Properties

The ability of a herbicide to control weeds selectively can depend on morphology (shape) and chemical variations between plant surfaces. Large, broad leaves disposed parallel to the soil surface are easier to hit with spray solutions applied by most field sprayers. Herbicide molecules are more

Table 13.2

Spray Retention by Different Species (Blackman *et al.*, 1958)

Species	Number of leaves	Height (cm)	ml water retained/ g shoot weight
White mustard	2	5–7	2.5
Sunflower	2	6	2.0
Flax	2	5	1.1
Pea	2	5–7	0.4
Barley	3	15–20	0.3

Table 13.3

Retention of Propanil by Wheat and Green Foxtail
Plants at the 3-leaf Stage of Growth
(Eberlein and Behrens, 1984)

Species	Propanil retained	
	mg/plant	mg/g fresh weight
Green foxtail	0.69	19.09
Wheat	1.20	2.49

likely to contact and remain on broad leaves of dicots than on grass leaves, which are often disposed perpendicular to the soil surface.

Velvetleaf, crabgrass, and some species of mallow have hairy leaf surfaces that prevent direct, quick contact of spray droplets with the leaf surface. But when a hairy surface becomes saturated, herbicide entry may be promoted because hairiness delays evaporation.

Because leaves are one of the principal entry points for herbicides, their structure and function are important. The primary leaf tissues are epidermal, mesophyll, and vascular. The epidermis is present on upper and lower leaf surfaces and consists of a single layer of interlocked cells with no chloroplasts. It is covered by the cuticle, which is often layered with waxes. These constitute a varnish-like layer or film that retards movement of water in and out of leaves.

All leaves have cuticles, a formidable barrier to herbicide entry, yet herbicides enter plants. Surface-active agents (surfactants) are used in some formulations to assist entry, and they often determine the amount of herbicidal activity obtained because of their influence on leaf surface penetration. Water is not compatible with many plant surfaces, especially those with thick or very waxy cuticles. Surfactants lower surface tension of liquid systems, increasing their tendency to spread and their ability to wet leaf surfaces. Surfactants aid penetration.

It is incorrect to assume that plants with thick, waxy cuticles absorb less herbicide or absorb the same amount more slowly than plants with thin cuticles. The reason is that cuticle hydration and composition are more important factors in herbicide absorption than cuticle thickness. Plant leaves growing in shade generally have thinner cuticles than those growing in full sun, and young leaves have thinner cuticles than old ones. Thinner cuticles are one reason, but not the only reason, young plants are more susceptible to herbicides than old plants.

Stomata appear to be obvious entry points, but most herbicides enter plants through leaf surfaces. Liquid spray droplets or volatile gases can enter stomata, but even after stomatal entry, herbicides must penetrate the thin cuticle present in substomatal chambers. Stomata vary in number, location, and size among different plant species, and while they can be located on upper and lower surfaces, most agricultural plants have the majority of stomata on lower surfaces. There may be as much as 10-fold variation among species in stomatal number.

Another barrier to entry through stomatal openings is surface tension of spray solutions. It is possible, but not very likely, for a droplet of a liquid with high surface tension to bridge a stomatal opening and not enter it. Surface tension is a more important determinant of the tendency to spread than it is of stomatal entry.

Often stomata are not open when herbicides are applied; they close during the heat of day and open during cool mornings and evenings. Thus, stomata may be closed when herbicides are applied. To achieve easy stomatal penetration, a herbicide spray must have low surface tension and high wetting power: a difficult combination.

2. Other Factors

The location of growing points or plant meristematic areas can determine herbicide selectivity. In grasses, growing points are usually at the bases of plants and are protected from foliar herbicides by surrounding leaves. In some plants, growing points are actually below the soil surface and not exposed to direct contact by foliar-applied herbicides. In contrast, broad-leaved plants usually have terminal, exposed growing points that may be more readily contacted and susceptible to herbicide action.

Selectivity can be obtained through herbicide placement. A herbicide can be applied to plant foliage, only to soil, only to soil and weeds between crop rows, or only over the crop row. A nonselective herbicide can be used selectively by controlling where it is applied. Selective placement can also be obtained by using granular herbicides that have little or no foliar activity because granules do not adhere to foliage.

The stage of plant growth at application is an important determinant of herbicide activity. It is a good generalization that seedling plants are more easily controlled by a herbicide than mature plants. An example of this was given earlier (2,4-D, Chapter 12).

3. Characteristics of Spray Solution

Composition of the spray solution is a very important aspect of selectivity and activity. A spray solution with little or no surfactant may have high surface tension and just bead up on a cuticular surface as water does on a newly waxed car. In this case, there is less opportunity for absorption because the contact area between the applied herbicide and the plant surface is limited. On the other hand, a spray solution with a surfactant decreases liquid surface tension, spreads out water droplets, increases surface coverage, and wets the surface, thereby promoting penetration. Frequently, nonphytotoxic crop oils are included in spray mixtures to promote herbicide penetration and activity. Diesel fuel is often included as an adjuvant in a water-based spray system for control of plants on rangeland because it promotes penetration of leaves and is phytotoxic.

Another, often uncontrolled, factor influencing herbicide activity is drop size. For a given amount of herbicide per unit area, activity usually increases as droplet size decreases (McKinley *et al.*, 1972). For a fixed droplet size, effective dosage can be increased equally well by increasing herbicide concentration in each drop or by increasing the number of drops per unit area. If a spray solution has 0.86 grams of 2,4-D per liter, three 400-micron drops per square centimeter will apply 64 times more spray volume and active ingredient per square centimeter than three 100-micron droplets (McKinley *et al.*, 1972). Drop size is difficult to control in most hydraulic sprayers but is fixed, in a narrow range, with controlled drop applicators.

B. ENVIRONMENTAL FACTORS

The influence of environmental factors on herbicide phytotoxicity is almost always related to differential absorption, translocation, or metabolism. These are affected by morphological characteristics imposed on plants by the environment. Altered plant susceptibility to herbicides can often be traced to environmental stress that alters a plant's ability to metabolize herbicides.

1. Moisture

If a herbicide molecule does not drift or volatilize, reaches its target, and is retained on the plant surface, its activity can still be affected by environmental factors. Herbicide users want to know the likely effects of weather (rain, snow, cold, hot, dry, wet, etc.) on herbicide performance. If the sun comes out immediately after application or even during application, as opposed to application on a gray, cloudy day, what effect does it have on a herbicide's activity and selectivity? Phenoxy acids formulated as esters are more fat-soluble than water-soluble. On a warm day, leaf cuticles may be more fluid and more readily penetrated by fat-soluble compounds such as esters. Therefore, warm days aid penetration and activity. It is a good generalization that the warmer it is, the better herbicide activity will be. For noncontact or soil-active herbicides, temperature at time of application is less important. Temperature influences a plant's metabolic rate and physiological activity. If a plant is rapidly metabolizing and photosynthesizing, it will translocate herbicides rapidly, enhancing their activity.

What if a herbicide is applied and then it rains? With phenoxy acids, penetration occurs within a matter of hours, so rain several hours after application does not affect activity. If, on the other hand, atrazine is applied to plant foliage, it penetrates poorly and rain will wash some off even if it rains as many as 7 days after application. The best recommendation for foliar herbicides is that they should be applied on warm, sunny days with little chance of rain within 24 hours after application. Activity of soil-applied herbicides may be enhanced by a light rain shortly after application that moves them into upper soil layers.

In general, high temperatures and low humidity are detrimental to cuticular absorption. Plants growing under these conditions may produce thicker, less penetrable cuticles or have thin, poorly hydrated cuticles that are not easily penetrated. Sprays dry rapidly, and water stress may cause stomatal closure. High relative humidity reduces water stress, delays drying, and favors open stoma. Plants sprayed with a herbicide under warm, dry conditions may die more quickly if they are moved to warm, moist conditions. Warm temperatures that are not excessive (above 100°F) usually promote herbicide penetration and action. Rain and hard winds before treatment may weather (break and crack) cuticle and more spray may be trapped and taken up by weathered leaves.

2. Temperature

Weed control is best when temperatures before spraying favor uniform plant germination and growth. High temperatures during application generally increase herbicide action by favoring more rapid uptake, but the effect may be offset by rapid drying of spray on leaf surfaces.

3. Light

Light is an important, but uncontrollable, environmental factor. It is essential for photosynthesis, but photosynthetic inhibitors do not have to be sprayed during the day. Many photosynthetic inhibitors are taken up by roots and can be sprayed at any time of day. Good light conditions may open stomata, increase photosynthetic rate, and increase transport of photosynthate and herbicide.

IV. PHYSIOLOGY OF HERBICIDES IN PLANTS

A. FOLIAR ABSORPTION

If a herbicide avoids all of the preceding problems and resides on a plant surface long enough, it must be absorbed for activity to follow. Very few herbicides are true contact materials that solubilize cuticles and membranes and thus do not have to enter plants to achieve activity.

Most herbicides must enter plants and reach an appropriate site of action before toxicity is expressed. Successful herbicide action requires herbicide absorption, translocation in the plant, and avoidance of detoxification prior to an attack at the molecular level on some process vital to plant growth.

No general description of the entire process of herbicide action is applicable to all herbicides, any more than such a description can be provided for all antibiotics or general pharmaceuticals. How does aspirin work? We do not know precisely. However, that does not mean we cannot use it intelligently.

Absorption of an herbicide can be regarded as passage through a series of barriers, any one of which may limit or prohibit action. With crops and weeds, functioning of such barriers can be the basis of selectivity. Modern herbicide formulations have been created with full knowledge of absorptive barriers, and while selectivity is most often explained by metabolism, absorption must still occur (see Section IV,D).

The terms symplast and apoplast are helpful when thinking about uptake and distribution. The essence of the concept of *symplast* is that all living cells of an organized, multicellular plant form a functionally integrated unit. The *apoplast* is the continuous nonliving cell-wall phase that surrounds and contains the symplast and the xylem that is part of the apoplast. The apoplast varies in composition from the highly lipid cuticle to aqueous pectin and cellulose cell walls. It is interposed between the symplast and the external environment. All herbicides that enter plants do so via the apoplast and usually bring about death by action on the symplast.

There are several barriers to apoplastic penetration. The role of each barrier varies with each herbicide–plant combination. These barriers include stomata, cuticle, epidermis, and cell wall.

1. Stomatal Penetration

As discussed earlier (Section III,A,1), stomatal presence, exposure, and distribution vary between plants and between plants of the same species grown in different environments. Stomata are an obvious port of entry but are not very important, because stomatal openings vary under field conditions and the time of maximum opening may be different from the time of application. Rapid drying of solutions (Table 13.1) also allows little time for stomatal penetration. Cuticular penetration is often easier and occurs regardless of stomatal presence or aperture size when herbicides are properly formulated and applied.

2. Cuticular Penetration

The cuticle is a waxy layer on the leaf surface, the thickness and absolute composition of which vary between species. Like stomata, it varies when plants of the same species are grown in different environments. Apart from root absorption (to be discussed) and some stomatal entry, cuticular penetration is the way most foliar herbicides enter plants. Cuticular entry is possible when stomata are closed and occurs under a range of environmental conditions. There are aqueous and lipid routes of entry through the cuticle. Both are available for simultaneous entry of herbicides and the relative rate of entry depends on the molecule entering and the environment.

Cuticles are somewhat open, spongelike material structures made up of a lipid frame with interspersed pectin (water-soluble) strands and possibly open pores. Pores can fill in a water-saturated atmosphere to provide an accessible water diffusion continuum. Herbicides concentrate as solution dries and gather in depressions, commonly over anticlinal walls, prior to absorption. Cuticular penetration is by diffusion through a water or lipid continuum. When a plant is under stress, pores fill with air, which acts as a barrier to water penetration, but the lipoidal routes are still available.

3. Fate of Foliar Herbicides

There are five possible fates of herbicides applied to plant foliage:

1. Volatilization from foliar surfaces and loss to the atmosphere
2. Retention on leaves in a viscous liquid or crystalline form
3. Penetration of the cuticle and retention there in lipid solution
4. Adsorption by the cuticle
5. Penetration of the cuticle

Penetration of the cuticle (number 5) is what is intended. Desirably, that is followed by penetration of the aqueous portion of the apoplast (epidermal cell walls) and migration via anticlinal walls to the vascular system. If a herbicide is not phloem mobile, it will remain in the apoplast and move with the transpiration stream to acropetal leaves. Some herbicides that move this way cannot cross the plasmalemma barrier and they translocate only acropetally in xylem. Many others cross easily to phloem. Because xylem translocation is much more rapid than phloem, the herbicide may appear to be translocated only or dominantly in xylem even though phloem translocation occurs.

Finally, after penetrating the cuticle, the aqueous phase of the apoplast, molecules are absorbed into the living cellular system (symplast) and translocated in phloem out of leaves in the assimilate stream. These molecules can become systemic and move throughout the plant to sites of high metabolic activity (e.g., meristematic regions). Many herbicides follow this route.

4. Advantages and Disadvantages of Foliar Herbicides

There are obvious advantages to foliar herbicides. Foliage is a readily available site of entry. There is often a high efficiency of foliar absorption, and treatments can be designed and scaled to control specific, observable weed problems. There are equally important disadvantages. Application timing is often critical because the herbicide may be most effective when applied at a certain stage of plant growth (e.g., difenzoquat or diclofop for control of wild oats). Some herbicides are not absorbed well by foliage, but are readily absorbed by roots. Wetting plant surfaces is difficult and weather conditions at the time of application affect performance. Herbicides control small plants better than large ones, but small plants do not have many leaves and contact and absorption may be inefficient. It often takes several days or even weeks for some plants (e.g., perennials) to grow enough foliage that good absorption and activity can be obtained.

B. ABSORPTION FROM SOIL

1. General

Some herbicides are directed at soil without any intention of hitting plant foliage. Most foliar herbicides are applied as broadcast sprays and much of the spray hits soil because it misses plant foliage. Because many herbicides are applied when plants are young, most of the soil surface is exposed. Thus, soil becomes an unavoidable target and repository for much of what is applied. Herbicide fate in soil becomes a significant determinant of performance and environmental effect (see Chapter 14).

2. Advantages and Disadvantages of Soil-Applied Herbicides

Application timing of soil-applied herbicides may be convenient and economical because it can be combined with other operations. The effectiveness of preplant or preemergence soil-applied herbicides is not dependent on stage of plant growth or physiological condition at time of application. Positional selectivity can be obtained by placing soil-applied herbicides at a particular depth relative to the crop plant or seed.

Soil-applied herbicides have important disadvantages. There is a tremendous dilution by soil and soil water following application and the amount available to plants is low. There is fixation by soil colloids (adsorption) that reduces the amount of herbicide available for plant absorption. Foliar herbicides are affected by weather conditions at application, whereas soil-applied herbicides are more affected by weather subsequent to application, especially dry conditions. There is often dependence on rainfall, irrigation, or soil incorporation for distribution and action. Persistent residues that may injure subsequent crops can occur after use of some soil-applied herbicides.

3. Root Absorption

It is generally conceded that herbicides enter roots via root hairs and the symplastic system, the same pathway that inorganic ions (plant nutrients) follow. Passive and active uptake occurs, but most uptake is passive with absorbed water and movement is with water in the apoplast. Active uptake involves respiration energy, oxygen, entry into protoplasts, and movement in the symplast. There is accumulation of herbicides at points of activity in the symplast and selectivity is expressed in the symplast. Phenylureas, sulfonylureas, triazines, and uracils are absorbed by roots and move upward apoplastically. Root absorption is highly dependent on a herbicide's lipophilicity.

4. Influence of Soil pH

For weak aromatic acids such as dicamba and 2,4-D, phytotoxicity increases as soil pH increases and reaches a maximum at pH 6.5 (Corbin *et al.,* 1971). The same is true for weak bases such as prometon and amitrole. Soil pH between 4.3 and 7.5 had no effect on phytotoxicity of picloram, weak aromatic acids, and the non-ionic herbicides dichlobenil and diuron. The conclusion is that no generalizations can be made about effects of soil pH on herbicide absorption (Corbin *et al.,* 1971). There is an influence and the effect of pH cannot be ignored, but there is no basis for predicting what it will be in every case.

Many soil-applied herbicides, including the triazines atrazine and simazine, the asymmetrical triazine metribuzin, the phenylurea linuron, and several of the sulfonylureas, show increased activity when soil pH is above 7.5. This is often seen on areas of exposed calcareous soil where more plant injury occurs and selectivity is reduced. This is because there is less herbicide adsorbed at high pH and more is biologically available.

When soil pH was raised from 5 to 7, soil microflora and degradation rate of EPTC increased and phytotoxicity was shortened 2 to 3 weeks. A similar increase in rate of degradation of EPTC was found when manure was added (Lode and Skuterud, 1983). Therefore, EPTC could be less effective on soils with high effective microbiological activity and high pH. These examples further illustrate the point that phytotoxicity is affected by soil pH, but no generalizations can be made.

C. Shoot vs Root Absorption

Different plants absorb herbicides at different sites. Grasses vary in seedling morphology, location of the mesocotyl, and depth of seed germination. Selectivity of diallate and triallate between wheat and wild oats is due to differences in location of the site of herbicide uptake (Appleby and Furtick, 1965). In wild oats, the mesocotyl elongates into herbicide-treated soil where herbicide is absorbed after seed germination. Wheat and barley have a short mesocotyl that does not elongate into the herbicide-treated zone. Depth of the herbicide zone in soil can be controlled by incorporation depth, and positional selectivity can be obtained when these herbicides are used to control wild oats in wheat or barley (Table 13.4). Root exposure has an effect, but wild oats survive. Coleoptile exposure results in plant death because of absorption by mesocotyls of emerging seedlings.

Parker (1966) confirmed these results and demonstrated preferential root or shoot absorption by sorghum with five herbicides (Table 13.5). Dichlobenil and trifluralin are dependent on root absorption, whereas EPTC and diallate depend on absorption by shoots of emerging seedlings.

Table 13.4

Dry Weight of Oat Seedlings Selectively Exposed to Diallate (Appleby and Furtick, 1965)

Plant part exposed	Dry weight (mg)
Coleoptile	0
Root	205.2
Coleoptile and root	0
Untreated control	303.8

Table 13.5

**Herbicide Dose Required to Cause 50% Reduction
in Root or Shoot Dry Weight of Sorghum
(Parker, 1966)**

| Herbicide | Equally effective concentration (ppm) | | Ratio |
	Root exposure	Shoot exposure	
Diallate	>8	2.5	1 : 0.33
Dichlobenil	0.055	1.25	1 : 23
EPTC	>16	0.8	1 : 0.05
Triallate	>4	4	1 : 1
Trifluralin	0.065	2.7	1 : 42

Triallate was equally effective on sorghum when absorbed through roots or shoots.

Studies with yellow nutsedge have shown that most tuber (often called nutlet) sprouts come from below 2 in. because tubers in the top 2 in. of soil often winter-kill. When tubers sprout, they develop a crown meristem about 1.5 to 2 in. below the soil surface; roots and new rhizomes arise from this crown. It is important to have herbicides in this area for absorption from soil. Many herbicides that have activity on nutsedge are soil incorporated at least 2 in. deep to ensure that crown meristems come in contact with the herbicide.

The area near or above the shoot of green foxtail is the primary area for herbicide uptake. Placement of soil-active herbicides in the top 1–2 in. of soil is essential for good control. Corn and sorghum have a different site of uptake because of deeper planting. This provides an opportunity to achieve selectivity through herbicide placement.

D. Absorption as a Determinant of Selectivity

Selectivity is a function of three factors: absorption, translocation, and metabolism. In some cases, absorption can explain why a herbicide affects one plant and not another.

Peas are tolerant to 2,4-D and tomatoes are susceptible because peas absorb 2,4-D for only 24 hours after exposure, whereas tomatoes absorbed greater quantities over 7 days (Fang, 1958). Wheat and corn absorb 2,4-D more slowly than beans, and the low rate of absorption into monocots is a factor in selectivity.

For many modern herbicides, absorption is not a major barrier to activity. Studies of herbicide selectivity frequently find metabolic ability or rate of application to be the defining difference between species. But not all differences in selectivity of new herbicides are due to metabolism. Differences in activity of glufosinate on tolerant barley and green foxtail (sensitive) were explained by differences in foliar absorption and translocation, but not by metabolism (Mersey *et al.*, 1990). Improved control of common milkweed and poor glyphosate activity on hemp dogbane were attributed to improved foliar absorption of glyphosate by milkweed when surfactants were used (Wyrill and Burnside, 1977).

A major determinant of herbicide selectivity is the plant's growth stage when the herbicide is applied. Some plants show maximum susceptibility in early seedling stages and greatly reduced susceptibility after fruiting. Much of this can be traced to absorption. Figure 13.3 shows the growth stages for wheat, oats, barley, and rye. Each of these crops is susceptible to growth regulator herbicides when they are applied during early stages. A growth regulator herbicide applied stem elongation has little effect. When

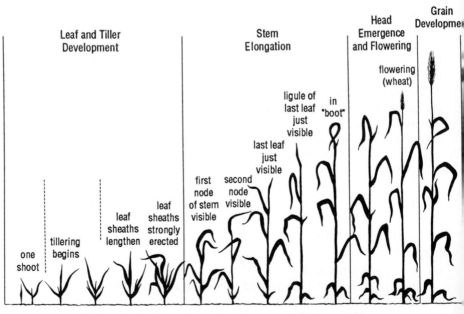

Figure 13.3   Growth stages in wheat, oats, barley, and rye.

they are applied during head emergence, susceptibility increases, but is not as high as it was in early stages. Not all of this can be explained by absorption, but much is due to greater absorption by young seedling plants and direct access to floral structures later. Susceptibility of small grains follows a consistent general pattern during various stages of growth. Thus, growth regulator herbicides should not be applied to small grain crops before tillers are formed. After tillering, susceptibility decreases and application is safe.

E. TRANSLOCATION

Translocation is important because, to be effective, most herbicides must move to sites of action. Translocation takes place through phloem and xylem, the transport systems in plants. It is common to find a direct correlation between foliar absorption and phloem transport and root absorption and xylem transport.

It is helpful to think of phloem translocation as movement from source to sink. Movement from source to sink often occurs with photosynthate transport from regions of high carbohydrate synthesis to regions of high use. Sources are points of entry of herbicides and sinks are sites of high metabolic activity where herbicides express their toxicity.

Herbicide movement in plants is determined frequently by patterns of photosynthate distribution and by the relative activities of sources and sinks. For example, movement from cotyledons and young leaves is predominantly to roots. From lower leaves of mature plants there is either no movement or movement to roots. From later formed leaves there is transport to roots and shoot tips or meristematic areas. From upper, mature leaves transport is to shoot tips, flowers, and fruits.

Herbicides that enter phloem can pass from it to xylem and are systemic. The reverse is rare but occurs. Herbicides that move symplastically and migrate to xylem can move up or down, whereas those that move only apoplastically (root uptake dominates) translocate only acropetally in the transpiration stream. Table 13.6 shows the primary translocation pathway for several herbicides (Ashton and Crafts, 1981). Rate of translocation for one herbicide varies between species and with different environmental conditions for one species. Many patterns are possible and no absolute generalizations can be made.

F. TRANSLOCATION AS A DETERMINANT OF SELECTIVITY

2,4,5-T is more mobile than 2,4-D in burcucumber (Slife *et al.,* 1962), but burcucumber is resistant to 2,4-D and susceptible to 2,4,5-T. Transloca-

Table 13.6

Mobility and Primary Translocation Pathways of Some Herbicides in Plants (adapted from Ashton and Crafts, 1981)

Free mobility			Limited mobility			Little mobility
Apoplast	Symplast	Both	Apoplast	Symplast	Both	
Chloroacetamides	Glyphosate	Amitrole	Chloroxuron	Phenoxy acids[b]	Endothall	Bensulide
Desmedipham		Dicamba	Diquat		Naptalam	Diphenyl ethers
Diphenamid[b]		DSMA	Fluridone[b]		Nitriles	DCPA
Methazole		MSMA			Phenoxy acids	Dinitroanilines
Napropamide[b]		Picloram			Propanil	
Norflurazon[b]		Glyphosate				
Phenmedipham		Imidazolinones				
Pronamide		Sulfonylureas				
Thiolcarbamates						
Triazines						
Uracils						
Ureas[a]						

[a] Except chloroxuron, limited apoplast.
[b] Translocation rate varies widely between species.

tion of 2,4-D is initially slow and there is no movement after 24 hours. Slow continual movement occurs in domestic cucumbers over 8 days. Burcucumber avoids 2,4-D injury by immobilizing it, whereas 2,4,5-T is translocated to sites of action.

Bean leaves absorb 2,4-D and it seldom moves elsewhere. It is strongly absorbed by roots, but moves in stems at low concentrations and not into leaves after root absorption (Crafts, 1966). Bean roots absorb 2,4-D, but translocate little into stems and none into leaves. After foliar absorption there is no apoplastic movement and no retransport. Barley leaves, on the other hand, absorb 2,4-D and translocate it symplastically but not apoplastically. Barley roots absorb it but transport very little. Thus, translocation may partially explain bean's susceptibility to 2,4-D and barley's low susceptibility.

Two-year-old white ash trees treated for 4 weeks with 10 ppm picloram in nutrient culture were only slightly injured, but young red maple, treated in the same way, died in 2 weeks (Mitchell and Stephenson, 1973). Rate of root uptake, acropetal translocation, and leaf accumulation was lower in red maple and would explain what happened except that red maple died. Foliar penetration was similar in both species and absorption could not explain selectivity. Picloram was metabolized at equal rates in both species and metabolism did not explain selectivity. Tolerance of white ash was not related to lower rates of uptake or faster metabolism. Red maple's high susceptibility was due to blockage of xylem by undifferentiated callus growth caused by picloram activity. Death was caused by lack of normal translocation and subsequent desiccation of leaves and stems. Picloram's activity prevented necessary translocation.

Other researchers asked why beans and peas differed in their tolerance to dimefuron (an oxadiazolinone) (Glasgow and Dicks, 1980). Beans are susceptible to dimefuron applied to roots, but peas are tolerant. Dimefuron (related to oxadiazon, but never commercialized) was translocated from roots to shoots in beans but not in peas. Beans are therefore tolerant to preemergence field applications because there is root absorption but no shoot absorption. When only roots are exposed, absorption is low and poor translocation explains selectivity.

G. HERBICIDE METABOLISM IN PLANTS

Modification of a herbicide's chemical structure usually eliminates phytotoxicity. This is not always true but is a good generalization. An example of the opposite case is the phenoxybutyric acid 2,4-DB, which, among other uses, controls broadleaved weeds in peanuts. It is chemically altered by plant metabolism through a process called beta-oxidation that produces the 2-carbon phytotoxic derivative, 2,4-D (Chapter 12).

Once a herbicide is absorbed by plants, it is susceptible to metabolism and loss of biological activity. The faster a herbicide is metabolized, the less there is available for translocation and activity at the site of toxic action. An example of plant metabolism is conversion of simazine to hydroxysimazine, a derivative with no herbicidal properties (Figure 13.4).

Many metabolic reactions occur in plants, but the most important are oxidation, reduction, hydrolysis, and conjugation (Hatzios and Penner, 1982). Plant metabolic reactions have been separated into three phases (Hatzios and Penner, 1982; Shimabukuro et al., 1981). Phase one includes nonsynthetic, generally destructive processes such as oxidation, reduction, and hydrolysis. Phase two reactions are conjugations that result in synthesis of a new molecule. Phase one reactions add OH, NH_2, SH, or COOH functional groups that usually change phytotoxicity, increase polarity, and a predisposition for further metabolism. Phase one reactions can be enzymatic or non-enzymatic. An example of the latter is photochemical reduction (detoxification) of bipyridyllium herbicides.

Phase two metabolism is conjugation that yields metabolites with reduced or no phytotoxicity, higher water solubility, and reduced plant mobility. Conjugations occur with glutathione, amino acids, and glucose and other sugars. Phase three metabolism is unique to plants because plants cannot excrete metabolites as animals can. Conjugated metabolites must be compartmentalized in plant cells or somehow removed from further metabolic activity. Herbicides become more water soluble as they are metabolized from phase 1 to 2 and they remain water soluble or become insoluble in phase 3. Phytotoxicity is reduced with each phase and herbicides metabolized to phase three are no longer toxic. Table 13.7 shows some reactions and the herbicide groups affected. A complete discussion of these reactions is beyond the scope of this text (see Ashton and Crafts, 1981; Corbett et al., 1984; Hatzios and Penner, 1982).

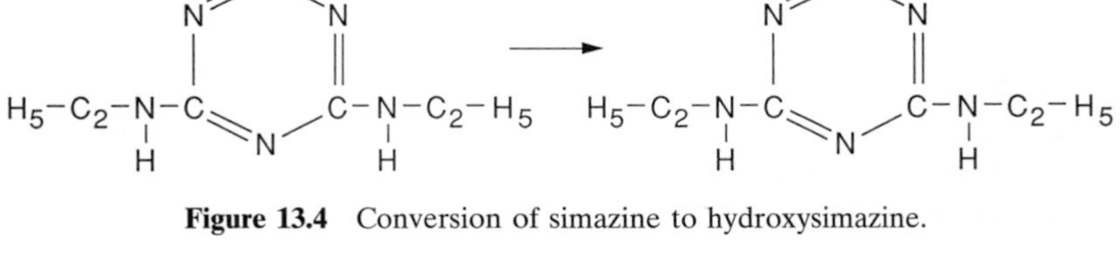

Figure 13.4   Conversion of simazine to hydroxysimazine.

Table 13.7

Plant Metabolic Reactions and the Herbicide Chemical Groups Affected

Chemical reaction	Affected chemical groups
Hydroxylation	Triazines, phenoxy acids, imidazolinones
Oxidation	Phenoxy acids
Decarboxylation	Benzoic acids, picolinic acids
Deamination	Ureas, dinitroanilines
Dethioation	Carbamothioates
Dealkylation	Dinitroanilines, triazines
Hydrolysis	Carbamates, sulfonylureas, imidazolinones
Conjugation with plant constituents e.g., glucosidation	Benzoic acids, imidazolinones

H. METABOLISM AS A DETERMINANT OF SELECTIVITY

Herbicide activity and selectivity are often directly attributed to differences in plant metabolism. For example, black currant is susceptible to 2,4-D and decarboxylates only 2% of applied 2,4-D. Red currant is tolerant and decarboxylates 50% of applied 2,4-D in the same time (Luckwill and Lloyd-Jones, 1960). The different rate of metabolism accounts for observed selectivity. Catchweed bedstraw is selectively controlled by MCPP, but not by MCPA, and there is no difference in absorption or translocation. Ten days after treatment, MCPP was not metabolized at all and MCPA was completely metabolized (Leafe, 1962). Rapid metabolism of MCPA explains its lack of effect, and no metabolism of MCPP leads to the plant's death.

Broadleaf plantain, common chickweed, and strawberry are resistant to phenoxy acids, especially 2,4-D. Dandelion, cucumber, soybean, pea, common lambsquarters, and wild buckwheat are moderately sensitive, and sunflower, mustards, and cotton are very sensitive to the same herbicides (Hatzios and Penner, 1982). These differences are explained by differences in rate of metabolism among the plants.

A portion of atrazine's selectivity can be explained by differential metabolism (Negi *et al.,* 1984). Data in Table 13.8 show the amount of atrazine remaining 10 days after preemergence application to eight different plants. Nonsusceptible species have a low concentration because they metabolize atrazine to an innocuous form. Species intermediate in susceptibility have a higher concentration than nonsusceptible species but a lower concentration

Table 13.8

**Amount of Atrazine in Shoots of 8 Plant Species 10
Days after Preemergence Application
(Negi *et al.*, 1964)**

Species	Susceptibility	ppm atrazine
Johnsongrass	none	19
Grain sorghum	none	8
Corn	none	10
Cotton	intermediate	222
Peanuts	intermediate	97
Oats	high	376
Soybean	high	322
Bean	high	227

than susceptible ones. Susceptible species, especially oats and soybeans, have the highest concentration and are, in part, susceptible because of their inability to metabolize atrazine. But beans and cotton differ by only 5 ppm, hardly a significant amount. The reason cotton is intermediate in its susceptibility is not related solely to metabolism. Atrazine accumulation in lysigenous (oil-bearing) glands of cotton is an isolating, protective mechanism. Higher concentrations exist, but the plant isolates atrazine, and a lower active concentration is present. This work illustrates the complexity of explaining selectivity.

Metabolism is the basis for differential atrazine tolerance among warm-season forage grasses (Weimer *et al.*, 1988). Big blue stem and switchgrass are not very susceptible to atrazine and yellow indian grass and side oats grama are susceptible in the seedling stage. Atrazine metabolism in big bluestem and switchgrass occurred primarily by glutathione conjugation. Conjugation by big bluestem and switchgrass occurred faster than *N*-dealkylation of atrazine in yellow indian grass and side oats grama. Differential tolerance to atrazine among these four grasses is due to the metabolic route by which atrazine is detoxified and the rate and type of metabolism that dominated in susceptible and resistant species.

Propanil has been used for selective control of green foxtail in hard red spring and durum wheat. Green foxtail moved into niches created when broadleaved species were controlled by growth regulator herbicides. Other research has shown that propanil was selective in rice because it is rapidly metabolized (Yih *et al.,* 1968a and b). Green foxtail retained more spray solution than wheat but less propanil (Table 13.3), and retention is important. Both plants had rapid absorption during the first 12 hours after treatment. Green foxtail absorbed about 10% more but differences in absorption after 48 hours did not account for selectivity. Because over 95% of applied propanil remained in leaves, translocation was not a major factor in selectivity.

Retention was important, but the most important determinant of propanil selectivity was metabolism (Table 13.9). Propanil was metabolized by wheat but not by green foxtail. Only 34% of the amount applied remained active 72 hours after application to wheat, whereas over 90% remained in green foxtail. Propanil is selective in wheat and rice through rapid metabolism, and wheat has the added advantage that it does not retain the amount of spray solution green foxtail does.

Many studies include absorption, translocation, and metabolism because it is generally recognized that all must be considered if selectivity is to be understood. Results of two studies are summarized next to illustrate their scope and complexity. Wilcut *et al.* (1989) studied selectivity of the sulfonylurea herbicide chlorimuron among soybean, peanut, and four broadleaved weeds. Absorption was similar in five species after 72 hours but lower in Florida beggarweed. There was slight symplastic and apoplastic translocation in all species. Peanut showed more tolerance with age because of reduced absorption by older plants and faster metabolism. Neither absorption nor translocation differences explained differential selectivity among the two crops and four weeds. Further experiments showed that tolerance was directly correlated with the amount of unmetabolized chlorimuron. Rate of metabolism was greatest in soybean and lowest in common cocklebur. After 24 hours, nearly

Table 13.9

Rate of Propanil Metabolism by Wheat and Green Foxtail (Eberlein and Behrens, 1984)

Hours after application	Percent applied propanil remaining	
	Wheat	Green foxtail
24	69.8	93.7
48	42.8	93.5
72	34.2	93.5

two times as much unmetabolized chlorimuron was found in the four weeds compared to the two crops. After 72 hours, 16.7% of the applied chlorimuron was present in soybean and peanut had 25.6%. Prickly sida, classified as intermediate in susceptibility, had 29.6% of the applied chlorimuron unmetabolized after 72 hours. Sicklepod and common cockle-bur, susceptible species, had 39.9 and 60.6%, respectively. Florida beggar-weed is susceptible to chlorimuron even though only 16.6% of the herbicide remained after 72 hours. This was equal to soybeans and should have made the weed tolerant if metabolism was the only factor. Florida beggarweed actually had over five times as much chlorimuron compared to soybean when chlorimuron was calculated as amount per gram dry plant weight. Its susceptibility occurred in spite of the fact that it absorbed less than half as much as soybean and seemed to metabolize rapidly. Chlorimuron concentration remained very high in Florida beggarweed even though the total amount was low (Wilcut *et al.*, 1989).

Field violet is controlled by terbacil, but only when it is applied to emerging seedlings with fewer than three leaves. Established plants with 12 leaves are not controlled by terbacil applied to control weeds in straw-berries. Doohan *et al.* (1992) demonstrated that field violet with 12 leaves absorbed less terbacil per gram fresh weight than 3-leaf plants. Young plants translocated twice as much terbacil to foliage after root uptake. Metabolism studies showed 79% of terbacil was still intact after 96 hours in 3-leaf plants, whereas in resistant 12-leaf plants only 40% of terbacil remained. Young plants were susceptible because although they absorbed less they translocated twice as much of what was absorbed and metabolized it more slowly than 12-leaf plants. A similar explanation was offered for the selectivity of fluroxypyr among four species. More fluroxypyr was recov-ered in susceptible wild buckwheat and field bindweed (about 70%) than in tolerant Canada thistle and common lambsquarters (about 30%) 120 hours after application (MacDonald *et al.*, 1994). Fifteen and 10% of applied fluroxypyr was translocated in Canada thistle and common lambsquarters, respectively, whereas 40% was translocated in the two susceptible species. Selectivity was due to limited translocation in tolerant species and more rapid metabolism (MacDonald *et al.*, 1994).

One of the most striking features of herbicides is selectivity: the ability to kill or affect the growth of one plant without affecting another. Factors that affect selectivity include the following:

1. Distribution as affected by drift, volatility, soil incorporation, and selective placement
2. Retention by plants as affected by leaf morphology, herbicide formulation, and the herbicide's chemical and physical properties

3. Absorption by plants as affected by site of uptake (root vs shoot), cuticle, weather, soil, and the herbicide's chemical and physical properties
4. Immobilization vs translocation in plants as affected by plant age, weather, the specific herbicide, herbicide formulation, and soil
5. Metabolism or molecular change of herbicides in plants and soil as affected by the herbicide and soil microorganisms
6. Plant age
7. Weather
8. Physiological factors, including translocation and inactivation without molecular change

In general, for maximum effectiveness the ideal herbicide should have the following properties:

1. Ability to enter plants at various sites
2. Ability to enter plants without local damage
3. Activity or ability to affect plant growth that is not confined to a particular stage of plant development or plant size
4. Ability to translocate in plants to appropriate sites of action
5. Metabolism or degradation to inactivity in target plants that is slow enough to permit full expression of activity
6. Moderate soil absorption to decrease leaching
7. Reasonable stability in soil, except for foliar-active, contact herbicides where soil persistence is of no consequence to plant action but may nevertheless have environmental consequences
8. A wide weed-control spectrum or specific activity against target weeds

There are no ideal herbicides. Some come close, but none meet all of these criteria. Herbicide selectivity means that all plants do not respond in the same way to all herbicides. Their use in agronomic and horticultural crops, lawn and turf, forestry, or aquatic sites is dependent on selective activity. Herbicide selectivity is dependent on morphological and metabolic differences between weed and crop. For most herbicides, selective action occurs over a relatively wide dose range. This gives users some assurance of selectivity and avoids catastrophe if small errors in calibration or application are made. The selective action and effectiveness of herbicides depends on differences in their toxicity at the cellular level. Selective action also depends on all the factors that influence the amount of herbicide that reaches sites of toxic action in cells.

For herbicides, dose is the most important determinant of selectivity. All herbicides have a recommended dose for particular tasks, and applicators need to know and apply the correct dose for each weed–crop situation.

THINGS TO THINK ABOUT

1. Can drift and volatility be eliminated?
2. How can drift and volatility be controlled?
3. Are drift and volatility current problems? Why? What are appropriate solutions?
4. How does plant morphology affect selectivity?
5. How can spray solutions be modified to affect selectivity?
6. How does weather affect performance of foliar-applied herbicides?
7. How does weather affect performance of soil-applied herbicides?
8. What can happen to foliar-applied herbicides after they contact plants?
9. Foliar- and soil-applied herbicides each have advantages and disadvantages. Compare and contrast them.
10. How do absorption, translocation, and metabolism interact to determine selectivity?
11. What are the phases of herbicide degradation in plants and what is the significance of each?
12. What factors determine a herbicide's selectivity?
13. What characteristics should a herbicide have to maximize activity?

LITERATURE CITED

Anonymous. 1996. Herbicide drift could be potent problem. *Successful Farming* **94**(2):13, February.

Appleby, A. P., and W. R. Furtick. 1965. A technique for controlled exposure of emerging grass seedlings to soil-active herbicides. *Weeds* **13**:172–173.

Ashton, F. M., and A. S. Crafts. 1981. "Mode of Action of Herbicides," 2nd ed., p. 34. Wiley, New York.

Blackman, G. E., R. S. Bruce, and K. Holly. 1958. Studies in the principles of phytotoxicity. V. Interrelationships between specific differences in spray retention and selective toxicity. *J. Exp. Bot.* **9**:175–205.

Brooks, F. A. 1947. The drifting of poisonous dusts applied by airplane and land rigs. *Agric. Eng.* **28**:233–239.

Corbett, J. R., K. Wright, and A. C. Baillie. 1984. "The Biochemical Mode of Action of Pesticides," 2nd ed. Academic Press, London. 382 pp.

Corbin, F. T., R. P. Upchurch, and F. L. Selman. 1971. Influence of pH on the phytotoxicity of herbicides in soil. *Weed Sci.* **19**:233–239.

Crafts, A. S. 1966. Comparative movement of labeled tracers in beans and barley. *Proc. Symp. Int. Atomic Energy Agency, Isotopes in Weed Control, Vienna,* pp. 212–214.

Derting, C. W. 1987. Wiper application, Chapter 14 *in* "Methods of Applying Herbicides" (C. G. McWhorter and M. R. Gebhardt, Eds.). Weed Sci. Soc. of America Monograph No. 4. Champaign, Illinois.

Doohan, D. J., T. J. Monaco, and T. J. Sheets. 1992. Effect of field violet (*Viola arvensis*) growth stage on uptake, translocation, and metabolism of terbacil. *Weed Sci.* **40:**180–183.

Eberlein, C. V., and R. Behrens. 1984. Propanil selectivity for green foxtail (*Setaria viridis*) in wheat (*Triticum aestivum*). *Weed Sci.* **32:**13–16.

Fang, S. C. 1958. Absorption, translocation, and metabolism of 2,4-D-1-^{14}C in peas and tomato plants. *Weeds* **6:**179–186.

Glasgow, J. L., and J. W. Dicks. 1980. The basis of field tolerance of field bean and pea to dimefuron. *Weed Res.* **20:**17–23.

Hartley, B. S., and I. J. Graham-Bryce. 1980. "Physical Principles of Pesticide Behavior," Vol. 2, Appendix 5, p. 93. Academic Press, New York.

Hatzios, K. K., and D. Penner. 1982. "Metabolism of Herbicides in Higher Plants," pp. 6–7. Burgess, Minneapolis.

Leafe, E. L. 1962. Metabolism and selectivity of plant growth regulator herbicides. *Nature* **193:**485–486.

Lode, O., and R. Skuterud. 1983. EPTC persistence and phytotoxicity influenced by pH and manure. *Weed Res.* **23:**19–25.

Luckwill, L. C. and C. P. Lloyd-Jones. 1960. Metabolism of plant growth regulators. I. 2,4-Dichlorophenoxyacetic acid in leaves of red and black currant. *Ann. Appl. Biol.* **48:**613–625.

MacDonald, R. J., C. J. Swanton, and J. C. Hall. 1994. Basis for selective action of fluroxypyr. *Weed Res.* **34:**333–34.

McKinlay, K. S., S. A. Brandt, P. Morse, and R. Ashford. 1972. Droplet size and phytotoxicity of herbicides. *Weed Sci.* **20:**450–452.

McWhorter, C. G. 1966. Sesbania control in soybeans with 2,4-D wax bars. *Weeds* **14:**152–155.

McWhorter, C. G. 1970. A recirculating spray system for postemergence weed control in row crops. *Weed Sci.* **18:**285–287.

Mersey, B. G., J. C. Hall, D. M. Anderson, and C. J. Swanton. 1990. Factors affecting the herbicidal activity of glufosinate-ammonium: absorption, translocation, and metabolism in barley and green foxtail. *Pestic. Biochem. Physiol.* **37:**90–98.

Mitchell, J. F., and G. R. Stephenson. 1973. The selective action of picloram in red maple and white ash. *Weed Res.* **13:**169–178.

Mullison, W. R., and R. W. Hummer. 1949. Some effects of the vapor of 2,4-dichlorophenoxy-acetic acid derivatives on various field crops and vegetable seeds. *Botan. Gaz.* **111:**77–85.

Negi, N. S., H. H. Funderburk, Jr., and D. E. Davis. 1964. Metabolism of atrazine by susceptible and resistant plants. *Weeds* **12:**53–57.

Parker, C. 1966. The importance of shoot entry in the action of herbicides applied to the soil. *Weeds* **14:**117–121.

Shimabukuro, R. H., G. L. Lamoureux, and D. S. Frear. 1981. Pesticide metabolism in plants: Principles and mechanisms, *in* "Biological Degradation of Pesticides" (F. Matsumura, Ed.). Plenum, New York.

Slife, F. W., J. L. Key, S. Yamaguchi, and A. S. Crafts. 1962. Penetration, translocation, and metabolism of 2,4-D and 2,4,5-T in wild and cultivated cucumber plants. *Weeds* **10:**29–35.

Wilcut, J. W., G. R. Wehtje, M. G. Patterson, T. A. Cole, and T. V. Hicks. 1989. Absorption, translocation, and metabolism of foliar-applied chlorimuron in soybeans (*Glycine max*), peanuts (*Arachis hypogaea*) and selected weeds. *Weed Sci.* **37:**175–180.

Weimer, M. R., B. A. Swisher, and K. P. Vogel. 1988. Metabolism as a basis for differential atrazine tolerance in warm-season forage grasses. *Weed Sci.* **36:**436–440.

Wyrill, J. B., and O. C. Burnside. 1977. Glyphosate toxicity to common milkweed and hemp dogbane as influenced by surfactant. *Weed Sci.* **25:**275–287.

Yih, R. Y., D. H. McRae and H. F. Wilson. 1968a. Mechanism of selective action of 3′,4′-dichloropropionanilide. *Plant Physiol.* **43:**1291–1296.

Yih, R. Y., D. H. McRae, and H. F. Wilson. 1968b. Metabolism of 3′,4′-dichloropropionanilide: 3-4-Dichloroaniline–lignin complex in rice plants. *Science* **161:**376–377.

Chapter 14

Herbicides and Soil

FUNDAMENTAL CONCEPTS

- There are three concerns about herbicides in soil:
  - (a) Equilibrium among the soil's gaseous, liquid, and solid phases
  - (b) Susceptibility to degradation
  - (c) Possible effects on soil flora and fauna
- Soil is a living medium with a vast adsorptive surface that plays a major role in determining a herbicide's activity and environmental fate.
- Several physical and chemical factors interact to determine a herbicide's fate in soil.
- Herbicides are degraded in the environment by soil microorganisms, nonenzymatic, and photochemical processes.
- When used in accordance with label directions, herbicides do not accumulate in the environment.

LEARNING OBJECTIVES

- To understand the affect of soil colloidal surfaces on a herbicide's activity and environmental fate.
- To know the physical and chemical factors that affect herbicide activity and performance in soil.
- To understand the importance of adsorption to a herbicide's fate in soil.

- To know the relationship between herbicide adsorption, leaching, volatility, and degradation.
- To understand the role of soil microorganisms in herbicide degradation.
- To understand the importance of chemical or non-enzymatic and photodegradation of herbicides.
- To understand the role of herbicides that persist in soil and their effect on weed management.

Independent of method of application, some of any herbicide reaches soil. Foliar applications may be washed off foliage to soil, and some herbicides are applied directly to soil. Interactions of herbicides and soil are an important part of herbicide use and environmental fate. Soil is not an instrument of crop production similar to a tractor, fertilizers, or pesticides. It is a complex, living, fragile, medium that must be protected because it can be destroyed. It is a living medium for plant growth and a myriad other biological and chemical activities. The thin mantle of soil on the earth's surface may be humankind's most essential and least appreciated resource. It is where our food grows.

There are three concerns about herbicides in soil. The first is the reciprocal equilibrium for exchange and distribution in the liquid, solid, and gaseous phases. The second is a herbicide's susceptibility to degradation and its rate of degradation. The third involves possible influences of herbicides on soil, soil fertility, and soil microorganisms.

After soil application, there is no immediate, direct contact between a herbicide and plant roots or emerging shoots. The physical processes of diffusion and mass flow of water bring herbicides to plant roots. These processes are necessarily weather-dependent (especially rainfall), and the dose that creates the biological response is a function of weather, soil properties, and rate of application. Some control of these factors is possible, especially in irrigated agriculture. An essential property of a soil active herbicide is activity over a fairly wide range of environmental conditions with reproducible reliability.

I. SOIL

Soil contains many heterogeneous organic and inorganic compounds. It is a dynamic system in which components are constantly displaced mechanically, or chemically or biochemically transformed. It contains gaseous, liquid, solid, and living phases. The solid phase, what is seen, is present in a finely distributed form that creates large surface areas. This is of great

Table 14.1

Increase of Soil Surface Area with Decreasing Particle Diameter

Size fraction	Particle diameter	Approximate surface
	(mm)	cm² per gram
Stones	>200	—
Coarse gravel	200–20	—
Fine gravel	20–2	—
Coarse sand	2–0.2	21
Fine sand	0.2–0.02	210
Silt	0.02–0.002	2,100
Clay	<0.002	23,000

importance to the soil behavior of herbicides. Table 14.1 shows how surface area increases with decreasing particle diameter. The fine, colloidal clay minerals with their large surface area determine behavior because of the properties of their surfaces rather than exclusively by the chemical composition of the particles. Small particles with huge surface areas play an important role in herbicide behavior in soil. Figure 14.1 (Dubach, 1972) shows the basic structure of kaolinite and montmorillonite, clays. The thin molecular layers of clay minerals are held together by chemical attraction to form a series of layers. In kaolinite, there is one silicon tetrahedral layer and one aluminum octahedral layer in a fixed lattice. Attraction between layers is

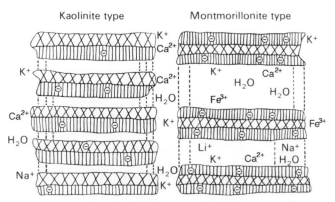

Figure 14.1 Structure of two clay minerals (Dubach, 1972).

so strong that water molecules and chemical ions cannot penetrate between the non-expanding, fixed layers. In montmorillonite clay, there are two silicon tetrahedral layers and one aluminum octahedral layer in an expanding lattice structure. Individual layers are held together weakly in a structure capable of expanding, and molecules and ions can penetrate between layers. Internal and external surfaces are available for chemical activity in expanding lattice clays, and only external surfaces are available in non-expanding (fixed) lattice clays.

The molecular lattice of clay colloids interacts with positively charged ions and molecules, from the soil solution, on its predominantly negatively charged surfaces. These molecules and ions are exchangeable between surfaces and soil solution. Most do not become permanently fixed to clay surfaces. The sum of negative charges, the cation exchange capacity, varies between clays (Table 14.2).

Soils also contain negatively charged organic colloids that have large internal and external surfaces, and an exchange capacity equal to or greater than that of expanding lattice clays (Table 14.2).

Herbicides in soil are subject to electrostatic attractive forces from soil colloidal surfaces and are thus adsorbed on surfaces. When adsorbed, it is difficult, often impossible, for them to be taken up by plants or microorganisms and they are at least partially protected from attack by microbes and non-enzymatic chemical reactions in soil. If a herbicide is sorbed by action at one moiety (part), an exposed or nonadsorbed portion can be susceptible to microbial or chemical attack, but the entire herbicide molecule cannot be absorbed by plants until it is desorbed from the soil surface.

There is a well-established negative correlation between a herbicide's soil activity and the soil's clay and organic matter content. Soil pH and soil water content are also important. In the field, rainfall, temperature, and clay and organic matter content (sorptive capacity) are important determi-

Table 14.2

**Comparison of Cation Exchange Capacities and Surface Area
for Three Clay Minerals and Soil Organic Matter**

Exchange surface	Exchange capacity (cmol (+)/kg)	Surface area (m^2/g)
Organic matter	100–300	500–800
Montmorillonite	100	600–800
Illite	30	65–100
Kaolinite	10	7–30

nants of activity, but each alone and all collectively can be affected by temporary, significant alterations of their interaction (e.g., drought). Great differences in a herbicide's activity are determined by whether application is on dry soil and rain falls afterwards, or whether soil is moist and there is or is not precipitation afterwards. Interaction of factors is important. Not all interactions are known, and results of all interactions cannot be predicted, but good generalizations, based on laboratory experiments, permit accurate prediction of field behavior. Laboratory experiments cannot duplicate the variability of the natural environment. Therefore, although predictions can be qualitatively accurate, they are not always quantitatively precise.

II. FACTORS AFFECTING SOIL-APPLIED HERBICIDES

A. PHYSICAL FACTORS

Five physical factors affect herbicides in soil: placement, volatility, adsorption, leaching, and soil moisture.

1. Placement

Some herbicides are taken up more readily by roots than by shoots and vice versa (see Chapter 13). This knowledge permits placement in soil to enhance or reduce uptake. Herbicides can be placed in or on soil to contact specific weeds or avoid crops. This sounds obvious and easy, but is difficult because control of movement after application is impossible.

2. Time of Application

Time of herbicide application can determine residual activity and soil persistence. Late summer or fall and early spring application yield good activity, but increase the possibility of leaching due to lower soil temperatures, reduced evapotranspiration, and a higher probability of heavy rainfall. Application when soil is dry may lead to no activity and extend soil life.

3. Volatility

Volatility affects location and a molecule's physical state, but does not cause chemical change or molecular degradation. All herbicides have a vapor pressure, although for many it is negligible. The vapor pressure of mothballs, gasoline, and ether is high and their scent is easy to detect.

Vapor pressure—the tendency to volatilize—increases with temperature and is measured in millimeters (mm) of mercury (Hg) at a specific temperature, usually 25°C. Volatilization of herbicides with low phytotoxicity does not create an obvious hazard, but may reduce environmental quality. Volatilization into the atmosphere of herbicides that are toxic to other plants or other species is undesirable.

Volatility occurs from soil or plant surfaces. Herbicides that volatilize from the soil surface move through the atmosphere, the easiest and most available route. Herbicides that volatilize in soil move laterally and toward the surface. Incorporation in soil decreases loss due to volatility and is required for many dinitroaniline and carbamothioate herbicides.

Application of herbicides to a dry soil surface followed by surface wetting and then hot, low-humidity, drying weather can move volatile herbicides to the soil surface and increase volatility. Some lateral and upward movement of volatile herbicides occurs after incorporation in soil and is desirable.

Measures should be taken to reduce or eliminate volatility. Formulation of phenoxyacetic acids as long-chain esters or complex ester chains with an ether linkage reduces volatility (see Chapter 13). Soil incorporation reduces volatility, by burial and enhancement of adsorption, and puts many herbicides near plant roots to enhance their activity. The relative volatility of some herbicides is shown in Table 14.3. The herbicides have been divided into high, medium, and low volatility groups according to their vapor pressures (in mm Hg at 25°C). Herbicides with high volatility have low vapor

Table 14.3

Relative Volatility of Some Herbicides

Volatility	Herbicides
High vapor pressure (10^{-2} to 10^{-4} mm Hg)	Most carbamothioates—butylate, EPTC Clomazone, dichlobenil, trifluralin Short-chain esters of phenoxy acids
Medium vapor pressure (10^{-5} to 10^{-6} mm Hg)	Alachlor, benefin, bromoxynil, butachlor, clopyralid, DCPA, dicamba, ethalfluralin, isopropalin, linuron, napropamide, oxyfluorfen, pendimethalin, pronamide, long-chain esters of phenoxy acids
Low vapor pressure ($<10^{-7}$ mm Hg)	Acetochlor, atrazine and most triazines, amitrole, bentazon, bromacil, cyanazine, diclofop, bipyridilliums, ethofumeste, fluazifop, fluometuron, glyphosate, hexazinone, most imidazolinones, oryzalin, picloram, sethoxydim, most sulfonylureas

pressure (10^{-2} to 10^{-4} mm Hg) and a high tendency to change state from liquid to gas at normal atmospheric pressure. High volatility is a caution, but not an automatic hazard. Most herbicides have low volatility (10^{-7} mm Hg or less).

4. Adsorption

Adsorption is a process of accumulation at an interface and is contrasted with absorption, or passage through an interface. In soil, clay, and organic matter, surfaces are interfaces between the solid soil surface and soil's gaseous and liquid phases. Through cation exchange and physical attraction, herbicides can be concentrated at adsorptive surfaces and removed from soil solution, from which plant uptake occurs. Adsorption is one of the most important mechanisms for reducing herbicide concentration in soil, and few herbicides completely escape adsorptive interactions. Manufacturers develop application rates to compensate for adsorption and to keep enough desorbed (in solution) for activity. The organic arsenicals, dipyridiliums, and glyphosate are adsorbed quickly and extensively and have no soil activity.

Adsorption affects movement and availability in soil and rate of degradation. It regulates degradation by soil microorganisms and chemical reactions. The adsorption–desorption equilibrium determines the amount adsorbed and the amount in solution and available for plant absorption. The equilibrium is the ratio of adsorbed herbicide to solution concentration and can be expressed mathematically given specific herbicide–soil combinations.

In adsorption, there are two factors to consider: strength of binding, and extent of binding. It is not true that the most extensively bound chemical will be the most strongly bound; both must be determined. Table 14.4 compares strength of adsorption for several common herbicides. The groups, from very strong to weak, were created by using each herbicide's K_{oc}. K_{oc}, expressed in ml/g, is the soil–organic carbon sorption coefficient. It is the herbicide's K_d (distribution coefficient) divided by the weight fraction of organic carbon in a soil:

$$K_{oc} = K_d/\text{weight fraction of organic carbon in soil.}$$

K_d, usually expressed in liter/kg or ml/g, is the ratio of sorbed to dissolved herbicide at equilibrium in a soil–water slurry:

$$K_d = \text{herbicide sorbed (mg/kg)/herbicide in solution } (\mu\text{m/liter})$$

These are standard measures available for most herbicides (Ahrens, 1994).

Table 14.4

Adsorption Strength for Several Herbicides

Adsorption strength	Herbicide
Very strong ($K_{oc} > 5,000$)	Benefin, bipyridilliums, bromoxynil, DCPA, diclofop, DSMA, fluazifop, glyphosate, MSMA, pendimethalin, prodiamine, oxyfluorfen, trifluralin
Strong (K_{oc} 600 to 4,999)	Bensulide, butachlor, cycloate, desmedipham, ethalfluralin, fluridone, napropamide, norflurazon, oryzalin, oxadiazon, pyridate, thiobencarb
Moderate (K_{oc} 100 to 599)	Alachlor, aciflurofen, amitrole, bensulfuron, butachlor, clomazone, dichlobenil, diuron, EPTC, fluometuron, glufosinate, isoxaben, quizalofop, most triazines, triasulfuron, vernolate
Weak (K_{oc} 0.5 to 99)	Acrolein, bentazon, bromacil, chlorsulfuron, clopyralid, dicamba, haloxyfop, hexazinone, imidazolinones, mecoprop, metribuzin, nicosulfuron, picloram, primisulfuron, sodium chlorate, sulfometuron, terbacil, tebuthiuron, tribenuron, triclopyr

Dalapon and TCA are no longer used. They were effective for control of grass weeds. Both existed as anions, with no positive charge. Neither was susceptible to cation exchange and both leached readily. Bipyridilium herbicides, on the other hand, are susceptible to cation exchange because they are cations and are adsorbed tightly and extensively by negatively charged surfaces. Imidazolinones and sulfonylureas are not adsorbed extensively or tightly. Their sorptive interactions are governed by soil pH and their sorption increases as soil pH decreases. At acidic pH, soil adsorption is higher because the molecules are more negatively charged. As pH becomes more basic, the molecules are neutral and not sorbed extensively.

Soils high in organic matter and clay require higher concentrations of herbicide for equal activity compared to soils low in clay and organic matter. Clays require more herbicide than sandy soils. High levels of organic matter and clay adsorb herbicides and residues can persist longer than they do in sandy soils. To illustrate, Table 14.5 shows the recommended change in trifluralin rate with increasing organic content of soil.

5. Leaching

Leaching is movement of a herbicide with water, usually, but not always, downward. It is of environmental concern because of the possibility of

Table 14.5

The Influence of Soil Texture and Organic Matter on Trifluralin Rate

| Soil texture | Pints of Treflan (EC) required per acre for | |
	Dry beans	Cotton in southeastern United States in spring
Coarse	1	1
Coarse with 2–5% organic matter	1.5	1.5
Medium	1.25–1.5	1.25–1.5
Fine	1.5–2	1.5–2
Fine with 2–5% organic matter	2	2
Soil with 5–10% organic matter	2	2–2.5

offsite movement and groundwater contamination. It can determine a herbicide's effectiveness by moving it into or out of its zone of action.

Leaching can be thought of as a chromatographic process where soil is the stationary phase and water the moving phase. Given the correct soil and herbicide information, leaching can be predicted mathematically. It is inversely related to percent organic matter and percent clay, and therefore to adsorption. The greater the adsorption of a herbicide and the adsorptive capacity of a soil, the less leaching will occur.

The extent of leaching is determined by the following factors:

(a) *Adsorptive interactions* between herbicide and soil.
(b) *Water solubility.* The greater a herbicide's water solubility, the greater the leaching potential.
(c) *Soil pH.* Sorption and leaching of imidazolinone and sulfonylurea herbicides is governed by soil pH. Adsorption increases as pH decreases; at low pH, more of the herbicides will be sorbed and leaching is reduced.
(d) pK_a. This is a measure of alkalinity and a property of the herbicide, not the soil. The higher the pK_a, the greater the leachability. At the pK_a, one-half of a molecule is neutral and one-half is ionized. For example, small amounts of acidic pesticides (phenoxyacetic acids, dicamba, picloram) are adsorbed on clay colloids when the pH equals the pK_a and molecular and anionic species occur in relatively equal amounts. For acidic herbicides, at pH levels above the pK_a. anionic species dominate and adsorption will be lower. When soil pH is below the pK_a, molecular species dominate and adsorption can increase. Basic

molecules such as the triazines have a high pK_a and adsorption is greatest at low pH.

(e) *The amount of water moving through the profile.* The more water present due to rainfall or irrigation, the more likely it is that leaching will occur.

(f) *Temperature.* In theory, leaching will be greater at higher temperature but this is very difficult to measure in the field.

If a herbicide is very water soluble, it is more likely to leach. But water solubility is not the sole determinant of leaching. A 4.5-inch rain or irrigation weighs about 1 million lbs. That is enough water to leach most herbicides out of the soil profile because their water solubility is greater than 1 ppm, but that does not happen. Even after 10, 15, or 20 inches of water as rain or irrigation, some herbicide remains in upper soil layers in spite of the fact that the water is capable of dissolving much more than was applied. The reason water does not leach out all of the herbicide is sorption. Table 14.6 shows the relative mobility of herbicides in soils. The table is approximately the inverse of Table 14.4. Adsorption and leaching are inversely related—the greater the adsorption, the lower the amount leached. Herbicides with weak adsorption will appear in mobility class 5 and those with very strong adsorption will be in mobility class 1.

In general, leaching is movement downward. It can occur laterally and upward. Upward movement occurs when a herbicide has been moved into the soil profile and there is subsequent movement of water upward by capillary action when there is a high rate of water evaporation from the soil surface. Herbicides can move up with evaporating water.

6. Interactions with Soil Moisture

If soil is wet and air is dry, plants transpire more. Roots absorb water from soil to replace transpired water and herbicides in soil move to roots by mass flow. More herbicide will be absorbed and activity will increase. Sometimes dry air and wind cause rapid foiliar water loss, and when not enough water is taken up by roots, plants wilt. Stomata then close, water movement in plants slows, and herbicide uptake decreases. Soil drying can increase soil adsorption and decrease root uptake.

Rainfall or irrigation is essential to move herbicides into the top few centimeters of soil where most weed seeds germinate. Some rain (perhaps an inch) may be essential to activate herbicides such as the triazines, which are taken up by roots. No rain for 10 to 14 days after application can cause weed control failures. Heavy rains, on the other hand, may move herbicides below zones of activity. Excess rainfall can leach herbicides through a zone of action unless they are adsorbed.

Table 14.6

Relative Mobility of Herbicides in Soil

Mobility class[a]				
5	4	3	2	1
Bromacil	Amitrole	Atrazine	Acifluorfen	Benefin
Clopyralid	Chlorsulfuron	Alachlor	Bensulide	Bromoxynil
Dicamba	2,4-D	Ametryn	Butachlor	DCPA
Haloxyfop	Metribuzin	Bensulfuron	Clomazone	Diclofop
Mecoprop	MCPA	Dichlobenil	Diuron	Difenzoquat
Picloram	Nicosulfuron	Fluometuron	EPTC	Diquat
Na chlorate	Tribenuron	Glufosinate	Imazapyr	Fluazifop
	Triclopyr	Isoxaben	Imazaquin	Glyphosate
		Prometon	Imazethapyr	MSMA
		Quizalofop	Linuron	Paraquat
		Simazine	Napropamide	Trifluralin
		Terbacil	Norflurazon	
		Triasulfuron	Oxyfluorfen	
			Prometryn	
			Propanil	
			Pyrazon	
			Siduron	

[a]Class 5 = very leachable, Class 1 = essentially immobile. Table compiled from Helling (1971) and published K_{oc} values in Ahrens (1994).

The effect of soil moisture cannot be generalized for all herbicides, crops, or application times. Pendimethalin controlled itchgrass in upland rice irrespective of soil moisture after preemergence application (Pathak *et al.*, 1989). When bentazon or 2,4-D was applied postemergence, either controlled purple nutsedge well only when soil moisture was above the demands of plant evapotranspiration (Pathak *et al.*, 1989).

B. CHEMICAL FACTORS

1. Microbial or Enzymatic Degradation

An enduring lesson of soil microbiology is that things in soil do not just rot or disintegrate. They are decomposed by active chemical processes. It is the large, heterogeneous microorganism population of soil that mediates much of the decomposition in soil. It consists of algae, fungi, actinomycetes, and bacteria. For soil degradation of herbicides, bacteria and fungi are most important.

Herbicides in the soil solution can be adsorbed by soil colloids or be degraded by microorganisms. Many herbicides provide a carbon source from which microorganisms derive energy. Some herbicides (perhaps most) are degraded as an incidental process as microorganisms attack soil organic material. Herbicide degradation is enhanced by warm, moist, aerobic conditions that favor microbial growth. Under similar temperature and moisture conditions, herbicide degradation occurs more rapidly in soils that are rich in organic material and have high microbial activity. In general, with high adsorptive capacity herbicides persist longer and are less available for microbial activity. Soil adsorption and microbial action combine with environmental conditions to determine rate of degradation.

Microbial degradation proceeds by many pathways, none of which are unique to herbicide degradation in soil. These include the following:

Dehalogenation	(loss of a halogen atom)
Dealkylation	(loss of a methyl or methylene group)
Decarboxylation	(loss of COOH, a carboxylic acid group)
Oxidation	(structural change by addition of oxygen)
Hydrolysis	(attack by water)
Hydroxylation	(addition of an OH group)
Ether cleavage	(breaking the R–O–R linkage)
Conjugation	(usually with a sugar or amino acid, sometimes with a protein)
Ring cleavage	(Breaking ring integrity)

In a few cases, decomposition leads to activity, but in most cases it results in loss of phytotoxic activity.

2. Chemical or Non-enzymatic Degradation

Several herbicides are degraded non-enzymatically by chemical reactions not mediated by soil microorganisms. Triazines are degraded commonly by hydroxylation and removal of chlorine at the 6-position that converts atrazine to the non-phytotoxic hydroxy derivative through a purely chemical, non-enzymatic process. Some soil microorganisms also degrade triazine molecules to their hydroxy form. In many other cases, herbicides are decomposed by non-enzymatic *and* enzymatic processes that work in concert and at different points on a single molecule.

The sulfonylurea herbicides are degraded in soil by simple hydrolysis if pH is acidic. As pH approaches neutrality, or under basic conditions, enzymatic degradation by microorganisms dominates. Degradation in acidic soils is more rapid because of the high rate of acid hydrolysis. Imazaquin (an imidazolinone) persists longer and is more active at low pH (5.1) (Marsh and Lloyd, 1996). Studies have been done to determine the influence of soil pH on herbicide degradation and activity. They show that no generalizations can be made about the influence of soil pH on the activity or rate of degradation of all herbicides. It is true that, for most herbicides, rate of degradation is slower as soil pH rises.

3. Photodegradation

Photodegradation, the effect of radiation on internal chemical bonds, is a form of chemical degradation. It is well established that many herbicides, particularly heterocyclic molecules (with carbon and nitrogen in a ring), and nitrogen-containing compounds undergo photodecomposition. Photochemical reactions have been reported for many herbicides, including phenoxy acids, dinitroanilines, propanil, and benzoic acids. Absorption of electromagnetic radiation at wavelengths between 290 and 450 nm influences excitation states of electrons and leads to bond rupture and can energize several common reactions, including oxidation, reduction, hydrolysis, substitution, and isomerization. Although there is no question that photodecomposition occurs, its importance as a determinant of activity or selectivity under field conditions is uncertain. Photo-oxidations are important environmental reactions because of the abundance of oxygen in air, soil, and water. Reactions can occur in a matter of hours and can affect any herbicide during its time in air or on an exposed surface. They are important especially for herbicides that remain in the atmosphere or move back into the air from the target. Once a compound is incorporated in soil, the importance of photodecomposition decreases.

III. SOIL PERSISTENCE OF HERBICIDES

Table 14.7 shows the rate range, crop of use, and expected soil life for several herbicides. Soil persistence is agriculturally important because residual herbicides control weeds over time, but may also injure present or succeeding crops. Soil residues can become residues in the edible portions of crops, contaminate water, and affect nontarget species. Soil residues may cause temporary or permanent effects on soil microorganisms. It would have enormous agricultural and environmental consequences if a herbicide

Table 14.7
Soil Persistence of Phytotoxic Activity and Use of Some Herbicides

Herbicide name (Common)	Herbicide name (Trade)	Major crop	Rate (lbs ai/A)	Soil persistence 1 wk	1 mo	2 mo	3 mo
Alachlor	Lasso	Soybeans, corn, dry beans	2–3			X	
Ametryn	Evik	Bananas, pineapple, sugarcane	4–8				X
Amitrole	Amitrol	Noncropland	2–10		X		
Ammonium sulfamate	AMS soil sterilant	Woody plant control	50–400			X	
Atrazine	Aatrex	Corn	1–3				X
Benefin	Balan	Lettuce, peanuts, tobacco, turf	1–1.5				X
Borates	Several	Sterilant	0.5–100 lb/sq. ft				X
Bromacil	Hyvar	Noncropland, citrus	1.5–25				X
Bromoxynil	Brominal	Small grains, alfalfa, corn	0.5–1	X			
Butylate	Sutan	Corn	3–4			X	
Chlorates	Several	Noncropland	1–8 #/sq rd				X
Chloroxuron	Tenoran	Soybeans	1–1.5				X
Chlorsulfuron	Glean	Wheat	1/6–1/2 oz ai/A			X	
Cyanazine	Bladex	Corn	1.2–2			X	
Cycloate	Ro-Neet	Sugarbeets	3–6			X	
DCPA	Dacthal	Turf	4–10				X
Diallate	Avadex	Wheat, barley, sugarbeets, potatoes	1.5–3.5		X		
Dicamba	Banvel	Small grains	0.5–3				X
Dichlobenil	Casoron	Aquatics, ornamentals	0.75–4				X
Diphenamid	Enide	Tomatoes	4				X
Diuron	Karmex	Cotton, alfalfa, orchards	.6–6.4				X

Common name	Trade name	Crop/Use	Rate (lb/A)				
Endothall	Endothal	Turf, aquatics, sugarbeet	1–2		X		
EPTC	Eptam	Potatoes, beans, alfalfa	2–6	X			
Glyphosate	Round-up	Contact–nonselective	1–4	X			X
Linuron	Lorox	Soybeans, corn, potatoes	0.5–1.5				X
MCPA	Several	Wheat, rice	0.25–1.5		X		X
Metribuzin	Sencor/Lexone	Soybean, potato, alfalfa	0.5–1.5			X	
MSMA	Several	Noncropland, cotton	2–4	X			
Paraquat	Paraquat	Desiccant, min. tillage, peanuts	0.25–1	X	X		
Phenmedipham	Betanal	Sugarbeets	1–1.5				
Picloram	Tordon	Brush, rangeland	0.25–1.5		X		X
Prometon	Pramitol	Sterilant	10–60				X
Prometryn	Caparol	Cotton	0.5–2.5			X	
Propachlor	Ramrod	Corn	3–6			X	X
Propanil	Rogue	Rice	3–6	X			
Pyrazon	Pyramin	Sugarbeets	2–4			X	
Simazine	Princep	Corn, orchards	2–4				X
Sodium chlorate	Several	Sterilant	0.5–2.25 lbs 100 sq ft				X
2,4-D	Several	Corn, turf, small grain	0.25–2	X			
Terbacil	Sinbar	Peppermint	0.8–3.2				X
Terbutryn	Igran	Wheat, sorghum	1.2–2.4			X	
Triallate	Avadex-BW	Small grains	1–1.5		X		
Trifluralin	Treflan	Cotton, soybeans, alfalfa	0.5–1				X
Vernolate	Vernam	Soybean, peanut, tobacco	2–4		X		
Weed oils		General use		X			

NOTE: Persistence and degradation of a particular herbicide vary with rate and with climatic, and edaphic conditions.

were released and then found to cause serious depression of activity or even death of nitrifying bacteria in soil. This has not happened because questions about possible effects on soil microorganisms are asked repetitively by manufacturers. Effects are found frequently and depression or stimulation of activity are both found. No registered herbicide has ever caused permanent damage to the soil microflora when it was used according to approved label directions.

Weed scientists, agriculturalists, and manufacturers must also know whether herbicides will accumulate in soil at a rate faster than their rate of dissipation. Under normal agricultural use patterns the answer is no (Figure 14.2; Hamaker, 1976)! The assumptions used to make this conclusion are that 1 lb of herbicide is applied annually at about the same time and the time it takes for half of it to degrade (Its half-life) is 1 year. After 1 year, 0.5 lb would remain and another pound would be applied. Continuing this sequence produces the sawtooth pattern in Figure 14.2. Residues will never exceed twice the annual rate of application, and therefore, except in very unusual circumstances (very cold, very dry, very long half-life), herbicides do not accumulate in soil with repeated annual use. All herbicides in use today have soil half-lives much shorter than 1 year and most herbicides are not used repetitively on the same field over several years. Therefore, the example is an extreme one, but illustrates the point well.

In most cases initial rate of degradation is independent of herbicide concentration in soil. If high rates degraded more slowly or if a herbicide was not degraded, residues would accumulate. Residues could accumulate if there were several applications in one growing season, but this is rare.

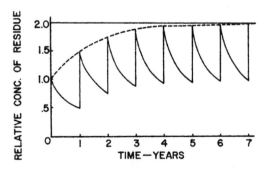

Figure 14.2 Residue pattern for a single, annual application and a half-life of 1 year (Hamaker, 1976).

When a herbicide is applied to soil, adsorption determines availability to plants and leachability. Absorption and translocation by plants lead to activity. Lability with respect to degradation in soil and susceptibility to physical processes in soil that do not degrade herbicides, but move them to different sites, must be understood. Herbicide persistence in soil is a problem, but it is essential for performance of some herbicides and for control of some weeds. Phytotoxicity that disappears too rapidly is not always good because weeds do not all emerge at once. Continued research will direct future use and take advantage of persistence when necessary and manage it where possible.

THINGS TO THINK ABOUT

1. How does the physical structure of soil affect herbicides?
2. How do a soil's chemical properties affect herbicides?
3. What is the role of adsorption in herbicide activity?
4. How do adsorption and leaching interact?
5. How do adsorption and volatility interact?
6. How do soil microorganisms affect herbicides in soil?
7. What are some non-enzymatic reactions that affect herbicides in soil?
8. What role does photodegradation of herbicides play?
9. What factors determine how long a herbicide persists in soil?
10. Do herbicides accumulate in the environment?

LITERATURE CITED

NOTE: Some of these references are cited in the text. Others are provided as sources of additional information.

Ahrens, W. H. (Ed.). 1994. "Herbicide Handbook," 7th ed. Weed Sci. Soc. Am., Champaign, Illinois. 352 pp.

Anonymous. 1972. Degradation of synthetic organic molecules in the biosphere: natural, pesticidal and various other man-made compounds. Proc. Conf., Nat. Acad. of Sci. 350 pp.

Cheng, H. H. (Ed.). 1990. "Pesticides in the Soil Environment: Processes, Impacts, and Modeling." Soil Sci. Soc. America, Series 2, Madison, Wisconsin. 520 pp.

Dubach, P. About 1972. Dynamics of herbicides in the soil. Ciba-Geigy, Ltd., Basel, Switzerland.

Garner, W. Y., R. C. Honeycutt, and H. N. Higg. 1986. "Evaluation of Pesticides in Groundwater," Amer. Chem. Soc. Symp. Series 315. Amer. Chem. Soc., Washington, D.C. 573 pp.

Goring, C. A. I., and J. W. Hamaker (Eds.). 1972. "Organic Chemicals in the Soil Environment," Vols. I and II. Dekker, New York.

Gould, R. F. (Ed.). 1966. "Organic Pesticides in the Environment," Advances in Chemistry Series, No. 60. Amer. Chem. Soc. Washington, D.C. 309 pp.

Grover, R. 1988. "Environmental Chemistry of Herbicides," Vol. I. CRC Press, Boca Raton, Florida. 207 pp.

Guenzi, W. D. (Ed.). 1974. "Pesticides in Soil and Water." Soil Sci. Soc. America, Madison, Wisconsin. 562 pp.

Hamaker, J. W. 1976. Mathematical prediction of cumulative levels of pesticides in soil, pp. 122–131 *in* "Organic Pesticides in the Environment" (R. F. Gould, Ed.), Adv. in Chem. Series 60. Amer. Chem. Soc., Washington, D.C.

Hance, R. J. (Ed.). 1980. "Interactions between Herbicides and the Soil." Academic Press, New York. 349 pp.

Helling, C. S. 1971. Pesticide mobility in soils. II. Application of soil thin-layer chromatography. *Soil Sci. Soc. Am. Proc.* **35**:737–743.

Kearney, P. C., and D. D. Kaufman (Eds.). 1976. "Herbicides: Chemistry, Degradation, and Mode of Action," 2nd ed., Vols. I and II. Dekker, New York.

Linn, D. M. (Ed.). 1993. "Sorption and Degradation of Pesticides and Organic Chemicals in Soil." Soil Sci. Soc. America Spec. Pub. No. 32. Madison, Wisconsin. 260 pp.

Marsh, B. H. and R. W. Lloyd. 1996. Soil pH effect on imazaquin persistence in soil. *Weed Technol.* **10**:337–340.

Pathak, A. K., S. Sankaran, and S. K. De Datta. 1989. Effect of herbicide and moisture level on *Rottboellia cochinchinensis* and *Cyperus rotundus* in upland rice. *Trop. Pest Man.* **35**:311–315.

Saltzman, S., and B. Yaron (Eds.). 1986. "Pesticides in Soil." Van Nostrand Reinhold, New York. 379 pp.

Sawhney, B. L., and K. Brown (Eds.). 1989. "Reactions and Movement of Organic Chemicals in Soil." Soil Sci. Soc. America Spec. Pub. No. 22. Madison, Wisconsin. 474 pp.

Chapter 15

Herbicide Application

FUNDAMENTAL CONCEPTS

- Sprayer calibration is one of the most important and most neglected aspects of herbicide application.
- Forward speed, pressure, and nozzle tip orifice size are the primary things that can be adjusted to change a sprayer's calibration.
- There are five basic types of pumps used on herbicide sprayers, each with its own advantages and disadvantages.

LEARNING OBJECTIVES

- To understand the fundamental importance of sprayer calibration.
- To know what can be adjusted to change a sprayer's calibration.
- To know the effect on calibration of each adjustment.
- To know the desirable features of a sprayer.
- To be familiar with calibration questions and to know how to solve simple calibration problems.

I. INTRODUCTION

It is important to understand the equipment required to apply herbicides properly. Although the equipment has changed over time, it remains basically the same (McWhorter and Gebhardt, 1987). More than 90% of all herbicides are still applied with hydraulic sprayers that contain the same four basic components available for more than 50 years: a tank, pressure regulator, pump, and spray nozzles. The conventional hydraulic sprayer continues to be the most acceptable and most widely used method of herbicide application. Great advances in herbicides and formulations have been made in recent years, but application technology has not advanced at the same pace. Most herbicides are still broadcast as an aqueous mixture from a hydraulic sprayer that uses simple nozzles to break the pressurized liquid stream into droplets. As long as the fuel, the herbicide, and the farmer's time were inexpensive and environmental contamination was not a major concern, a cheap method of herbicide application was appropriate. These conditions have changed, and more effort is now being expended to improve hydraulic sprayers and herbicide application.

In most cases herbicides are applied as broadcast sprays to an entire area, whether the area is an entire field or a band over the crop row. Not all of the area sprayed may have weeds, but it is all sprayed. This means that herbicide is applied where there are no weeds. This has been efficient because it has been easier and cheaper to spray an entire area, and the technology to spray just the weeds has not been available. There has been no way to detect each weed. Yet weed scientists know that weeds usually exist in patches in a field, not as uniform stands, and spraying the entire field is not necessary. New research ideas (Felton, 1990) make it possible to apply one herbicide to one species and another herbicide to a second species in one pass across a field. Weed species are detected because the green leaf tissue of each species differs in reflectance. Microprocessors turn the sprayer on only when weeds are sensed. The system reduces total spray, herbicide use, and cost, does not waste herbicide, reduces environmental presence, and reduces the likelihood of off-target movement and nontarget effects. When morphological and foliar reflectance characteristics of different species are incorporated, specific weed control will be possible. These advances combined with global positioning system (GPS) technology will allow a grower or commercial applicator to know and remember where species are and be very precise with herbicide applications.

A few years ago there was great interest in controlled droplet applicators (CDA technology), but it has waned in recent years. In principle CDA technology produces droplets over a narrow size range. The principle holds for low-volume applications and CDAs are quite effective for drift reduction (see Chapter 13, Section II,A,1).

Herbicides are applied as granules with applicators capable of being calibrated. Granule application can often be conveniently combined with crop planting. Because of its exclusively foliar activity, glyphosate led to the development of wiper application. It can be applied through nylon ropes that act as wicks. The ropes are saturated with glyphosate, but do not drip the herbicide on nontarget species. Weeds emerging above a crop canopy receive a lethal dose of glyphosate when wiped by the rope. Rollers covered with shag carpet have been used, but they have been largely replaced by the rope wicks.

Each kind of herbicide applicator can be calibrated with the same basic technique. The applicator is driven over a known area and output is measured, or output is measured for a certain time with the applicator stationary. Special devices are available to assist with calibration by direct reading during spraying or while stationary. No technique is difficult or complex, but each takes time before herbicide application.

II. THE REASON FOR CALIBRATION

Even with sophisticated, specialized knowledge of herbicide chemistry, mechanism of action, application timing, rate of application, selectivity, and activity, herbicides may fail to control the weeds they should control, achieve desired crop selectivity, and leave undesirable environmental residues. A major reason for failure is not lack of knowledge about how the herbicide acts, but rather that herbicides are frequently not applied properly. A Nebraska study (Reichenberger, 1980), found that two of every three pesticide applicators made application errors due to inaccurate calibration, incorrect mixing, worn equipment, or failure to read and understand the product label. These mistakes caused over- and under-application, which cost farmers between $2 and $12 per acre in added chemical expense, potential crop damage, and lost weed control. When results were extrapolated to the entire United States, a billion-dollar application blunder was made each year. Other studies of farmers' sprayers have shown similar problems (Ozkan, 1987).

It is not totally inaccurate to say that a major problem with agricultural chemicals is the people who apply them. In spite of all the specialized research and technology required to develop and market a herbicide, the end result is often dependent on decisions made by a user, just prior to use. These decisions, often made quickly are frequently wrong. The reason more accidents have not occurred is that herbicides are developed to be reasonably idiot-proof, but they are not completely so; not all mistakes are tolerable.

Because of application blunders and concern for human and environmental safety, government regulation of herbicides will increase. No legislative body can enforce a law against stupidity, but all can pass laws that make penalties for stupidity greater and encourage use of reasonable intelligence. Such laws become more likely when reasonable intelligence is not the norm.

The metallic salts, the first selective herbicides, were applied at 100+ lb/A in at least 100 gallons of water per acre. Some may also have been applied in relatively low volume by brushing or wiping (Gebhardt and McWhorter, 1987). Invention of the compressed air sprayer in the early 1900s improved application (Gebhardt and McWhorter, 1987) but did not reduce the amount of herbicide required for weed control. Early weed sprayers were high-volume sprayers with wooden tanks. Later sprayers, capable of applying lower volumes, had steel tanks. Those sprayers and modern sprayers have basically the same parts–a tank, pressure regulator, pump, and nozzles. Today, 90% of all herbicides are applied with low-pressure ground sprayers drawn by a tractor (Felton, 1990). Herbicides are also sprayed by airplane and with large, self-propelled ground implements.

Spraying may be followed by soil incorporation to reduce or control volatility, put the herbicide in position to maximize plant uptake, and promote control of emerging seedlings or root uptake. Failure to incorporate well is a frequent reason for poor herbicide performance. Power roto-tillers are the best incorporation implements, but are not used on most farms. Disking is probably the most common incorporation technique and works best if done twice, with the second pass at right angles to the first. A single disking produces zones of high herbicide concentration and other areas with virtually no herbicide because of the tendency of the disk to ridge soil.

Herbicides can be applied by injection into water flowing in furrows or ditches and through sprinklers. This technique, called herbigation, is effective for herbicides taken up by plant roots from soil, but is not effective for all herbicides.

III. SPRAYER CALIBRATION

A. CALIBRATION QUESTIONS

Calibration of equipment prior to application is the most important step in any herbicide application. However, it is also the step most often estimated, forgotten, or ignored. Sprayers should be calibrated before every major spraying operation because it is an essential step in any successful program of chemical weed control. Any time a herbicide is applied with a hydraulic sprayer, regardless of the herbicide, crop, or weed, the same questions should be answered:

1. How much herbicide should be added to the spray tank?
2. How much water is being applied per acre?
3. How many acres should a tank of spray solution cover?

The answers are found by calibrating the sprayer. For each chemical weed control task, it is necessary to know the exact amount of herbicide applied. Too little will give poor weed control and the weeds not controlled may reduce crop yield. Too much wastes money, may damage the crop, and may leave environmental residues that can make crops unsalable and soil unusable.

B. General Aspects of Calibration

Adjustable factors that determine calibration and application rate are speed, pressure, nozzle size and type, herbicide concentration, and boom height. The role of each will be discussed briefly.

1. *Speed.* Forward speed for tractor-pulled ground sprayers should be held constant, between 4 and 6 mph. Some ground sprayers are designed and calibrated to travel and apply herbicides successfully at higher speeds.

2. *Pressure.* Pressure should be held constant, normally between 30 and 40 lb/sq. in. Higher pressures produce smaller droplets and increase drift hazards. Lower pressures may not give uniform application.

3. *Nozzle size and type.* With constant pressure, larger nozzle orifices deliver a greater amount of herbicide to a given area. Orifice shape determines spray pattern, and accurate calibration is impossible if nozzles are worn. In general it is accepted that any nozzle that produces droplets between 100 and 400 microns will produce good results over a range of conditions unless application volume is very low (less than 5 gallons per acre) or very high (greater than 100 gallons per acre).

4. *Herbicide concentration.* Changing the amount of herbicide in the spray tank changes rate of application. However, within the dose range recommended for crops, the amount of herbicide does not affect calibration: the amount of liquid applied per unit of ground covered. Viscosity of the spray mixture is important and influenced by the type of herbicide and its concentration. Wettable powder suspensions have higher viscosity than water- and oil-based solutions. If calibration is with water and a wettable powder suspension is applied, the amount of herbicide may be lower than expected. Calibration ideally includes wetting agents, oils, and other adjuvants.

5. *Boom height.* Boom height is not often adjusted, but should be considered. Improper height adjustment affects nozzle pattern overlap and changes rate of application. Height is important for band application because it determines band width. Boom height affects the amount of herbicide applied, but not the sprayer's calibration. It affects where the herbicide goes, but not the volume applied by each nozzle tip.

C. THE SPRAYER

Because rate of application must be changed for different tasks, a properly equipped sprayer has a pressure regulator and exchangeable nozzle tips. Other features found in the best crop sprayers are these:

1. Ease of filling
2. Ease of cleaning, which usually implies a bottom drain
3. Easy-to-regulate output from the tractor
4. A line with a valve for attachment of a single-nozzle spray gun
5. Allowance for different boom lengths and adjustable boom height
6. Adequate agitation
7. Ruggedness and durability

Preparation for spraying requires equipment adjustment to provide the intended pattern and rate over the target. Before calibrating and spraying, one should determine if all sprayer parts function normally. All nozzle tips and screens, including line and suction strainers, should be cleaned before the first spraying operation. All nozzle tips must be the same size for ground sprayers. (Nozzle tips are often different sizes on airplane sprayers to compensate for wind vortices.) If nozzle tips are dirty, they should be cleaned using an old toothbrush or paintbrush—never a metal brush, wire, or a jackknife. Plugged and worn tips should be discarded.

For most weed spraying, a pump that produces 30 to 60 lb/sq. in. pressure is adequate. For high volumes (e.g., for brush control) higher pressures are required. The advantages and disadvantages of pumps commonly found on sprayers are shown in Table 15.1.

There are many types and sizes of nozzles. Flat fan nozzles spray an oval pattern, preferred for broadcast spraying. For band spraying where uniformity across the band is important, the even fan, with its rectangular pattern, is preferred. Cone nozzles are the choice for insecticides because they provide better plant coverage. Hollow cones are for low volumes and solid cones for high volumes. Boomless sprayers with a broad jet nozzle are not recommended for herbicides because of nonuniform coverage.

Nozzle tips are selected to control volume per acre and coverage. Most manufacturers code their nozzles, and the manufacturer's guide reveals the nozzle's intended output. Manufacturers' catalogs also indicate the relationship between nozzle spacing, spray angle, boom height, and output.

Discharge rate for a nozzle orifice varies approximately as the square root of pressure. For example, if a nozzle is discharging 0.5 gal/min at 30 psi, it would require about 120 psi to increase the output to 1 gal/min. For this reason, changing nozzle orifice size is more practical than changing pressure to increase or decrease output. Also, higher pressures increase

Table 15.1

Advantages and Disadvantages of Pumps Suitable for Herbicide Sprayers

Pump type	Advantages	Disadvantages
Centrifugal	Adaptable for use with all sprayers. Handles large volumes. Low wear with abrasive materials well adapted to wettable powders. Quiet.	Medium to low pressure (not over 60 psi). Not self-priming. Requires higher speed and will not operate well off tractor PTO.
Piston	High pressure. Adaptable for use with all sprays. Long-lasting and low-maintenance.	May give uneven pressure. Requires a cushioning chamber. Expensive. Large and difficult to mount directly on PTO or in small space.
Gear	Inexpensive. Pressure up to about 100 psi. Compact.	Low volume. Wears rapidly with wettable powders. Noisy. Short life.
Roller	Medium price. Medium volume. Compact. Most common weed sprayer pump.	Pump life short if (nylon) is run dry. Requires high speed. Wears rapidly with wettable powder suspensions. OK, if rollers changed every 200–300 acres.
Rubber impeller	Low price. Low pressure between 15 and 30 psi. Quiet. Compact.	Will wear rapidly if impeller is run dry. Will not stand sprays with high oil content. Low volume.

the number of fine droplets, which increases drift hazard. Low pressures may not produce a desirable spray pattern.

Nozzle tips are made of six materials (brass, nylon, plastic, stainless steel, hardened stainless steel, and ceramic) that vary in cost and durability. The initial cost of hardened stainless steel tips is high, but their great durability more than compensates for their cost. Brass tips are usually the cheapest and the least durable. Plastic tips can be more durable than brass and less expensive than hardened stainless steel. Ceramic tips are encased in plastic to prevent shattering with impact. They are expensive, but very durable. All nozzle tips increase in flow as a function of the square root of the time in use (Reichard *et al.,* 1991).

All sprayers should have screens and strainers to keep spray mixtures clean, protect sprayer parts, and prevent operator frustration from frequent stopping to clean nozzle tips. A 50 to 80 mesh screen is recommended for suction and in-line strainers. Nozzle tip screen sizes vary but are usually 50 or 100 mesh. All nozzle screens should be the same size. Screens should have enough capacity to handle output from the pump and not plug with the spray solution or suspension being used.

Herbicides are usually mixed in water prior to spraying, but oil may be used as a carrier. The amount of agitation required is dependent on the specific herbicide formulation and the carrier. Hydraulic agitation can be accomplished by bypassing some of the spray solution from the discharge side of the pump back to the tank. Sprayer pumps have a greater capacity than is required to maintain pressure to the boom. The excess, diverted through a bypass back to the tank, is sufficient to mix emulsions and solutions. Bypass agitation is inadequate for wettable powders and some emulsions. A pressurized agitator is employed and it requires a separate line from the discharge side of the pump to provide vigorous hydraulic agitation. Mechanical agitation is also common and adequate.

Sprayer calibration would do much to eliminate application blunders and ensure proper application of herbicides to achieve desired goals with minimal environmental effect. Calibration will not eliminate off-target movement and nontarget effects, but it will reduce them. For a complete discussion of herbicide application equipment, readers are referred to Ozkan (1995).

D. CALIBRATION PROBLEMS

The end result of sprayer calibration is knowing the number of gallons of spray solution applied per acre. That is essential information, but it is not the only thing that must be known before spraying can commence. The applicator must know how much of the product to be applied should be put in the spray tank (Question 1 in Section III,A). The easy calculations are essential if the correct amount is to be applied. The herbicide label is the source of information on how much of the product should be applied to accomplish the intended purpose. The label does not reveal how much of the product to put in a spray tank when the sprayer will be applying a certain number of gallons on each acre. The following questions illustrate the calculations that might follow calibration.

1. Pete has 200 acres of Pinto beans. He wants to apply 2 lb ai/A of the emulsifiable concentrate (EC) formulation of trifluralin (4EC). His sprayer has a 300-gallon tank with a 40-foot boom and he used 10 gallons to spray 363 feet.
   (a) How much trifluralin is needed per tank?
   (b) How many full tanks will be needed to spray 200 acres?
   Answers: (a) 5 gallons, (b) 20 tanks.
2. Zeke's spray tank holds 200 gallons and it will spray 20 acres. He is applying Aatrex™ 80% wettable powder to his corn field at

0.75 lb ai/A. After he began to spray, he quit for awhile to watch "Days of Our Lives" and have tea with his wife. After the TV show, he measured and found he had used 45 gallons of spray solution.

(a) How much formulated Aatrex should he put in the tank as he refills it?

(b) He must also agitate the mixture thoroughly. Why?

Answers: (a) 4.2 pounds. (b) Because a wettable powder forms a suspension, not a solution, and it will settle out on standing.

3. A farmer is going to plant 1,500 acres of sugarbeets, and is going to apply 3 lb ai/A of preplant incorporated cycloate. The sugarbeet rows are 24 inches apart and a 4-inch band of cycloate formulated as a 6 lb ai/gal emulsifiable concentrate will be applied over each row. The spray tank holds 400 gallons. Nozzles are spaced 24 inches apart on the boom and the calibrated sprayer delivers 20 gallons per acre.

(a) How much of the formulated cycloate is needed for each full tank?

(b) How many acres of land will he have sprayed when he has driven over 100 acres?

Answers: (a) 10 gallons, (b) 16.7.

4. A farmer wants to apply triallate for wild oat control in her spring wheat. She has decided on a rate of 1.25 lb ai/A and triallate is formulated with 4 lb ai/gallon. She must treat 520 acres and her 250-gallon sprayer is calibrated to deliver 18 gallons per acre.

(a) How much formulated triallate is needed for the whole job?

(b) How much formulated triallate should be added as the tank is filled?

Answers: (a) 162.5 gallons, (b) 4.3 gallons.

5. My neighbor has a tree that blocks my southern view. I plan to acquire my view (with my neighbor's agreement) by applying SPIKE™ (tebuthiuron), a soil sterilant, to kill the tree. The rooting area is a circle with a 15 foot 8 inch radius. SPIKE is a 5% ai granular formulation and the lethal application rate for woody species is 7 lb ai/A.

(a) How many pounds of tebuthiuron granules must be used to kill the tree?

Answer: 2.5 pounds.

6. The weed district's sprayer has a 33-foot boom and travels 5 mph. It delivers 20 gallons in 3 minutes.

   (a) How many acres can be treated in one hour? **NOTE:** There are 43,560 square feet in one acre.

   (b) How many miles of roadside can be sprayed with one full 500-gallon tank?

Answers: (a) 20 A/hour, (b) 6.25 miles.

7. During a routine check of a spring wheat field, a developing infestation of common lambsquarters and wild buckwheat was noticed. The state weed control handbook recommended a tank mixture of 0.125 lb ai/A of dicamba and 0.5 lb ai/A of 2,4-D. A full 350-gallon spray tank will treat one 35-acre field.

   (a) How much dicamba should be added to the tank as it is filled?

   (b) How much 2,4-D should be added to the tank as it is filled?

Answer: (a) 1.1 gallons of dicamba, (b) 2.9 gallons of 2,4-D.

8. The aerial applicator is going to apply Gramoxone™ to 750 acres of chemically fallowed wheat stubble. The airplane's spray tank holds 50 gallons and the plane applies 3 gallons per acre. Gramoxone has 2 lb of active ingredient per gallon. The Gramoxone label recommends addition of a nonionic surfactant at $\frac{1}{2}$ of 1% of the total spray volume.

   (a) How many acres can be sprayed with one tank of spray solution?

   (b) How much herbicide is needed per full tank?

   (c) How much surfactant is required per tank full?

Answers: (a) $16\frac{2}{3}$ acres per tank, (b) 4.2 gallons, (c) 0.25 gallons.

9. Lasso™ (alachlor) and Aatrex™ (atrazine) will be applied to 480 acres of corn. Alachlor will be applied at 2 lb ai/A, and 1.25 lb ai/A of 90% water-dispersible atrazine will be applied. While driving 660 feet at 3.5 mph, the sprayer with a boom of 17 nozzles spaced 20 inches apart delivers 12.5 gallons. The sprayer's tank holds 250 gallons.

   (a) What is the calibration, the GPA?

   (b) How much of each herbicide is needed for each full tank?

   (c) How much of each formulated herbicide will be required for the 480 acres?

Answers: (a) 29.1 gpa; (b) 4.3 gallons of Lasso per tank and 12 lb of Aatrex per tank; (c) 240 gallons of Lasso and 667.2 pounds of Aatrex.

10. In the suburban United States, a lawn is infested with dandelions. Before entertaining his boss, your client feels compelled to seek advice on ridding the lawn of the dandelions. The lawn is 60 feet by 181.5 feet (small, but beautifully landscaped). The

recommendation is 0.75 lb ai/A of a 2 lb/gallon formulation of 2,4-D. The groundskeeper will use a backpack sprayer with an 8-foot boom, and he will empty the 2-gallon tank in 363 feet.
(a) How much 2,4-D is needed per tankful?
(b) How much actual 2,4-D is needed for the entire lawn?
Answers: (a) 94.6 ml, (b) 355 ml or 0.75 pints.

THINGS TO THINK ABOUT

1. What is known when a sprayer is calibrated?
2. What happens to calibration when speed, pressure, and nozzle tip orifice size increase? Decrease?
3. What is the easiest thing to change to make a small (less than 5 gallons per acre) change in a sprayer's output?
4. What is the easiest thing to change to make a large change in a sprayer's output?

LITERATURE CITED

Felton, W. L. 1990. Use of weed detection for fallow weed control. Proc. Central Great Plains conservation Tillage Symp., pp. 241–244.

Gebhardt, M. R., and C. G. McWhorter. 1987. Introduction to herbicide application technology, Chapter 1 *in*: "Methods of Applying Herbicides" (C. G. McWhorter and M. R. Gebhardt, Eds.), Monograph 4. Weed Sci. Soc. of America, Champaign, Illinois.

McWhorter, C. G., and M. R. Gebhardt (Eds.). 1987. "Methods of Applying Herbicides," Monograph 4. Weed Sci. Soc. of America, Champaign, Illinois. 358 pp.

Ozkan, H. E. 1987. Sprayer performance evaluation with microcomputers. *Appl. Engineering Agric.* **3**:36–41.

Ozkan, H. E. 1995. Herbicide application equipment, pp. 155–216 *in* "Handbook of Weed Management Systems" (A. E. Smith, Ed.) Dekker, New York.

Reichard, D. L., H. E. Ozkan, and R. D. Fox. 1991. Nozzle wear rates and test procedures. *Trans. ASAE* **34**:2309–2316.

Reichenberger, L. R. 1980. Chemical application: The billion-dollar blunder. *Successful Farming,* April. 5 pp.

Chapter 16

Herbicide Formulation

FUNDAMENTAL CONCEPTS

- All herbicides are formulated.
- A formulation is a physical mixture of several or one herbicide(s) and inert ingredients, which provides effective and economical weed control.
- The goals of formulation are to improve biological efficacy and to put the herbicide in a physical form convenient for use.
- There are eight basic types of herbicide formulations.

LEARNING OBJECTIVES

- To know the goals of formulation.
- To understand why users can assume herbicide mixtures are homogenous.
- To know the physical characteristics of different herbicide formulations.

I. INTRODUCTION

If an herbicide has excellent biological activity and can be produced safely at reasonable cost, it may never reach the market if it cannot be formulated to retain or enhance its biological activity. All herbicides are formulated, which means they are combined with a liquid or solid carrier so they can be applied uniformly and transported, and still perform effectively. A variety of formulations have been designed for particular methods of application, to gain increased selectivity, to facilitate use, or to increase effectiveness. A formulation is a physical mixture of one or several herbicides and inert ingredients that provides effective, economical weed control. It is important to note that a formulation is a physical, not a chemical, mixture. That is, no chemical reactions are designed into the mixture. Each formulation has one or several biologically active chemicals. The second issue of Volume 10 (1996) of the journal *Weed Technology* lists 134 separate herbicides formulated into more than 204 products with different trade names. The same issue of the journal also lists 112 formulations with two or more active ingredients.

Formulations include inert ingredients that can be chemically or biologically active but have no herbicidal activity. Effectiveness is a clear, objective judgment. The product performs as labeled or it does not. Whether or not a formulation is economical, as defined earlier, is often a subjective judgment.

Formulation chemists have two primary goals. The first is to improve biological effectiveness by altering vapor or liquid mobility in soil, changing resistance to breakdown, or improving ability to penetrate biological surfaces. The second, equally important, goal is to place the herbicide in a physical form convenient for users and appropriate for intended uses. A formulation should also be convenient to use, as inexpensive as possible, and have good shelf life. Exactly how the formulator does all of this involves art and science. Many aspects of formulation are trade secrets and there are no complete texts on the art of herbicide formulation.

Formulation can be crucial to success. The phytotoxicity of trichloroacetic acid was first discovered by a Du Pont scientist and the ammonium salt was patented. The Dow Chemical Company patented the sodium salt–a different chemical, and thus a different formulation. The sodium salt was the successful herbicide.

The pesticide chemical industry is dominated by a few large companies engaged in a wide range of activities. Each company may buy or synthesize several hundred to several thousand new chemical structures annually. Each major group in a company examines compounds obtained or synthesized by another group to see if they have activity specific to the group's interests. Thus, the agricultural group will obtain chemical compounds from groups interested in many other things. Agricultural groups also employ organic synthesis chemists to prepare new compounds based on known chemical groups or observed structure–activity relationships.

Biologists, in the pesticide development group, receive a few to several hundred grams of a compound and preformulate it so it can be applied to plants to determine biological activity. Chemists receive small samples and determine several physical and chemical properties, such as melting point, boiling point, rate of hydrolysis, vapor pressure, specific gravity, solubility, and susceptibility to ultraviolet degradation. Inherent biological activity is determined by biologists who work closely with chemists.

During development, inert ingredients are selected to be formulated with a potential herbicide. The formulation chemist determines the physical properties of the formulation and is concerned about compatibility of ingredients with the potential herbicide and its container.

When a herbicide is marketed, a farmer or custom applicator will move it from a point of distribution to a point of use or further sale where it may stored under a variety of conditions. The formulation chemist must be concerned about stability of herbicides that will be stored in a shed where things freeze in the winter and where the temperature might be 120°F at noon in the summertime. A formulation that spontaneously combusts or explodes at temperatures above 100°F is not acceptable. Formulations that freeze without renewal of activity on thawing are not acceptable. Formulators must be concerned about storage at various temperatures for different times and the effects of storage conditions on stability of active ingredients, on performance, and on wettability and dispersion. Chemists can modify formulations to affect solubility, volatility, and phytotoxicity.

When a user selects a herbicide and mixes it for spraying, an assumption is made that is rarely thought about. Most herbicides are added to a volume of carrier, usually water, and mixed in a spray tank. It is assumed that the mixture is homogenous and that any volume of water taken from a spray tank contains the same amount of herbicide as any other volume of water. With variable water hardness (presence of divalent cations, e.g., calcium, magnesium, iron), suspended particulates, and microbial activity, however, the formulation chemist cannot assume homogeneity. Homogeneity must be tested for and assured.

When a formulation is added to water in a spray tank, foaming can occur. Some foaming is inevitable if a surfactant is included, but excessive foaming is undesirable. It leads to imprecise determination of volume, and potential environmental contamination as foam exits the tank. Control of foaming is a formulation problem.

Formulation chemists must be concerned about spray solution viscosity because it affects flow patterns, particle size, and weed control. The primary goal of formulation is to maintain or improve biological efficacy. A homogenous formulation that is stable at all possible temperatures to which it will be exposed, does not foam, and maintains appropriate viscosity, but that, in the process of formulation, loses its phytotoxicity, is a failure.

If a formulation performs satisfactorily in spray equipment and has passed all of the preceding tests, then chemists and biologists are concerned with how it interacts with the weed. Among the questions asked are these:

1. What percent is retained on the weed?
2. What is the residual nature of the herbicide on the weed?
3. Does the herbicide penetrate and translocate?
4. Does the herbicide form crystals on the plant surface?
5. What is its biological efficacy?
6. What is the site and mode of action?
7. What is its persistence in the environment?
8. What are the possibilities of effects on nontarget organisms through drift, soil residues, residues in water, and presence in other parts of a food web?

Formulation chemists are included in manufacturing decisions. Formulators must achieve the highest practical degree of efficacy at the lowest cost. Formulation chemists discuss formulations with chemical engineers to see if a formulation that works can be manufactured economically. The formulation chemist is also concerned about packaging. A package that is sensitive to moisture is obviously inappropriate. One does not want a package that will be corroded by its contents. Thus, a formulation must maintain biological efficacy and be compatible with containers that will inevitably be stored under diverse conditions.

II. TYPES OF HERBICIDE FORMULATIONS

There are two general types of herbicide formulations: liquid and dry. However, nothing as complex as herbicide formulation should be divided into two simple categories. Liquid formulations include solution concentrates, emulsifiable concentrates, and flowables. Dry formulations include wettable powders, dry flowables, and granules.

A. Liquid Formulations

1. Solution Concentrate

A solution concentrate is a herbicide dissolved in a solvent system designed to provide a concentrate soluble in a carrier, usually water. If a herbicide is immiscible with water, a one-phase solution containing the herbicide, one or more emulsifiers, and one or more solvents can be made to force or bridge a herbicide into solution or very fine suspension. The basic requirements for making a solution concentrate are dependent on active ingredient solubility. The herbicide must be soluble in a small enough quantity of solvent to make packaging and shipping economical. The concentrate must be completely and rapidly soluble in water at all temperatures and concentrate–carrier ratios likely to be encountered when it is used. Usually, solution concentrates require little formulation and have a high concentration of active ingredient. Some acidic salts are formulated as solution concentrates, but few herbicides are soluble enough and also capable of being stored at very high or low temperatures as a water solution. The formulation is not widely used.

2. Emulsifiable Concentrate

An emulsifiable concentrate consists of the herbicide dissolved in an organic solvent with sufficient emulsifier added to create an oil-in-water emulsion when the concentrate is added to water. Salts of acidic herbicides that are soluble in water and could be formulated as solution concentrates are commonly formulated as emulsifiable concentrates because the herbicide may react with metallic ions in water, precipitating the active ingredient and clogging spray equipment. These formulations are used when the active ingredient may not enter plant foliage readily because of high water surface tension or evaporation that leaves herbicides on foliage and results in no activity.

Other herbicides have low water solubility but can be dissolved in an organic solvent (e.g., xylene) and mixed with water to form an emulsion. An emulsion is a mixture in which one liquid is suspended in another (e.g., fat globules in milk). In herbicide emulsions, water (the carrier) is the continuous phase and oil globules (solvent plus technical herbicide) are dispersed in it. This is called an oil-in-water (O/W) emulsion. Oil-soluble esters of acid herbicides and other herbicides, such as the carbamothioates, dinitroanilines, and some alphachloroacetamides, are formulated in this way. Because phases may separate, an emulsifying agent is added to keep the dispersed phase (herbicide) in suspension. This combines the two liquids without direct contact between them, and adverse reactions between the chemicals are not likely. Most agricultural emulsifiable concentrates consist of 60 to 65% (by weight) of herbicide dissolved in 30 to 35% organic solvent, with 3 to 7% of an appropriate emulsifier added to create an oil/water emulsion when the concentrate is added to water. They may contain a small amount of emulsion stabilizer and a surfactant selected to permit appropriate interaction with water or plant surfaces.

Most manufacturers use the highest possible concentration, but 4 lb/gal is common. Almost everyone knows that if there are 4 lb in a gallon, there is 1 lb in a quart, 0.5 lb in a pint, etc., and pint or quart measures are easy to find in the United States.

Emulsifiable concentrates, when added to water, form an emulsion that is opaque or milky. Thus, if one sees a milky or opaque herbicide mixture, it is reasonable to conclude it was made from an emulsifiable concentrate. These concentrates usually penetrate waxy foliage better than other formulations. The solvent, in some cases, is also phytotoxic and aids herbicide activity. Emulsifiable concentrates can be applied in hard water without adverse reactions and are less apt to be washed off foliage by rain or irrigation. Herbicides formulated in this way evaporate more slowly from plant surfaces. The formulations are easy and inexpensive to make and easy to measure and handle. They were the first method attempted for many herbicides.

Advantages of emulsifiable concentrates include low price and ease of handling and transport; also, agitation is not required, and nozzle plugging is rare. One disadvantage is that the formulants (solvents and other ingredients) are concentrated and may be toxic; overapplication is a potential problem. A formulation challenge is to minimize hazards to machines and people from required toxic formulants.

3. Invert Emulsions

In an invert emulsion (W/O), oil is the continuous phase and water the dispersed phase. A common example is mayonnaise. An invert's primary advantage is drift reduction because they are more viscous and produce large drops. They are used in formulation of phenoxy acid herbicides for rangeland and industrial weed control. Inverts are nearly always applied in a large volume of diesel fuel or another low-grade petroleum product to aid plant absorption. Special emulsifiers are required and the formulation is usually expensive.

4. Flowable Concentrate

Flowable concentrates can be thought of as liquid extensions of wettable powders. They are concentrated aqueous dispersions of herbicides that are insoluble, or nearly so, in water. Not many of these formulations are available. They contain little or no organic solvent, but do include clays similar to those used in wettable powders, some oil, water, an emulsifying agent, and a suspending agent. These approach other liquid formulations in ease of dispersion in water and ease of measurement and do not require vigorous agitation. They are more difficult to make and have been used in lieu of wettable powder formulations. The entire system can gel and become unusable, or the system can become solid with the oil portion rising to the surface.

5. Encapsulated

Encapsulated formulations enclose dry or liquid herbicide molecules in microscopic, porous polymer (plastic) capsules that are sprayed in water suspension. After application the capsule releases the herbicide slowly. Rate of release can be controlled so timed release can be achieved. They offer the advantage of timed release through a longer portion of the crop season. With most formulations, maximum availability occurs at application, which may coincide with maximum crop susceptibility and the lowest weed population. These formulations attempt to change that. They are water applied, mix easily, and will not freeze.

B. Dry Formulations

1. Dusts and Dry Powders

Dusts are finely powdered, free-flowing, dry materials used to provide extensive surface coverage. They are relatively easy to formulate, but no herbicides are formulated as dusts because of the drift potential. Some herbicides are so soluble in water that they require little formulation. Spontaneity of solution can be a problem, but few herbicides have such water solubility and none are formulated as dry powders.

2. Wettable Powders

Wettable powders are finally divided (dustlike) solids that are easily suspended in water. When herbicides are insoluble in water or oil solvents, the formulation chemist may turn to wettable powders. They are formulated by impregnating the active ingredient in or on an inert material such as a clay and adding a wetting and dispersing agent. The wetting agent wets the active ingredient when it is mixed with water. Dispersing agents disperse the finely ground particles when mixed with water. A wettable powder with 50% active ingredient may contain 42% clay, 2% wetting agent, 2% dispersing agent, 4% inert ingredients, and 50% active herbicide. Because wettable powders form suspensions, not solutions, they will settle without continued agitation in the spray tank. These formulations typically have less foliar activity than liquid formulations. Because they are suspended, finely divided solids, their abrasive action can wear pumps and spray nozzle tips and frequent calibration is required. To aid dispersion and ensure homogeneity, wettable powders should be mixed in a thick slurry before mixing with water in the spray tank. Most of the triazines, phenylureas, uracils, and members of several other herbicide groups have been formulated as wettable powders. A major problem with this formulation is the difficulty of measuring weight in the field.

Wettable powders can present an inhalation hazard to those measuring them or mixing them in water. Vigorous agitation in the spray tank is required. They are the most abrasive of formulations to nozzle tips and pumps, and frequent nozzle plugging is a problem.

3. Granules

Some herbicides can be formulated as granules: solid materials with 2 to 10% active ingredient. The cost per unit of herbicide is high. Granules are not applied in water or oil carriers and there is less drift hazard. They are not sprayed, but applied as solid granules that tend to fall off plant foliage with little or no damage to plants via foliar uptake. Granular formulations are restricted to herbicides with soil activity. Equipment required for application can be inexpensive and application can be combined with other field operations. However, uniform application is more difficult and granules may be moved by wind or water after application.

Granular technology combines 5 to 10% ai with 1 to 2% surfactant; the balance is carrier. Carriers must be available in a uniform size range, free of dust and fine particles, and their structure must not be destroyed with repeated handling. The granule must have sufficient adsorptive capacity to take up and hold the active ingredient. Granules are bulky to handle, costly to ship, and expensive per unit of active ingredient.

A mycoherbicide (see Chapter 10) has been formulated as a unique kind of granule (Connick et al., 1991). The product is not commercially available but has been called "Pesta." Appropriate fungal propagules were entrapped in a matrix of wheat gluten. A dough prepared from wheat flour, filler, fungus, and water was rolled into a thin sheet (the process for preparation of pasta), air-dried, and ground into granules. The fungus was the active agent that grew and sporulated on the wheat granules after application to soil. Acceptable control of four broadleaved species has been obtained (Connick *et al.*, 1991).

4. Dry-Flowable and Water-Dispersible Granules

Dry-flowable and water dispersible granule formulations combine granule and wettable powder technology. A wettable powder resembles flour and a granule is a large particle. A dry flowable is dustless, small, dry particles that flow and are measured by volume rather than weight prior to mixing in water and spray application. They offer the convenience of measurement of liquid formulations, reduce the mess of liquid formulations, and retain the advantages of solid formulations. Many herbicides are now formulated as dry flowables.

5. Water-Soluble Packets and Effervescent Tablets

Water-soluble packets reduce mixing and handling hazards by eliminating direct contact with the formulation. The package, containing the formulated herbicide, dissolves when placed in water. Some agitation is required to mix the formulation in the spray tank. These packets are usually small and are very appropriate for small landholders in the developing world.

Effervescent tablets resemble Alka-Seltzer™ tablets when mixed in water. The tablet, usually palm-sized, can be used whole or broken into pieces. Some agitation is required to mix the formulation in the spray tank.

III. SURFACTANTS AND ADJUVANTS

Surfactants, surface-active agents, do many things in formulations, including increasing wettability and spreadability, enhancing phytotoxicity, and increasing penetration. Their effects are due to their ability to increase wetting of the target surface and enhance penetration. It has been shown that one of the things surfactants do is reduce the energy required to absorb herbicides across cuticle and exterior leaf membrane barriers. Surfactants may be part of the purchased formulation (e.g., glyphosate) or added to the spray tank prior to use if recommended on the label (e.g., Gramoxone).

An adjuvant is something added that may or may not be phytotoxic. One example is addition of a surfactant to promote foliar activity, spreading, sticking, or absorption. There are safeners or protectants available for use with specific carbamothioate or alphachloroacetamide herbicides, which extend their range of selectivity. Spray modifiers are available in several forms to reduce drift or promote spreading and sticking. If foaming is a problem, there are antifoam agents. Nitrogen has been found to enhance activity of some herbicides when added to the spray tank. Ammonium sulfate is used as an adjuvant and increases herbicidal activity in some cases. The herbicide label should always be consulted as the one reliable guide on use of surfactants or adjuvants with any herbicide.

THINGS TO THINK ABOUT

1. Why are all herbicides formulated?
2. What is an acceptable definition for each type of formulation?
3. What are the advantages and disadvantages of each type of herbicide formulation?
4. What is a surfactant, and what do surfactants do in herbicide formulations?
5. What is an adjuvant, and what do adjuvants do in herbicide formulations?

LITERATURE CITED

Connick, W. J., Jr., C. D. Boyette, and J. R. McAlpine. 1991. Formulation of mycoherbicides using a pasta-like process. *Biological Control* **1**:281–287.

Chapter 17

Herbicides and the Environment

FUNDAMENTAL CONCEPTS

- Herbicides are chemical molecules that do not occur naturally in the environment, but they all are not inherently dangerous when used properly.
- Herbicides control weeds and manage vegetation in situations where no other method is as efficient.
- Herbicide performance is measured by activity, selectivity, and soil residual behavior.
- Herbicide resistance is an important, manageable aspect of herbicide use.
- There are positive and negative interactions that occur whenever a herbicide is used in the environment.
- Science can measure risk, but safety is a normative political judgment.

LEARNING OBJECTIVES

- To understand how activity, selectivity, and residue characteristics determine a herbicide's environmental interactions.
- To understand how intended herbicide use causes positive and negative environmental effects.
- To know what herbicide resistance and tolerance are and to know how they can be managed.
- To know about enhanced herbicide degradation in soil.

- To know that herbicides and plant pathogens interact and how this affects herbicide use.
- To understand the energy relationships of herbicide use.
- To appreciate the complexity of herbicide interactions with humans and the environment.
- To understand how the LD_{50} and perception of risk affect herbicide use.
- To know rules for safe use of herbicides.

Herbicides are chemicals that do not occur naturally in the environment. That does not make them inherently evil or dangerous. It signals a need for caution and should encourage attention to possible detrimental effects that can be prevented by intelligent use. This chapter presents information on harmful and beneficial aspects of herbicides. It is not an exhaustive treatise on herbicide–environment or herbicide–human interactions. Additional sources of information have been cited at the end of the chapter. It is incorrect to assume that all environmental interactions or effects of herbicides are negative or harmful. Some are, some are not. Examples of both will be presented to encourage understanding of and clear thought about herbicides and the environment.

I. HERBICIDE PERFORMANCE

Performance–the fact of weed control–is the reason herbicides are used. They work. Therefore, a positive aspect of herbicide–environment interaction is vegetation management and weed control. Herbicides control undesirable plants–weeds–in many places, and that is an advantage to farmers and others with weed problems and to all the rest of us, the consumers, who benefit from reduced food costs because of reduced production costs.

Herbicides, used in agriculture, solve one aspect of the problem of weeds. Weeds also cause aesthetic pollution when they interfere with enjoyment of our world. For example:

1. Bicycle riders think the world would be a better place without puncturevine and sandbur. Both bear seeds with sharp, durable spines that easily puncture tires.
2. Some people are very allergic to poison ivy and poison oak. They suffer bouts of itching and discomfort and cannot tolerate either in their yard or garden. Herbicides help clear our immediate environment of these unwanted plants.
3. Plants that cause allergies can be managed by herbicides. Thousands of people suffer from "hay fever" caused by weeds.

4. Many weeds and other common plants are poisonous when consumed. Some poisonous weeds are larkspur, monkshood, spotted water hemlock, nightshades, buttercups, poison hemlock, and jimsonweed. Other plants that are poisonous when ingested are lily-of-the-valley, oleander, wild cherries, rhubarb, foxglove, iris, and sadly, a romantic one, mistletoe.
5. Not too many homeowners love crabgrass or dandelions in their lawns. Herbicides are the only technique available to control these pests quickly, easily, and inexpensively. Herbicides should be combined with proper fertilization, mowing, and water management.
6. Few fishermen fish where it is hard to see the water because of aquatic vegetation. Proper weed management of aquatic sites may include herbicides because of their ability to selectively control aquatic plants without polluting water.

No one advocates herbicides for all situations where some plant bothers people who decide it is a weed and should be controlled. The foregoing is an illustration of places where herbicides, perhaps uniquely, provide a way to control weeds. Weeds often exist in places where no other control technology is appropriate.

Herbicide performance is measured in terms of activity, selectivity, and residual characteristics. Activity is reflected in the rate used to control weeds. How much is needed is another way of asking how active the herbicide is. Selectivity (Chapter 12) determines what plants are affected and not affected. It determines the crops or cropping systems in which a herbicide can be used. Soil residual characteristics (Chapter 14) determine how much of the herbicide resides in soil to control weeds over time and possibly affect the next year's crop.

Each of these traits is affected by environmental factors including wind, rain, air and soil temperature, light, humidity, soil texture (adsorptive capacity), soil pH, and other plants. These are the givens, albeit complicated ones, of herbicide use.

II. ECOLOGICAL CHANGES

A. EFFECTS OF HERBICIDE USE

Weed control with herbicides concerns weed scientists, ecologists, and other scientists because of intentional ecological alteration. It is commonly accepted that one reason annual grasses have become the dominant weed problems in wheat and barley is that 2,4-D so successfully controlled annual broadleaved weeds. Failure or lack of preventive programs of field and seed sanitation contributed to development of the annual grass problem. Without 2,4-D's success it is unlikely that annual grasses would ever have developed into the dominant weed problems in small grains. Use of 2,4-D yielded all the benefits of good weed control–improved yield, ease of harvest, lower production cost, etc. It also yielded an unintentional, but predictable, ecological alteration–a major vegetation shift and the ecological opportunity for different weeds to succeed. It seems anomalous that a successful weed control technique could create a problem. In fact, any technology creates and solves problems at the same time. Predicting what problems will be created is a much more difficult task than observing what problem has been solved.

In the U.S. Pacific Northwest, continuous wheat is a common cropping system. Phenoxy acid herbicides have been used on some fields for 20 years, leading to changes in the weedy vegetation (Table 17.1) (McCurdy and Molberg, 1974). Each of the herbicides in Table 17.1 is active against broadleaved species, but they are not equally active against all species. Because the work in Table 17.1 was completed in the early 1970s, it is possible that the poor control of redroot pigweed may have been the development of resistance to the herbicides (see Section II,B of this chapter).

A similar situation, although not well documented, is the invasion of corn fields by yellow nutsedge after several years of successful use of atrazine for weed control in corn. Nutsedge is not affected by atrazine used for weed

Table 17.1

Percent Weed Reduction in Wheat Fields Treated Annually for 20 Years with a Phenoxy Acid Herbicide (McCurdy and Molberg, 1974)

Weed species	Herbicide		
	2,4-D amine	2,4-D ester	MCPA
Stinkweed	97	94	98
Russian thistle	88	58	35
Common lambsquarters	90	85	86
Wild buckwheat	32	54	51
Wild tomato	52	53	23
Redroot pigweed	55	15	30
Total all weeds	86	83	69

control in corn, and it moved into a niche opened by atrazine's successful control of other weeds. Atrazine's success created opportunities for invasion by crabgrass, witchgrass, fall panicum, shattercane, and wild proso millet. None are controlled by normal use rates of atrazine.

The phenomenon of vegetation shifts is not limited to annually cropped fields (Table 17.2). Herbicides are used in orchards to eliminate broadleaved weeds and encourage a grass groundcover. This makes other orchard maintenance activities easier and facilitates harvest. After 4 years and six separate herbicide applications of specific herbicides in an apple orchard, there were few patches of bare ground, but the soil was not barren (Schubert, 1972). Continued application of the same herbicide or herbicides that affect the same weeds encourages unaffected genera because those susceptible to the herbicides are controlled.

The lesson of these data is that herbicide rotation is a good idea. Continued use of a single herbicide for many years on one field *will* change the nature of the weedy flora and may complicate weed management.

On Black Mesa in western Colorado, the butyl ester of 2,4-D was used to control pocket gophers (*Thomomys talpoides*) (Tietjen *et al.,* 1967). In large doses, the herbicide is toxic to many mammals. When used to control weeds, although the dose is insufficient to kill pocket gophers, their population was controlled. Pocket gophers live by consuming small, broadleaved forbs that were abundant on Black Mesa. The 2,4-D reduced the forb population from 77% (South Crystal Gulch) or 63% (Myers Gulch) to 9%, and the pocket gophers' food supply was nearly eliminated (Table 17.3). Use of 2,4-D closed the pocket gophers' grocery store and they had to move to a new neighborhood or face starvation. Eventually the forb population

Table 17.2

Plant Genera Encouraged after Successive Annual Applications (Brown, 1978, after Schubert, 1972)

Genera encouraged	After successive annual application of
Rumex	2,4,5-T, simazine, diuron, or terbacil
Plantago	2,4,5-T amine, diuron, or monuron
Polygonum	2,4,5-TP, simazine, or diuron
Convolvulus	Simazine, diuron, or terbacil
Rubus	Simazine, diuron, terbacil, or dichlobenil
Cerastium	Dalapon or amitrole

Table 17.3

Pre- and Postspray Forb Composition in Two Locations
(Tietjen et al., 1967)

	Composition of forbs		
Location	Prespray	Postspray	After 5 years
		(%)	
South Crystal Gulch	77	9	44
Myers Gulch	63	9	10

recovered, but after 5 years it had not achieved prespray levels. A detailed analysis of vegetative composition on Grand Mesa, Colorado, 3 years after 2,4-D was applied showed that slender wheatgrass increased with a corresponding decrease in broadleaved species (Turner, 1969).

Santillo et al. (1989a) thought glyphosate, a contact, broad-spectrum herbicide, would affect small forest-dwelling mammals by altering vegetation structure and cover and by reducing plant and insect food resources. Glyphosate applied at 4 lb ai/A controlled 75% of nonconiferous, brushy plants. Insectivore and herbivore species were less abundant for the 3 years of observation after glyphosate application. Omnivores were equally abundant in treated and control areas. The difference in small-mammal abundance paralleled herbicide-induced reductions in invertebrate species and plant cover. The total number of birds was lower on clear-cut areas treated with glyphosate (Santillo et al., 1989b).

Careful study might find a negative ecological effect any time a herbicide is applied. These findings must be balanced against net gains in production of useful crops and reductions in labor required to produce crops. It is also true that herbicide-caused depressions in community diversity may be small and transitory. Three herbicides effective against broadleaved species were applied to control spotted knapweed in Montana (Rice et al., 1992). Weed control was 84 to 90% 2 years after application. There was a small decline in community diversity 1 year after spraying, but diversity increased relative to areas with spotted knapweed 2 years after spraying. The data suggest that all herbicide-treated areas had greater diversity 3 years after spraying. Aggressive, perennial weeds such as spotted knapweed tend to form nearly monocultural communities. Tebuthiuron enhanced rangeland diversity, increased forage production for livestock grazing, improved wildlife habitat, and protected against watershed erosion in studies in new Mexico and Wyoming (Olson et al., 1994). Controlling weedy plants with herbicides that have no other harmful environmental effects (e.g., leaching, drift, hazard to nontarget species) is wise vegetation management.

B. HERBICIDE RESISTANCE

Herbicide resistance is defined as the decreased response of a species' population to a herbicide (LeBaron and Gressel, 1982). It is "survival of a segment of the population of a plant species following an herbicide dose lethal to the normal population" (Penner, 1994). Resistance is contrasted with tolerance, or the natural and normal variability of response to herbicides that exists within a species and can easily and quickly evolve (LeBaron and Gressel, 1982). Tolerance is characterized by "survival of the normal population of a plant species following an herbicide dosage lethal to other species" (Penner, 1994). The terms are not always clearly distinguished and often are used as synonyms. The ecological aspect is the shift of the population to the resistant biotype. The weed species does not change; the ability to control does.

For many years, weed scientists knew that insects developed resistance to insecticides and more of the same insecticide did not solve the problem nor would new insecticides or combinations help much. Weed scientists assumed that weeds could become resistant to herbicides, but that it was not likely to be a major problem for several reasons, the more important of which were these:

1. Weeds, even annuals, have a long life cycle compared to insects.
2. Weeds are not as environmentally mobile as insects.
3. There was a wide range of herbicides in use, and they had several different modes of action. Insecticide resistance, it was assumed, was based on continued exposure to chlorinated hydrocarbons or organophosphate materials. The two groups had different modes of action, but all members of each group shared a mode of action.
4. Crop rotations offered the possibility of using different herbicides in a field.
5. Cultivation and other cultural techniques were used in the same field herbicides were used in and would kill resistant weeds. It was assumed that integration of methods was common.
6. There is, and it was assumed always would be, a large soil seed reserve.
7. Resistant species will probably be less competitive and will not survive well.

These were all logical but bad assumptions, because herbicide resistance developed and is a serious problem for weed scientists. The time for development of resistance has proven to be short. It was first reported for common groundsel that was found to be resistant to atrazine and simazine in 1968 after the herbicides had been applied once or twice annually for 10 years (Ryan, 1970). In 1986, more than 50 weeds were resistant to triazines (National Research Council, 1986) and more than 107 resistant biotypes had evolved around the world. In 1990, 55 weeds were resistant to triazine herbicides in 31 U.S. states, 4 Canadian provinces, and 18 other countries (LeBaron and McFarland, 1990). Weeds are now resistant to 14 other classes of herbicides. Several species have evolved cross-resistance to more than one herbicide. Since 1982 the number of resistant weeds has more than tripled and the land area involved has increased 10 times. Multiple resistance has been observed and occurs when resistance to several herbicides results from two or more distinct resistance mechanisms occurring in the same species. Annual ryegrass in Australia is resistant to all herbicides in seven herbicide families, and blackgrass in Europe is resistant to herbicides in eight different chemical families. In general, but not always, there are enough alternative herbicides and other control measures (e.g., rotation, tillage) to manage resistant weeds effectively, but when multiple resistance occurs, chemical options are eliminated and weed scientists must resort to biological principles to create weed management systems.

The triazines are broad-spectrum herbicides that inhibit photosynthesis and quickly kill a high percentage of emerged seedlings. Because they persist in soil, they continue to kill weeds that emerge after application, and there is a long time when susceptible plants are not present to compete with resistant ones. It was believed that the soil seed reservoir, unique to plant populations, would slow the appearance of resistance because of the small percentage (2–10%) that germinates in one year. The large seed reserve slowed, but could not prevent, expression of resistance.

The sulfonylureas and imidazolinones are active at fractions of an ounce per acre, often persistent in soil, and have a specific mechanism of action. They have important advantages and have replaced some herbicides with multiple sites of action and different soil persistence. Resistance to some of these herbicides has developed in as little as 3 years (Gressel, 1990).

Assume that in the first year of herbicide application there was only 1 resistant weed in a population of 100 million. It would not be noticed, or if it was, anyone would assume it had emerged after herbicide application or had been missed. If 100 million plants were spread, uniformly, over a large field, there would be a small population on each square foot. It is reasonable to assume that a herbicide would achieve 90% control. The resistant weeds would not be noticed after the first year (Table 17.4). It is likely that the resistant population would not be noticed for several years. It would take a person with unusual powers of observation and a keen

Table 17.4

**Development of a Population of Resistant Weeds
with Repeated Use of a Single Herbicide
(Gressel, 1990)**

Year	Susceptible population	Resistant population
1	100,000,000	1
2	10,000,000	4
3	1,000,000	16
4	100,000	64
5	10,000	256
6	1,000	1024
7	100	4026

knowledge of weeds to notice 256 weeds in a large field. If one assumed the 100 million weeds were all in a 50-acre field, there would be 46 weeds per square foot. That is a dense population, but it is more likely one would be delighted with the excellent weed control achieved rather than notice a few escapes mixed with other weed species the herbicide did not control. Another way to look at the same problem is to note that concomitant with 90% population reduction of the susceptible species, the resistant species might increase by a factor of 4 each year. If it was a 10-acre field (435,600 square feet), with 90% control and an annual quadrupling of the resistant population, Table 17.5 shows what would happen. Farmers and weed scientists must anticipate and prevent these problems.

It is incorrect to assume that resistance will occur with all herbicides, although there are several examples (Table 17.6). It is most likely to occur where some or all of the following factors are present:

1. The herbicide has a high degree of control of the target species. It is very active and efficient.
2. The weed's seed has a short life in the soil seedbank.
3. The herbicide has long soil persistence.
4. The herbicide is used frequently–annually for many years or more than once per year for several years.
5. Annual herbicide rotation is not practiced.
6. The herbicide has a single site of action.
7. The herbicide's rate is high.
8. Herbicides are not mixed in a crop.

Table 17.5

**Development of a Resistant Weed Population in a
10-Acre Field**

Year	Resistant population	Susceptible population
1	1	4,356,000
2	4	435,600
3	16	43,560
4	64	4,356
5	256	436
6	1,024	44
7	4,096	4
8	16,384	1
9	65,356	0
10	130,712	0

It is equally incorrect to assume that the phenomenon of resistance is the death knell for herbicides. Resistant weeds are not superweeds and are often less fit ecologically than their susceptible relatives. It is important to recognize that resistance is possible and to determine the reasons for it. Identification of the cause and mechanism of action of resistance was one impetus for the intentional use of biotechnology to transfer resistance to crops.

The reasons cited at the beginning of this section are good reasons that herbicide resistance will remain a less important phenomenon than it has become with insects and insecticides. Crop rotation and herbicide rotation for different weed problems are important techniques to combat the problem and should be used in integrated weed management systems. If the same herbicide, herbicides from the same chemical family, or those with the same mechanism of action are used on the same land for several successive years, development of resistance is more likely. Integration of crop rotation and mechanical control in weed management, rather than relying on herbicides to solve all problems, is an important part of the answer to the problem of resistance.

C. ENHANCED SOIL DEGRADATION

Because of the nature of the crop grown and the weed problem, some soils have been treated with the same herbicide several years in succession.

Table 17.6

Examples of Weeds Resistant to Herbicides (LeBaron, 1990 and other sources)

Herbicide	Resistant weed(s)	Location
Paraquat	Horseweed	Japan
	Perennial ryegrass	U.K.
	American black nightshade	Florida
Atrazine	Redroot pigweed	Several
	Common ragweed	Pennsylvania
	Downy brome	Five states
	Common lambsquarters	Several
	Velvetleaf	Maryland
	Kochia	Twelve states
	Wild buckwheat	Germany and Pennsylvania
	Common groundsel	Several
	Black nightshade	Several
	Barnyardgrass	Several
	Annual bluegrass	Several
MSMA and DSMA	Common cocklebur	N. and S. Carolina
Sulfonylureas	Kochia	Eight states
	Prickly lettuce	Idaho
	Russian thistle	Five states
Trifluralin	Palmer amaranth	S. Carolina
	Green foxtail	Canada
Bromoxynil	Common lambsquarters	Germany
2,4-D	Musk thistle	New Zealand
	Canada thistle	Hungary
	Wild carrot	Ontario
Picloram	Yellow starthistle	Idaho
Pyrazon	Common lambsquarters	Europe
Diuron and linuron	Redroot pigweed	Hungary
	Horseweed	Hungary
Amitrole	Annual ryegrass	Australia
	Annual bluegrass	Belgium
Bromacil	Redroot pigweed	Hungary
	Smooth pigweed	Hungary

This has led to enhanced degradation that was first reported for EPTC, a carbamothioate herbicide, in New Zealand (Rahman *et al.,* 1979). Since then, several cases of enhanced or accelerated degradation in soils repeatedly treated with carbamothioate or phenoxy acid herbicides have been reported. In Section II,B, the problem of weeds becoming resistant to herbicides was described: The herbicide works less well with time. This is the opposite problem: Microorganisms responsible for herbicide degradation become more capable of degrading the herbicide, and weed control decreases. The precise mechanism is one of four (Gressel, 1990):

1. The soil could be enriched in a population of a rare or minor microorganism that increases because of the herbicide and rapidly degrades it.
2. Repeated application of the herbicide could select microorganisms from existing populations that degrade the herbicide more rapidly.
3. It is well known that substrates are capable of inducing enzymes in microorganisms. The presence of the substrate (the herbicide) could induce enzymes that rapidly degrade the herbicide *or* induce mutations in microorganisms so they are more capable of degrading the herbicide.
4. Finally, it is possible that when the herbicide is present with other soil chemicals, rapid degradation is promoted. This, co-induction, is related to the presence of another compound or compounds that may not be degraded.

The phenomenon of enhanced degradation has not eliminated use of susceptible herbicides. It has encouraged development of alternative control strategies and new chemicals designed to inhibit rapid degradation. Both techniques have been successful, and enhanced degradation is real but rare.

D. INFLUENCE OF HERBICIDES ON SOIL

Most of any herbicide application reaches the soil, and soil–herbicide interactions are important (see Chapter 14). A related question is, do herbicides damage soil or any of its living components? It would be tragic if a herbicide were approved for use that destroyed an important decomposer organism or affected the nitrogen cycle. This has not happened, and it is not likely to happen, because of careful and continuing evaluation of herbicides, and all other pesticides, by the manufacturer and the U.S. Environmental Protection Agency before approval for use (see Chapter 18). The approval process cannot detect all possible environmental interactions, because often scientists do not even know what questions to ask until after an observation has been made. One should not assume that pesticide use in the United States is one large experiment where no problems are anticipated or addressed until after an observation has been made. Nature is more complex, and the present level of understanding does not permit anyone to anticipate or ask every question that nature may reveal as technology develops.

With normal use rates, the quantity of herbicide applied to soil is too small in relation to total soil volume to have any detectable influence on a soil's physical or chemical state. Research has shown that tilling soil has limited benefit other than weed control, and it has the negative effects of breaking weed-seed dormancy and enhancing soil erosion. Without herbicides, investigation of the effects of tillage would have been impossible because of excessive weed growth.

Part of the research on any candidate herbicide is a determination of its effects on soil microorganisms. Nearly all investigations show a positive or negative effect. Reactions such as nitrification are often suppressed, but, at field use rates, suppression is not permanent. Because of large populations, short reproductive cycles, and great adaptability to environmental insult, microorganism populations are very resilient.

Metham, a soil fumigant applied to seedbeds to control weeds and plant pathogens, is a general biocide and can decimate a soil's microorganism population. But one of the most difficult things to do in the laboratory is to keep soil sterile. Microorganisms are ubiquitous and sterility, while easy to obtain, is almost impossible to maintain with exposure to air or water.

E. HERBICIDE–DISEASE INTERACTIONS

One of the facts of ecology has become almost a cliché: In the natural world, you cannot do just one thing. Everything is connected to everything else and it is impossible to tinker with one environmental parameter without affecting others. All possible effects of an environmental intervention cannot be determined in advance, yet one must act. Food must be grown and weeds must be managed. All environmental effects of food production and weed management are not known, but we cannot stop either to determine all possible effects.

Some herbicides promote plant diseases and others reduce disease incidence. Herbicides predisposed 20 hosts (crops and weeds) to higher disease levels in cases involving 20 pathogens (Altman and Campbell, 1977). One of the earliest reports was after herbicide use in peanuts. Where herbicides had been used, peanuts were larger and more vigorous and the effect was seen in the absence of weeds. This work involved dinitro and other phenolic herbicides that are no longer used, and it was proposed, and later proven, that these herbicides inhibited growth and vigor of parasitic and pathogenic fungi that affected peanuts. Sugarbeets grown in nematode-infested soil and treated with tillam (a carbamothioate) had a higher level of nematode infestation 6 years later than those grown in soil not treated with tillam (Altman et al., 1990). It has also been reported that the carbamothioate herbicide cycloate enhanced cyst development on sugarbeet roots (Altman *et al.,* 1990). Soil residues of chlorsulfuron increased take-all (*Gaeuman-nomyces graminis* var. *tritici*) and *Rhizoctonia,* root diseases of barley and wheat, and yield was reduced (Rovira and McDonald, 1986). The soil-applied herbicide trifluralin alters the fusarium disease syndrome in beans.

It was not intuitively obvious that these interactions should occur. They are an examples of the fact that, in nature, one cannot do just one thing. Herbicide–disease interactions are another element in the equation that must be considered in weed management systems. The data are not available to predict if there will be an interaction and, if so, what kind it will be for all herbicide, crop, and disease combinations. The possibility exists and must be considered. A few examples are cited in Table 17.7.

III. ENERGY RELATIONSHIPS

In 1974, Nalewaja projected that if all the corn grown in the United States were to be weeded by hand in 6 weeks, it would take more than 17 million people working 40 hours per week. That was more than 4 times the number of workers then employed on U.S. farms. The job had to be done in 6 weeks because of the early critical period for weed control. If weeding was not done during the critical period, yield would decrease. There is not enough labor available to weed the U.S. corn crop. It is doubtful that even if people were available they would be in the right place at the

Table 17.7
Examples of Herbicide–Disease Interactions (Katan and Eshel, 1973)

Organism	Disease and crop	Herbicide
Diseases promoted		
Rhizoctonia solani	Damping off—cotton	Trifluralin
Helminthosporium sativum	Seedling disease—barley	Maleic hydrazide
Fusarium oxysporum	Wilt disease–tomato	Maleic hydrazide, Dalapon
Alternaria solani	Early blight—tomato	2,4-D
Diseases suppressed		
Cercosporella herpotrichoides	Foot rot–wheat	Diuron
Fusarium oxysporum	Wilt disease–tomato	Propham, TCA
Alternaria solani	Early blight—tomato	Maleic hydrazide, Dalapon

right time and be willing to weed corn. Hand weeding or hoeing are not among life's desired occupations.

Yet the corn has to be weeded. Not all acres need the same amount of weeding, but all need some. Years ago, animals were substituted for hand labor as hand labor became scarce and more expensive, and later tractors replaced animal power. If the tractors on U.S. farms were not used, others have estimated it would take 61 million horses and mules to replace them. There are not that many work animals available in the United States, and the land required to grow feed for them would reduce available cropland. The shift from hand labor to animals reduced human labor. The change to tractors added more petroleum energy to agriculture. The trend has continued, and agriculture has become more mechanized and chemicalized. Nitrogenous fertilizers and pesticides are highly dependent on petroleum energy for manufacturing and distribution. The purpose of this section is to discuss, with reference only to the use of herbicides for weed control, U.S. agriculture's dependence on petroleum energy. Is energy use for herbicide manufacture and application efficient? How does use efficiency compare to other methods of weed control? The quick answer is that herbicides do not demonstrate an overdependence on energy and their efficiency compares well with other methods of weed control.

A study of the economic relationships for weed control techniques in six experiments on corn showed that when weeds were controlled by hand labor there was a net loss (Nalewaja, 1974). When appropriate herbicides were used there was a net profit (see Table 11.3). Table 11.4 shows similar data for cotton. Of course, all costs are higher now than in 1976. The relationship among the techniques is still valid even though absolute costs have changed.

The data in Table 11.5 show the 1974 energy relationships for weed control in corn (Nalewaja, 1974). Land was plowed, disked, and prepared in the conventional manner and the comparison is only for weed control. The data show an energy advantage for hand labor, but it is not significantly better than herbicide. Both are more advantageous than cultivation with a tractor and cultivator. Corn yield with herbicide use and hand labor was nearly identical. Table 17.8 shows energy costs for several weed control practices. Some equipment requires more energy than others, and energy costs for herbicides increase directly with rate, although application cost is constant. Hand labor is not the cheapest way to weed crops in terms of energy expended. The sulfonylureas and imidazolinones use fractions of an ounce per acre and energy costs compare favorably with any other method of weed control. The energy cost of mechanical tillage and cultivation do not compare favorably with herbicides.

Table 17.8

**Energy Inputs per Performance for Various Weed Control Practices
(Nalewaja, 1974)**

	Energy input				
Method	Gas	Indirect machine	Hand labor (kcal/ha)	Herbicide	Total
Hand labor			53,800		53,800
Field cultivator	120,800	60,400	170		181,370
Tandem disk	93,100	46,500	220		139,820
Rod weeder	26,000	13,000	170		39,170
Rotary hoe	19,700	9,800	120		29,620
Row cultivator	36,700	18,300	310		55,310
Rotary tiller	262,300	131,100	930		394,330
Herbicide					
0.5 kg/ha	8,100	4,000	70	13,600	25,770
1.0 kg/ha	8,100	4,000	70	27,200	39,370
2.0 kg/ha	8,100	4,000	70	54,400	66,570
4.5 kg/ha	8,100	4,000	70	108,700	120,870

Most U.S. cropping systems replace human and animal energy with petroleum energy in the form of fuel for tractors and other machines, manufacture of nitrogen and other fertilizer, and water pumping for irrigation, transportation, and pest control. U.S. agriculture is energy based, but it is not the major energy-consuming sector of the U.S. economy: Farm production consumes only 3% of total U.S. energy. Petroleum energy has been substituted for hand labor and animal power, and chemical energy substitutes for mechanical energy. But herbicides and the cost of application are not a significant portion of the energy cost of producing crops (Table 17.9) (Pimentel and Pimentel, 1979). The energy used for herbicides ranges from 0.1% of total energy expended to produce oranges to 27.3% for soybeans. The mean for the 19 crops is 6.8%, and this is a reasonable picture of energy use for weed control in U.S. crops.

Is the level of energy expenditure excessive? There is no clear answer. Many argue that the business of agriculture is to produce food at a reasonable cost to the consumer and profit to the grower. Agriculture's business is not to produce energy, but it must use it efficiently and responsibly. Herbicides are an efficient use of energy, in view of the energy costs and efficiency of alternative methods of weed control.

Table 17.9

Energy Inputs for Herbicides in Several Crops (Pimentel and Pimentel, 1979)

Crop	Location	Herbicide as percent of total energy Rate (kg/ha)	For herbicides (%)
Corn (grain)	U.S.	2	3.1
Corn (silage)	NY	2.5	4.0
Wheat	U.S.	0.5	1.8
Oats	U.S.	0.2	0.9
Rice	Philippines	0.6	2.4
Rice	CA	11.2	7.7
Rice	Japan	7	9.7
Sorghum	U.S.	4.5	8.4
Soybean	U.S.	5	27.3
Dry bean	U.S.	4	14.6
Peanuts	GA	16	14.6
Potato	NY	18	11.2
Apple	U.S.	2	1.1
Orange	U.S.	0.2	0.1
Spinach	U.S.	2	1.6
Tomato	CA	2	1.2
Brussels sprouts	U.S.	10	12.4
Alfalfa	OH	0.2	0.8
Hay	U.S.	1	5.8
		Average	6.8

The weed control and management techniques used by U.S. agriculture could be more efficient and conserve more energy. Weed scientists are developing more effective integration of weed control techniques that use less energy. Herbicide rates are decreasing and energy use for weed control will decrease. It cannot go to zero because agriculture and weed management require energy. Agriculturalists and the general society will participate in the debate over what form the energy will take and how much is needed. There is no question that agriculture can become more efficient and use less of the total U.S. energy supply. How this will be achieved is not clear. Some production systems use far less energy (Table 11.7). To move quickly toward these systems the United States would sacrifice food production for energy; presently not a good trade.

IV. THE EFFECTS OF HERBICIDES ON PEOPLE

A. GENERAL EFFECTS

It is difficult to assess the environmental toxicity of herbicides to people. A determination of effects is confounded by at least four factors that affect the accuracy of any determination:

1. The changing character of the environment and our attitude toward it
2. The changing character of the population
3. The changing character of the problem
4. The changing character of public health responsibility

Everyone wants a protected and protecting environment, but we must also have a productive environment. At least part of the debate about herbicides and the environment centers on differing views of the appropriate balance among these factors and how to achieve it. The discussion always includes one or more of three concepts:

1. Toxicity: The inherent capability of something to cause injury
2. Risk or hazard: The probability that injury will occur
3. Safety: The practical certainty that injury will not occur

Science can measure toxicity and estimate risk, but cannot measure or determine safety. Safety is judged by people and mandated in legislative acts. Sometimes safety is not a scientific question, but a political question. It is true that one of the biggest problems of herbicide use is misuse. It is not the only problem. Herbicides are toxic to other forms of life and they move in the environment. These things can cause major environmental effects and affect ecological relationships. When a herbicide is put in the environment, it is incorrect to assume that it will affect weeds and not do anything else. That is sloppy ecological thinking and poor science. Weed management is basically an ecological problem of relationships between weeds, other plants, nontarget organisms, and the environment. These can be studied and, when detrimental effects are discovered, changes can be made in use patterns or use can be discontinued.

In the past those who worked with herbicides knew about ecological relationships but did not ask the right questions about the effect of herbicides on ecological systems. Today the right questions are being asked and environmental effects are examined, in depth, with great care. (Because good questions are asked and answered, however, we should not assume all problems have been solved and no future environmental effects or ecological disruptions will occur because of herbicide use.) Good questions asked about all herbicides include the following:

1. What are possible effects on public health?
2. Will domestic animals be affected?
3. Will products for human consumption be contaminated (e.g., meat, milk, fruits)?
4. Will beneficial natural predators or parasites be affected?
5. What is the likelihood of resistance developing?
6. Will honeybees or wild bees be affected and will pollination be reduced?
7. Will there be crop damage?
8. Will ground or surface water be contaminated?
9. Will there be negative affects on fish, wild birds, mammals, microorganisms, or invertebrates?

Reasons for problems with herbicides in the world's developing countries include a limited capability to predict environmental hazards coupled with the dominant attitude that, though effects might be real, they would be minimal. It was true in the United States, and is still true in most of the world's developing countries, that possible or real environmental hazards of herbicides are given low priority. Many of the world's developing countries do not have good environmental policies. An important and debated reason for problems in developing countries is the lack of suitable alternatives to herbicides.

It is often helpful when thinking about complex issues to consider related perhaps, more familiar matters as examples. Two will be discussed briefly.

B. THE CASE OF FLUORIDES

Fluorine is poisonous and is an element in some herbicides. Its most famous, and in some circles, still controversial, use is as an intentional additive to drinking water to prevent tooth decay. It is added to drinking water at about 1 ppm, and debate has rarely focused on its efficacy. The debate is about its toxicity to people who must drink water to survive and should not be poisoned because they cannot avoid drinking water.

The determining factor in fluoride toxicity is the same as it is for the toxicity of any chemical: dose or concentration. A 150-lb man will become ill if he ingests 0.25 mg of fluorine in one day. The same man will become very sick if he ingests 1 g and he will die if he ingests 4 to 8 g. At the prevailing level of fluorine in U.S. drinking water, a 150-lb man would have to drink more than 42 gal of water containing 1 ppm to ingest 1 g–more than 3 bathtubs full. The man would die from water intoxication long before he would be affected by fluorine toxicity.

C. 2,4,5-T

2,4,5-T controls of a wide-range of broadleaved and woody plants. It has been used selectively in crops, on home lawns, in forests, and in rice. When the United States was engaged in the Vietnam War, a 2,4,5-T ester was used in combination with a 2,4-D ester as Agent Orange, to eliminate unwanted vegetation. When 2,4,5-T is manufactured, temperature control is required to minimize formation of an undesirable, nonphytotoxic contaminant, 2,3,7,8-tetrachloroparadioxin. This substance is one member of a family of compounds know as dioxins and is a potent teratogen. A teratogen causes terata, or birth defects, when pregnant women are exposed. This dioxin also causes chloracne, a skin condition characterized by blisters and irritation. There was never any debate about whether the dioxin contaminant in 2,4,5-T was a teratogen or caused chloracne. Part of the concern and debate ensued because of the unknown level of exposure of Vietnam-era servicemen and American women to the contaminant.

The *Pesticides Monitoring Journal* reports surveys of pesticide levels found in the American food supply. In one report they surveyed 24,000 food samples and 3 contained measurable quantities of 2,4,5-T: 2 in milk and 1 in meat. Reported on a whole-milk and fresh meat basis the average 2,4,5-T content was 0.006 ppm for all 24,000 samples: The 2,4,5-T content was 7.5×10^{-7} ppm or roughly equal to 1 mg in 133,000 metric tons. Based on other studies, the presumed maximum nonteratogenic dose of 2,4,5-T with the dioxin contaminant for a 130-lb pregnant woman is 1.26 grams daily. At the observed level of 2,4,5-T in the nation's food supply, a 130-lb pregnant woman could have consumed 170 million tons of food per day for 9 days without fear of teratogenic effects on a fetus.

Many people want absolute assurance of safety and any level of risk of a deformed child is too high, if the risks can be eliminated. Certainly the risk of exposure to a herbicide can be minimized; however, many think that even an infinitesimally small risk is too large and should be eliminated.

A Harvard scientist[1] disputed the 2,4,5-T toxicity theory and calculated the risks associated with spraying 2,4,5-T. If a person applied 2,4,5-T with a backpack sprayer 5 days a week, 4 months a year, for 30 years, the chances of developing a tumor would be 0.4 per million. Other risks of developing a tumor are larger (Table 17.10).

[1]Wilson, R. Cited in the Pesticide Pipeline. Colorado State Univ. XIV (7) July 1981.

Table 17.10

Risks of Developing a Tumor

Activity	Chance of developing a tumor (per million)
Smoking cigarettes	1,200
Being in a room with a smoker	10
Eating ¼ pound of charcoal-broiled steak per week	0.4
Drinking 1 can of diet soda with saccharin per day	10
Drinking milk with aflatoxin or eating 4 tablespoons of peanut butter/day	10
Drinking 1 can of beer/day	10
Sunbathing	5,000

D. SUMMARY

The data required to resolve these questions are difficult to obtain and are controversial. Yet decisions have to be made. Weeds must be controlled. Informed debate is best, but debaters should understand that such decisions, when made, will be based in part on factual information and in part on perceptions or other factors that may have no basis in scientific fact.

Table 17.11 shows data on the risk of death associated with certain human activities which many people do voluntarily. These data are presented not to provide a conclusion or judgment about herbicides and the environment or herbicides and human welfare. Such statements can be found in several of the references cited in the Literature Available section at the end of this chapter; in most cases, thought is required. It cannot be overemphasized that the end of these debates is usually a value judgment, not a clear decision based solely on scientific fact. It may be true that chances of getting cancer are increased by 1 in 1,000,000 by consuming Miami drinking water for a year. Residents and visitors in Miami must decide what, if anything, they propose to do about the scientific evidence. How does one judge the importance of the facts to life? This is a question that must be dealt with by those who consider the problems and advantages of herbicides in the environment.

There is another point of view that should be considered when thinking about herbicides and the environment. The United States is a rich country that can afford to ask and answer difficult environmental questions. We can afford to make decisions that favor environmental protection over

Table 17.11

Acts That Increase the Risk of Death by 0.000001 (1 Chance in 1 Million)

Act	Hazard
Smoking 1.4 cigarettes/day	Cancer, heart disease
Drinking $\frac{1}{2}$ liter of wine/day	Cirrhosis of liver
Spending 1 hour/day in a coal mine	Black lung disease
Spending 3 hours/day in a coal mine	Accident
Living 2 days in New York or Boston	Air pollution
Traveling 6 minutes by canoe	Accident
Traveling 300 miles by car	Accident
Traveling 10 miles by bicycle	Accident
Flying 1000 miles by jet	Accident
Flying 6000 miles by jet	Cancer from cosmic radiation
Living 2 months in Denver on vacation from N.Y.	Cancer from cosmic radiation
Living 2 months in average stone or brick home	Cancer from natural radioactivity
One chest X-ray in a good hospital	Radiation cancer
Living 2 months with a smoker	Cancer, heart disease
Eating 40 tablespoons of peanut butter	Liver cancer from aflatoxin-B
Consuming Miami drinking water for 1 year	Cancer from chloroform
Drinking 30 12 oz cans of diet soda (at one time)	Cancer from saccharin
Living 5 years in the open at the site boundary of a nuclear power plant	Radiation cancer
Living 150 years within 20 miles of a nuclear plant	Radiation cancer
Eating 100 charbroiled steaks (all at one time)	Benzpyrene-induced cancer

productivity. Poor countries may not choose, or cannot afford, to put productivity second. Most of the world's people are poor, hungry, landless, and lack formal education. They do not have access to reasonable health care, or a hope for a brighter future. If one is hungry, one has only one need—food. Obtaining or producing food is the only goal and environmental questions, if thought of, are unrealistic obstacles that may stand in the way of food production. One can argue that without consideration of environmental questions, such as whether or not the risk of soil, personal, or environmental contamination from herbicides is acceptable, long-term food production may be at risk. However, if one is hungry, only the short-term goal of obtaining sufficient food is important. Attitudes toward herbicides may be very different when herbicides are perceived, correctly or incorrectly, primarily as a way to produce food rather than primarily as environmental risks.

V. HERBICIDE SAFETY

How safe are herbicides? It is a simple question, but a definitive answer is hard to find because each answer may have a bias that should be understood. A reasonable response to the question is, Compared to what? Compared to some things, herbicides are dangerous, but compared to others they are safe. Most people believe herbicides, and all other pesticides, are very dangerous. They are regarded as poisons: things that are not safe to use or be around. In fact, they are poisons. If they were not poisonous to something, they would not be useful as herbicides. The suffix "-icide" comes from the Latin *caedere,* to kill.

Answers to safety questions are complex. The questioner usually expects a factual response, not an opinion, but answers are nearly always composed of fact *and* opinion. Some respondents are vested, automatically, with authority and veracity by a questioner, but that does not change the fact that most answers are part opinion.

There are facts about herbicide safety; not all answers are entirely opinions. The dermal and oral LD_{50} (lethal dose at which 50% of test population dies) for all herbicides is known. Necessary safety precautions during use and storage are well known. The agrichemical industry knows and avoids uses that create problems. The U.S. Environmental Protection Agency and state agencies regulate herbicide use and users.

A. PERCEPTION OF RISK

Science can measure risk and determine the probability of occurrence of a defined risk. Science cannot measure safety. Safety is a normative personal or political judgment. Judgment of safety is not, and cannot become, a scientific decision. Science plays a role in creating the data on which many judgments and decisions are based, but scientists, through the scientific process, cannot determine what should be done about their data.

Something may be described as unsafe because it is found through experiment and observation (the methods of science) to increase the risk of undesirable consequences. For example, motorcycle riding without a helmet can be fatal when an accident occurs. Scientists can measure the risk (the likelihood) of a fatal accident from riding a motorcycle without a helmet. Parents may decide not to buy a child a motorcycle, insurance companies may charge high premiums, and legislative bodies may pass laws requiring helmets because of the scientific evidence. Scientists may agree with these actions, but science does not create them.

People perceive risk in different ways depending on where they live, how rich or poor they are, what their options are, their level of education, their friends, the scientific evidence they are aware of, what they read, etc. Perception of risk may differ from the facts as determined by scientific study. Table 17.12 is from a frequently cited study that shows how three different groups judged the risk of several common things. It is obvious that not all share the same perception of risk. In addition to reporting how people in various groups perceive risk, Table 17.12 also shows the actual number of deaths from the hazard. Neither actual deaths nor perceptions of risk are an adequate way to decide what to do.

It is not the purpose of this section to debate the question of herbicide safety, but rather to frame a perspective from which the debate can proceed. The annual U.S. death rate from motor vehicle accidents is more than 40,000 people. People properly conclude that automobiles are dangerous. Yet people drive too fast, without seatbelts, and often with alcohol. Because they think they are in control, many people do not think automobiles are as dangerous as they are. The danger is there, but as long as people think they are in control, they think they control the risk. With pesticides the U.S. death rate is about 30 per year, but pesticides are regarded as more dangerous than the data show they really are. This is because they are seen as an uncontrolled, involuntary risk.

More Americans die from bee stings (150/year), aspirin (200/year), and falls (7,700/year) than from pesticides. There are 2,000 to 3,000 cases of pesticide poisoning each year in the United States from voluntary and involuntary exposure, but only a few deaths. There are several thousand cases of pesticide poisoning and many more deaths in the world's developing countries each year.

There should be no debate about whether or not herbicides can be hazardous to humans. They are toxic to people and will poison and may kill, if not used properly. The last phrase is the key: if not used properly. Many prescription pharmaceuticals, household cleaning agents, aspirin, automotive fuel, and other common products are dangerous, if not used properly. Their inherent toxicity does not change with use, but the possibility of danger increases with improper use. Stupidity does not increase the inherent toxicity of anything, but it increases risk.

B. RULES FOR SAFE USE OF HERBICIDES

There are rules for safe use of herbicides that also do not change inherent toxicity, but make accidents and the expression of toxicity less likely. The rules are simple, obvious, and often overlooked. A sample set of rules is shown below.

Before use
1. Keep away from children.
2. Purchase the right herbicide.
3. Read the label.
4. Follow label directions.
5. Label equipment so cross-contamination is avoided.

Storage
1. Keep in a locked storage place.
2. Never store any herbicide in anything other than its original container.
3. Store outside the residence and away from food, feed, seed, or fertilizer.
4. Protect liquids from freezing.

Handling
1. Keep away from children.
2. Mix in a well-ventilated area, preferably outside.
3. Do not inhale spray or dust. Wear a protective mask when needed.
4. Never smoke, eat, or chew while spraying or handling.
5. Wash with soap and water and change clothing immediately if the herbicide is spilled on skin or clothing.
6. Wear loose-fitting clothing.
7. Wear rubber gloves and rubber boots.
8. If herbicide is swallowed, call a physician or the nearest hospital poison control center at once.
9. If herbicide is splashed in eyes, flush with clean water immediately and call a physician.
10. If symptoms of illness occur during or shortly after handling or use, call a physician or your nearest hospital poison control center.

Application
1. Look out for children.
2. Be aware of the hazards of drift and volatility.
3. Be aware of other people in the area.
4. Do not contaminate wells, cisterns, other water sources, nontarget crops, or animals.
5. Apply at proper time and rate.

Table 17.12

Actual Risk and the Perception of Risk by Three Groups

Rank order of actual risk	Activity	U.S. deaths/year	Perceived risk		
			League of women voters	College students	Business and professional club members
1	Smoking	150,000	4	3	4
2	Alcohol	100,000	6	7	5
3	Automobiles	50,000	2	5	3
4	Handguns	17,000	3	2	1
5	Electric power	14,000	18	19	19
6	Motorcycles	3,000	5	6	2
7	Swimming	3,000	19	30	17
8	Surgery	2,800	10	11	9
9	X-rays	2,300	22	17	24
10	Railroads	1,950	24	23	20
11	Private aviation	1,300	7	15	11
12	Construction	1,000	12	14	13
13	Bicycles	1,000	16	24	14

14	Hunting	800	13	18	10
15	Home appliances	200	29	27	27
16	Fire fighting	195	11	10	6
17	Police work	160	8	8	7
18	Contraceptives	150	20	9	22
19	Commercial aviation	130	17	16	18
20	Nuclear power	100	1	1	8
21	Mountain climbing	30	15	22	12
22	Power mower	24	27	28	25
23	High school and college football	23	23	26	21
24	Skiing	18	21	25	16
25	Vaccinations	10	30	29	29
26	Food coloring	*	26	20	30
27	Food preservatives	*	25	12	28
28	Pesticides	*	9	4	15
29	Prescription antibiotics	*	28	21	26
30	Spray cans	*	14	13	23

*Not available.
Source: Adapted from Slovik, 1982.

6. Do not contaminate food and water containers, including those for livestock.

After use

1. Always dispose of empty containers so they pose no hazard. Puncture containers to prevent reuse.
2. Wash and change to clean clothing.
3. Be sure the person who washes contaminated clothing is aware of the contamination.
4. Clean equipment soon after use.

The precaution concerning children is repeated for good reason. Children often do unexpected things and adults have to be prepared. The other precautions for herbicide use are not difficult to understand. Most are just common sense. If poisoning occurs, treat it seriously and take the victim to a physician or hospital promptly. It is always a good idea to take the pesticide container along. Do not move victims who are in shock without treating for shock. Often, doing nothing except removing the victim from any possibility of further poisoning is better than doing something, if you are not sure what is correct.

C. THE LD$_{50}$ OF SOME HERBICIDES

The LD$_{50}$ is a good indicator of relative toxicity and safety. It is not the only measure of safety. The LD$_{50}$ helps deal with toxicity compared to other things. It is a measure of acute, not chronic, oral toxicity. All values are expressed in milligrams (mg) per kilogram (kg) of body weight. If the LD$_{50}$ is multiplied by 0.003, it is converted to ounces (oz) per 180-lb man. The value would be different for a woman or for someone with greater or lesser weight.

The U.S. Environmental Protection Agency has classified all pesticides into four groups based on their toxicity (Table 17.13). Table 17.14 shows the LD$_{50}$ values of some herbicides and a few common chemicals. It is wrong to assume that because two things have the same LD$_{50}$ they are equally toxic. This is because routes and likelihood (the chance) of exposure differ.

In one 5-oz cup (about 150 ml) of roasted and brewed coffee there is about 85 mg of caffeine. Instant coffee has 60 mg and there is about 3 mg in decaffeinated coffee. To be poisoned by coffee, a 180-lb man would have to drink 7.62 gal all at once–192.5 cups, nonstop. That is impossible, and it is very unlikely that anyone will die from acute caffeine intoxication.

Table 17.13

Pesticide Toxicity Classes Based on LD_{50}

Toxicity class	Signal words[a]	LD_{50}	Toxic amount
I	Danger—poison + a skull and crossbones	5 6–49	A taste to 7 drops 8 drops to a teaspoonful
II	Warning—may be fatal	50–499	1 teaspoonful to 1 tablespoonful or 1 ounce
III	Caution	500–4,999	1 ounce to 1 pint
IV	Caution	5,000–14,999	1 pint to 1 quart

[a]Signal words must appear on pesticide label.

The LD_{50} of ethyl alcohol is 4,500 mg/kg, and a 180-lb man would have to consume a little more than 0.8 pints to be acutely poisoned. This is not commonly done, but it is possible, and people do die from alcohol intoxication.

The LD_{50} as a measure of toxicity of anything is valuable, but its use must be tempered with knowledge of exposure, route of administration, rate, time, and physiological factors. It is a useful indicator of toxicity, but not a perfect one.

THINGS TO THINK ABOUT

1. How do weeds interfere with human activities?
2. What ecological changes can occur after herbicide use?
3. How can ecological change, created by herbicides, be prevented or managed?
4. What is herbicide resistance?
5. How important is herbicide resistance?
6. Where is resistance most likely to occur?
7. Is resistance equally likely for all herbicides?
8. How can resistance be managed?
9. How do herbicides influence soil?
10. Is herbicide use in U.S. agriculture energy intensive?
11. Are herbicides always harmful to people?
12. Why is the debate about the environment, herbicides, and people so complex?
13. What is the LD_{50} and how is it used?
14. How can herbicides be used safely?

Table 17.14

The LD$_{50}$ of Some Herbicides and a Few Other Chemicals

Common name	LD$_{50}$ (mg/kg) Technical[a]	Common name	LD$_{50}$ (mg/kg) Technical[a]
Herbicides			
Acetochlor	2,148	Dichlorprop	800
Acifluorfen	1,540	Diclofop methyl ester	557–580
Acrolein	29	Difenzoquat	617
Alachlor	930–1,350	Diquat	230
Ametryn	1,160	Diuron	3,400
Amitrole	>5,000	DSMA	1,935
Atrazine	3,090	Endothal	38–51
Benefin	>5,000	EPTC	1,652
Bensulfuron	>5,000	Ethalfluralin	>10,000
Bensulide	770	Ethofumesate	>6,400
Bentazon	1,100	Fenoxaprop	3,310
Bifenox	>5,000	Fluazifop-P butyl ester	4,096
Bromacil	175	Flumetsulam	>5,000
Bromoxynil	440	Fluometuron	6,416
Butachlor	2,000	Fluridone	>10,000
Butylate	4,659	Fomesafen	1,250–2,000
Chlorimuron	4,102	Glufosinate	2,170
Chloroxuron	3,700	Glyphosate—isopropylamine salt	>5,000
Chlorsulfuron	5,545	Haloxyfop	337
Clomazone	2,077	Hexazinone	1,690
Clopyralid	4,300	Imazamethabenz	>5,000
Copper sulfate	470	Imazapyr	>5,000
Cyanazine	334	Imazaquin	>5,000
Cycloate	3,200	Imazethapyr	>5,000
2,4-D acid	764	Isopropalin	>5,000
2,4-D dimethylamine	>1,000	Isoxaben	>10,000
2,4-D isooctyl ester	1,045	Linuron	1,254
DCPA	>10,000	MCPA acid	1,160
Desmedipham	>10,250	MCPB	680
Diallate	1,050	Mecoprop	650
Dicamba	1,028	Metolachlor	2,877
Dichlobenil	4,460	Metribuzin	1,090

(continues)

Table 17.14 (*Continued*)

Common name	LD$_{50}$ (mg/kg) Technical[a]	Common name	LD$_{50}$ (mg/kg) Technical[a]
Metsulfuron	5,000+	Sulfometuron	>5,000
MSMA	1,800	Tebuthiuron	644
Napropamide	>5,000	Terbacil	1,255
Naptalam	1,770	Thifensulfuron	>5,000
Nicosulfuron	5,000+	Triallate	1,100
Norflurazon	9,000	Triasulfuron	>5,000
Oryzalin	>5,000	Tribenuron	>5,000
Oxadiazon	>5,000	Triclopyr	713
Oxyfluorfen	>5,000	Trifluralin	>5,000
Paraquat	112–150	Triflusulfuron	>5,000
Pendimethalin	>5,000	**Other chemicals**	
Phenmedipham	>8,000	Aspirin	750
Picloram K+ salt	>5,000	Caffeine	200
Primisulfuron	5,050	DDT	87
Prodiamine	>5,000	Diazinon	66
Prometon	4,345	Ethyl alcohol	4,500
Prometryn	4,550	Gasoline	150
Pronamide	>16,000	Kerosene	50
Propachlor	1,800	Methyl parathion	9
Propanil	1,080	Nicotine sulfate	83
Pyrazon	2,200	Paradichlorobenzene	>1,000
Quizalofop	1,670	Parathion	3
Sethoxydim	2,676–3,124	Phorate	1
Siduron	>7,500	Pyrethrins	200
Simazine	>5,000	Sodium chloride (table salt)	3,320
Sodium chlorate	5,000	Water	25 ml/kg

[a]Herbicide values are for the acute oral LD$_{50}$ in adult rats for technical-grade herbicide (=95% pure).

LITERATURE CITED

Altman, J., and C. L. Campbell. 1977. *Ann. Rev. Phytopath.* **15:**361–386.
Altman, J., S. Neate, and A. D. Rovira. 1990. Herbicide–pathogen interactions and mycoherbicides as alternative strategies for weed control, pp. 241–259 *in* "Microbes and Microbial Products as Herbicides" (R. E. Hoagland, Ed.). Amer. Chem. Soc., Washington, D.C.

Brown, A. W. A. 1978. "Ecology of Pesticides," p. 325. Wiley, New York.

Gressel, J. 1979. Will weeds develop resistance to herbicides. *Weeds Today* **10**(2):26–27.

Gressel, J. 1990. Synergizing herbicides. *Rev. Weed Sci.* **5**:49–82.

Katan, J., and Y. Eshel. 1973. Interactions between herbicides and plant pathogens. *Res. Rev.* **45**:145–177.

LeBaron, H. M., and J. Gressel. 1982. "Herbicide Resistance in Plants," p. xv. Wiley Interscience, New York.

LeBaron, H. M., and J. McFarland. 1990. Herbicide resistance in weeds and crops: An overview and prognosis, pp 336–352 *in* "Managing Resistance to Agrochemicals: From Fundamental Research to Practical Strategies" (M. B. Green, H. M. LeBaron, and W. K. Moberg, Eds.), ACS Symp. Ser. **421**. Amer. Chem. Soc. Washington, D.C.

McCurdy, E. V., and E. S. Molberg. 1974. Effects of the continuous use of 2,4-D and MCPA on spring wheat production and wheat populations. *Can. J. Plant Sci.* **54**:241–245.

Nalewaja, J. D. 1974. Energy requirements for various weedcontrol practices. *Proc. N. Cent. Weed Control Conf.* **29**:19–23.

National Research Council. 1986. "Pesticide Resistance, Strategies and Tactics for Management, pp. 11–70. Natl. Acad. Press, Washington, D.C.

Olson, R., J. Hansen, T. Whitson, and K. Johnson. 1994. Tebuthiuron to enhance rangeland diversity. *Rangelands* **16**:197–201.

Penner, D. 1994. Herbicide action and metabolism, pp. 37–70 in "Turf Weeds and Their Control." Amer. Soc. of Agron. and Crop Sci. Soc. of Amer. Madison, Wisconsin.

Pimentel, D., and M. Pimentel. 1979. "Food, Energy and Society, pp. 69–98. E. Arnold, London.

Rahman, A., G. C. Atkinson, and J. A. Douglas. 1979. Eradicane causes problems. *N. Z. J. Agric.* **139**(3):47–49.

Rice, P. M., D. J. Bedunah, and C. E. Carlson. 1992. Plant community diversity after herbicide control of spotted knapweed. USDA/Forest Service. Intermountain Res. Paper No. INT-460. 7 pp.

Rovira, A. D., and H. J. McDonald. 1986. Effects of the herbicide chlorsulfuron on *Rhizoctonia* bare patch and take-all of barley and wheat. *Plant Dis.* **70**:879–882.

Ryan, G. F. 1970. Resistance of common groundsel to simazine and atrazine. *Weed Sci.* **18**:614–616.

Santillo, D. J., D. M. Leslie, Jr., and P. W. Brown. 1989a. Responses of small mammals and habitat to glyphosate application on clearcuts. *J. Wildlife Management* **53**:164–172.

Santillo, D. M., P. W. Brown, and D. M. Leslie, Jr. 1989b. Response of songbirds to glyphosate-induced habitat change on clearcuts. *J. Wildlife Management* **53**:64–71.

Schubert, O. E. 1972. Plant cover changes following herbicide applications in orchards. *Weed Sci.* **20**:124–127.

Slovik, P. 1982. Perception of risk. *Science* **236**:280–285.

Tietjen, H. P., C. H. Halvorsen, P. L. Hegdal, and A. M. Johnson. 1967. 2,4-D herbicide, vegetation and pocket gopher relationships on Black Mesa, Colorado. *Ecology* **48**:635–643.

Turner, G. T. 1969. Responses of mountain grassland vegetation to gopher control, reduced grazing, and herbicide. *J. Range Management* **22**:377–383.

LITERATURE AVAILABLE

Adams, T., and J. Tryens (Eds.).1988. "The Pesticide Crisis: A Blueprint for States." Natl. Center for Policy Alternatives, Washington, D.C. 92 pp.

Avery, D. T. 1995. "Saving the Planet with Pesticides and Plastic: The Environmental Triumph of High-Yield Farming." Hudson Institute, Indianapolis, Indiana. 432 pp.

Avery, D. T. 1997. Saving the planet with pesticides, biotechnology and European farm reform. British Crop Protection Conference–Weeds, pp. 1–16.

Barrons, K. C. 1981. "Are Pesticides Really Necessary?" Regnery Gateway, Chicago. 245 pp.

Bender, J. 1994. "Future Harvest: Pesticide-Free Farming." Univ. Nebraska Press, Lincoln, Nebraska. 159 pp.

Bosso, C. J. 1987. "Pesticides and Politics: The Life Cycle of a Public Issue. Univ. Pittsburgh Press., Pittsburgh. 294 pp.

Bull, D. 1982. "A Growing Problem: Pesticides and the Third World Poor." Oxfam, Oxford, U.K. 192 pp.

Carson, R. 1962. "Silent Spring." Several editions available.

Colburn, T., D. Dumanoski, and J. P. Myers. 1996. "Our Stolen Future: Are We Threatening Our Fertility, Intelligence, and Survival? A Scientific Detective Story." Penguin Books USA, New York. 306pp.

Conroy, M. E., D. L. Murray, and P. M. Rosset. 1996. A cautionary tale: Failed U.S. development policy in Central America, *in* "Food First Development Studies" (P. M. Rosset, Ed.). Lynne Rienner, Boulder, Colorado. 211 pp.

Doyle, J. 1986. "Altered Harvest: Agriculture, Genetics , and the Fate of the World's Food Supply." Penguin, New York. 502 pp.

Fletcher, W. W. 1974. "The Pest War." Halsted Press, New York. 218 pp.

Graham, F., Jr. 1970. "Since Silent Spring.: Houghton Mifflin, Boston. 333 pp.

Green, M. B. 1976. "Pesticides–Boon or Bane." Westview Press, Boulder, Colorado. 111 pp.

Gunn, D. L., and J. G. R. Stevens. 1976. "Pesticides and Human Welfare." Oxford Univ. Press, Oxford, U.K. 278 pp.

Marco, G. J., R. M. Hollingworth, and W. Durham. (Eds.). 1987. "Silent Spring Revisited." Amer. Chem Soc., Washington, D.C. 214 pp.

McMillan, W. 1965. "Bugs or People." Appleton-Century, New York. 228 pp.

Murray, D. L. 1994. "Cultivating Crisis: The Human Costs of Pesticides in Latin America." University of Texas Press, Austin, Texas. 177 pp.

Perring, F. H., and K. Mellanby (Eds.). 1977. "Ecological Effects of Pesticides." Academic Press, New York. 193 pp.

Pimentel, D., and H. Lehman. (Eds.). 1993. "The Pesticide Question: Environment, Economics, and Ethics." Chapman and Hall, New York. 441 pp.

Regenstein, L. 1982. "America the Poisoned." Acropolis Books, Washington, D.C. 414 pp.

Rudd, R. L. 1964. "Pesticides and the Living Landscape." Univ. Wisconsin Press., Madison, Wisconsin. 320 pp.

Sheets, T. J., and D. Pimentel (Eds.). 1979. "Pesticides: Contemporary Roles in Agriculture, Health, and the Environment. Humana Press, Clifton, New Jersey. 186 pp.

Van den Bosch, R. 1978. "The Pesticide Conspiracy." Univ. California Press, Berkeley, California. 226 pp.

Waggoner, P. E. 1994. "How Much Land Can Ten Billion People Spare for Nature?" Council for Agricultural Science and Technology, Task Force Report No. 121. 64 pp.

Wargo, J. 1996. "Our Children's Toxic Legacy: How Science and Law Fail to Protect Us from Pesticides." Yale Univ. Press, New Haven, Connecticut. 380 pp.

Weir, D., and M. Shapiro. 1981. "Circle of Poison: Pesticides and People in a Hungry World. Inst. for Food and Development Policy, San Francisco. 96 pp.

Whelan, E. 1985. "Toxic Terror." Jameson Books, Ottawa, Illinois. 348 pp.

Whitten, J. L. 1966. "That We May Live." Van Nostrand, Princeton, New Jersey. 251 pp.
Whorton, J. 1974. "Before Silent Spring: Pesticides and Public Health in Pre-DDT America."
    Princeton Univ. Press., Princeton, New Jersey. 288 pp.
Wright, A. 1990. "The Death of Ramon Gonzalez: The Modern Agricultural Dilemma. Univ.
    Texas Press, Austin, Texas. 337 pp.

Chapter 18

Pesticide Legislation and Registration

FUNDAMENTAL CONCEPTS

- The pesticide registration process is complex, mandatory, and founded in state and federal legislation.
- The U.S. Environmental Protection Agency is the federal organization charged with administration of federal pesticide laws, and it is responsible for pesticide registration.

LEARNING OBJECTIVES

- To understand the purpose and complexity of Federal pesticide laws.
- To be aware of the protection the regulatory process provides the U.S. public.
- To become familiar with the basic steps of pesticide registration under U.S. law.
- To understand the role of pesticide regulation.

I. THE PRINCIPLES OF PESTICIDE REGISTRATION

Most of the world's nation states have some sort of pesticide[1] registration procedure. In some less developed countries, procedural and data require-

[1]Federal and state pesticide regulations apply equally to all classes of pesticides. The general term pesticide will be used in most places in this chapter instead of the specific term herbicide.

ments are few to nonexistent, primarily because of fiscal constraints but also because of unawareness of the need. Countries that regulate pesticides share the goal of providing protection from adverse effects of pesticides and gaining the benefits of pesticide use (Snelson, 1978). These objectives are achieved through registration and control of the pesticide label. Registration enables a government to exercise control over use, claims about performance, label directions and precautions, packaging, and advertising to ensure proper use and environmental and human protection. In general, the process protects the public's interest and the manufacturer's rights (Snelson, 1978).

It is apparent from the questions raised in the popular media that the public does not have adequate knowledge of the intricacies of pesticide registration and the laws that govern the process. This chapter describes some general aspects of U.S. pesticide registration but it is not intended to be a complete description of the process.

In the world's developed countries, pesticides have been subject to some kind of governmental regulation for more than 75 years. The public is aware of pesticides and fearful because of mistakes that have occurred. The Alar scare of the early 1990s is an example of the public's fear. The public does not know about the intricate and continually reviewed procedures necessary for registration of a pesticide prior to use. Registration is a complex process that should not be confused with registering a dog or a car. It is not simply recording ownership and paying a nominal fee. Registration of pesticides means compliance with legal requirements that establish a regulatory process that demands proof of safety and, more rarely, proof of efficacy. Different nation states establish registration processes that conform to their needs. The system in the United States is among the most complex and successful. It is not perfect: There are many complaints about it from the general public and environmental groups who are concerned that protection is not sufficient, and from manufacturers who find the process slow, expensive, and unnecessarily cautious. The United Kingdom used to work with a voluntary approval scheme wherein a consensus was reached among the manufacturer, government, and users about appropriate regulation. That scheme was abandoned in the mid-1980s because of pressure from the European Economic Community (EEC) for uniform standards. The U.K. scheme and the EEC process now resemble the procedures followed in the United States, including provisions that regulate advertising, storage, application, and crop use. Many nations follow the standards put forth by the UN/FAO Codex Committee on Pesticide Residues that establishes maximum residue limits (MRLs) for pesticides in food. The CODEX also guides countries on safety regulations for use, storage, and analysis of pesticides.

This chapter describes the history of pesticide legislation and registration in the United States and general procedures that must be followed. Other descriptions are available (Harrison and Loux, 1995; Keller *et al.*, 1982).

II. FEDERAL LAWS

The Food and Drug Act of 1906 was the first U.S. law that dealt with pesticides, and it was administered originally by the U.S. Department of Agriculture (USDA). Its purpose was "to halt the exposure of the general public to filthy rotten food, adulteration, substitution, and misleading claims." There were several cases of arsenic poisoning in England from imported U.S. apples, and these stimulated passage of the law.

The first U.S. federal law that directly involved pesticides was the Insecticide Act of 1910. It was passed to stop unethical persons from merchandising ineffective or adulterated products and was specifically aimed at Paris Green, lead arsenate, and other insecticides and fungicides. The law, administered by USDA, introduced a labeling requirement that mandated an ingredient statement and the manufacturer's name. The early 1900s were a period of slow development of pesticides and there was little public concern about them because they were generally considered useful. The law did not cover evaluation of hazards of the pesticides it controlled, and chemical analysis for crop residue was the most important enforcement procedure. The Insecticide Act protected the public against the possible loss of crops or damage to property from pesticide use, but there was no assurance that pesticides were not health hazards.

The Federal Food, Drug, and Cosmetic Act was passed in 1938 to gain more control of adulteration, misbranding, and substitution of food, drugs, and cosmetics and to ensure the integrity and safety of food moving in interstate commerce. It was originally enforced by the Federal Security Agency, which was abolished in 1953 when the Department of Health, Education, and Welfare (HEW) was created. At the present time, the law is administered by the Department of Health and Human Services, and the Environmental Protection Agency (EPA), created in 1970, has responsibility for setting tolerances for pesticides. The need for such a law was the increase in use of potentially adulterating chemicals in food, drugs, and cosmetics. Manufacturers were required to prove safety and usefulness. The Federal Security Agency and HEW established safe levels of residues. This procedure required a health agency to make agricultural decisions (on usefulness) and was cumbersome.

III. FEDERAL INSECTICIDE, FUNGICIDE, AND RODENTICIDE ACT

Pesticide development during and after World War II created the need for stronger laws. USDA, supported by the pesticide industry, developed the Federal Insecticide, Fungicide, and Rodenticide Act of 1947 (FIFRA). The law retained the key portions of the Insecticide Act of 1910, but extended the principle that a pesticide formulation should meet proper standards. No other federal law had authority over the pesticide and its labeling.

The FIFRA added two new ideas to pesticide regulation. The first was that all pesticides intended for shipment in interstate commerce must be registered with the U.S. Secretary of Agriculture before shipment. The second stipulation was that the USDA was given control over all precautionary statements on the pesticide label. USDA was empowered to review the public presentation of safety procedures so important to proper use. This law also placed the burden of proof of use and safety on the manufacturer. These provisions stopped shipment of untested or improperly labeled products in interstate commerce by requiring that labeling be adequate and that all labels be approved (registered) by USDA. Withholding registration was an effective way of stopping shipment of untried pesticides. The USDA could withhold registration until data were provided to prove the pesticide would give the degree of pest control claimed or implied on the label. Labels could also be withheld pending submission of adequate evidence of human safety from appropriate studies. The act had several specific and new items:

1. Protection of the user from physical injury or economic loss.
2. Protection of the public from injury. Previous laws had only protected the purchaser of the product from injury.
3. The manufacturer had to prove that the pesticide was effective for its intended use.
4. A pesticide was defined and limited to economic poisons that were defined as "any substance or mixture of substances intended for preventing, destroying, repelling, or mitigating any insects, rodents, fungi, weeds, and other forms of plant or animal life or viruses except viruses on or in living man or animals which the Secretary of Agriculture shall declare to be a pest."

The major public protection came from strict control over every feature of labeling. FIFRA had no control over the user of the product. It left the user with the responsibility to read and heed the label and avoid misuse and environmental contamination.

Questions of coverage of economic poisons not specifically defined under the FIFRA of 1947 arose. The law was amended in 1959 to include nematocides, plant regulators, defoliants, and desiccants as economic poisons. Residues of these compounds in or on food or feed crops were now regulated. This broadened scope did not include adequate protection for fish and wildlife, and the law was further amended to include pesticides sold for control of moles, birds, predatory animals, and other nonrodent pests. It also included certain plants and viruses when they are injurious to plants, domestic animals, or to man. Thus, it was possible to regulate pesticides designed to control specific things. The regulations were expanded to include the following:

1. Mammals, including but no limited to dogs, cats, moles, bats, wild carnivores, armadillos, and deer
2. Birds, including but not limited to starlings, English sparrows, crows, and blackbirds
3. Fishes, including but not limited to the jawless fishes such as sea lamprey, the cartilaginous fishes such as sharks, and the bony fishes such as carp
4. Amphibians and reptiles, including but not limited to poisonous snakes
5. Aquatic and terrestrial invertebrates including but not limited to slugs, snails, and crayfish
6. Roots or other plant parts growing where they are not wanted
7. Viruses other than those in or on living man or other animals

One might ask who was enamored of armadillos and got them included. To ask the question misses the point that the law was being expanded in scope consistent with Congressional interpretation of the public's desire for environmental safety.

A. AMENDMENTS

The Miller Pesticide Amendment, or PL-518, amended the Federal Food, Drug, and Cosmetic Act of 1938. It was passed in 1954 to correct cumbersome enforcement procedures in the 1938 law. It was formulated by the Delaney Committee of the House of Representatives that was formed to investigate chemicals in food and cosmetics. The committee decided that a better way to establish tolerances on food crops was required and assumed the initiative to formulate a way. Congressman Miller of Nebraska formulated the recommendations in a bill known as the Pesticide Chemicals Amendment to the Federal Food, Drug, and Cosmetic Act of 1938. A pesticide chemical was defined as "any substance which alone, in chemical combination, or in formulation with one or more other substances, is an economic poison as defined by the FIFRA of 1947, as now or as hereinafter amended and which is used in the production, the storage, or the transportation of raw agricultural commodities." This definition through use of the term economic poison related the Miller amendment to FIFRA.

The amendment established new procedures for obtaining tolerances. The HEW Secretary was charged with establishing tolerances or maximum allowable limits of pesticide residues in or on raw agricultural commodities moving in interstate commerce. A raw agricultural commodity was defined as "any food in its raw or natural state including all fruits in a washed, colored, or otherwise treated state in their unspoiled form prior to marketing." This formalized the establishment of tolerances in the Federal Food and Drug Administration (FFDA) of HEW. The USDA had to certify to the FFDA that the chemical for which a petition for tolerance had been filed would be useful for the purposes described. The USDA had to express an opinion as to whether the tolerance requested reasonably reflected the residues likely to remain on the crop when the pesticide was used as directed. This change assigned agricultural functions to the USDA, an agricultural agency, and health functions to the FFDA under HEW, a health agency. Tolerances were obtained from FFDA and use clearance from USDA. Prior to filing a tolerance petition, the chemical must have been registered under the FIFRA. All registration functions are now handled by the U.S. Environmental Protection Agency (EPA).

The 1968 Color Additive Amendment subjected all color additives to the provisions for food additives and the 1964 Seed Coloring Amendment subjected all seed colorings to the provisions for food additives. The most controversial amendment to the Federal Food and Drug Control Act (FFDCA) was the so-called Delaney Cancer Clause, included with the 1958 food additive amendment. It has been widely and hotly debated since it was incorporated into law. It said "no additive is deemed safe if found to induce cancer when ingested by man or animal or if it is found after tests which are found appropriate for the evaluation of the safety of food additives to induce cancer in man or animals." Much of the controversy concerning the use and misuse of pesticide chemicals has centered on the Delaney Cancer Clause. It is important to note that it says nothing at all about dose, nor does it mention a particular length of time within which cancer must be induced. It applies only to processed food, not raw agricultural commodities. After years of debate many (but not all) were pleased when the 38-year-old Delaney clause was removed in the 1996 Food Quality Protection Act signed into law by President Clinton on August 3. Since the Delaney clause was enacted (1958), chemical analytical technology has progressed so that it is now possible to routinely detect part per billion (ppb) or trillion (ppt) amounts that are well below what could be detected in 1958 and pose no known human health hazard. The standard of reasonable certainty is now defined, in part, as "no more than a one-in-one-million chance of getting cancer after a lifetime of exposure." Replacement of Delaney standards with new standards has not eliminated concern about health issues. The new standards may be just as tough or tougher than the old, widely discussed standard.

IV. THE ENVIRONMENTAL PROTECTION AGENCY

In December 1970 the Environmental Protection Agency (EPA) was created when the entire pesticide regulations division of the USDA and, somewhat later, the pesticide office of the FFDA came to the EPA office of pesticides. This office contained the pesticides registration and enforcement divisions. There were five sections in the pesticides regulation division: efficacy, chemistry, human safety, fish and wildlife safety, and registration. The first four conducted scientific reviews prior to registration.

The pesticides enforcement division had a petitions control branch that reviewed chemistry and toxicology. In addition, they had the following groups: inspection services and imports, case review and development, control officer and prosecutions, and field enforcement staff.

V. FEDERAL ENVIRONMENTAL PESTICIDE CONTROL ACT

On October 21, 1972, President Nixon signed into law the Federal Environmental Pesticide Control Act (FEPCA). This law made many changes as an amendment to the FIFRA of 1947, which is still the primary law. The new law was designed to protect man and the environment and extended federal regulation to all pesticides, including those manufactured and used within a state.

Responsibility for use and misuse was now lodged with the pesticide applicator. In addition, no pesticide could be registered or sold unless its labeling was designed to prevent injury to man and any unreasonable effects on the environment. Future label evaluation by EPA had to consider the public interest, including benefits from pesticide use. However, under the 1947 FIFRA ultimate responsibility for pesticide use and misuse was borne by the manufacturer who prepared the label. Under FEPCA, the amended FIFRA, manufacturers still must establish safety and efficacy, but responsibility rests with the user if failure to follow label directions results in human, or environmental harm. Violators can be prosecuted under civil or criminal law.

FIFRA required the registration of pesticides moving in interstate commerce. The amended FIFRA requires registration of all pesticides, regardless of their point of manufacture or use. All registered pesticides are classified for restricted or general use. General use or unclassified pesticides can be used by anyone in accordance with label directions. The restricted category includes all pesticides that demonstrate the potential for harm to human health or the environment even when used according to the label. The restricted classification must appear on the front label of the pesticide package. The FEPCA requires certification of commercial applicators that involves demonstration, on a written examination, of a minimum level of knowledge and competence about pesticides. A commercial applicator is one who may use or supervise the use of restricted-use pesticides. Private applicators are certified by participation in approved training programs. A private applicator is a certified applicator who uses or supervises use of a restricted-use pesticide for purposes of producing any agricultural commodity on property owned or rented by the individual or an employer or on property of another person (if such application is without compensation other than trading personal services). Under the amended FIFRA, nonessentiality is not a sufficient reason to deny registration. This means that if one pesticide is already available for a specific use, registration cannot be denied to a new product.

The FEPCA strengthened EPA enforcement procedures. All pesticide producing establishments must be registered and regular reports of sales and production are required. A pesticide's registration can also be revoked by EPA. States may impose more stringent pesticide regulations than the amended FIFRA. In the past, some states had no pesticide laws, but the FEPCA required all to have them or federal regulations would apply.

EPA can cancel registration after 5 years even when continued use is requested by the manufacturer. EPA also has the power to reclassify, suspend, or cancel a pesticide if it causes unreasonable adverse effects on the environment when used as directed. Unreasonable adverse effects include any unreasonable risk to man or environment. The decision must be based on the economic, social, and environmental costs and benefits of pesticide use. If a use (or all uses) of a pesticide is suspended (taken off the market) and later canceled, the law provides indemnities to the manufacturer and other owners. Such indemnities are designed to protect the manufacturer who has met all legal requirements and may suffer large monetary losses when new knowledge demonstrates that continued use of the product may be hazardous.

VI. PROCEDURAL SUMMARY

In 1988, the FIFRA was again amended. The principal focus of the amendments was to ensure that previously registered pesticides measure up to current scientific and regulatory standards. The procedure for pesticide registration can be summarized in two statements. First, the manufacturer must file with the Office of Pesticide Programs, Pesticides Registration Division (PRD), of the EPA for registration of a pesticide. The EPA has a total of 90 days, segmented into two 45-day periods, to determine the completeness of the application (first 45-day period) and the usefulness of the compound as requested on the label and to comment on the data. Second, the manufacturer must file with the Hazard Evaluation Division of the EPA for a tolerance or for an exemption from tolerance. EPA has 90 days to render an opinion on this petition. They can recommend a tolerance exemption, a petition withdrawal, a petition amendment, or a petition rejection. After the Hazard Evaluation Division has granted a tolerance, the Registration Division may register the label. In practice, these deadlines are often extended because of requests for additional information or submission of inadequate data.

Petitions must be supported by prescribed information, including the identity and composition of the pesticide chemical, methods of analysis, complete information on use, full reports of investigations made on residues produced, and toxicity information. During preregistration, all other involved federal agencies can express an opinion regarding use and registration of a pesticide. These agencies include the Forest Service, the U.S. Department of the Interior, the Bureau of Land Management, and other conservation and wildlife interests (including private interests). These organizations cannot accept or reject a chemical, but their opinions are of great value to decision makers. Other federal agencies are involved in pesticide use but not in registration. They include the National Research Council, which promulgates information on safe pesticide use; the Occupational Safety and Health Administration, which protects workers who handle pesticides; and the Federal Aviation Administration, which regulates aspects of safety for aerial pesticide application.

A certificate of usefulness may or may not be issued. If it is denied, the pesticide will not be registered and approved for use. If it is issued, the Tolerance Division has 90 days to act on a petition for tolerance and issue residue regulations in the Federal Register. If a tolerance is not established, the pesticide may fail to be registered. It is possible to obtain either without the other, but both are necessary for registration.

Under existing federal law, the EPA will register a pesticide only when the following criteria are met (Harrison and Loux, 1995):

1. The pesticide's composition must be such as to warrant the claims proposed by the registrant.
2. The proposed label must conform to FIFRA requirements.
3. The pesticide must perform its intended function without unreasonable adverse environmental effects.
4. The pesticide must not cause unreasonable adverse environmental effects when used in accordance with accepted practices.

In each case the burden of proof is on the manufacturer (registrant). The EPA may waive the requirement to prove efficacy on the assumption that manufacturers will not be so foolish as to risk their reputation by marketing a product that does not do what they claim it will do.

A manufacturer may apply to EPA for an experimental use permit (EUP) before a pesticide is granted full registration. Such permits are usually granted for 1 year and crops may be used or destroyed as determined by EPA when the permit is granted. EUPs permit the manufacturer to sell the product while gathering performance information under field conditions to support full registration.

Section 18 of FIFRA permits EPA to authorize use of a pesticide before full registration if an emergency condition can be established. Permits are granted only when the weed (or other pest) problem is urgent and nonroutine, no other registered herbicides will provide effective control, and no other control measures are economically or environmentally feasible (Harrison and Loux, 1995).

A final procedural matter relates to the ability of states to regulate the sale or use of a federally registered pesticide under Section 24c of FIFRA. State regulatory agencies may register a federally approved active ingredient or product for a special local need that is not part of the EPA-approved label.

VII. TOLERANCE CLASSES

All pesticides fall into one of four tolerance classes.

A. EXEMPT

Some pesticides are exempt because there is no known human or animal health concern. They are generally recognized as safe (GRAS) and do not require a tolerance. These pesticides include lime ($CaCO_3$), sodium chloride, sulfur, sodium carbonate, acetic acid, ammonium chloride, calcium hypochlorite, potassium chloride, sulfur dioxide, and *Bacillus thuringensis*.

B. ZERO

If, because of toxicological characteristics, the PRD of EPA decides it is not in the public interest to accept any detectable residue of a given chemical, it can establish a zero tolerance that means none of the chemical is permitted in any crop. It is not possible to register a pesticide with a zero tolerance for use on food crops. This is a regulatory position and applies to pesticides even when no manufacturer ever applied for a tolerance. Zero tolerance used to mean that when used according to label directions, no detectable residue would remain. However, more sensitive methods of detection invalidated the concept. Today, no one knows what level might be detectable tomorrow. Parts per trillion are not uncommon, and smaller amounts can be found. As of 1966, a finite tolerance must be established, but zero can still be used. EPA applies the zero tolerance to pesticides, but they then refuse to register them for use on food crops.

C. FINITE

A finite tolerance is used when chemical residues are known to exist; this is the tolerance under which most pesticides are now registered. Any raw agricultural commodity moving in interstate commerce and found to have pesticide residues over the stated amount is subject to seizure by the FFDA. Before a finite tolerance can be obtained, 2-year feeding studies on at least two species of mammals (usually rats and dogs) are used to establish no-effect levels.

D. NON-FOOD-CROP REGISTRATION

When a pesticide is applied to soil many days before planting and has been proven to decompose or metabolize rapidly into natural substances, or is used in a way that presents no possibility of its remaining at harvest, it may be considered as a non-food-crop use. Such a chemical can then be registered without establishing a finite tolerance. Herbicides in this group could be applied to parking lots, but there could be reentry restrictions. Under this registration, range and pasture are considered food crops even though they are not consumed by humans, because cattle or sheep are consumed by humans. Seed treatments have often been put in this group, as have applications of pesticides to dormant crops when the pesticide is known to disappear rapidly. Persistent compounds, on the other hand, require an established tolerance even if applied preplant or during the dormant season.

No-residue registration was eliminated for pesticides in 1970. The concept was that if no residue could be detected with the best analytical method available at the time, a compound could be registered under the no-residue provision. The problem is that detection methods have improved so much that a compound that originally could not be detected by methods sensitive to a part per million can now be detected in parts per trillion or less. In 1967, no-residue registrations were gradually converted, on petition of the manufacturer, to finite tolerances. If a manufacturer did not request a finite tolerance, the pesticide's uses were canceled.

VIII. THE PROCEDURES FOR PESTICIDE REGISTRATION

A complete registration petition contains a great deal of information necessary for full consideration of benefits and risks. At a minimum, the petition must contain the following:

(a) A statement of active and inert ingredients in the product, chemical and physical properties of the compound in the formulation, the complete quantitative formula of the product, its environmental stability, and known impurities in the formulation.
(b) Five copies of the proposed label, including:
  1. Brand name
  2. Complete chemical name and physical and chemical properties
  3. Ingredients statement plus samples of the chemical and its formulation(s)
  4. Directions for use, specifying crops or sites intended for treatment
  5. Amount(s) to be used
  6. Timing of application
  7. Any precautions or limitations on use
  8. Warning statement for protection of nontarget species
  9. The antidote in case of human consumption
  10. Warning to keep out of reach of children
  11. Manufacturing details
  12. Net weight statement
  13. Restricted vs general use statement

(c) Full description including the data of scientific tests used to determine effectiveness and safety.

(d) Complete toxicity report of tests on lab animals, and the methods for obtaining the data. At a minimum, such tests must include items 1 to 11 below. Studies are often expanded to include data on oncogenicity, spermatogenicity, aspects of mutagenicity, and other risk-related factors research may identify.

    1. Two-year rate feeding study to determine reproductive and carcinogenic effects
    2. Eighteen-month mouse feeding study to determine reproductive and carcinogenic effects
    3. Two-year dog feeding study
    4. Dominant lethal mutagenic possibilities in mice
    5. Teratogenic study in rabbits
    6. Three-generation rat reproduction study and reproduction studies in chickens
    7. Meat residue in cows, chickens, and swine; milk residue in cows; egg residue
    8. Ninety-day rat feeding study to determine mammalian metabolites
    9. Twenty-one day subacute toxicity in rabbits, and oral and dermal LD_{50} in rats
    10. Eye and skin irritation
    11. Tests to determine effects on two species of fish and quail

(e) Results of tests on the amount of residue remaining and the description of the analytical methods. This is extremely critical for tolerances. Tolerances are set on the amount of residue remaining and not on the highest figure permissible from a health standpoint. EPA is interested in residue data on the crops and animals on which the pesticide will be used and in the soil (or other portion of the environment) that will be treated with the pesticide.

(f) Practical methods for removing residues exceeding any proposed tolerance, including a description of the method.

(g) Proposed tolerance for the pesticide and supporting reasons for the level requested.

(h) Reasonable grounds in support of the petition, including a summary of data in the entire petition and a summary of benefits when used in agriculture.

If there are no food residues and if no other residue exists, and if the Pesticide Regulation Division of EPA concludes that the pesticide is safe and conforms to the manufacturer's claims, then the Fish and Wildlife Service of the Department of the Interior and the Hazard Evaluation Division of EPA are notified of the intent to register. These agencies and others can concur or reject the petition. If they reject, a reevaluation must occur. When uses will result in residue at harvest of crops or slaughter of animals, registration is subject to the requirements of the Miller Amendment. In seeking to register such a compound, a petition proposing a tolerance or exemption must be submitted to EPA. It must provide information on the pesticide, its use, and reports of safety tests. The safety information must include results of animal susceptibility experiments, tests on residues, and the analytical methods employed, practical methods for removing residues that exceed proposed tolerances. Table 18.1 summarizes the process.

Table 18.1

A Summary of the U.S. Pesticide Registration Procedure

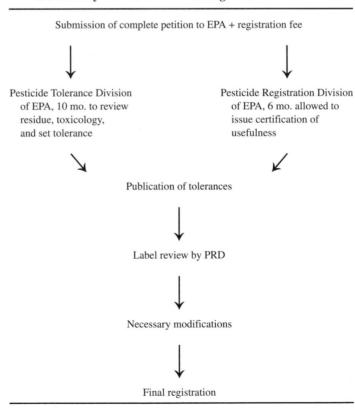

Submission of complete petition to EPA + registration fee

Pesticide Tolerance Division of EPA, 10 mo. to review residue, toxicology, and set tolerance

Pesticide Registration Division of EPA, 6 mo. allowed to issue certification of usefulness

Publication of tolerances

Label review by PRD

Necessary modifications

Final registration

The FIFRA, through its registration and enforcement features, provides the primary public protection against improperly labeled or adulterated pesticides. The law is also the primary effort to protect the health of users and consumers against potential adverse effects of pesticides. At the present time, there are at least 20,000 to 25,000 products made from one or more of 600 pesticidal chemicals. The Pesticides Registration Division of EPA

(1) certifies the chemical is useful for the use for which the label is requested, and
(2) expresses an opinion as to whether the tolerance requested reasonably reflects the residues likely to remain on the treated crop.

The burden of proof is always on the applicant. The Pesticide Tolerances Division of EPA establishes tolerances or maximum allowable limits of pesticide residues in or on raw agricultural commodities in interstate commerce. The government's role does not necessarily stop at this point. EPA and state officials collect unregistered pesticides, look for misbranded or adulterated products, and can take legal action against offenders. The FDA also has a program of monitoring the nation's food supply for pesticide residues and a program of environmental monitoring for pesticide residues. This program was established under the Federal Food, Drug, and Cosmetic Act of 1938. Under this program, the FDA's pesticide program sampled more than 8,000 U.S. food products in 1991. No detectable pesticide residue was found in 64% of the samples. In all samples taken, fewer than 1% had over-tolerance or violative residues, fewer than 1% had residues of pesticides for which there was no tolerance on the food crop, and 34% had detectable, but nonviolative, residues.

States, through their departments of agriculture, environmental agencies, or other designated bodies, have the power to register and regulate the intrastate use of pesticides. States are allowed to register additional uses of federally registered pesticides for special local needs (24-C registrations). States cannot invoke any regulations that are less stringent than the amended FIFRA, but can impose additional requirements on the registrant. Pesticides formulated and distributed within a state must be registered by the Federal EPA. State registrations cannot be obtained if EPA has already denied FIFRA registration for a particular use.

THINGS TO THINK ABOUT

1. Why does a nation or state bother to register pesticides?
2. What federal agency governs pesticide registration in the United States?

3. What federal acts govern pesticide registration in the United States?
4. What does it mean when a pesticide is registered?
5. Why is the pesticide label important?
6. What is the significance of the Miller pesticide amendment to the FIFRA?
7. What things must a manufacturer prove to register a pesticide?
8. What are the tolerance classes under which a pesticide may be registered in the United States?
9. What information must be included with a petition for registration, and who bears the responsibility for preparing the information?

LITERATURE CITED

Harrison, S. K., and M. M. Loux. 1995. Chemical weed management, pp. 101–153 in "Handbook of Weed Management Systems" (A. E. Smith, Ed.). Dekker, New York.
Keller, J. J., et al. (Eds.). 1982. "Pesticides Guide: Registration, Classification, and Application." J.J. Keller and Assoc., Neenah, Wisconsin.
Snelson, J. T. 1978. The need for and principles of pesticide registration. *FAO Plant Prot. Bull.* **26**:93–100. **NOTE:** This issue contains several articles on pesticide registration.

Chapter 19

Weed Management Systems

FUNDAMENTAL CONCEPTS

- A weed management system can be designed for any crop–weed situation, but more research must be done to integrate available weed management techniques.
- There are six logical steps that are fundamental to all complete weed management systems.

LEARNING OBJECTIVES

- To understand the logical steps that are part of complete weed management systems.
- To know how to combine weed management and control techniques into a weed management system.
- To understand the design and implementation of weed management systems for a few crops and cropping situations.

I. INTRODUCTION

Weed control is a process of reducing weed growth to an acceptable level. To date there is no clear definition of what weed management is. It is an evolving concept. When defined it will include the dictionary sense of the term—"taking charge of and directing" the growth of weeds. It will also include handling carefully, a rarer definition of management that includes the concept of husbandry, that is, to manage economically and conserve.

Weed management has been defined as: "An environmentally sound system of farming using all available knowledge and tools to produce crops free of economically damaging, competitive vegetation" (Fischer *et al.,* 1985). This definition lacks the specificity of an economic threshold and advocates no yield loss (i.e., no economic damage). It could be interpreted as advocating a high level of weed control because it mentions crops free of economically damaging, competitive vegetation.

Fryer (1985) defined weed management as the "rational deployment of all available technology to provide systematic management of weed problems in all situations." Unfortunately, there is no agreement on what is rational or systematic. It is also not good to include the word defined in the definition, as if everyone already knew what it meant.

Weed management will be a systematic approach to minimize weed impacts and optimize land use, and it will combine prevention and control (Aldrich, 1984). It will emphasize minimizing the effect of weeds, but probably not eliminating them. Weeds will be accepted as a normal and manageable part of the agricultural community, albeit a part to learn to fight and live with. The objective of effective weed management will be to manipulate the crop–weed relationship so that growth of the crop is favored over that of the weeds.

When research has provided an adequate base for weed management systems, it will include the following components:

1. Incorporation of ecological principles
2. Use of plant interference and crop–weed competition
3. Incorporation of economic and damage thresholds
4. Integration of several weed control techniques including selective herbicides
5. Supervised weed management, frequently by a professional weed manager employed to develop a program for each crop–weed situation

Systems will be designed to prevent or eliminate the probability that weed problems will develop and to anticipate problems that are likely to occur. Systems will manage weed problems that, if ignored, will reduce yield. Ecological considerations will include natural weed mortality, inter- and intraspecific competition, plant density, and genetic manipulation. The last may develop populations more susceptible to control techniques such as tillage and herbicides. Successful weed management will also incorporate precise timing of cultural practices such as tillage to maximize benefit, and careful selection of rate and application time of herbicides, something that is already done.

Weed scientists will develop Integrated Weed Management Systems (IWMS) (Shaw, 1982) for crops *and* specific weeds or weed complexes in crops. These systems will demand integration of the whole agricultural system, not just its parts, and will consider three choices (Weiner, 1990):

1. Maximum short-term yield
2. Maximum sustainable yield
3. Maximum yield sustainability (minimum risk)

Maximum yield sustainability characterizes Third-World agriculture: systems dominated by low, stable (barring environmental disasters such as floods or drought), long-term production. Weiner (1990) and Jackson (1984) suggest that choosing maximum short-term yield "requires high input costs, high environmental costs, and high nutrient and capital fluxes." Both authors advocate low input, environmental cost, nutrient, and capital flux agricultural systems. They suggest that option 2 is the most desirable. Option 2 is most likely to produce systems that integrate the fewest inputs to achieve the desired result. Results will always be important. One measure of the results of a weed management system is how well it manages the problem in the year it is first used. But I suggest that the best measure of results of any weed management system is whether or not it reduces the likelihood of future problems. A primary question is: will there be fewer seeds or vegetative propagules after the management system is imposed than before? If the answer is yes, it is probably a good system. Weed management systems will reduce weed problems, not eliminate them. The goal may best be described as stabilization of populations at a low level through management techniques that are economically and environmentally sound.

This chapter illustrates some principles and available components of weed management systems. Each system is incomplete, partly because the research base is still developing, and partly because few management systems will ever be fully complete and fixed for all time. Weed problems will evolve as they always have and management systems must be dynamic. This chapter will not include or discuss every weed problem in every crop or cropping situation. It will describe weed problems in several general situations and illustrate the techniques that can be integrated in management systems. The chapter is not a weed control guide or how-to manual. A longer discussion of weed management systems is available (Smith, 1995). It is important that students know that weed management systems neither stand alone nor are imposed in isolation. They are part of agriculture and landscape management. Each must mesh with soil conditions, tillage practices, economic and political realities, and social and other aspects of plant culture. Instructors and students are urged to apply the principles developed in this chapter to weed management situations important in their region.

II. A METAPHOR FOR WEED MANAGEMENT

The necessity for weed management occurs when a place is selected for planting. The history of the place is important as a weed management system is developed. Past cropping sequences and weed control methods reveal the kinds of weeds to be expected. The way in which the soil and seedbed have been prepared will be important. Plowing the field exposes a different population of weed seed than disking or chiseling. The kind and timing of irrigation influence weed species. If the land has been observed carefully and edges, ditches, and fences have been kept clean of new sources of weed infestation, the problem is different than if field sanitation has never been practiced. Past and present insect and disease management must be integrated with weed management, and they influence each other. Everyone does some of these things, but few pay attention to their effect on the weed problem or the weed management program. More research is needed to determine specific effects of each necessary practice on weeds.

Weed management systems can be compared to a carpenter's toolbox. Almost everybody recognizes and knows something about the use of a hammer, a screwdriver, or a tape measure. But in a good carpenter's toolbox there is a whole assortment of other things, the use of which is unknown to noncarpenters. Their purpose is a mystery. Many people would be pleased to have the toolbox of a good carpenter and would quickly use several tools. Other tools would be used later as knowledge developed or after someone explained their purpose and use. Still other tools might remain interesting but unused. The purpose and use of some tools will be obvious. Others will look familiar but their use won't be obvious. There will probably be many with no clear purpose, and one may even wonder who decided they should be in the box.

There are some features that will be common to all weed management systems–all the toolboxes. Each weed management toolbox should have three compartments: weed prevention, weed control, and weed eradication. The prevention compartment will have the tools used to keep a weed species from occurring in a previously uninfested area. The compartment labeled control will be the largest, and many tools that belong there will also be found in one of the other compartments. Control tools are things used to reduce the weed population so land can be used productively. The eradication compartment is the smallest one, or, if not the smallest, it is the least used. The tools in it are not more complex than others, but they require great persistence and just do not seem to work as well as others do. They are designed for complete removal of a weed species from an area.

Dewey *et al.* (1995) proposed that noxious weed management could be regarded as forest managers think of wildfire management. The area of range and forest land infested with noxious weeds grew from 2.5 million acres in 1985 to 8.4 million in 1994 and is projected to grow to 10 million in 2000. Dewey *et al.* (1995) regard weeds as a raging biological wildfire. All aspects of wildfire and weed management are similar except two. The first dissimilarity is that wildfires spread more rapidly than weeds, even though their patterns of spread are similar. The second, more important, difference is that forest managers never fail to see a wildfire. Fires cannot be ignored; weeds can be. Weeds do not obviously destroy things and because the occupied area increases slowly, they can be and often are, ignored until they dominate large areas. Then people ask, "Why didn't someone do something?" It is too late then for other than expensive, time-consuming attempts to control. Action needs to be taken when fires and weeds begin. The next section describes the logical steps for developing a weed management system to take the necessary action.

III. THE LOGICAL STEPS OF WEED MANAGEMENT

Most toolboxes and good tools come with a set of instructions on purpose and use. Weed management tools also come with instructions. Presently, instructions are general, but they will improve and become more specific as knowledge expands. At a minimum the instructions for nearly all areas and all cropping systems will include seven logical steps[1]:

A. PREVENTION

The first, most obvious, and perhaps the most frequently omitted step is prevention. Weeds that do not appear because clean seed is planted, machines are cleaned, and new cattle are separated (see Chapter 9, Section I) do not have to be controlled. Early detection is part of weed prevention. Detecting a new weed does not prevent its arrival, but quick action can prevent its spread. Preventive action can be as simple as bending over and pulling the weed and removing it from the site.

[1]I am indebted to Dr. K. G. Beck, Assoc. Prof., Dept of Bioagric. Sci. and Pest Management, Colorado State Univ. for the insights of these management steps.

B. Mapping

An accurate map of weed infestations must be made before a good management program can commence. Problems must be defined, by species, and located in the field or area to be managed before solutions are proposed. No patient wants to be treated for unspecified, unknown illnesses. An accurate medical diagnosis is expected, before treatment. Weed management must be as specific. The best weed management systems will be designed to control specific weeds in specific places.

It is not feasible to map every weed. Major weed species should be known, and those species likely to become problems (e.g., perennials, hard-to-control annuals) should be located and the size of the infestation specified. Early detection of new weeds is part of mapping and prevention.

C. Prioritization

Money, time, technology, or labor are often lacking. It is not possible to do everything. The best weed managers will know as much as possible about the weed problems and select those to be managed. Those species that pose the greatest threat to intended land use should receive highest priority.

D. Development of an Integrated Weed Management System

When it is determined that management is necessary, one begins to look closely at tools in the box to see what is available and if a particular tool or set of tools will be best for the weeds to be managed. The best weed management systems will not select a single technique. All appropriate tools will be examined, and an integrated approach including two or more tools or strategies will be selected. Integration will consider cultural methods such as grazing management, fertility, irrigation practices, seeding rate, and use of competitive cultivars. Mechanical methods include tillage before and during the crop's growth, mowing, burning, flooding, and mulching. Biological and chemical control will also be considered. All of these methods will be in the weed management toolbox (see Chapter 9). Some tools are numerous, more apparent, or easier to grab and are used more frequently. Chemical control is the method of choice in many situations. Other tools always seem to wind up on the bottom of the box and are not often selected. This could be because they are not as easy to use, or because the knowledge of how to use them is not available. For example, soil fertility influences weeds and should be considered in weed management. The knowledge of how to manipulate fertility in weed management systems is not abundant.

Figure 19.1 is a conceptual model of weed management systems. It is a glimpse into the weed management toolbox. After completing the preliminary steps shown in Figure 19.1, control options are selected to develop a system to manage weeds. When the methods are selected, weed managers must ask what can be done and what should be done? Some things that are possible may not be desirable. For example intensive tillage might increase soil erosion, or intensive herbicide use might pollute water or harm nontarget species. The weed management system should be integrated with other aspects of management (e.g., insect or disease control, fertility) of the crop, landscape, or area on which weeds are to be managed.

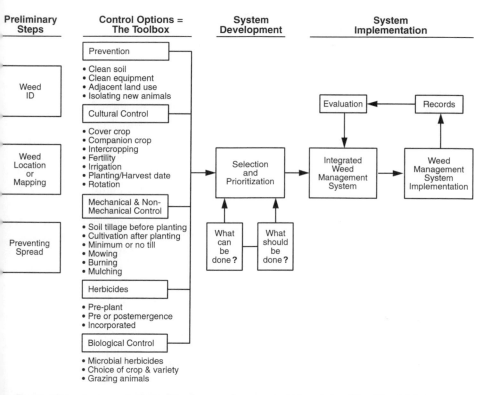

Figure 19.1 A conceptual model of a weed management system—The Weed Management Toolbox.

E. IMPLEMENTATION OF SYSTEMATIC WEED MANAGEMENT

With a map of the weeds and a management plan, managers can begin a program. Not everything has to be accomplished in one season or with one technique. The program is systematic, meaning that timely, planned management will be done over several seasons. The manager may decide to begin with control of perimeter infestations to prevent their spread and eliminate sources of future problems. Another approach could be spot treatment of developing perennial infestations. Based on available techniques, their cost, and adaptability to the situation, the manager can choose among several courses of action (Dewey *et al.*, 1995). If an infestation is small but the weed is very aggressive and likely to spread quickly, eradication is often the best choice. A weed could be a serious threat but not easy to control. In this case it could be contained or confined with other tools to be employed if the weed threatens a crop. In some cases managers should opt to do nothing except monitor the weed regularly and evaluate management options.

F. RECORD-KEEPING AND EVALUATION

Records of what was done and its success must be kept (Figure 19.1). Good records allow the manager to repeat successes and learn from mistakes. Evaluation should be continuous and not just a week or month after a control was done. Evaluation over 2 or 3 years is required to measure success. For example, students are required to write final examinations at the end of classes. A better measure of learning might be an examination 3 to 5 years after completing a class. Such an examination would measure knowledge retention. It is the same with a weed management program. Success over time is more important than success in the short run. A bad mistake will be made when a system is developed and the user assumes it works but does not do regular evaluation to verify the system's success.

G. PERSISTENCE

Successful weed management is not achieved after application of one control technique. The soil seedbank and new sources of infestation demand continued attention. Vigilance is the price of success. Weed management is not a one-step (one control) process. Methods must be integrated over time to reduce or eradicate some weeds and manage other populations to reduce yield loss and crop injury.

IV. WEED MANAGEMENT PRINCIPLES IN SIX SYSTEMS

There are as many weed management systems as there are cropping systems and weed complexes. The number of systems may equal the number of weed managers. Each manager will put unique touches on any system, but each system shares some general characteristics. The systems to be described should be thought of as generic management systems. It is assumed that weed identification, mapping, preventive measures, record keeping, and evaluation are part of each system. How these things are done will differ with each system. Several research programs are now investigating sampling methods to determine the easiest, most efficient, and cost effective way to sample an area for weeds. The following examples focus on identifying possible components of integrated management systems. The examples are not intended to be complete, readily adaptable, prescriptive management systems. They are intended to be a basis for discussion of such systems and their development.

A. SMALL GRAIN CROPS

1. Prevention

Preventive strategies, the first phase of weed management in small grain crops, including wheat, oats, barley, and rye, are not complex; they are the basis of good farming practices. The first preventive step is identical for all crops–plant weed-free seed. Custom combines and other itinerant machines are sources of weed seed and should be cleaned before leaving a farm and before moving from an area of known contamination to a weed-free area. Competitive weeds on field edges and roadsides should be managed because they are sources of new field infestations. Trucks or wagons used to transport grain should be covered to prevent wind dissemination of weed seed from uncleaned grain.

2. Mechanical Methods

When preparing land for small grains, there is a wide choice of techniques. Traditionally, soil was plowed, but that is no longer as common. The sequence and kind of preplanting tillage is influenced by soil type and condition, prevailing weather, implements available, and grower preference. Preplant tillage might mean plowing 8–16 in. deep, disking or field cultivating up to 6 in. deep, surface tilling 1–3 in. deep, or direct planting with no tillage at all. Each of these practices and its timing affects the type, presence, and abundance of weeds. No cultivation or a shallow, noninverting cultivation increase the incidence of perennial weeds and decrease annuals (especially broadleaved species).

No-till methods tend to increase the incidence of annual grass weeds such as wild oats, bromes, annual bluegrass, volunteer rye, and jointed goatgrass. Plowing and disking help prevent spread of perennials, but neither alone will control Canada thistle effectively. Plowing is 10–20% more effective than shallow cultivation or disking for control of perennials, but it returns previously buried roots and rhizomes to the surface where, if they are not desiccated rapidly, they will grow. After plowing or without plowing, early cultivation of land to be planted to small grains stimulates germination of annual weed seeds, and seedlings then can be controlled by subsequent cultivations.

Cultivation of stubble in fields from which a small grain has been harvested can aid control of perennial grasses and prevent some annual weeds from producing seed. If stubble cultivation is done at the wrong time or with weeds that survive cultivation, the weed problem can worsen. For example, stubble cultivation soon after harvest could bury wild oat seed and reduce loss through natural causes (e.g., cold weather). Seedlings of winter annuals such as downy brome easily survive shallow, noninverting tillage, and partial burial.

Fallow (noncropping), or fallow combined with tillage, is an effective weed management technique. Seedlings can be eliminated with cultivation, and more than one cultivation may be required to control most germinated seed and emerged seedlings.

3. Cultural

Farmers, for many reasons, want to plant early, and date of sowing affects weed management. The earlier a crop is planted, the less time is available for weeding of any kind before planting, and this may increase the chance that weeds will germinate and grow with the crop. Seeding winter wheat at a higher rate reduces competitiveness of slender foxtail or wild oats. Increasing crop density by higher seeding rate or by narrowing row width tends to increase competitiveness of wheat and other cereals against spring grass weeds.

Delaying winter wheat planting until emerged weeds can be killed by a light tillage is an effective, inexpensive weed management technique. On the other hand, a quick-emerging, vigorous, dense crop stand is also an important weed management technique. For example, early planting of spring grains may allow crop development before foxtails germinate. Small grains are normally planted in 7-in. rows with adequate rain or irrigation and in about 14-in. rows on dryland. This accepted agronomic practice is usually not changed for weed management reasons even though it affects weeds and their control.

Correct seedbed preparation for the soil and cultural system; cultivar selection; use of high-quality, clean seed; and careful attention to optimum fertility to produce rapid emergence of vigorous crop plants contribute to weed management. Many farmers save seed of small grains from year to year to save money. With no or poor cleaning, weed seed can be planted (see Chapter 5).

As Italian ryegrass density increased, wheat yield decreased and semi-dwarf cultivars had lower yield than tall cultivars with the same density of Italian ryegrass (Appleby *et al.*, 1976) (Table 19.1). In Canada, it was shown that green foxtail is more competitive in semi-dwarf than in tall spring wheat cultivars (Blackshaw *et al.*, 1981). These data point out the importance of cultivar selection, crop canopy, and crop competitiveness. Semi-dwarf cultivars have a more open crop canopy, permit more light to reach weeds, and allow Italian ryegrass to be more competitive. Cultivars are not chosen for weed management, but their influence should not be ignored. Unfortunately, the basis for cultivar competitiveness is too poorly understood and it is a tool that cannot yet be used in weed management systems.

Table 19.1

Yield of Four Wheat Cultivars Grown with Three Densities of Italian Ryegrass (Appleby *et al.*, 1976)

Cultivar height	Ryegrass plants (plants/sq. yd.)	Wheat yield (lb/A)
Tall	0	3,096
	33.4	2,520
	82.8	2,232
Tall	0	3,924
	33.4	2,925
	89.4	2,709
Semidwarf	0	3,042
	32.6	2,214
	80.3	1,908
Semidwarf	0	3,465
	36.8	2,565
	82.8	2,115
LSD at $p = 0.05$		423

Crop rotation breaks a weed's life cycle by altering the crop it must associate with. It demands use of weed management techniques adapted to different crops. Rotation to another crop effectively manages winter annual grasses in winter grains or summer annual grasses in spring grains. Each crop has its own set of cultural practices that create habitats for certain weeds. Changing crops changes available habitats and weeds.

Fallow weed management must be done in winter wheat–fallow systems. Weeds such as downy brome, jointed goatgrass, and volunteer rye use moisture during fallow and the seed produced easily infests the next wheat crop. Their life cycles are similar to that of winter wheat, so inclusion of a spring crop (e.g., barley or corn) is a useful management tool. The spring crop may permit use of herbicides that control the problem weeds but cannot be used in wheat. Adding summer annual crops to a winter wheat–fallow rotation lengthens the time before the next wheat crop reduces the annual weed problem, and increases seed mortality.

Fertilizer placement is a regular practice in row crops but not in small grains, where it may have potential as a weed management practice. Placing nitrogen fertilizer in the crop seed row, away from weeds, achieved a small reduction in seed production of volunteer rye and jointed goatgrass. Greater reduction was obtained when fertilizer placement was integrated with an increased wheat seeding rate and a taller, more competitive cultivar (Anderson, 1994).

4. Biological

There are few biological weed control techniques available for use in small grains. Use of an endemic anthracnose disease to control Northern jointvetch in rice is one instance (see Chapter 10). When developed, these agents will have to be integrated with other weed management techniques.

5. Chemical

Herbicides used in small grains are generally safe, efficient, and profitable but, like other methods, when used alone, they will not solve all weed problems. Herbicides must be part of a total weed management program. For maximum effectiveness, herbicides should be applied when weeds are young and have not yet affected crop yield.

Information on proper application of any herbicide is critical to successful use. Always read and follow label directions. If herbicides are mixed, follow label directions. Combining herbicides with different actions and activity can improve the weed control spectrum. When mixed, the rate of one or both herbicides may be reduced.

For the many weed problems in small grains, there are herbicides or herbicide combinations that provide good control and crop safety if they are properly applied at the right time and rate. Most postemergence, foliar-absorbed herbicides require actively growing weeds for maximum effectiveness. Weed growth is reduced by cool temperatures and drought. Herbicides with soil activity are less affected by temperature, but their activity is reduced in dry soil. A key to successful weed control with herbicides is early use when weeds are most susceptible.

Managers should know what weed is to be controlled before selecting a herbicide. The herbicide should be applied uniformly with a properly calibrated sprayer. Reading the part of the herbicide label specific to each crop–weed situation is essential. Table 19.2 summarizes the attributes of several herbicides that can be used in small grains. It is not a table of recommendations. Specific label recommendations must be consulted and followed because approved uses vary from place to place.

The following section is a brief discussion of some of the common herbicides labeled for use in small grains in the United States.

Metsulfuron (Ally) is most effective when absorbed through foliage, but is absorbed by roots. It can be applied to wheat after the two-leaf stage and before jointing. A surfactant is always required. Because it persists in soil, there may be rotational cropping restrictions. Its soil degradation slows as soil pH increases, and it should not be used on soils with a pH above 8.

Chlorsulfuron (Glean) provides broad-spectrum control of many annual broadleaved weeds and suppression of some grasses. It has foliar activity but is most effective after root uptake. It can be applied after the two-leaf stage and before the boot stage of wheat and, as with metsulfuron, a surfactant is needed for optimum postemergence effectiveness. It persists in soil, there may be rotational cropping restrictions, and it should not be used if soil pH is above 8. There are problems with weed resistance to chlorsulfuron and its use has been restricted because of the rapid development of resistance.

Thifensulfuron (Harmony) is usually mixed with tribenuron (Express) and sold as Harmony Extra. It has postemergence activity on many broadleaved weeds and wild garlic It can be applied after wheat has two leaves but before the third node is detectable. Broadleaved weeds should be less than 4 in. tall and wild garlic plants less than 8 in. It has little or no soil residual activity and no rotational restrictions beyond 2-months.

Bromoxynil (Buctril) controls annual broadleaved weeds and has complete selectivity in small grains. It has contact activity and can be applied from the two-leaf up to the early boot stage, but is most effective when applied before full tillering while weeds are small. Because it is taken up

Table 19.2

Herbicides for Small Grain Crops

Herbicide	Rate range/A	Time of application and comments
Metsulfuron	0.1 oz	From 2-leaf stage until just before boot. In durum wheat from 4-leaf stage to boot.
Chlorsulfuron	0.17–0.33 oz	From 2-leaf to boot.
Thifensulfuron	0.5–0.6 oz	Winter wheat after 2-leaf stage and before the 3rd node is detectable. Spring wheat—from 2-leaf stage but before first node is visible.
Triasulfuron	0.0131–0.0262 lb	Postemergence in wheat, barley, or fallow and preemergence in wheat. Surfactant required.
Tribenuron	0.17–0.33 oz	From 2-leaf stage but not after first flag leaf is visible.
Dicamba	0.09–0.13 lb	In winter wheat apply after dormancy breaks in spring and before wheat begins to joint. Winter wheat is more tolerant than spring wheat. In spring wheat apply after 3 leaves and before wheat has 5 leaves. Dicamba is best used in combination with other broadleaved herbicides.
	0.09 lb	Rate for spring barley.
Bromoxynil	0.25–0.5 lb	Apply postemergence until just before boot.
MCPA	0.25–0.5 lb	Apply to seedling weeds after tillering has begun (about 5 leaves) and before the boot stage.
2,4-D	0.5–0.75 lb	Apply to seedling weeds after tillering has begun (about 5 leaves) and before the boot stage. Often injures winter wheat.
Clopyralid/Stinger/Reclaim	0.06–0.25 lb	Postemergence broadleaved weed control and Canada thistle suppression.
Diclofop/Hoelon/Iloxan	0.75–1.25 lb	Postemergence control of wild oats and some annual grasses in the 1–3 leaf stage.
Difenzoquat/Avenge	0.63–1 lb	Postemergence control of wild oats in fall-seeded wheat and spring barley.
Triallate/Avadex	1.25 lb	Pre- or postseeding control of wild oats.
Imazamethabenz/Assert	0.18–0.47 lb	Postemergence control of wild oats and many Brassicaceae.

by foliage and poorly translocated, thorough coverage is important and late application reduces the amount of exposed weed foliage.

Dicamba (Banvel) gives postemergence control of many broadleaved weeds that are difficult to control with 2,4-D. In winter wheat, it can be applied after full tillering up to the prejoint stage. In spring wheat, it must be applied before the 5-leaf stage.

2,4-D is absorbed through foliage and roots. It performs best when the plants are not stressed. Ester formulations readily penetrate plant foliage, particularly under cool, dry conditions. Amine formulations are water soluble and do not penetrate foliage as well, but have less potential for crop injury. 2,4-D is applied after full tillering and before jointing. Small grains can be injured with 2,4-D, MCPA, or dicamba if application is made too early or too late. 2,4-D and MCPA can be used after the soft dough stage as a harvest aid, without risk of crop injury, but the benefit to yield is minimal. Small-grain tolerance of MCPA is similar to that of 2,4-D, and their weed control spectra are similar but not identical. Clopyralid (Stinger/Reclaim) is often applied with 2,4-D or MCPA postemergence to control many annual and perennial broadleaved weeds. It broadens the spectrum of control and suppresses Canada thistle. Translocation from treated foliage is rapid in phloem after nearly complete absorption. It persists in soil and can affect some rotational crops.

Triallate, diclofop, difenzoquat, and imazamethabenz-methyl control wild oats, but each is unique. Triallate (Avadex) is effective on wild oats and suppresses annual brome grasses. It can be used preplant or preemergence and is absorbed only through the coleoptile node of the shoot, not through the root of wild oats. It must be incorporated to prevent volatility loss and to place it in the right soil zone for uptake by wild oats. The small grain must be seeded below the area of incorporation.

Diclofop (Hoelon/Illoxan) controls wild oats and some other annual grasses if applied when the weeds have one to three leaves. It does not need to be incorporated. Difenzoquat (Avenge) gives postemergence control of one- to three-leaf wild oats in spring and winter wheat. Application rate should be increased to 2 quarts per acre as wild oats' density increases. Imazamethabenz (Assert) provides postemergence control of wild oats and a few other grasses and several mustards in wheat and barley. Wild oats must be in the one- to four-leaf stage and wheat has good tolerance, up to jointing, of rates between 0.3 and 0.5 pounds active ingredient per acre. A surfactant or crop oil is recommended to improve absorption. It can be applied in water or in nonphosphorus liquid fertilizer. Growth is inhibited within a few hours of application, but chlorotic and necrotic symptoms do not appear for 1 to 2 weeks. Activity on target weeds is only postemergence but it persists in soil and can affect rotational crops.

B. Corn and Row Crops

One of every four U.S. crop acres grows corn, and it has been selected as representative of the many row crops grown in the United States. Soybeans, dry beans, sugarbeets, cotton, tobacco, sorghum, peanuts, vegetables from broccoli to zucchini squash, and potatoes each have their own, unique weeds, and thus weed management requirements and solutions. Because they are all annual row crops, however, they share some weed management principles.

Corn, which most of the rest of the world calls maize, will be used to emphasize the shared principles of weed management. Less than 10% of the U.S. corn crop is eaten by humans directly. Most is fed to animals, and much of the U.S. crop is exported. Corn has more than 3,000 uses in more than 1,200 food items ranging from corn syrup to margarine. Other uses include paper production, plastics, cleaning agents, cosmetics, additives for pesticides, and ethanol production for fuel.

These crops are called row crops because they are planted in rows from 20 to 30 or more in. apart. Small grain crops are also planted in rows, but the rows are narrow and mechanical inter-row cultivation is impossible. Rows were invented because of the necessity of cultivating for weed control. Weed management is only one reason crops are planted in rows. Rows make planting and harvesting easier, and modern equipment demands straight rows. Manufacturing facilities concentrate human power, talent, and resources in factories for mass production. Agriculture requires a different spatial geometry, and the advantages of concentration are limited. Spacing in rows is required for optimum yield per unit area for all row crops, and yield is not increased if plants become too crowded.

1. Prevention

Weed prevention strategies are similar for most crops. There is nothing sophisticated or mysterious about them. Most practices are just common sense and should be incorporated in all good weed management system. The reader is referred to Chapter 9, Section II, for a brief discussion of preventive practices.

2. Mechanical

Not too many years ago, it was standard practice to moldboard plow the preceding fall or in the spring before planting corn. Plowing has not been abandoned as a weed management/seedbed preparation technique, but its use is diminishing. Plowing controls emerged weeds and buries weed seeds while it brings other seeds to the surface where they can germinate. Like many practices that affect seeds, it controls and encourages weeds at the same time. Plowing is usually followed by other tillage operations to prepare the seedbed. Disking and harrowing break down clods and make crop planting easier with traditional seed drills and corn planters. They also create ideal conditions for germination of weed seeds with, or just before, the crop. Plowing is a soil-inverting operation, but chiseling is not, and fewer weed seeds are brought to the surface by the latter operation. Tillage operations, subsequent to plowing, are shallower and encourage germination of shallowly buried seeds, but not those deeper in the profile. The effects of the two kinds of tillage on weed populations were discussed in Chapter 9.

The use of conservation or reduced tillage has expanded greatly in the past decade, and interest in adoption of some form of reduced tillage has expanded even faster. These systems range from surface disking to break up the residue of the preceding crop to no tillage at all and planting directly into the preceding crop's residue with specially designed no-till planters. In the first year or two after no-tillage is begun, weed problems decrease dramatically, but without careful management, weeds can increase in subsequent years. For example, in one experiment with monoculture corn and conservation tillage, all plots were weed-free the first year. In the second and third years, fall panicum dominated, and smooth pigweed dominated in the fourth and fifth years, reaching densities of 85% of total plot area (Coffman and Frank, 1992). The authors related the change of weed flora to continuous use of certain herbicides. Fall panicum dominated in plots treated with atrazine and a carbamothioate herbicide, and a triazine-resistant biotype of smooth pigweed assumed dominance in plots treated with atrazine plus cyanazine (Coffman and Frank, 1992).

Ridge-till systems are used to reduce soil erosion and the need for herbicides in some corn–soybean rotations. A special ridge-till cultivator makes ridges over the crop row during the final, summer cultivation of either crop. The ridges are disturbed at harvest and during spring planting are leveled by moving some soil into the furrows. Immediately after smoothing or "knocking off" the ridges, the crop is sown on the remainder of the ridge and the ridges are gradually rebuilt, during cultivation, as the crop matures. The system was most effective in Minnesota when corn and soybeans were rotated (Forcella and Lindstrom, 1988a). The ridges crack and weeds emerge when corn is grown continuously.

Ridge tillage in corn–soybean rotations is not without problems. Knocking off ridges controls many weeds, and ridging soil during the summer encourages germination of numerous weed seeds that can produce seeds to infest the next crop. Conventionally tilled corn had about two-thirds fewer weed seeds than ridge-tilled corn because of the large seed production by weeds that germinated when ridges were rebuilt (Forcella and Lindstrom, 1988b). Ridge tillage cannot be successful without herbicides to control late-emerging weeds.

Studies in Indiana evaluated no-tillage, moldboard-plow, and chisel-plow systems in three rotation systems, each of which included corn (Martin *et al.*, 1991). Net incomes for no-till systems on all farms were lower than incomes for moldboard or chisel plow systems because of slightly lower yields and higher herbicide costs. In general, farm incomes were higher with moldboard as opposed to chisel plow systems. It is important to note that these studies were done on highly productive, flat, well-drained soils that are not highly eroded. A different situation for any system will probably yield different results. Agriculture and agricultural research is definitely site specific.

After land preparation, corn can be cultivated mechanically, one or more times, with various types of implements ranging from a straight shank to several different duckfoot-shaped tools. A rotary hoe can also be used to kill weeds between rows. Cultivation can move soil into corn rows and cover emerging seedling weeds, but it cannot till in the row. Newer cultivation implements operate close to the crop row and make herbicide banding attractive. If all weeds between rows are controlled mechanically, then herbicide quantity and cost can be reduced by applying the herbicide only in a band over the crop row. This requires more application skill and accuracy than broadcast application. Postemergence flaming has been used selectively in corn and cotton, but has never been used widely.

3. Cultural

Crop rotation can be a profitable and useful weed management technique. Corn is often grown in monoculture or in a limited rotation with soybeans or another row crop. Rotating to a small grain or hay crop, or both, in succession, often results in reasonable yields, but lower than those in a corn–soybean rotation (Helmers, 1986). Rotational possibilities are limited by land, climate, market opportunities, and the availability of suitable rotational crops. Rotating corn with other crops is critical to managing weeds in systems attempting to reduce herbicide use. Introduction of crops with different life cycles and cultural practices deters growth of summer annual weeds with a life cycle similar to that of corn; their growth would be encouraged by continuous or frequent corn crops. Rotation reduces annual grass weeds in corn in the Central Great Plains of the United States (Wicks and Smika, 1990). Many annual grasses germinate in May and set seed by late August before corn is harvested. As mentioned earlier, rotating to winter wheat changes the times of tillage and crop presence and disrupts the life cycle of annual weeds.

4. Biological

At present, there are no biological control agents used routinely in corn or other row crops. Because row crops usually have several weeds rather than just one, the specificity of a biological control organism may not fit the weed control need.

5. Chemical

The main weed control technique that has to be integrated in corn weed management systems is chemical control. Table 19.3 summarizes attributes of some of the herbicides available for use in field corn. Some, but not all, of these herbicides may also be used in sweet corn or popcorn. Table 19.3 is not a table of recommendations. Label and local recommendations must always be consulted before using any herbicide. The dominant herbicide families used for weed control in corn are triazine (atrazine), chloroacet-amide (alachlor, metolachlor), and carbamothioate (EPTC, butylate). Herbicides in these families may be used alone, but are most often used in combinations. Some states' recommendations include more than 30 soil-applied and a separate list of more than 30 postemergence herbicides or herbicide combinations for weed control in corn. About one-third of the soil-applied treatments combine atrazine with a chloroacetamide or a carba-mothioate. Perhaps a half-dozen soil-applied herbicides are single applications of a chloroacetamide herbicide. The greatest variety of herbicides is found in the postemergence group. Much progress has been made in discovery of postemergence annual grass herbicides, including nicosulfuron, primisulfuron, and rimsulfuron and thifensulfuron, applied in combination. Each of these herbicides (Table 19.3) is applied at very low rates, and they have solved postemergence annual grass control. Other new herbicides (e.g., halosulfuron, flumiclorac, and flumetsulam) control many annual broad-leaved and grass weeds, including some (such as velvetleaf) that are not controlled well by other herbicides.

6. Integrated Strategies

Simulation models for weed management in corn (King *et al.,* 1986; Lybecker *et al.,* 1991) suggest that flexible weed management strategies, based on control variables, outperform fixed or prescriptive weed management programs. The variable used for deciding what and how much herbi-

Table 19.3

Partial List of Herbicides Used in Corn

Herbicide	Usual rate range (lb ai/A)	Time of application and comments
Alachlor/Lasso	2–3 lb	Can be applied preplant and shallowly incorporated, or postplant but prior to crop emergence. A minimum of 10 gallons of total spray volume per acre ensures good coverage. It is available in several formulations, including a microencapsulated, slow-release formulation. Often applied in combination with atrazine, cyanazine, or dicamba. It controls several annual grasses.
Atrazine/Aatrex	1.5–3 lb	Apply preplant incorporated, pre- or postemergence before weeds are 1.5 in. tall. Early application preferred. Lower rates on sandy soils and higher rates are for clay soil or soil with greater than 2.5% organic matter. Soil residues can injure succeeding, susceptible crops. Effective on a wide range of annual broadleaved species. Used in combination with alachlor, metolachlor, or cyanazine.
Bromoxynil/Buctril	0.25–0.38 lbs	Good spray coverage is required because of foliar uptake. Applied to weeds that emerge before corn planting until just prior to corn emergence. Can be applied postemergence—follow label directions. Controls several broadleaved species not controlled well by 2,4-D.
	0.5	For chemigation only.
Butylate/Sutan	4–6 lb	Both trade products are a combination of butylate and a safening agent (different in each case). Apply preplant and incorporate 3 to 4 in. deep as soon as possible after application. Delay reduces weed control especially on moist soil. Short soil persistence. May control some perennial grasses, including nutsedge. Controls annual grasses and broadleaved weeds.
Cyanazine/Bladex	1.5–2 lb	Applied preplant, pre- or postplant before crop emergence. Shallow incorporation may be required for activation if there is no rain or irrigation within 6 days of application. Does not normally persist into the next year. Controls annual grasses and broadleaved weeds.
Dicamba/Banvel	0.25–0.5	Apply 0.5 lb from spike to 5-true-leaf stage of corn for broadleaved weed control. Later apply lower rate up to when corn is 3 feet tall or before tassels emerge. Drop nozzles to avoid spraying corn foliage must be used for late applications.
2,4-D amine or ester	0.5–0.75 lb	Apply to small, emerged broadleaved weeds in corn that is 6 to 12 inches tall. 2,4-D makes corn brittle and cultivation or wind soon after may damage it, and the chance is greater at temperatures above 85°F, but its action is often enhanced when it is warm. The lower rate is recommended for ester formulations.

EPTC/Eradicane extra	4–6 lb 3–6 lb	Eradicane is a combination of EPTC and a safener or antidote to protect corn. Eradicane extra includes a different safener and another chemical to extend EPTC's soil life. Controls several annual grasses and some broadleaved species. It must be applied preplant and incorporated immediately to prevent loss by volatility.
Flumetsulam/Scorpion/ Broadstrike	4 oz	Scorpion is a combination of flumetsulam and clopyralid for control of annual broadleaved weeds. Broadstrike is flumetsulam combined with metolachlor.
Flumiclorac/Resource	4–6 oz	Postemergence, annual broadleaved weeds
Glufosinate/Liberty		Broad spectrum weed control, but only in cultivars genetically engineered for resistance to glufosinate.
Halosulfuron/Permit	0.66–1.33 oz	Postemergence, annual broadleaved weeds
Imazethapyr/Pursuit	4 oz	Pursuit-resistant corn cultivars are required. Postemergence control of many annual broadleaved and grass weeds and shattercane.
Linuron/Lorox	1–1.5 lb	Used only postemergence and directed at the base of the corn plant. Injures corn foliage if sprayed on it. It is not incorporated. Controls some germinating broadleaved and annual grass weeds. A surfactant is recommended.
Metolachlor/Dual	1.5–3 lb	Action is similar, but not identical, to alachlor. Used preplant or premergence to control a wide range of annual grass and broadleaved weeds. Rate varies with soil type. Combined with atrazine or cyanazine.
Metribuzin/Sencor	0.25–1 lb	Broadleaved weeds in corn less than 8 in. tall.
Nicosulfuron/Accent Rimsulfuron + Thifensulfuron/ Combined as Basis	0.25–1.5 oz 0.33 oz	Used postemergence to control previously hard-to-kill grasses such as shattercane, johnsongrass, foxtails, sandbur, and wild proso millet. A crop oil concentrate increases activity.
Pendimethalin/Prowl	1–1.5 lb	Incorporation is not mandatory, but may enhance activity. Used postemergence and often mixed with atrazine or cyanazine. Controls several annual grasses.
Primisulfuron/Beacon	0.38 + 0.38 lb 0.76 lb	Application early and again any time before tassel emergence controls previously hard-to-kill grasses such as shattercane and johnsongrass. Single application.

cide to use is weed seed number in soil. The models require knowledge of losses due to specific weed densities, percent emergence of weed seed from the soil seedbank, and the efficacy of each herbicide against each weed. The models do not consider the effect of weed escapes on the next year's crop. A flexible strategy lowered total herbicide cost and the quantity of herbicide used, increased postemergence herbicide use, decreased preemergence herbicide use, and increased the farmer's gross profit margin (Lybecker *et al.*, 1991, 1992). The models were developed for irrigated corn and have not been transferred to rainfed corn areas. To date, the models do not incorporate mechanical methods of weed control. They herald a new era of weed management when decisions will be informed by knowledge of weed seed in soil and the efficacy of different control measures. Weed management decisions have often been made on the basis of what someone thought the problem was going to be, and have therefore been prophylactic rather than directed at a specific problem.

C. Turf

Desirable turf-grass species vary with climate, rainfall, and intended use. They are usually divided into cool- and warm-season species. The United States has about 25 species that can be used for turf (Vengris and Torello, 1982). They are usually perennials that do well with continuous close mowing. Cool-season species grow best during cool (60 to 75°F), wet conditions in the spring and fall and may become partially dormant in hot, summer months. Warm-season species grow vigorously during hot, dry times when temperatures are above 80°F. Vengris and Torello (1982) list 108 weeds that invade turf. Those that occur most frequently in the United States and Canada are crabgrass, dandelion, annual bluegrass, common chickweed, plantains, and prostrate knotweed, but there are many others. Of the 108 common turf weeds, 17 are perennial monocots, 11 annual monocots, 44 perennial dicots, 29 annual dicots, 4 winter annual dicots, and 3 are biennials. Weed control on many turf sites is principally elimination of broadleaved species and because of the variety of herbicides available, the task is not difficult.

Turf is no different from any other crop in the sense that prevention is an essential component of weed management. Preventive practices are used by turf managers when turf is established and during its life. Turf grasses are particularly poor competitors during establishment, and elimination of weeds by thorough tillage prior to planting or use of preplanting herbicides is important. Imported topsoil is almost always contaminated with weed seeds and vegetative propagules, and delaying planting turf species until some of these seeds germinate and can be controlled is wise. Planting the correct turf grass or grass mixture is an obvious preventive strategy. Vigorous emergence and growth reduces weed growth. Grass seed must be of high quality (high percent germination), free of weed seed, and sown at the right time for the climate. Cool-season grasses do best when planted in the fall or early spring, and warm-season grasses do best when planted in spring or early summer.

Weeds in established turf can often be traced to windblown seed sources and poor management practices. Weeds most easily invade cool-season grasses such as Kentucky bluegrass, ryegrass, or fescue when they are mowed less than $1\frac{1}{2}$ to 2 in. On the other hand, weeds most easily invade stoloniferous and rhizomatous bermudagrass turf when it is mowed too high (above 1 in.). Too little water may stress turf grasses and drought-tolerant broadleaved weeds will invade. Too much water will create ideal conditions for establishment of annual and winter annual grasses. Most turf is fertilized with nitrogen to keep it vigorous and maintain desired color. Fertilization appropriate to the climate and turf species helps prevent weed invasion. Prevention can also be practiced by controlling turf wear from constant use of certain spots. Change of traffic or play patterns helps maintain a vigorous turf and prevent weed invasion.

The oldest method of weed control, hand pulling, is still appropriate and common in turf. For a home lawn it is efficient, even if not pleasant. For golf courses and public lawns it is inefficient and not economical. In home lawns, hand weeding will eventually control even the most persistent perennials for which there are no other selective control techniques. Tillage is not appropriate in established turf unless extensive renovation is undertaken. Scarification or vigorous raking is used to thin stoloniferous grasses and control some broadleaved species. If overdone it thins turf and allows weed invasion. Aeration is used in many climates to reduce compaction and stimulate vigorous turf growth thereby reducing weed competition.

Because turf is valuable, fumigation, although expensive, may be a desirable preplanting control strategy. Herbicides are used commonly in turf, but are most appropriate after all preventive management techniques have been employed. There are at least 35 different herbicides that can be used in turf. Not all herbicides can be used with all turf species. For example, five herbicides (benefin, bensulide, DCPA, oxadiazon, and pronamide) control crabgrass and annual bluegrass in cool-season, perennial turf grasses. All of them can be used on bermudagrass turf, except that some hybrids and fine-leaved varieties can be injured (Elmore, 1985). Only bensulide can be used on bentgrass turf. Local recommendations must be consulted before application. There are no selective herbicides for removal of coarse-leaved, perennial grass weeds from perennial turf species. This is a major unsolved weed-control problem. It can be accomplished by spot application of a translocated, nonresidual, nonselective herbicide such as glyphosate, but it will kill other plants it contacts. The most common herbicide used in turf is 2,4-D alone or in combination with another growth regulator herbicide to control broadleaved species. It injures seedling grasses less than 4 to 6 weeks old, but controls a range of annual and perennial broadleaved species without injuring most turf grasses. Because of the nature of plant growth and the translocation pattern in perennials, fall application is often the most effective. As in other crops, application when the crop is growing vigorously and the weeds are young is best.

Warm-season grasses in warm climates present a unique weed control situation. These grasses commonly are dormant during the winter, and herbicides such as paraquat that would desiccate the bermudagrass foliage if applied when it is actively growing can be used to control cool-season grasses and annual broadleaved weeds that grow when the grass is dormant. Atrazine has been used in St. Augustine grass, although it would kill cool-season turf grasses. Postemergence application of organic arsenical herbicides (MSMA, DSMA) selectively controls weedy annual grasses in warm-season turf grasses. Local recommendations and turf managers should always be consulted prior to herbicide choice and application.

D. PASTURES AND RANGELAND

Pastures and rangeland cover more than 40% of the world's agricultural acres. These are diverse habitats that exist in all topographies and climates, and over most soil types. The desirable vegetation is equally diverse, ranging from the short-grass prairies of the mid-United States, to the oak/pine associations found in western states, to designed, planted, irrigated pastures of irrigated and rainfed areas. With the exception of intensively managed, planted pastures, rangeland and pasture may have more than 100 species per square mile. These areas are often very large, and hilly or mountainous terrain make access for mechanical or chemical weed control difficult if not impossible.

Controlled burning has been a common management technique to reduce competition from woody species and competition for water. It has some serious environmental drawbacks (smoke pollution, potential erosion of bare soil) and use has been reduced. It is often followed, on large areas, by reseeding, commonly by air. Biological control with grazing animals–managed grazing–is a desirable weed management technique. Goats are particularly good browse animals, but they must be carefully managed so they do not compete with cattle or sheep. Some of the major successes in biological control of weeds with insects have been achieved on rangeland through use of *Cactoblastis cactorum* to control prickly pear and *Chrysolina quadrigemina* to control St. John's wort (see Chapter 10). Mechanical removal with bulldozers and by chaining are used on rangeland, but are expensive, and results are temporary. Mowing is a good technique for control of weeds in pastures, but not on large areas of rangeland. Growth regulator herbicides are used to control woody species such as big sagebrush or greasewood on rangeland.

A major problem in the arid western states is the perennial, herbaceous weed leafy spurge (see Chapter 10). Successful management has been accomplished when techniques have been integrated. Sheep or goats (biological control) will graze the weed early in the growing season to release desirable grasses from leafy spurge competition and make the spurge more susceptible to fall-applied herbicides. A total of seven insects have been released in the United States for biological control of leafy spurge. The leafy spurge hawkmoth (*Hyles euphorbiae*) eats leafy spurge leaves and flowers during its caterpillar stage (Harris *et al.*, 1985). A root- and stem-boring beetle (*Oberea erythrocephala*), imported from Italy, was established in Montana and North Dakota (Lehninger, 1988). The beetles puncture stems and lay eggs. Larvae bore into stems and move to roots where they mature and exist on carbohydrate root reserves. Adult *Aphthona* spp. beetles feed on leaves and the larvae bore into roots and destroy vascular structure while feeding. Grass infested with leafy spurge will be favored by fertilization and irrigation if either is economically feasible (cultural control). These strategies reduce competition and permit efficient grazing by animals such as cattle that do not eat leafy spurge. In the fall leafy spurge can be sprayed with picloram or picloram plus 2,4-D (chemical control). Neither herbicides nor grazing animals have greatly affected vitality or future performance of biological control insects. It is a certainty that this integration of methods will not eliminate leafy spurge in one season, but it will keep the population at a level that permits effective land use. Persistence, defined as continued use of several techniques and continued evaluation, will be required.

Perennial weeds such as Canada thistle are controlled better when herbicides and mowing are combined (Beck and Sebastian, 1993). Mowing improves pastures and stresses perennial weeds that may then be more susceptible to herbicides. The value of combining herbicides and mowing is illustrated well by control of the exotic invader tropical soda apple in Florida (Mullahey *et al.*, 1996). The annual weed was first found in 1981, but was not identified as an important invader until 1990. It now infests more than 500,000 acres of Florida pastureland, where it flowers and sets seed throughout the year. It has been found in five other southern states.

Control has been best when plants are mowed or chopped 60 days prior to spraying with the growth regulator herbicide triclopyr. Mowing three times 60 days apart gave 83% control after 180 days. Mowing or chopping 60 days prior to triclopyr application was 93 to 100% effective, 180 days after herbicide application. Further mowing is not required, but spraying escaped plants is recommended. Cattle ranchers are urged to isolate new cattle and monitor cattle movement between pastures, because cattle eat the fruit and seed easily passes through their digestive system to reinfest pastures. Cattle isolation is another step in an integrated weed management system.

E. PERENNIAL CROPS

Perennial crops grow for several seasons and are then rotated (e.g., alfalfa), or grow for several years (e.g., apples or almonds) after which the trees are removed and another orchard begun on the same site. A diverse group of weeds succeeds in perennial crops, including annuals, biennials, and perennials favored by perennial culture. Some perennial crops, such as alfalfa, peppermint, asparagus, or strawberries, are not cultivated mechanically, and without good crop competition and weed management, perennials can invade and succeed. Cultivation can be a part of a weed management program in tree fruits and nuts that have wide rows and low crop density. Weed management options are limited because the crop's longevity precludes use of rotation and, in some crops, mechanical tillage. Cover crops and mulching are feasible in perennial crops and should be incorporated in weed management planning. Biological control must be chosen based on the weeds present and cannot be prescribed for all perennials.

1. Prevention

Vigilance is a prerequisite for a good weed management program. The manager must be aware of sources of weed infestation and take appropriate action to prevent invasion. In perennial cultures these could include screening of irrigation water to prevent import of weed seed, careful selection of clean mulch material, composting of manure to kill weed seed before spreading, mowing to prevent seed production, careful selection of adapted crop cultivars to maximize competitiveness, and planting weed-free seed or seedling stock. The last two can only be done at planting, an opportunity that should not be lost. Site selection is a weed management technique because perennial weeds are favored in perennial crops; sites without them should be selected, when possible, for initial planting. Annual weeds will be present on almost any site, and some control can be achieved by preplant tillage, just as it can be before annual crops are planted. Perennials are not controlled easily by tillage and avoiding them is always good planning.

2. Cultural

Timing of planting is a cultural control and preventive technique. Planting should be done when a quick-emerging (or quick-establishing, in the case of transplants), vigorous crop is ensured. For example, alfalfa planted in the fall in southern California becomes better established and is a better competitor with weedy spring grasses than spring- or summer-planted alfalfa (Mitich, 1991). Planting time varies with climate and environmental conditions, but its role in weed management should not be ignored.

Irrigation timing is an important cultural practice that influences weeds. Barnyardgrass and yellow foxtail establish readily when alfalfa is stressed before or during harvest and water is applied when there is little alfalfa growth to shade soil. When water is applied near cutting, weed invasion is reduced.

Grazing animals on perennial cropland contributes weeds in manure, and overgrazing always encourages weed invasion. Grazing animals control some annual weeds.

For some short-duration (3–5 year) perennials (e.g., alfalfa), planting with a nurse or cover crop is a useful weed management technique. Crop yield may be reduced in the first year, but subsequent crops have lower weed populations.

Cover crops, groundcovers, or grassed, mowed alleyways are part of good orchard management in many perennial row crops and fruit orchards. Groundcover species adapted to local environments should be selected based on local recommendations. Cover crops and groundcovers compete directly with weeds, but should not compete with the crop. They may also have allelopathic effects. Regular mowing of grassed areas changes ecological relationships and affects weed populations. In orchards with grassed inter-row areas, mowing is done between tree rows and tree rows may be weeded with herbicides.

3. Mechanical

For alfalfa, peppermint and similar crops that are not planted in wide rows and eventually cover the soil, cultivation is not possible. In tree crops, clean cultivation is a widely practiced weed management technique that precludes grassed alleyways or groundcovers. Clean cultivation is common in many nut orchards and is usually combined with chemical methods of weed control. It is a desirable management technique but increases the risk of soil erosion from water or wind.

4. Chemical

Herbicides for perennial crops are as diverse as the crops. Because this is a book of principles rather than recommendations, the several herbicides available for perennial crops will not be listed. Local recommendations should be consulted for each crop. Many persistent herbicides, including dinitroanilines (peppermint, spearmint, sugarcane), triazines (asparagus, alfalfa, citrus fruits, nuts, pineapple, sugarcane), and uracils (peppermint, spearmint, pineapple), are approved for use in perennial crops. Decisions on herbicide use must be based on the weeds to be controlled and how herbicides affect other weed management strategies (especially incorporation of a permanent groundcover). Herbicides are valuable tools in these crops and should be regarded as part of the overall weed management program, not the entire program.

F. AQUATIC WEED SITES

Detailed recommendations for aquatic weed control can be found in McNabb and Anderson (1985) and in the complete manual of aquatic weed management with herbicides from the Department of the Interior (Hansen *et al.,* 1983). This section presents an integrated view of several available techniques.

It is seldom necessary or desirable to remove all vegetation from water. The aquatic weed manager must decide if complete eradication is desirable or if some level of control is more appropriate. It may be possible to control one especially troublesome or dominant species and leave others undisturbed. Control can be infrequent by mechanical or chemical means, and it can be just removal of excessive growth for part of a season without actually killing any plants.

1. Classification of Weeds

A brief introduction to aquatic weeds was given in Chapter 3, Section D. It used the usual, and simple, classification into free-floating, submersed, and emersed weeds. Although those are useful divisions, the aquatic world is more complicated. A good explanation of the complexity is offered by McNabb and Anderson (1985), who subdivide the usual categories to provide more information about habit of growth and plant type (Table 19.4). The aquatic weed manager must know exactly what weed is to be controlled or managed and its method of reproduction. Algae reproduce asexually by cell division. Completely submerged aquatic plants reproduce by fragmentation, vegetatively by rhizomes and runners, and by specialized submerged buds (turions) and tubers. Submersed plants that have some floating leaves such as American pondweed reproduce by seed, as do emersed plants. Free-floating plants (e.g., waterhyacinth) reproduce by seed and asexually or clonally by fragmentation, and some reproduce by spores and clonally (e.g., *Salvinia*). With dual modes of reproduction, some weeds cannot be managed by preventing seed production as is the case with terrestrial annuals. If the manager does not know the plant's growth habit and method(s) of reproduction, poor or no control may result from improper choice of methods.

2. Prevention

Preventive strategies depend on knowledge of the factors affecting growth of aquatic vegetation. These include light, nutrients, water depth, water flow rate, the growth medium (water or soil) and its nutritional status, dissolved gases, and temperature. Although the last two are important, there is little that can be done to change them. Light can be managed by

Table 19.4

Classification of Aquatic Weeds (McNabb and Anderson, 1985)

Type of plant	Growth habit	Examples
Algae	Unicellular or microscopic colony	Phytoplankton
	Free-floating attached to substrate	Diatoms
	Filamentous—green,	Cladophora
	Colonial—attached or floating	Spirogyra
	Blue-green	Nostoc
Vascular plants	Completely submerged	Sago pondweed, hydrilla
		Eurasian watermilfoil
		Some mosses
	Submersed with floating leaves	Waterlilies
		American pondweed, arrowhead
	Emergent	Cattail, bulrush
		Several grasses
	Free-floating	Waterhyacinth
		Duckweed, azolla
		Salvinia (a fern)

control of water depth. Water management to control weeds by reducing water depth through intentional drawdown manages some weeds effectively. It has little effect on floating vascular plants or algae, but will aid control of rooted species. It is not permanent, because other weeds adapt to new water levels, but it can help manage current weed problems. Watermilfoil, arrowhead, and water lilies can be managed by drawdown, but most pondweeds, cattails, and rushes are not affected. Drawdown can be done at any time, but most irrigation structures were not designed to facilitate the technique and it is not used widely. If the manager understands the biology of the weed to be controlled, drawdown can be timed to stop production of reproductive structures. The opposite technique, ponding or deepening water, can be used to manage some aquatic species.

Water depth can also be affected by turbidity. Turbid, or more nearly opaque, water provides less light to submerged species. If turbidity can be tolerated, it can be created by stirring or intentional incorporation of silt or other soil particles. Turbidity can also be created by fertilization to promote algal "blooms" or abundance. This technique can cause other problems because algae may be toxic or otherwise undesirable. Fertilizer stimulates algal growth that shades plants that root underwater and do not emerge. A bloom must be maintained through a growing season to achieve control. Careful monitoring is required. More commonly, nutrients from surrounding fields or other sources encourage growth of weedy plants and worsen the weed problem. Dredging or reshaping a pond to remove shallow areas reduces light on the edges and reduces growth of submerged or emersed weeds.

Moving irrigation water inevitably brings weeds with it (see Chapter 5, Section II,C). Prevention of water movement to ponds and lakes is nearly impossible. Animals, birds, and humans transport seeds and vegetative reproductive organs to water and, with the exception of humans, cannot be prevented easily, if at all.

3. Mechanical

Mechanical methods of weed control adapted for use on aquatic sites frequently require large, specially adapted machinery. Aquatic weeds can be mowed with floating mowers, but these are expensive and there is a problem of disposal of the mowed, smelly vegetation. Repeated mowing is required and the method is not adapted to large areas. As is true for terrestrial, perennial weeds, mowing may release dormant, vegetative buds and actually worsen the weed problem unless it is done frequently or integrated with another method. It is only appropriate for rooted plants.

Floating plants could be collected by a large mower that collected what it mowed (Figure 19.2), and they would be controlled, but it would be an incidental benefit. It is similar to physical removal of vegetation. Removal is a good technique, especially for floating plants. However, the biomass of aquatic plants is often quite high and large equipment is needed to collect the vegetation; the method is only appropriate for large lakes or straight waterways. Vegetative reproducers will quickly repopulate an area, and disposal of collected biomass is a problem.

Physical disruption of rooted plants by chaining or dredging is a good technique for straight irrigation ditches. It immediately reduces clogging by weeds, but the relief is only temporary for plants that grow back from severed roots. Dredging or reshaping a ditch or pond can be more effective if roots and vegetative reproductive organs are removed, but it is expensive.

Burning is used in much of the western United States each spring to remove plant residue from irrigation ditches. It is more for sanitation and good housekeeping than for weed control. Temperatures are not high enough to kill buried seeds or vegetative organs. If young seedlings are emerging, they will be controlled, but the main benefit is cleanup.

Figure 19.2 A weed mower for aquatic weeds.

4. Biological

The same criteria for success of a biological control organism apply to aquatic and terrestrial environments (see Chapter 10). *Agasicles hygrophila,* a South American flea beetle, has been used to clear southern U.S. waterways of alligatorweed, a free-floating plant. *Cercospora rodmanii* has shown great promise for control of waterhyacinth in tropical and semitropical waterways (Strobel, 1991). Although *Agasicles* has been very successful, there are no other examples of a widely adapted, successful insect or pathogen control for aquatic plants. Other organisms have been used for control for aquatic plants. They include the sea manatee, a large aquatic mammal, and two fish, the grass carp or white amur (*Ctenopharyngodon idella*) and tilapia (*Tilapia melanopleura*). These eat aquatic vegetation but are generally nonselective; however, that is not regarded as a serious disadvantage. Survival and reproduction are problems with any biological control organism. The manatee has no known enemies except man and a pathogen that reduced their population in 1996. Manatees survive well in warm waters, but do not reproduce well.

Fish provide an alternative crop in many aquatic environments. Tilapia are intolerant of cold water and are therefore adapted only to subtropical and tropical climates. The grass carp reproduces well in cold water and can grow to marketable size, but its reproductive ability limits its acceptability in many places (it is not approved for use in California; McNabb and Anderson, 1985). On the other hand, failure to reproduce keeps a population under control and prevents escape of an otherwise desirable control organism.

5. Chemical

Aquatic plants are susceptible to a wide range of herbicides, but there are few herbicides available for use in water. However, herbicides are the most common method of weed control in the United States. These apparently contradictory statements are explained by the multiple uses that water has and the fact that it moves. Herbicides offer the same advantages in the aquatic environment that they do in field crops. They are selective, easy to apply, act quickly, are relatively inexpensive, and can be used where other methods do not work well. Nevertheless, their use is limited to defined aquatic sites and geographic areas. Most limitations on herbicide use are because water has multiple uses. Aquatic herbicides are applied directly to emerged plants or to water. Some herbicides can be applied to exposed soil after water is drawn down or removed. In all cases, water's multiple uses must be considered. Table 19.5 shows some herbicides that can be

Table 19.5

Some Herbicides for Aquatic Weed Control

Plant type	Herbicide	Restrictions
Algae—all types	Copper sulfate	Kills trout
	Endothall	Waiting period for fish and domestic use
Sumerged and submersed plants	Endothall	Waiting period for fish and domestic and livestock consumption
	Diquat	Minimum 2-week wait for all uses
	2,4-D	Cannot be used in irrigation, stock, or domestic water
Emergent	Glyphosate	Foliar application only; cannot be applied to water
Floating plants	Diquat	Minimum 2-week wait for all uses

used in water and includes a brief comment on use restrictions. Label and local recommendations must be consulted prior to use. There are very few herbicides for aquatic use, and restrictions limit use of approved herbicides.

McNabb and Anderson (1985) developed a decision-making scheme for integration of methods. The first question in all cases is "What will be the use of the water?" The answer determines control options, and other questions can only be asked after the first has been answered.

McNabb and Anderson (1985) divide water into three categories: industrial, potable or recreational, and irrigation. Their scheme integrates all control methods through a series of logical questions regarding the possibility of nutrient control, drawdown of water level, and control of downstream flow through ponding or temporarily holding water. Their scheme is presented for one situation in Table 19.6.

It is obvious that compared to terrestrial crops, herbicide choices for aquatic weed control are limited. Some aquatic weeds can be managed by skillful combination of preventive and control techniques in an integrated system.

G. SUMMARY

There are few fully integrated weed management systems. Each developing system must be adapted to local environmental, economic, and farming realities, and therefore, no single system will be appropriate for a crop everywhere it is grown. For many years, herbicides will continue to be

Table 19.6

Decision Scheme for Control of Attached Submersed Weeds where Water is Potable

Potable water

        Can nutrients be reduced?

Yes		No
Can water be held temporarily?		Is drawdown possible?
Yes	No	Excavate
		Dredge
Compatible herbicide	Herbicide with potable	Foliar herbicide for plants
Excavate	use	with floating leaves
Dredge	Mechanical removal	
Line pond bottom	Herbivorous fish	**Yes**
Foliar herbicide	Reduce light	
Soil incorporated herbicide	Plant beneficial	Soil-incorporated herbicide
	competitive plants	Line bottom

important components of most weed management systems. Their use may be reduced as integrated systems become more common and effective. Research is being done to develop effective integrated weed management systems that minimize cost, optimize weed control and are sustainable with changing economic conditions. Although these systems will not solve all weed management problems, they will stabilize weed populations at a low level by employing an array of control techniques.

Systems will evolve and change over time because of failure to prevent invasion by new weeds, development of resistance to one or more control techniques, and development of a population not controlled and, in fact, favored by a given management system. Weeds will never be eliminated. They can be managed.

THINGS TO THINK ABOUT

1. What are the basic weed management techniques that should be considered for weed management systems?
2. Describe the components of a good weed management system.
3. Can you design a weed management system for a crop of your choice?
4. What things other than weeds must be considered in the design of a weed management system?

5. What information is essential to create better weed management systems?
6. Explain the role of biological control in present weed management systems.
7. Explain the role of herbicides in present weed management.
8. Explain the role of mechanical methods in weed management systems.
9. How can cultural control techniques be incorporated in weed management systems?

LITERATURE CITED

Aldrich, R. J. 1984. "Weed–Crop Ecology: Principles in Weed Management." Breton Publishers, North Scituate, Massachusetts.

Anderson, R. L. 1994. Management strategies for winter annual grasses in winter wheat, pp. 114–122 in Proc. MEY Wheat Management Conf. PPI and FAR (L. S. Murphy, Ed.). Denver, Colorado.

Appleby, A. P., P. D. Olsen, and D. R. Colbert. 1976. Winter wheat yield reduction from interference by Italian ryegrass. *Agron. J.* **68:**463–466.

Beck, K. G. 1990. Weed management on rangeland and pastures. Service in Action, Colorado State Univ. No. 3. 105. 2 pp.

Beck, K. G., and J. R. Sebastian. 1993. An integrated Canada thistle management system combining mowing with fall-applied herbicides. *Proc. West Soc. Weed Sci.* **46:**102–104.

Blackshaw, R. E., E. H. Stobbe, and A. R. W. Sturko. 1981. Effect of seeding dates and densities of green foxtail (Setaria viridis) on the growth and productivity of spring wheat (Triticum aestivum) *Weed Sci.* **29:**212–214.

Coffmann, C. B., and J. R. Frank. 1992. Corn–weed interactions with long-term conservation tillage management. *Agron J.* **84:**17–21.

Elmore, C. L. 1985. Ornamentals and turf, pp. 387–397 in. "Principles of Weed Control in California." Thomson, Fresno, California.

Fischer, B. B., E. A. Yeary, and J. E. Marcroft. 1985. Vegetation management systems, pp. 213–228 in. "Principles of Weed Control in California." Thomson, Fresno, California.

Forcella, F., and M. J. Lindstrom. 1988a. Movement and germination of weed seeds in ridge-till crop production systems. *Weed Sci.* **36:**56–59.

Forcella, F, and M. J. Lindstrom. 1988b. Weed seed populations in ridge and conventional tillage. *Weed Sci.* **36:**500–503.

Fryer, J. D. 1985. Recent research on weed management: new light on an old practice, Chapter 9 in "Recent Advances in Weed Research" (W. W. Fletcher, Ed.). Gresham Press, Old Working, Surrey, U.K.

Hansen, G. W., F. E. Oliver, and N. E. Otto. 1983. "Herbicide Manual." Water Resources Tech. Pub. U.S. Dept. of Interior. Bur. Reclamation, Denver, Colorado. 346 pp.

Harris, P., P. H. Dunn, D. Schroeder, and R. Vormos. 1985. Biological control of leafy spurge in North America, pp. 79–82 in "Leafy Spurge Monograph No. 3" (A. K. Watson, Ed.). Weed Sci. Soc. Am., Champaign, Illinois.

Helmers, G. A. 1986. An economic analysis of alternative cropping systems for East Central Nebraska. *Am J. Alternative Agric.* **1:**153–158.

Jackson, W. 1984. Toward a unifying concept for an ecological agriculture, pp. 209–221 in "Agricultural Ecosystems" (R. Lowrance, B. R. Stinner, and G. J. House, Eds.). Wiley, New York.

King, R. P., D. W. Lybecker, E. E. Schweizer, and R. L. Zimdahl. 1986. Bioeconomic modeling to simulate weed control strategies for continuous corn (Zea mays). *Weed Sci.* **34:**972–979.

Leininger, W. C. 1988. Non-chemical alternatives for managing selected plant species in the western United States. Colo. State Univ. Coop. Ext. Ser. Pub. No. XCM-118. 40 pp.

Lybecker, D. W., E. E. Schweizer, and P. Westra. 1991. "Computer Aided Decisions for Weed Management in Corn," pp. 234–239. West. Agri. Econ. Assoc., Portland, Oregon.

Lybecker, D. W., E. E. Schweizer, and P. Westra. 1992. Reducing herbicide loading in corn with weed management models. Abstracts, Div. Environ. Chem., Amer. Chem. Soc., San Francisco.

Martin, M. A., M. M. Schreiber, J. R. Riepe, and J. R. Bahr. 1991. The economics of alternative tillage systems, crop rotations, and herbicide use on three representative East-Central corn belt farms. *Weed Sci.* **39:**299–307.

McNabb, T., and L. W. J. Anderson. 1985. Aquatic weed control, pp. 440–455 in. "Principles of Weed Control in California." Thomson, Fresno, California.

Mitich, L. W. 1991. Alfalfa, pp. 232–237 in. "Principles of Weed Control in California." Thomson, Fresno, California.

Shaw, W. 1982. Integrated weed management systems. *Weed Sci.* **30,** Supp. 2:2–12.

Smith, A. E. (Ed.). "Handbook of Weed Management Systems." Dekker, New York. 741 pp.

Strobel, G. A. 1991. Biological control of weeds. *Sci. American* July: 72–78.

Vengris, J., and W. A. Torello. 1982. "Lawns–Basic Factors, Construction and Maintenance of Fine Turf Areas," 3rd. ed., pp. 133–167. Thomson, Fresno, California.

Weiner, J. 1990. Plant population ecology in agriculture, pp. 235–262 in "Agroecology" (C. R. Carroll, J. H. Vandermeer, and P. M. Rosset, Eds.). McGraw-Hill, New York.

Wicks, G. A., and D. E. Smika. 1990. Central great plains, pp. 127–157 in "Systems of Weed Control in Wheat in North America" (W. W. Donald, Ed.). *Weed Sci. Soc. Am.,* Champaign, Illinois.

Chapter 20

Weed Science: The Future

FUNDAMENTAL CONCEPTS

- Weed science has a rich, productive history, and weed management holds great promise for the future.
- There are many research areas in weed science that will create a new and challenging future.

LEARNING OBJECTIVES

- To know the promising areas for weed science research.
- To know that weed science is an evolving discipline.
- To understand that research opportunities in weed science and the science's potential contributions to food production and alleviation of world hunger are great.
- To understand the opportunities and problems related to use of biotechnology in weed science.
- To understand the opportunities and problems of transgenic crops in weed science.

Those who have read this far know the end is near, but the end of the book is not the end of the story. This book has presented a short, accurate history of weed science and described the present situation. Comments about the future are more difficult. The Danish humorist Victor Borge said, "Prophecy is a difficult business, especially of the future." However, when the past is known and the present is understood, it is easier to glimpse the future. What follows will be a glimpse, because the future cannot be predicted accurately, although some of it can be seen. Weed science is young, but it has an enviable record of achievement, and a good future.

The practice of weed control was a recognition of necessity by farmers who had been controlling weeds long before herbicides were invented. The advent of herbicides changed the way weed control was done, but did not change its fundamental purpose: to maintain or improve yield of desirable species. Herbicides replaced human, animal, and mechanical energy with chemical energy. No other method of weed control, before herbicides, was as efficient at reducing the need for labor, or as selective. People with hoes could discriminate between weeds and crops: they could weed selectively. Mechanical and cultural methods were not selective enough. Herbicides prevented weeds, reduced the growing population, and selectively removed weeds from crops. Other methods could do these things, but not as well or as easily. Weed control in the world's developed countries now depends on herbicides as the primary technology. This situation will prevail well into the 21st century.

I. RESEARCH NEEDS

There are at least three important problems that have hindered, and may continue to hinder, progress in weed science. The first problem is the assumption that anyone can control weeds. Those who make this assumption understand neither the complexity of weed problems nor their solutions. They do not know how much specialized, sophisticated knowledge is required to control weeds correctly. Those who have studied weeds and their control know how wrong this assumption is.

The second problem is that weeds are such steady components of the environment. They lack the appeal and urgency of sudden, serious infestations of other pests. Other pests *are* serious, but that does not mean they are more serious than weeds or deserve more attention. If a problem is always there, it will not receive the attention or funding that new, obvious, but perhaps no more serious problems receive. A good analogy can be drawn with world hunger. Famines receive the greatest attention, but persistent, widespread hunger and malnutrition are the most serious aspects of world hunger.

The third problem is a lack of people and research funds. A great deal of research is done by large herbicide development and sales corporations. It is good work, but it is inevitably oriented toward sales. Research on weed biology, ecology, seed dormancy, and other problems that lead to basic understanding rather than immediate control is done by too few scientists. Public funding of agricultural research is not growing. There continues to be a rapid increase in the research capability of private-sector corporations, and the publicly funded agricultural research and development system is and, it seems, will remain a minority component. Private-sector research increasingly induces the nature of public-sector research, rather than the opposite (Buttel, 1985). Thus, weed science has few scientists, and too few of those are working on long-term biological approaches to weed management.

This chapter will deal with the future of weed science in terms of research needs rather than in terms of what will be accomplished. The chapter is a narration of what ought to be done. It may not be exactly what will be done, because research does not always follow a straight path and other developments may change what ought to be done. For example, restrictive environmental legislation that reduced herbicide availability could rapidly change the way agriculture is practiced. A description of research needs is a safer prophetic stance. Describing what ought to be done, instead of describing the situation several years hence, reduces the possibility that the prophet will be wrong.

A. Weed Biology

1. Weed Biology and Seed Dormancy

Weed scientists know that dependence on herbicides is equivalent to treating the symptoms of a disease without actually curing the disease, but there has been little choice. Agriculture would be far better served if weed scientists learned how to control weed seed dormancy and seed germination so weed problems could be prevented, rather than controlled as they appeared. No one knows enough about weed seed dormancy, and much research remains to be done to know enough.

Empirical herbicide testing has made it easy to control weeds without studying their biology. Any attempt to control must know what weeds are to be controlled and where they are growing. That is, control is not blind. There is an object to be controlled, and it is known. With herbicides, however, it has been necessary to know something, but not much, about the weed. In general, herbicide development has not been based on knowledge of weak points in a plant's life cycle or has exploited specific physiological knowledge for control purposes. The safest approach has been to aim for complete control of weeds in a crop.

As knowledge grows, however, scientists find that some weeds are not injurious in some crops and control is not necessary. Some plants now considered weeds may be beneficial and should not be controlled. To acquire this knowledge, the entire life cycle of a weed, including knowledge of the dormancy of its seeds and vegetative reproductive organs, must be known. A marvelous series of projects could be developed on the biology of perennial weeds. These projects should include those mentioned by Wyse (1992):

Regulation of weed seed dormancy
Regulation of bud dormancy of perennials
The development and loss of reproductive propagules of perennials
Weed population genetics
Modeling of crop-weed systems

A number of research programs are oriented toward modeling crop-weed systems. The available models take several forms. One combines effective use of several tillage implements with computer technology to find weed control strategies that minimize herbicide use. Models routinely include knowledge of the size of the soil seedbank, rate of seed emergence, and seedling survival. It is only logical to assume that control and management methods will improve as knowledge of weed biology and seed dormancy improves.

2. Weed-Crop Competition and Weed Ecology

Much of the basic information required to develop computer-based models (see Section D, this chapter) of weed-crop systems that lead to best use of available weed control techniques has come, and will continue to be derived, from weed biology and ecology research. What plants compete for is known, but when competition is most severe or whether competition for a resource is equally intense at various stages of crop and weed development are not known. The old, but still-used, period threshold concept of weed competition (Dawson, 1965) suggests that weed competition is time dependent. Weeds in the crop early in the season are less detrimental than those that compete with the crop later on. This principle led to timely use of herbicides and other techniques for weed management. Some crop cultivars are more competitive than others, and this needs to be considered in developing weed management systems and as a basis for cooperative work with plant breeders.

Weed populations change with time, and the reasons are beginning to be understood. One reason is the development of weeds resistant to some herbicides, often after less than 5 years' use in one field (see Chapter 11). Active research on why resistance occurs is coupled with development of techniques to combat it. The chances of selection for resistance are increased when a persistent herbicide with a single site of action is used for several years. When resistance has developed, it has not led to totally unmanageable weed populations because other weed control techniques (e.g., cultivation or crop rotation) are available. Other reasons weed populations change with time are found in the study of weed ecology; a field still in its infancy. Understanding why populations change and the weed management implications of population shifts is important to development of weed management systems.

Even the casual observer of the world of weeds will recognize that weed problems change. The weeds most difficult to control in most crops today are not those that were discussed and worked on 10 or 20 years ago. That is evidence that weed scientists have developed successful solutions to weed problems. It is also evidence that nature abhors empty niches. When successful control efforts have reduced the population of a species, they have inevitably left space unoccupied and resources unused. Other species move into empty niches created by successful weed control.

Solutions to this dilemma take two forms. The first solution is to reduce the attractiveness of the niche. Farmers typically overprovide for crops. Fertilizer placement and precise recommendations have reduced surplus nutrients, but whole fields are irrigated and light cannot be controlled. If water could be placed as precisely as fertilizer and only what was needed was provided, the attractiveness of the niche and the number of potential invaders could be reduced. This is a preventive approach to weed management. The second approach has an element of prevention. No one has looked carefully, but some of the important problem weeds of the next decade are already in the fields or lurking on the edges. If they were identified and their weedy potential determined, weed scientists could try to predict which ones would be the successful invaders and they could be controlled or managed before they invaded. More basic biological-ecological knowledge is essential to either approach.

B. ALLELOPATHY

Allelopathy could be discussed with herbicide technology (the following section). It is a well-known plant phenomenon that has been studied for a long time, but has not been commercially exploited (see Chapter 7). It is intriguing that plants have natural chemical defenses that could be discovered and exploited as herbicides. It is one of those things that seems to be forever just beyond our grasp. Maybe there has been too little research. Maybe the responsible chemicals are very common in nature, not selective enough, or not active enough for commercial use. Maybe some observations are not really allelopathy at all. The lure remains. It will take time and good research to discover if a natural herbicide (an allelochemical) can be found and developed for commercial use.

C. BIOLOGICAL CONTROL

What if the agrichemical industry had never developed? Just suppose that someone had discovered an organism in 1944 that selectively controlled annual broadleaved weeds in small grains. Would we use as many herbicides as we now do? Would our environmental problem be what to do with mutating organisms rather than polluting chemicals? Interesting questions make nice debate topics, but they do not always solve problems. Biological control is still in its infancy compared to other control methods (see Chapter 10). Its theoretical potential, unrealized in all but a few cases, has had little effect on integrated weed management systems for agronomic or horticultural crops. Future research may discover specific biological control organisms and combinations of organisms that are effective and safe and can be integrated with other methods of weed management in crops. There are not many now.

D. WEED CONTROL AND BIOECONOMIC MODELS

Rotation of crops is an effective way to reduce weed competition and will be a more important part of future weed science research. Future rotations will include use of smother crops and green manure crops to keep land covered and protected from wind and water erosion more of the year. It is well known that rotations reduce soil erosion and manage weeds, but the research has not been done so that one can say exactly what weeds will be managed effectively and how rotations can be used to greatest advantage. Rotational research will have to include research on competitiveness of crop cultivars and weeds.

Cultivation has been part of weed management for years, but it has often been employed without knowing its full effects in integrated management systems. Cultivation will be part of many weed management systems, but its potential encouragement of soil erosion will have to be managed. Research will have to be done to determine the effect of cultivation timing with different types of implements on specific weeds. Much of the work will emphasize modeling of the crop–weed system to determine optimum timing for different weed management techniques during the crop's life (Wyse, 1992).

In the early 1980s, two decision-aiding, computer-based models were available. By the early 1990s, 21 mostly crop-based decision-aiding software models for weed control management were available to researchers and farmers (Mortensen and Coble, 1991).

Models used to predict crop yield losses due to weed competition (Weaver, 1993) have been divided into two groups. The first and largest group (type a) is based on soil weed-seed populations and selects mostly preemergence control strategies (dominantly herbicides) based on efficacy and optimizing weed control. The second group (type b) selects mostly postemergence control methods based on cost/benefit consideration after considering possible crop losses based on estimates of soil seedbank and weed seedling populations. Type b models include estimation of economic benefit by considering herbicide or other weed control costs and potential crop loss. In the early 1990s, 16 type a and 5 type b models were available. Models of either type can choose only preemergence methods *or* combine those with postemergence methods. The latter are not applied until weed problems are defined by field observations. Available models will be refined and other models will be developed. Some scientists suggest the ideal is a mechanistic weed–crop competition model that considers and responds to changing environments. The essential knowledge of weed biology to construct such models is not yet available (Schweizer *et al.,* 1996). In most models, weed density is the sole variable used to estimate crop yield loss. Future models will incorporate a variable for relative time of crop and weed(s) emergence, crop density, climate variation, method and amount of fertilizer applied, and weed density (O'Donovan, 1996). Rather than just predicting crop loss, as important as that is, future models will also enable realistic monitoring of weed population development and long-term implications of failure to control, or, to put it another way, the long-term implication of seed production by uncontrolled weeds (O'Donovan, 1996).

As more biological knowledge of weeds becomes available, as weed seed sampling techniques improve, and as models improve, they will be used by decision makers. Weed management decisions guided by models will lead away from prophylactic control methods. Profitability and weed control efficacy will continue to be important criteria. As new knowledge is incorporated, computer-based weed management decision-aid models will provide greater assurance of achieving profitability and appropriate long-term weed management. Some of the knowledge required to develop better models is shown in Table 20.1.

E. Herbicide Technology

Herbicides are the most successful weed control technology ever developed. They are selective, are not too expensive, are fairly easy to apply,

Table 20.1

Some Examples of Knowledge Required to Develop Improved Weed Management Systems and Decision Aid Models (Buhler *et al.*, 1997)

Management goal	Research needed
Management decision aids = models	Relationship of the size of the weed seedbank to final weed population
	Emergence rate of individual weed species
	Determination of economic optimum thresholds for control
	Interaction of management practice and weed seed production
	Effect of weed density on control
Prediction of seedling emergence	Mechanisms of dormancy
	Determination of interaction of environmental conditions, seed germination, and dormancy
Effect of management on weed seed	Effect of crop rotation on weed seedbank size
	Effect of living and dead mulches
	Rate of seed predation and decay
	Rate of seedling mortality
	Light requirement for seed germination
	Role of tillage and cultural practices

and have persistence that can be managed; also, many formulations and kinds are available. In spite of their many advantages, however, herbicides are far from perfect even in the eyes of their staunchest advocates. We do not have ideal herbicides. The ideal herbicide would be all of the following:

(a) Effective on a spectrum of weeds, or able to control a single species selectively
(b) Very selective in at least one of the major crops
(c) Nontoxic to other species, not just to humans
(d) Persistent in soil but not beyond the period of intense weed competition
(e) Easily degraded to innocuous substituents by soil microorganisms or nonenzymatic soil processes
(f) Applied postemergence to avoid the prophylactic nature of preemergence use
(g) Active at very low rates (mg or g/ha = ounces or fractions of an ounce per acre)
(h) Nonleachable and nonvolatile

There are some herbicides that meet nearly all of these criteria, but none are perfect, and much remains to be done to develop better herbicides.

1. Study of Plant Biochemistry and Physiology

It is known that dichlobenil and isoxaben affect a plant's ability to synthesize cellulose. However, neither was developed to do so. It is serendipitous that they interfere with a specific plant process. There are four ways to discover new herbicides (Beyer, 1991; Evans, 1992):

Method 1. *Random selection of chemicals submitted to targeted biological screening tests.* This method relies on carefully developed, targeted biological screening techniques to detect chemicals with activity and selectivity. Candidate chemicals are obtained from a company's several divisions or purchased. This is often referred to as the "blue sky" approach (Evans, 1992).

Method 2. *Screening of chemical derivatives of herbicides with known activity.* Once activity is found in method 1, derivative or analog development ensues and further screening attempts to gain more activity or greater or different selectivity.

Method 3. *Development of leads taken from natural products that display biological activity.* Nature can be viewed as an intense arena of complex chemical activity. Allelopathy is only one arena of chemical activity. Biologists and chemists can find clues in this chemistry that could lead to successful herbicides. There have been few successes (Evans, 1992). Natural products are inherently complex and the chemicals frequently have insufficient potency or the wrong activity or selectivity to provide strong leads for chemical synthesis (Evans, 1992).

Method 4. *Rational design based on biochemical principles and knowledge of plant physiology and biochemistry.* The most intellectually attractive concept is biorational design. The organism to be controlled is considered as a target, and an enzyme essential to its survival is chosen for direct attack. Candidate chemicals are developed to inhibit an essential plant function, based on biochemical knowledge of the target site. To date there has been little success, but the technique is advancing rapidly (Evans, 1992).

Scientists know how to use all four techniques. In practice the first two have provided almost all presently available herbicides. Method 1, in which a large number of chemical compounds is screened for possible activity, is the most common method of herbicide discovery. The screening process includes structural relatives of compounds with known activity and chemical structures with unknown activity.

Method 4, biorational design, will become increasingly feasible and useful. Combination of biological performance, toxicological properties, and environmental behavior with computer analysis will make this a powerful technique for herbicide development. The broad screen (method 1) has been very successful, but the herbicide industry would like to base future screening programs on greater understanding of plant biochemistry and physiology (method 4). It is a certainty that future screens will more precisely target specific plant processes as the biochemistry of those processes becomes known.

2. Rate Reduction and Precise Application

Sulfonylurea and imidazolinone herbicides (see Chapter 12) reduced herbicide rates by an order of magnitude, from kilograms to grams per hectare (pounds to ounces per acre). That is a desirable direction from an economic and environmental perspective. Further reductions are possible when there is more information on threshold levels for phytotoxic activity. Most herbicides are now broadcast over the entire target area. Some are banded over the plant row. These methods require enough herbicide to satisfy the soil's adsorptive requirements and must account for dilution in soil. When knowledge of precise thresholds is combined with precise application, a further reduction in rate will be achievable. The essential question is, what is the minimum amount that can be applied to achieve the desired control level?

3. Soil Persistence and Controlled Soil Life

It is at once a major problem and a significant advantage that herbicides persist in soil. Excessive persistence may affect succeeding crops, lead to contamination of ground or surface water, or cause undesirable residues in crops. Some persistence is good because it gives weed control over time and avoids the need for repeated herbicide applications or for using other weed control methods in the crop. Regulating or controlling soil life is a desirable goal for future herbicide development. If nonpersistent herbicides could be given a few weeks of persistence and the soil life of herbicides with long persistence but desirable activity and selectivity could be shortened, it would be good. If these things are achieved it will probably be through controlled-release formulations that have already had some success (e.g., encapsulated alachlor). The technology for these formulations is a special challenge in the complex soil environment.

4. Safeners for Herbicides

This technology has already been developed for acetamide and carbamothioate herbicides (see Chapter 12). It may be developed for other herbicides unless, as seems likely, it proves easier to achieve the same thing through genetic engineering (see Section I,E,6, below). The principle is simple. If a chemical can be found that interferes with herbicide activity in a crop in which the herbicide is normally not selective—selectivity will be achieved. It sounds easy. Empirical screening has been used to find these chemicals but the search would be easier if the herbicide's precise mode-of-action was known. The task is confounded by the necessity of achieving selectivity in crops and not losing activity on weeds.

5. Formulation Research

A few decades ago, only a few formulations were available. Most herbicides were emulsifiable concentrates, solution concentrates, or wettable powders. A few granular formulations were available, but application technology limited their use. Today, formulation chemists have reduced dust, foaming, and storage problems, made handling easier and safer, and improved efficacy. Users can choose from all the previous formulations and several improved ones. Further improvement will occur to make formulations safer and easier to use. Controlled-release formulations could reduce volatility, leaching, and use rates and increase selectivity. They could give nonpersistent herbicides a desirable soil life without fear of residual carry-over to the next crop.

6. Biotechnology and Herbicide-Tolerant Crops

More than 25 companies now have research programs to incorporate herbicide tolerance in crops (so called transgenic crops). Success has been achieved with about 15 herbicides, and most of the work has focused on major crops: corn, soybean, wheat, rice, cotton, and tobacco (Duke *et al.,* 1991). The technology for agricultural crops was introduced in the mid-1980s. From then to 1994, more than 1,500 approvals for field testing of a wide range of transgenic organisms were granted, and 40% of them were for herbicide tolerance (Hopkins, 1994). Because the technology has been available for some time, it is not unreasonable to ask why this section is included in a chapter on the future of weed science, rather than in the chapter on herbicides. The reason is that although we know what has been done with herbicide-resistant crops, the technology is so new and changing so rapidly that we do not know what might be done. That is, the direction of research is clear, the destination is not. We cannot be sure what new possibilities will be discovered as the technology of herbicide resistance continues to develop rapidly.

There are three physiological mechanisms for natural or induced tolerance or resistance to a herbicide:

1. Reduced sensitivity at a molecular site of action
2. Increased metabolic degradation
3. Avoidance of uptake or sequestration (hiding) after uptake (Duke
    et al., 1991)

Each of these has potential use in development of resistance in crops. Several examples of presently available, herbicide-resistant commercial cultivars are shown in Table 20.2. Roundup Ready™ soybeans and canola have achieved commercial success in the United States and Canada. Other glyphosate-resistant crops (e.g., corn and cotton) are being developed by Monsanto.

Criticism of herbicide-resistant crops is common and is usually related to all or some of four perceived risks:

1. *Public health.* Concern has been expressed about water or food
    contamination from increased herbicide use. Additional concern
    centers on use of herbicides in crops that do not metabolize the
    herbicide, meaning that the unaltered herbicide could be consumed
    by people.

Table 20.2

Development and Release of Herbicide-Tolerant Crops (Duke *et al.,* 1991; Hopkins, 1994)

Herbicide	Crop	Released in
Atrazine	Canola	Canada
Bialaphos	Corn, soybean	
Bromoxynil	Cotton, tobacco	
Glufosinate	Alfalfa, sugarbeet, tobacco	
	Oilseed rape	Canada—1995
	Corn	U.S.—1996
Glyphosate	Poplar, tomato	
	Canola	Canada—1995
	Soybean	U.S.—1995
	Cotton	U.S.—projected 1997
	Corn	U.S.—projected 1998–99
	Sugarbeet and oilseed rape	U.S., Canada—projected 2000+
Imidazolinones	Canola	Canada
	Corn	U.S.—1992
Sulfonylureas	Tobacco, canola	
	Soybean	U.S.—1993
2,4-D	Tobacco	

2. *Environmental concern* related to increased herbicide use.
3. *Social concern* related to the following:
    (a) Fear that the technology will favor large farms and lead to loss of more small farms and small-scale farmers
    (b) Fear that the cost of food production and thus food cost to the consumer will rise
4. *Weed control concerns* related to the following:
    (a) *Development of herbicide resistance.* Herbicide resistance among weeds may become more widespread because of continued use of a herbicide to which a crop is resistant.
    (b) *Resistant gene flow to sexually compatible plants.* This is acknowledged as one of the greatest potential risks of introducing any genetically engineered (transgenic) crop variety. The risk is transfer of desired herbicide resistance from the crop to a weed where undesirable resistance persists by natural selection. It is worth noting that this has happened when genes from herbicide-resistant canola moved to a nonweedy relative in the mustard family and then to wild mustard in a short time.[1] The risk may be especially high where the crop and weed are closely related and can interbreed: for example, red rice and rice or johnsongrass and grain sorghum.
    (c) *Resistant crop plants becoming hard-to-control volunteer weeds.* This has not been shown, but Keeler (1989) urged caution and pointed out the example of wild proso millet, which emerged as a weed in the 1970s, after more than 200 years in which proso millet had been successfully cultivated in North America without becoming a weed. Keeler (1989) used wild proso millet to emphasize how much we do not understand about weed evolution.

The quite legitimate concerns of epistasis and pleiotropy must also be recognized. They are similar to the concern in point 4c above, but not identical. Epistasis is the suppression of gene expression by one or more other genes, and pleiotropy is defined as a single gene exerting simultaneous effects on more than one character. In short, one of the rules of ecology may apply: you cannot do just one thing. When science manipulates a genome, any genome, specific outcomes are intended, and even when these are achieved, other, unplanned, things may also occur. Genetic engineering, with the best intention to do a good thing, may do unexpected things that could be good or bad.

[1]*Denver Post,* April 14, 1996, and *The New York Times,* March 7, 1996.

Another common critique of herbicide-resistant crops is that the technology will promote the use of herbicides (item 2 above), not decrease it, while continuing to develop what many view as an unsustainable, intensive monocultural agriculture. It is also suggested that herbicide-resistant crops will reinforce farmers' dependence on outside, petroleum-based, potentially polluting technology. An associated concern is that there is no technical reason to prevent a company from choosing to develop a crop resistant to a profitable herbicide that has undesirable environmental qualities such as persistence, leachability, or harm to nontarget species. It is undoubtedly true that nature's abhorrence of empty niches will mean that other weeds will move into the niches created by removal of weeds by the herbicide in the newly resistant crop. In other words, herbicide resistance will solve some, but not all, weed problems. Weeds not susceptible to the herbicide to which the crop is resistant will appear. Weeds are not conscious, but they seem to be clever.

Development of herbicide-resistant crops is proceeding rapidly, and there are important advantages that provide good reasons for their development. Many argue that the technology will provide lower-cost herbicides and better weed control. These are powerful arguments in favor of the technology because both can lead to lower food costs for the consumer. It is also true that herbicide-resistant crops are providing solutions to intractable weed problems in some crops. Glyphosate resistance has been created in several crops. Glyphosate is an environmentally favorable herbicide, and therefore, many argue, it is better to use it in lieu of other herbicides that are not environmentally favorable.

An important argument in favor of the technology is that it has the potential to shift herbicide development away from initial screening for activity and selectivity and later determination of environmental acceptability to the latter occurring first. Resistance to an herbicide that is environmentally favorable, but is not selective enough in any crops or in a major crop that its development would be profitable, could be engineered and the herbicide's usefulness could be expanded greatly. This has important implications for minor crops (e.g., vegetables, fruits) where few herbicides are available because the market is too small to warrant the cost of development. If resistance to a herbicide already successful in a major crop (e.g., cotton) could be engineered into a minor crop, manufacturers and users would benefit.

Biotechnology was discussed by Christiansen (1991), a self-acknowledged outsider, and his view is quoted here to present an alternative view of this research area:

> I think it would be a pity if the power of the use of mutants and mutation
> to uncover and describe physiology and development were limited, in the hands

of weed scientists, to the isolation and description in yet another species of yet more genes that confer resistance to yet another herbicide. To this outsider, it seems that the central issue for weed science is understanding the nature of weeds: What makes a weed a weed? [See Chapter 8, this book.] How can weeds consistently come out ahead when matched up against the finest commercial varieties my plant-breeding colleagues develop? Weeds persist, they spread, and they out compete the crop plants, reducing yields when left uncontrolled. The nature of this "competitive ability" that weeds possess seems an interesting target for research and an appropriate target for analysis through generation of mutants.

Transgenic crops are a controversial research area that is developing rapidly (see Hileman, 1995, for a summary of the controversy). It is not the purpose of this text to analyze the controversy in depth. A complete book (Duke, 1996) is available, as are articles too numerous to mention here. Much more work will be done and discussed, but it is important to realize that the technique is already widely promoted, accepted, and used.

7. Perennial Weeds

There are some herbicides that control perennial broadleaved or grass species. Perennial members of the *Cyperus, Cynodon, Sorghum, Cirsium, Convolvulus,* and *Eltrygia* genera are not generally controlled selectively by available herbicides. Few herbicides that are sufficiently active on perennial weeds are also selective enough to be used in most crops, and several persist in soil. It is relatively easy to control the emerged shoots of perennial weeds, but far more difficult to ensure translocation of an adequate amount of the herbicide throughout the extensive root, root runner, rhizome, or stolon system of a perennial weed. Selectivity is a greater problem than activity. A continuing problem for turf managers is selective control of coarse-leaved perennial grasses (e.g., quackgrass or tall fescue) from fine-leaved turf species (e.g., Kentucky bluegrass).

8. Aquatic Weeds

The difficulties of controlling aquatic weeds are related to their habitat, not their life cycle. Seventy percent of the earth's surface is covered with an interconnected water system. All waters flow to the sea, and all contaminants can be carried along as dissolved solutes or adsorbed to eroded soil. Almost all water contamination is, in the minds of most people, unacceptable. Because of heightened and enlightened concern about environmental quality, its unacceptability has resulted in legislation to prevent, control, and punish water pollution. Only nine herbicide active ingredients can be used in aquatic systems. Present and future herbicides must be compatible with all other actual and potential uses for water. This is a most difficult requirement, and extensive, expensive research will be required if acceptable weed management techniques are to be developed. In the aquatic environment, it is likely that proof of safety will be demanded by an anxious public. Reasonable assurance or reasonable doubt will not suffice. Because of these appropriate concerns, the aquatic environment is a likely site for development of nonchemical control techniques.

9. Parasitic Weeds

Parasitic weeds are not major problems in the United States, but they are in the world's developing countries. Few selective herbicides or other control techniques are available. Their unique habitat was discussed in Chapter 3. Research has not focused on them because they are very difficult problems, and because herbicide development and developers have concentrated on large-acreage crops of the developed world. It takes only a little experience with parasitic weeds to recognize how devastating they can be and how large crop yield losses can be. Third-World farmers stop growing susceptible crops, abandon fields, and are often defeated by parasitic weeds. Developing reliable, affordable management techniques for parasitic weeds would be a significant scientific contribution and a major contribution to development of agriculture in the world's poor countries.

10. Packaging and Labeling

Major herbicide manufacturers have developed safe packaging and are concerned about personal safety of users. Systems that minimize human exposure are available and will become more common. Manufacturers and users are working together to minimize the hazards of herbicide handling and container disposal. Safe, efficient herbicide delivery systems will be needed, and they will demand adaptations in formulations and application methods.

Herbicide labels and use instructions in the world's developed countries are explicit, are readily available, and contain adequate instructions for all approved uses. Sadly, the same is not true in much of the rest of the world. Manufacturers are well aware of the problem. Labels have to be developed that are clear, simple, adequate to the task, but not too complex. When potential users may be illiterate, clear instructions are imperative. It is a difficult challenge to create instructions that combine the need for clarity and accuracy in view of the growing complexity of herbicide chemistry and use.

11. The Agricultural Chemical Industry

Innovation and progress in herbicide development depend on the agrichemical industry, where consolidation has led to domination by a few large, world-oriented European and U.S. companies. The industry is growing less rapidly than it once did because of greater market saturation in the developed countries and limited prospects for quick expansion in the large, diffuse market in the world's developing countries. The development of no-tillage and minimum tillage, other evolving soil tillage practices, and the changing weed spectrum in crops will inevitably affect the herbicide market.

After 1986, when Du Pont bought Shell's pesticide division, the agricultural groups of Dow and Eli Lilly merged their agricultural chemical divisions into a new company—DowElanco. The French firm Rhone-Poulenc bought Union Carbide's agrichemical business, which had previously absorbed AmChem (which previously had purchased the herbicide business begun by Sherwin-Williams Paint Co.). The American division of Britain's Imperial Chemicals Co. (ICI) bought Stauffer Chemical group from Cheeseborough-Pond, and the Swiss company Sandoz bought VS Crop Protection (Velsicol). The most recent merger was the combination of Sandoz Agrochemical and Ciba to form Novartis. To the outsider it seemed that boards of directors must have said, let's get big or get out. There has been no negative effect on the availability of herbicides. It remains for historians to ascertain the effect of these mergers on agriculture. There is a downward price pressure due to patent expirations, market saturation, and competition.

The cost of development for a single herbicide (which must pay for all failures) is estimated to be greater than $40 million. The agricultural chemicals business is not one for the timid. Costs of herbicide development are increasing and new, profitable herbicides are rarer and more expensive. Biorational approaches that consider biological efficacy, toxicology, and environmental interactions, and presumably save money, will dominate design of new herbicides. These approaches will be combined with more specialized chemical, biological, and safety testing to determine mode-of-action, use, economic benefit, and environmental acceptability.

The green or environmental movement has helped create ever more restrictive herbicide legislation and regulation. Some feel there is more law than science in decisions that affect herbicide use and development and govern environmental decisions. Regulations that are too restrictive suppress herbicide development.

These brief comments on the agrichemical industry are included to give the reader insight into the business side of agriculture, which is often not noticed by those who control weeds. Decisions in the business world may have a much greater effect on future weed management programs and the direction of weed science than anything that occurs in a research laboratory.

F. ENGINEERING RESEARCH

The preceding section described research needs for herbicides. No matter how selective or active, herbicides have to be applied in the environment to be successful. The herbicide in the package is interesting, but not functional. The Weed Science Society of America published a monograph on applying herbicides (McWhorter and Gebhardt, 1987). The monograph points out that over 90% of all herbicides applied annually in the United States were sprayed using sprayers with the same four basic components tank, pressure regulator, pump, and nozzles that were on sprayers when herbicides were first sprayed. The technology has evolved and application equipment is more precise, more durable, and more flexible than older pieces of equipment. But must herbicides be sprayed? Is there a better way? Some of any sprayed herbicide never hits the target. Can that spray be recovered and reused, or at least handled so it does not reside in the environment without fulfilling its intended purpose? Low-volume and ultra-low-volume application techniques are available, but not widely used. There is potential for decreasing the volume of spray required. Sprayers have been developed that can sense weed presence and just spray where weeds are. Other sprayers can sense variations in soil type and adjust herbicide concentration to account for differences. These techniques will be pursued and aided by global positioning technology that will permit repeated application precisely on weed patches. The goal is to improve application accuracy and operator safety while maintaining environmental quality and protecting crop yield.

Herbicide application is not the only area in which engineers can contribute to improved weed management. The effects of tillage over time on weed populations is not known. There is room for improvement in methods of mechanical control of annual and perennial weeds in crops. Cultivators are available to weed the entire area between crop rows and leave only a narrow band over the crop row. This is a vast improvement over a person with a hoe or an animal-drawn cultivator. Further research will reveal even more specialized tillage and cultivation methods for weeds in crops.

Tillage research will also show that much tillage is unnecessary and, in fact, complicates weed management. Tillage exposes weed seeds to sunlight and encourages germination (see Chapter 9). No-till methods leave seeds buried and prevent or inhibit germination. The best tillage for weed management may be none. That does not mean that weeds will not be present. No-till creates a niche for weeds that do well without tillage. It will change, but not eliminate, the need for weed management.

G. Vegetation Management and Integration of Methods

In science, as in most human activities, movements occur, directions change, and progress may result. Some movements are called bandwagons. Each has associated words and phrases that define and identify it. Some call these buzzwords. Each movement makes its contribution to the parade of ideas and contributes to the general cacophony of competing ideas. Some ideas assume a position at the head of the line. It may be too soon to tell if integrated weed management will be just another interesting but temporary movement, or if it is assuming a position of centrality and leadership. Integrated pest management (IPM) has endured as a good alternative to previous pest management systems. It has been challenged because from 1968 to 1992, when U.S. interest in IPM grew steadily, pesticide use in U.S. crops increased 125% (Gardner, 1996). Many see IPM as a buzzword that has not changed pest control. If integrated weed management systems are to endure, change will be required. The direction and scope of change may determine the enduring success of the concept and practice of integrated weed management. Historians tell us, after the fact, what things have endured and why, and why other ideas were only temporary phenomena. Judgments from the present are often flawed because one is so close that subjectivity dominates. The perspective of time is often a prerequisite to objectivity.

Will integrated weed management be regarded as just another buzzword that, to quote Andy Warhol's remark about people, had its 15 minutes of fame? The idea of integrated weed management makes so much sense that it is likely to endure. It is not perfect, but it is better than anything else we have. The evidence in this text is sufficient to demonstrate that weed management systems for crops are incomplete. They are developing, but research gaps preclude defining complete systems (see Table 20.1). Here are a few good research questions:

1. What seeds emerge first from complex soil seedbanks?
2. What percentage of seeds, by species, emerge each year?
3. What is the precise percent control for different weed control techniques for different weeds at different growth stages?

4. If soil seedbank composition is known, can the weed complex be predicted for a crop?
5. How do weed management techniques affect other pests and other pest control techniques?

Integrated control cannot limit its focus to weeds. To be successful, the focus should be the total vegetation complex or better–habitat management, rather than weed control in a crop. Perhaps it is best to say that the scale of concern must change. Sustainable integrated weed management systems will extend concern to environmental quality and future generations. These are large-scale concerns, demanding large-scale thought. Until now, small-scale concerns have dominated, including how to control weeds in a crop this year. Small-scale thought has been enough for such individual concerns. Now, however, large thoughts are needed for large systems. Everything needs to be integrated to have a complete crop management system. That is so easy to say. It won't be easy to do. It *is* necessary.

H. OTHER RESEARCH

There are several other research areas that should be included in planning for weed science. These are less well developed than those that preceded, but may be just as important. They include the following:

1. What is the value, and what are the advantages and disadvantages, of monoculture?
2. What is the role of companion cropping and regular inclusion of cover crops in weed management?
3. What are the long-term effects of soil erosion after regular plowing and cultivation? These are all too apparent in the brown color of a country's rivers. Weed scientists were not too concerned with long-term effects when the science was developing. Weeds decreased crop yield: a detrimental long-term effect. The vision did not extend much farther. Solving the weed problem was a sufficient challenge. Any technology, used for enough time, has demonstrable environmental and social effects. A longer-term view will help reveal these effects and compel their consideration before widespread use is achieved.
4. Weed science must begin to work more closely with economists who can ask, What does it cost and what is it worth? Several reports that do this are cited in this text. What is it worth to do the work to develop a more competitive cultivar, to deplete the soil seedbank, to have assurance of 80% weed control? What will it be

worth to be able to predict weed problems? No one knows, but the answers are important to complete weed management systems.

II. POLITICAL CONSIDERATIONS

Weed scientists and most people engaged in agriculture are not, by nature or choice, good politicians. Most agriculturalists consider a career as a politician to be more noble than being a Mafia don or shoplifter, but probably about on a par with those folks who call you, just at supper time, to persuade you your carpets need cleaning.

However, the failure or inability to consider the fact that we live in a political world and are affected by it is a prescription for disaster. Political considerations affect our lives daily. A major political creation, in many countries, is cheap food, especially in urban areas. Most enjoy the benefit of this, often unstated, government policy. It affects the way we practice agriculture and manage weeds. If the government removed itself from agricultural policy making and the market, weed management systems would have to change to fit a new system.

Concern about environmental pollution is a fairly recent thing on the political scene. It was not too long ago that belching smokestacks meant prosperity and progress, rather than environmental pollution and corporate irresponsibility. For example, a study commissioned by the American Farm Bureau, an organization noted for its defense of agriculture (King, 1991), showed that only 15% of the American public was in favor of abolishing pesticide use in agriculture. However, 66% of the people surveyed thought pesticide use should be limited in the future, and 38% thought farmers were using more pesticides than they had in the past. Such information and concern has political meaning and consequences. Such data are ignored or dismissed only by those who willfully ignore the effects of political action. Political acts change many things, and agriculture has to recognize and work in a political milieu or suffer the consequences of regulation by those who do.

III. CONCLUSION

The American author and farmer Wendell Berry (1981) has written often and eloquently about problems facing American agriculture and about their solutions. He advocates solving for pattern. Berry says that "to the problems of farming, then, as to other problems of our time, there appear to be three kinds of solutions." The first kind causes a ramifying series of new problems. The only limitation on the new problems is that they "arise beyond the purview of the expertise that produced the solution." That is, those who are encumbered by the new problems are not those who devised solutions for the old problem. This kind of solution shifts the burden away from those who created the problem.

The second kind of solution is one which immediately worsens the problem it is intended to solve. These are often quick-fix solutions that take the form of questions such as, What herbicide will kill the weed? Adopting this kind of problem solving leads to the need for more quick-fix solutions. Everyone who has tried to fix something is familiar with this kind of solution. What was tried first didn't work, and some study (but perhaps little knowledge) revealed that loosening another bolt or screw would do it. Alas, loosening that screw was the wrong thing to do, because it loosened other things, and suddenly parts are everywhere and neither the source of each part nor a way to fit them back together is known. These solutions are common.

The third and most desirable solution creates a ramifying series of solutions. Parts do not fly off in all directions; they fit together. These solutions make, and keep, things whole. For Berry (1981), a good solution is one that acts constructively on the larger pattern of which it is a part. It is not destructive of the immediate pattern or the whole. People who devise the best solutions recognize the pattern in which they must fit and work to create a set of solutions that maintains the essential pattern. Good solutions solve for the whole, not for a single goal or purpose.

Those who will create the next generation of weed management systems for simple and complex weed problems will do well to remember Berry's admonition as they search for the best solutions: Know the whole system and devise solutions that create more solutions that maintain the pattern or make it better.

THINGS TO THINK ABOUT

1. Why are herbicides the primary weed control technique in the world's developed countries?
2. What are the major problems that impede progress in weed science?
3. Can you identify several important, future research problems?
4. Why is understanding weed-crop competition crucial to the future of weed science and weed management?
5. Why are studies of seed dormancy and seed germination so important to weed management?

6. What will be the future role for bioeconomic models of weed-crop competition, and what parameters should new models incorporate?
7. What are the characteristics of an ideal herbicide?
8. Why are perennial weeds so hard to control?
9. Why are parasitic weeds so hard to control?
10. What is the problem with herbicide use for aquatic weed management?
11. Why is there concern about packaging and labeling of herbicides?
12. What has been the recent evolutionary trend in the agrichemical industry?
13. Why are herbicides sprayed?
14. In what areas can engineering research contribute to weed management?
15. Why are political considerations important to weed science?
16. What is solving for pattern, and how does it relate to weed science and weed management?

LITERATURE CITED

Berry, W. 1981. Solving for pattern, pp. 134–145 in "The Gift of Good Land." N. Point Press, San Francisco.

Beyer, E. M., Jr. 1991. Crop protection—meeting the challenge. Proc. British Crop Prot. Conf.—Weeds **1**:3–22.

Buhler, D. D., R. G. Hartzler, and F. Forcella. 1997. Implications of weed seed bank dynamics to weed management. Weed Sci. **45**:329–336.

Buttel, F. H. 1985. The land-grant system: A sociological perspective on value conflicts and ethical issues. J. Agric. and Human Values **2**:78–95.

Christiansen, M. L. 1991. Fun with mutants: Applying genetic methods to problems of weed physiology. Weed Sci. **39**:489–495.

Dawson, J. H. 1965. Competition between irrigated sugar beets and annual weeds. Weeds **13**:245–249.

Duke, S. O. (Ed.). 1996. "Herbicide-Resistant Crops: Agricultural, Environmental, Economic, Regulatory, and Technical Aspects." CRC-Lewis, Boca Raton, Florida. 420 pp.

Duke, S. O., A. L. Christy, F. D. Hess, and J. S. Holt. 1991. "Herbicide Resistant Crops." Comment from CAST. No 1991-1. Council for Agric. Sci. and Technol. 24 pp.

Evans, D. A. 1992. Designing more efficient herbicides. Proc. First Int. Weed Control. Cong. **1**:34–41.

Gardner, G. 1996. IPM and the war against pests. WorldWatch, March/April P21–27.

Hileman, B. 1995. Views differ sharply over benefits, risks of agricultural biotechnology. Chem. & Eng. News, August 21 pp. 8–17.

Hopkins, W. L. 1994. Pp. 157–159 in "Global Herbicide Directory," 1st ed. Ag. Chem. Info. Services, Indianapolis, Indiana.

Keeler, K. H. 1989. Can genetically engineered crops become weeds? Biotechnology **7**:1134–1139.

King, J. 1991. A matter of public confidence. Agric. Eng. **72(4):**16–18.

McWhorter, C. G., and M. R. Gebhardt (Eds.). 1987. "Methods of Applying Herbicides," Monograph 4. Weed Sci. Soc. of America, Champaign, Illinois. 358 pp.

Mortensen, D. A., and H. D. Coble. 1991. Two approaches to weed control decision-aid software. Weed Technol. **5:**445–452.

O'Donovan, J. T. 1996. Computerised decision support systems: Aids to rational and sustainable weed management. J. Plant Sci. **76:**3–7.

Schweizer, E. E., D. W. Lybecker, and L. J. Wiles. 1996. Important biological information needed for bioeconomic weed management models. Adv. Soil Sci., in Press.

Weaver, S. 1993. Simulation of crop-weed competition: Models and their application, pp/ 3–11 in "Symp. on Weed Ecology, Expert Comm. on Weeds, Edmonton, Canada."

Wyse, D. L. 1992. Future of weed science research. Weed Sci. **6:**162–165.

Appendix A

List of Crop and Other Plants Cited in Text (Alphabetized by Common Name)

Common name	Scientific name
Alfalfa	*Medicago sativa* L.
Almonds	*Prunus dulcis* (Mill.) D.A. Webb
Apple	*Malus* spp.
Artichoke	*Cynara scolymus* L.
Ash, white	*Fraxinus americana* L.
Asparagus	*Asparagus officinalis* L.
Bahiagrass	*Paspalum notatum* Fluegge
Banana	*Musa paradisiaca* L.
Barley	*Hordeum vulgare* L.
Bean	*Phaseolus* spp.
Bean, fava = broad bean	*Vicia faba* L
Bean, lima	*Phaseolus limensis* L.
Bean, mung	*Vigna radiata* (L.) Wilczek var. radiata
Bean, wild winged	*Psophocarpus palustris* Desv.
Beet, red	*Beta vulgaris* L.
Begonia	*Begonia* spp.
Bentgrass	*Agrostis* spp.
Bermudagrass	*Cynodon dactylon* (L.) Pers.
Big bluestem	*Andropogon gerardii* Vitm.
Blackberries	*Rubus allegheniensis* Porter
Blueberries	*Vaccinium angustifolium* Ait.
Bluebunch wheatgrass	*Agropyron spicatum* (Pursh) Scribn. & Sm.

Common name	Scientific name
Blue grama	*Bouteloua gracilis* Lag.ex
Bluegrass, Kentucky	*Poa pratensis* L.
Broadbean	*Vicia faba* L.
Broccoli	*Brassica oleracea*–Italica group
Bromegrass, smooth	*Bromus inermis* Leyss
Brussels sprouts	*Brassica oleracea*–Gemmifera group
Buckwheat	*Fagopyrum esculentum* Moench.
Buffalograss, common	*Buchloe dactyloides* (Nutt.) Engelm
Cabbage	*Brassica rapa* L.
Cabbage, Chinese	*Brassica rapa* L.
Canola = rapeseed	*Brassica napus* L.
Carnation	*Dianthus caryophyllus* L.
Carrot	*Daucus carota* L.
Cassava = manioc	*Manihot esculenta* Crantz
Celery	*Apium graveolens* L.
Centro	*Centrosema pubescens* Benth
Cherry	*Prunus* spp.
Cherry, wild	*Prunus* spp.
Chickpeas	*Cicer arietinum* L.
Clover	*Trifolium* spp.
Clover, crimson	*Trifolium incarnatum* L.
Clover, red	*Trifolium pratense* L.
Clover, subterranean	*Trifolium subterraneum* L.
Clover, white	*Trifolium repens* L.
Coconut	*Cocos nucifera* L.
Coffee	*Coffea arabica* L.
Corn = maize	*Zea mays* L.
Cotton	*Gossypium hirsutum* L.
Cowpea	*Vigna unguiculata* (L.) Walp.
Cranberries	*Vaccinium macrocarpon* Ait.
Crownvetch	*Coronilla varia* L.
Cucumber	*Cucumis sativus* L.
Currant, black	*Ribes nigrum* L.
Currant, red	*Ribes sativum* Syme
Douglas fir	*Pseudotsuga menziesii* (Mirbel) Franco
Fescue	*Festuca* spp.
Flax	*Linum usitatissimum* L.

Common name	Scientific name
Foxglove	*Digitalis* spp. (esp. *D. purpurea*)
Grama, side-oats	*Bouteloua curtipendula* (Michx.) Torr.
Grapes	*Vitis* spp.
Groundnut = peanut	*Arachis hypogaea* L.
Guayule	*Parthenium argentatum* A. Gray
Hops	*Humulus lupulus* L.
Indian grass, yellow	*Sorghastrum nutans* (L.) Nash
Iris	*Iris pseudacorus* L.
Jute	*Corchorus olitorius* L.
Lentil	*Lens culinaris* Medik.
Lettuce	*Lactuca sativa* L.
Lily-of-the-valley	*Convallaria majalis* L.
Macadamia	*Macadamia* spp.
Maize = corn	*Zea mays* L.
Maple	*Acer* spp.
Maple, red	*Acer rubrum* L.
Manzanita	*Arctostaphylos pungens* Kunth
Meadowsweet	*Filipendula ulmaria* (L.) Maxim
Melon, egusi	*Citrullus lanatus* L.
Millet, common, or proso	*Panicum miliaceum* L.
Millet, foxtail	*Setaria italicum* (L.) Beauv.
Millet, pearl	*Pennisetum glaucum* L.
Mistletoe	*Arceuthobium* spp.
Narcissus	*Narcissus poeticus* L.
Oak	*Quercus* spp.
Oats	*Avena sativa* L.
Oleander	*Nerium oleander* L.
Onion	*Allium cepa* L.
Orange	*Citrus sinensis* (L.) Osb.
Orange, Satsuma	*Citrus nobilis* Lour.
Orchardgrass	*Dactylis glomerata* L.

Common name	Scientific name
Pangolagrass	*Digitaria decumbens* Stent
Peanut = groundnut	*Arachis hypogaea* L.
Peppermint	*Mentha piperita* L.
Peppers	*Capsicum* spp.
Pigeon pea	*Cajanus cajan* (L.) Millsp.
Pine	*Pinus* spp.
Pine, Ponderosa	*Pinus ponderosa* Dougl.ex P. & C. Laws.
Pineapple	*Ananas comosus* (L.) Merr.
Poinsettia	*Euphorbia pulcherrima* Willd.ex Kl.
Potato, sweet	*Ipomoea batatas* Lam
Potato, white	*Solanum tuberosum* L.
Psoralea	*Psoralea* spp.
Rapeseed = canola	*Brassica napus* L.
Raspberries	*Rubus* spp.
Red clover	*Trifolium pratense* L.
Rhodendron	*Rhododendron* spp.
Rhubarb	*Rheum rhaponticum* L.
Rice	*Oryza sativa* L.
Rye, common	*Secale cereale* L.
Ryegrass	*Lolium* spp.
Safflower	*Carthamus tinctoris* L.
Salvia	*Salvia* spp.
Sanfoin	*Onobrychis viciifolia* Scop.
Snapbean	*Phaseolus vulgaris* L.
Snapdragon	*Antirrhinum majus* L.
Sorghum	*Sorghum bicolor* (L.) Moench
Soybean	*Glycine max* (L.) Merr.
Spearmint	*Mentha spicata* L.
Spinach	*Spinacia oleracea* L.
Spiraea = spirea	*Spirea* spp.
Squash	*Cucurbita* spp.
Squash, zucchini	*Cucurbita maxima* L.
St. Augustine grass	*Stenotaphrum secundatum* (Walt.) Ktze.
Strawberries	*Fragaria vesca* L.
Sudangrass	*Sorghum sudanense* (Piper) Stapf
Sugarbeet	*Beta vulgaris* L.
Sugarcane	*Saccharum officinarum* L.
Sunflower	*Helianthus annuus* L.

Common name	Scientific name
Switchgrass	*Panicum virgatum* L.
Timothy	*Phleum pratense* L.
Tobacco	*Nicotiana tabacum* L.
Tomato	*Lycopersicon esculentum* Mill.
Trillium	*Trillium* spp.
Walnut, black	*Juglans nigra* L.
Watermelon	*Citrullus lanatus* (Thunb.) Matsum. & Nakai
Wheat	*Triticum aestivum* L.
Wheatgrass, crested	*Agropyron desertorum* (Fisch. ex Link) Schult.
Wheatgrass, slender	*Agropyron trachycaulum* (Link) Malte ex H.F. Lewis

Appendix B

List of Weeds Cited in Text
(Alphabetized by Common Name)

Common name	Scientific name
Ageratum	*Ageratum* spp.
Ageratum, tropic	*Ageratum conyzoides* L.
Alder	*Alnus* spp.
Alder, red	*Alnus rubra* Bong.
Alkaliweed	*Cressa truxillensis* H.B.K.
Alligatorweed	*Alternanthera philoxeroides* (Mart.)Griseb.
Amaranth, Palmer	*Amaranthus palmeri* S.Wats.
Amaranth, slender	*Amaranthus gracilis* Desf.
Amaranth, spiny	*Amaranthus spinosus* L.
Amaranth	*Amaranthus* spp.
American dragonhead	*Drachocephalum parviflorum* Nutt.
Arrowgrass	*Triglochin maritima* L.
Arrowhead	*Sagittaria sagittifolia* L.
Aspen = quaking aspen	*Populus tremuloides* Michx.
Azolla = pinnate mosquitofern	*Azolla pinnata* R. Brown
Barberry, European	*Berberis vulgaris* L.
Barley, foxtail	*Hordeum jubatum* L.
Barley, meadow	*Hordeum brachyantherum* Nevski
Barnyardgrass	*Echinochloa crus-galli* (L.) Beauv.
Bedstraw, catchweed	*Galium aparine* L.
Beggarticks, hairy	*Bidens pilosa* L.
Bent, creeping	*Agrostis stolonifera* L.
Bermudagrass	*Cynodon dactylon* (L.) Pers.

Common name	Scientific name
Bindweed	*Convolvulus* spp.
Bindweed, field	*Convolvulus arvensis* L.
Bindweed, hedge	*Calystegia sepium* (L.) R.Br.
Bitterbush	*Eupatorium oderatum* L. = *Chromolaena odorata* (L.) R.M. King & M. Robinson
Bittercress	*Cardamine* sp.
Blackgrass = slender foxtail	*Alopecurus myosuroides* Auds.
Bluegrass	*Poa* spp.
Bluegrass, annual	*Poa annua* L.
Bluegrass, Kentucky	*Poa pratensis* L.
Bluegrass, roughstalk	*Poa trivialis* L.
Bouncing Bet	*Saponaria officinalis* L.
Brackenfern	*Pteridium aquilinum* (L.)Kuhn
Brazilian peppertree	*Schinus terebinthifolius* Raddi.
Bristly starbur	*Acanthospermum hispidium* DC.
Brome	*Bromus* spp.
Brome, downy	*Bromus tectorum* L.
Brome, Japanese	*Bromus japonicus* Thunb. ex Murr.
Brome, smooth	*Bromus inermis* Leyss.
Brome, soft	*Bromus mollis* L.
Broomrape	*Orobanche* spp.
Buckeye, Ohio	*Aesculus glabra* Willd.
Buckthorn, European	*Rhamnus cathartica* L.
Buckwheat, wild	*Polygonum convolvulus* L.
Buffalobur	*Solanum rostratum* Dun
Bulrush	*Scirpus* spp.
Burcucumber	*Sicyos angulatus* L.
Burdock, common	*Arctium minus* (Hill)Bernh.
Bursage, skeletonleaf	*Ambrosia tomentosa* Nutt.
Buttercup	*Ranunculus* spp.
Buttercup, creeping	*Ranunculus repens* L.
California sagebrush	*Artemesia californica* Less.
California chapparral = whiteleaf sage	*Salvia leucophylla* Greene
California peppertree	*Schinus molle* L.
Camelthorn	*Alhagi pseudalhagi* (Bieb.) Desv.
Carpetgrass	*Axonopus compressus* (Swartz)Beauv.
Catnip	*Nepeta cataria* L.

Common name	Scientific name
Cat's ear, spotted	*Hypochoeris radicata* L.
Cattail, common	*Typha latifolia* L.
Celosia	*Celosia argentea* L.
Charlock	*Brassica* spp.
Charlock, yellow	*Brassica arvensis* Ktze = *Sinapis arvensis* L.
Chickweed, common	*Stellaria media* (L.) Vill.
Chickweed, mouseear	*Cerastium vulgatum* L.
Chickweed, star	*Stellaria pubera* Michx.
Chicory	*Cichorium intybus* L.
Chokecherry	*Prunus virginiana* L.
Cinquefoil, oldfield	*Potentilla simplex* Michx.
Cinquefoil, sulfur	*Potentilla recta* L.
Cladophora	*Cladophora* spp.
Coat buttons	*Tridax procumbens* L.
Cocklebur	*Xanthium* spp.
Cocklebur, common	*Xanthium strumarium* L.
Coffee senna	*Cassia occidentalis* L.
Cogongrass	*Imperata cylindrica* (L.)Beauv.
Coltsfoot	*Tussilago farfara* L.
Coontail	*Ceratophyllum demersum* L.
Corn cockle	*Agrostemma githago* L.
Cottonwood	*Populus* spp.
Crabgrass	*Digitaria* spp.
Crabgrass, large	*Digitaria sanguinalis* (L.)Scop.
Crabgrass, southern	*Digitaria ciliaris* (Retz.)Koel
Crazyweed	*Oxytropis* spp.
Cress	*Lepidium sativum* L.
Cress, hoary	*Cardaria draba* (L.)Desv.
Crotalaria, showy	*Crotalaria spectabilis* Roth
Crowfootgrass	*Dactyloctenium aegyptium* (L.)Willd.
Crupina, common	*Crupina vulgaris* Cass.
Dallisgrass	*Paspalum dilatatum* Poir.
Dandelion	*Taraxacum officinale* Weber in Wiggers
Dayflower	*Commelina* spp.
Dock	*Rumex* spp.
Dock, broadleaf	*Rumex obtusifolius* L.
Dock, smooth	*Rumex* lanceolatus Thunb.

Common name	Scientific name
Dock, curly	*Rumex crispus* L.
Dodder	*Cuscuta* spp.
Dodder, field	*Cuscuta campestris* Yuncker
Dodder, smallseed	*Cuscuta planiflora* Tenore
Dropseed	*Sporobulus* spp.
Dropseed, sand	*Sporobulus cryptandrus* (Torr.) Gray
Duckweed	*Lemna* spp.
Dwarf mistletoe	*Arceuthobium vaginatum* M. Bieb.
Dyer's woad	*Isatis tinctoria* L.
Eclipta	*Eclipta prostrata* L.
Elodea, Western	*Elodea nuttallii* (Planch.) St. John
Eveningprimrose, common	*Oenothera biennis* L.
False flax	*Camelina* spp.
Fescue, Tall	*Festuca arundinacea* Schreb.
Fiddleneck	*Amsinckia intermedia* Fisch. & Mey.
Field violet	*Viola arvensis* Murr.
Fireweed	*Epilobium angustifolium* L.
Flatsedge, rice	*Cyperus iria* L.
Flaxweed = flatseed falseflax	*Camelina alyssum* (Mill.) Thell
Flixweed	*Descurainia sophia* (L.) Webb. ex Prantl
Florida beggarweed	*Desmodium tortuosum* (Sw.) DC.
Florida pusley	*Richardia scabra* L.
Foxglove, common	*Digitalis purpurea* L.
Foxtail	*Setaria* spp.
Foxtail, giant	*Setaria faberi* Herrm.
Foxtail, green	*Setaria viridis* (L.)Beauv.
Foxtail, slender	*Alopecurus myosuroides* Huds.
Foxtail, yellow	*Setaria glauca* (L.)Beauv.
Fringerush, globe	*Fimbristylis miliacea* (L.) Vahl
Fumitory	*Fumaria officinalis* L.
Galinsoga	*Galinsoga* spp.
Galinsoga, smallflower	*Gallinsoga parviflora* Cav.
Garlic, wild	*Allium vineale* L.
Geranium, spotted	*Geranium maculatum* L.
Goldenrod	*Solidago* sp.
Goosefoot, oakleaf	*Chenopodium glaucum* L.

Common name	Scientific name
Goosegrass	*Eleusine indica* (L.) Gaertn.
Gooseweed	*Sphenoclea zeylandica* Gaertn.
Greasewood	*Sarcobatus vermiculatus* (Hook.) Torr.
Groundcherry	*Physalis* spp.
Groundsel	*Senecio* spp.
Groundsel, common	*Senecio vulgaris* L.
Guayule	*Parthenium argentatum* A. Gray
Guineagrass	*Panicum maximum* Jacq.
Gumweed, curlycup	*Grindelia squarrosa* (Pursh) Dunal
Halogeton	*Halogeton glomeratus* (Stephen ex. Bieb.) C.A. Mey
Heath	*Erica scoparia* L.
Heliotrope	*Heliotropium* spp.
Hemlock, poison	*Conium maculatum* L.
Hemp dogbane	*Apocynum cannabinum* L.
Hempnettle, common	*Galeopsis tetrahit* L.
Henbane, black	*Hyoscyamus niger* L.
Henbit	*Lamium amplexicaule* L.
Hoary alyssum	*Berteroa incana* (L.)DC.
Horsetail, field	*Equisetum arvense* L.
Horseweed	*Conyza canadensis* (L.)Cronq.
Hydrilla	*Hydrilla verticillata* (L.f.)Royle
Indian tobacco	*Lobelia inflata* L.
Itchgrass	*Rottboellia cochinchinensis* (Lour.) W.D. Clayton
Jerusalem artichoke	*Helianthus tuberosus* L.
Jimsonweed	*Datura stramonium* L.
Johnsongrass	*Sorghum halepense* (L.)Pers.
Jointvetch, northern	*Aeschynomene virginica* (L.)B.S.P.
Junglerice	*Echinochloa colonum* (L.)Link
Karibaweed	*Salvinia molesta* Mitch.
Kikuyugrass	*Pennisetum clandestinum* Hochst. ex Chiov.
Klamath weed	See St. John's wort
Knapweed, diffuse	*Centaurea diffusa* Lam.
Knapweed, spotted	*Centaurea maculosa* Lam.

Common name	Scientific name
Knapweed, Russian	*Acroptilon repens* (L.) DC.
Knotweed, Japanese	*Polygonum cuspidatum* Seib. & Zucc.
Knotweed, prostrate	*Polygonum aviculare* L.
Kochia	*Kochia scoparia* (L.) Schrad.
Kudzu	*Pueraria lobata* (Willd.) Ohwi
Ladysthumb	*Polygonum persicaria* L.
Lambsquarters, common	*Chenopodium album* L.
Lambsquarters, slimleaf	*Chenopodium leptophyllum* (Moq.) Nutt. ex S.Wats.
Lambsquarters, netseed	*Chenopodium berlandieri* Moq.
Lantana	*Lantana* spp.
Lantana, largeleaf	*Lantana camara* L.
Larkspur	*Delphinium* spp.
Leafy spurge	*Euphorbia esula* L.
Lettuce, prickly	*Lactuca serriola* L.
Locoweed	*Astragalus* spp.
Mallow, bull	*Malva nicaeensis* All.
Mallow, common	*Malva neglecta* Wallr.
Mallow, venice	*Hibiscus trionum* L.
Maple, bigleaf	*Acer macrophyllum* Pursh
Marestail	*Hippuris vulgaris* L.
Marigold, corn	*Chrysanthemum segetum* L.
Marigold, wild	*Tagetes erecta* L.
Marijuana	*Cannabis sativa* L.
Mayweed (prob. dogfennel)	*Eupatorium capillifolium* (Lam.) Small
Meadow fescue	*Festuca pratensis* Huds. = *F. elatior* sensu hitchc.
Meadow foxtail	*Alopecurus pratensis* L.
Medic, black	*Medicago lupilina* L.
Medusahead	*Taeniatherum caput-medusae* (L.)Nevski
Melon, wild	*Cucumis melo* L.
Mesquite	*Prosopis* spp.
Milkweed, common	*Asclepias syriaca* L.
Milkweed, showy	*Asclepias speciosa* Torr.
Milkweed, whorled	*Asclepias verticillata* L.
Millet, foxtail	*Setaria italica* (L.)Beauv.
Millet, wild proso	*Panicum milaceum* L.

Common name	Scientific name
Milograss, wild	*Panicum* spp.
Mimosa, catclaw	*Mimosa pigra* L.
Mistletoe, dwarf	*Arceuthobium vaginatum* M. Bieb.
Monkshood	*Aconitum napelus* L.
Monochoria	*Monochoria vaginalis* (Burm.f.)Kunth
Moonflower, purple	*Ipomoea alba* L.
Morningglory	*Ipomoea* spp.
Morningglory, pitted	*Ipomoea lacunosa* L.
Morningglory, tall	*Ipomea purpurea* (L.)Roth
Mugwort	*Artemesia vulgaris* L.
Mullein	*Verbascum* spp.
Mullein, common	*Verbascum thapsus* L.
Multiflora rose	*Rosa multiflora* Thunb. ex Murr.
Mustard	*Brassica* spp.
Mustard, white	*Sinapis alba* L.
Mustard, wild	*Sinapis arvensis* L.
Mustard, black	*Brassica nigra* (L.)W.J.D. Koch
Mustard, tumble	*Sisymbrium altissimum* L.
Needle-and-thread grass	*Stipa comata* Trin. & Rupr.
Nettle, burning	*Urtica urens* L.
Nettle	*Urtica spp.*
Nightshade	*Solanum* spp.
Nightshade, American black	*Solanum americanum* Mill.
Nightshade, Eastern black	*Solanum ptycanthum* Dun
Nightshade, black	*Solanum nigrum* L.
Nightshade, hairy	*Solanum sarrachoides* Sendtner
Nightshade, silverleaf	*Solanum elaeagnifolium* Cav.
Nostoc	*Nostoc* spp.
Nutsedge	*Cyperus* spp.
Nutsedge, purple	*Cyperus rotundus* L.
Nutsedge, yellow	*Cyperus esculentus* L.
Oak	*Quercus* spp.
Oak, gambel	*Quercus gambelii* Nutt.
Orach, halberdleaf	*Atriplex patula* var. *hasta*ta (L.) Gray
Orchardgrass	*Dactylis glomerata* L.
Orobanche	*Orobanche* spp.
Oxalis	*Oxalis* spp.
Oxalis, buttercup	*Oxalis pes-caprae* L.

Common name	Scientific name
Panicum, fall	*Panicum dichotomiflorum* Michx.
Panicum, Texas	*Panicum texanum* Buckl.
Paragrass	*Brachiaria mutica* (Forsk.)Stapf
Parthenium ragweed	*Parthenium hysterophorus* L.
Paspalum, sour	*Paspalum conjugatum* Bergius
Pennycress, field	*Thlaspi arvense* L.
Peppergrass = greenflower pepperweed	*Lepidium densiflorum* Schrad.
Pepperweed, Virginia	*Lepidium virginicum* L.
Pepperweed, field	*Lepidium campestre* (L.)R.Br.
Pigweed	*Amaranthus* spp.
Pigweed, prostrate	*Amaranthus graecizans* auctt., non L.
Pigweed, smooth	*Amaranthus hybridus* L.
Pigweed, tumble	*Amaranthus albus* L.
Pigweed, redroot	*Amaranthus retroflexus* L.
Pimpernel, blue	*Anagallis coerulea* Nathh.
Pineapple-weed	*Matricaria matricarioides* (Less.) Port.
Plantain	*Plantago* spp.
Plantain, buckhorn	*Plantago lancolata* L.
Plantain, broadleaf	*Plantago major* L.
Plantain, common	*Plantago* spp.
Poison hemlock	*Conium maculatum* L.
Poison ivy	*Toxicodendron radicans* (L.)Ktze
Poison oak	*Toxicodendron toxicarium* (Salisb.)Gillis
Pokeweed, common	*Phytolacca americana* L.
Pondweed	*Potamogeton* spp.
Pondweed, American	*Potamogeton nodosus* Poir.
Pondweed, sago	*Potamogeton pectinatus* L.
Prickly pear cactus	*Opuntia littoralis* (Engelmann) Cockerell
Prickly pear, plains	*Opuntia polyacantha* Haw.
Prickly pear, spreading	*Opuntia vulgaris* Mill.
Prince's feather	*Polygonum orientale* L.
Puncturevine	*Tribulus terrestris* L.
Purple loosestrife	*Lythrum salicaria* L.
Purslane, common	*Portulaca oleracea* L.
Quackgrass	*Eltrygia repens* (L.)Nevski
Quaking aspen	*Populus tremuloides* Michx.

Common name	Scientific name
Rabbitbrush, gray	*Chrysothamnus nauseosus* (Pallas)Britt.
Rabbitbrush	*Chrysothamnus* spp.
Radish, wild	*Raphanus raphanistrum* L.
Ragweed	*Ambrosia* spp.
Ragweed, common	*Ambrosia artemisiifolia* L.
Ragweed, giant	*Ambrosia trifida* L.
Ragweed, Western	*Ambrosia psilostachya* D.C.
Redrice	*Oryza sativa* L.
Red sprangletop	*Leptochloa filiformis* (Lam.)Beauv.
Redtop	*Agrostis gigantea* Roth
Reed, common	*Phragmites australis* (Cav.)Trin. ex Steud.
Reed canarygrass	*Phalaris arundinacea* L.
Rocky Mountain beeplant	*Cleome serrulata* Pursh
Rush	*Scirpus* spp.
Russian olive	*Elaeagnus angustifolia* L.
Rye, common	*Secale cereale* L.
Ryegrass	*Lolium* spp.
Ryegrass, annual	Probably–*Lolium multiflorum* Lam.
Ryegrass, Italian	*Lolium multiflorum* Lam.
Ryegrass, perennial	*Lolium perenne* L.
Ryegrass, poison	*Lolium temulentum* L.
Ryegrass, rigid	*Lolium rigidum* Gaudin
Sacramento thistle	*Cirsium vinaceum* (Woot. and Standl.)
Sage, Mediterranean	*Salvia aethiopis* L.
Sage, red	*Salvia splendens* F. Sellow ex Roem.
Sage, Syrian	*Salvia syriaca* L.
Sagebrush	*Artemisia* spp.
Sagebrush, big	*Artemisia tridentata* Nutt.
Sagebrush, fringed	*Artemisia frigida* Willd.
Sagebrush, sand	*Artemisia filifolia* Torr.
Sakhalin knotweed = S. knotgrass	*Polygonum sachalinense* F. Schmidt ex Maxim.
Saltbush	*Atriplex* spp.
Salt cedar = tamarisk	*Tamarix ramosissima* Ledeb.
Saltgrass	*Distichlis spicata* (L.) Greene
Salvinia	*Salvinia molesta* Mitch.
Sandbur	*Cenchrus* spp.

Common name	Scientific name
Sandbur, longspine	*Cenchrus longispinus* (Hack.) Fern.
Scotch broom	*Cytisus scoparius* (L.) Link
Sedge, smallflower umbrella	*Cyperus difformis* L.
Senna, coffee	*Cassia occidentalis* L.
Sensitiveplant	*Mimosa pudica* L.
Sesbania, hemp	*Sesbania exaltata* (Raf.) Rtdb. Ex A.W.Hill
Shattercane	*Sorghum bicolor* (L.) Moench
Shepherd's-purse	*Capsella bursa-pastoris* (L.) Medik.
Sicklepod	*Senna obtusifolia* (L.)
Sida	*Sida* spp.
Sida, prickly	*Sida spinosa* L.
Signalgrass	*Bracharia* spp.
Skeletonweed, rush	*Chondrilla juncea* L.
Smartweed	*Polygonum* spp.
Smartweed, marshpepper	*Polygonum hydropiper* L.
Smartweed, Pennsylvania	*Polygonum pensylvanicum* L.
Smooth cordgrass	*Spartina alterniflora* Loisel.
Snakeweed, broom	*Gutierrezia sarothrae* (Pursh) Britt. & Rusby
Sorghum-almum	*Sorghum almum* Parod.
Sorrel, green	*Rumex acetosa* L.
Sorrel, red	*Rumex acetosella* L.
Sowthistle, perennial	*Sonchus arvensis* L.
Sowthistle, annual	*Sonchus oleraceus* L.
Speedwell, corn	*Veronica arvensis* L.
Speedwell, Persian	*Veronica persica* Poir.
Speedwell, slender	*Veronica filiformis* Sm
Spirogyra	*Spirogyra* spp.
Sprangletop	*Leptochloa* spp.
Spreading dayflower	*Commelina diffusa* Burm. f.
Spurge, garden	*Euphorbia hirta* L.
Spurge, prostrate	*Euphorbia supina* Raf.
Spurge, toothed	*Euphorbia dentata* Michx.
Spurred anoda	*Anoda cristata* (L.) Schlecht.
St. Augustine grass	*Stenotaphrum secundatum* (Walt.) Ktze.
St. John's wort	*Hypericum perforatum* L.
Starthistle, yellow	*Centaurea solstitialis* L.

Common name	Scientific name
Stinkgrass	*Eragrostis cilianensis* (All.) E. Mosher
Stinkweed	*Pluchea camphorata* (L.) DC.
Stranglerviine	*Morrenia odorata* (H. & A.) Lindl
Sumac, smooth	*Rhus glabra* L.
Sunflower, common	*Helianthus annuus* L.
Sweetclover	*Melilotus* spp.
Sweetclover, yellow	*Melilotus officinalis* (L.) Lam.
Tamarisk = salt cedar	*Tamarix ramosissima* Ledeb.
Tansy ragwort	*Tanacetum vulgare* L.
Tansymustard, pinnate	*Descurainia pinnata* (Walt.) Britt
Tares = common vetch or	*Vicia sativa* L.
darnel	*Lolium temulentum* L.
Thistle, artichoke	*Cynara cordonculus* L.
Thistle, bull	*Cirsium vulgare* (Savi) Tenore
Thistle, Canada	*Cirsium arvense* (L.) Scop.
Thistle, musk	*Carduus nutans* L.
Thistle, Russian	*Salsola iberica* Sennen & Pau
Thistle, yellowstar	*Centaurea solstitialis* L.
Toadflax	*Linaria* spp.
Toadflax, dalmation	*Linaria dalmatica* (L.) Mill.
Toadflax, oldfield	*Linaria canadensis* (L.) Dumont
Toadflax, yellow	*Linaria vulgaris* Mill.
Tomato, wild	*Solanum triflorum* L.
Torpedograss	*Panicum repens* L.
Tree cactus	*Opuntia megacantha* Salm-Dyck
Tropical soda apple	*Solanun viarum* Ounal
Unicornplant	*Proboscidea louisianica* (Mill.) Thellung
Velvetgrass, common	*Holcus lanatus* L.
Velvetgrass, German	*Holcus mollis* L.
Velvetleaf	*Abutilon theophrasti* Medikus
Vervain, prostrate	*Verbena bracteata* Lag.&Rodr.
Vetch	*Vicia* spp.
Vetch, common	*Vicia sativa* L.
Vetch, hairy	*Vicia villosa* Roth

Common name	Scientific name
Waterhemlock	*Cicuta* spp.
Waterhemlock, spotted	*Cicuta maculata* L.
Waterhemlock, Western	*Cicuta douglasii* (DC.) Coult & Rose
Waterhyacinth	*Eichhornia crassipes* (Mart.) Solms
Waterlettuce	*Pistia stratiotes* L.
Water lily	*Nymphaea* sp.
Watermilfoil, Eurasian	*Myriophyllum spicatum* L.
White campion	*Silene alba* (Mill.)E.H.L.Krause
Wild cane	*Saccharum spontaneum* L.
Wild carrot	*Daucus carota* L.
Wild cherry	*Prunus* spp.
Wild oat	*Avena* spp.
Wild oat	*Avena fatua* L.
Wild oat	*Avena sterilis* L.
Wild oat, winter	*Avena ludoviciana* Durieu
Wild onion	*Allium canadense* L.
Willow	*Salix* spp.
Witchgrass	*Panicum capillare* L.
Witchweed	*Striga asiatica* (L.) Ktze.
Wood sorrel	*Rumex* spp.
Yarrow	*Achillea millefolium* L.
Yellow rocket	*Barbarea vulgaris* R. Br.
Yucca	*Yucca* spp.
Yucca, Great Plains	*Yucca glauca* Nutt. ex Fraser

Glossary

Absorption Process by which herbicides are taken into plants, by roots, or foliage (stomata, cuticle, etc.) (*see* Adsorption)

Acid equivalent (ae) The theoretical yield of parent acid from an active ingredient in acid-based herbicides

Acropetal Toward the apex; generally upward in shoots and downward in roots (*see* Basipetal)

Active ingredient (ai) Chemical(s) responsible for herbicidal effects

Adjuvant Ingredient that facilitates or modifies the action of the principal ingredient; an additive

ADP (adenosine diphosphate) Adenosine-derived ester formed in cells converted to ATP for energy storage

Adsorption Chemical or physical attraction of a substance to a surface; can refer to gases, dissolved substances, or liquids on the surface of solids or liquids (*see* Absorption)

Aliphatic Derived from straight-chain hydrocarbons

Allelopathy Adverse effect of one plant or microorganism on another caused by release of a chemical from living or decaying organisms

Allopatry Occurring in separate, widely differing geographic areas (*see* Sympatry)

Annual Plant that completes its life cycle in one year (i.e., germinates from seed, grows, flowers, produces seed, and dies in the same season); examples: redroot pigweed, common ragweed, mustards, foxtails, large crabgrass

Antidote (1) Chemical applied to prevent or reduce phytotoxic effect of a herbicide; (2) a substance used in medicine to counteract poisoning

Apoplast Continuous, nonliving, cell wall phase that surrounds and contains the symplast

Aquatic weed Weed that grows in water; there are three kinds: (1) submersed–grow beneath the surface (e.g., sago pondweed, elodea, watermilfoil); (2) emersed–grow above the water (e.g., cattails and water lilies); and (3) floaters–float on the surface (e.g., waterhyacinth)

Aromatics Compounds derived from the hydrocarbon benzene

ATP (adenosine triphosphate) Adenosine-derived nucleotide; primary source of energy through conversion to ADP

Band application Application to a continuous restricted band such as in or along a crop row, rather than over the entire field

Basal treatment Herbicide applied to the stems of woody plants at and just above the ground

Basipetal Toward the base; generally downward in shoots (*see* Acropetal)

Bed Narrow flat-topped ridge on which crops are grown with a furrow on each side for drainage of excess water, or area in which seedlings or sprouts are grown before transplanting

Biennial Plant that completes its growth in two years: The first year it produces leaves and stores food; the second year it produces fruits and seeds (e.g. wild carrot, bull thistle)

Biological control Controlling a pest with natural or introduced enemies

Blind cultivation Cultivating before plant emergence

Broadcast application Application over an entire area rather than only on rows, beds, or middles

Broadleaved plants In general, used as an antonym to grass plants

Brush control Control of woody plants

Calibration Series of operations to determine the amount of solution (volume) applied per unit area of land and the amount of pesticide to add to a known volume of diluent

Carrier Liquid or solid material added to a chemical compound to facilitate its application (usually water, but diesel oil has been used with water for brush control)

Cation exchange capacity Total exchangeable cations a soil can adsorb; expressed as moles or millimoles of negative charge per kilogram soil (or other exchange material, e.g., clay)

Chlorosis Loss of green foliage color

Clay (1) Soil consisting of particles <0.002 mm diameter; (2) soil textural class: soil containing >40% clay, < 45% sand, and < 40% silt

Cleistogamous Having small, unopened, self-pollinated flowers

Compatibility Quality of two compounds that permits them to be mixed without effect on the properties of either

Competition Active acquisition of limited resources by an organism that results in a reduced supply and consequently reduced growth of other organisms

Concentration Amount of active ingredient in a given volume of diluent; recommendations for concentration of herbicides should be on the basis of weight or volume of active ingredient or product per unit volume of diluent

Contact herbicide Herbicide that kills primarily by contact with plant tissue rather than as a result of translocation

Cotyledon First leaf, or pair of leaves, of the embryo of seed plants

Defoliator or defoliant Causes foliage to fall from plants

Desiccant Promotes dehydration of plant tissue and may lower moisture content of seeds to facilitate harvest

Dicot Abbreviation of dicotyledon; preferred term to describe broad-leaved plants

Diluent Liquid or solid to dilute an active ingredient in the preparation of a formulation

Dioecious Having male and female reproductive organs on separate plants; literally = two houses

Directed spray Application to minimize the amount of herbicide applied to the crop; usually accomplished by setting nozzles low with spray patterns intersecting at the base of plants just above the soil line

Dormant Condition in which seeds or other living plant organs are not dead but do not grow; state of suspended animation

Dormant spray Chemical application in winter or very early spring before plants have begun active growth

Drift Movement as a liquid

Ecosystem Ecological entity composed of the biotic community and nonliving environmental phases functioning together in an interacting system

Edaphic Of or pertaining to soil

Emergence Appearance of a plant above the soil

Emersed plant Rooted or anchored aquatic plant that grows with most of its stem tissue above the water surface

Emulsifiable concentrate (EC) Single-phase, liquid formulation that forms an emulsion when added to water

Emulsifier Material that facilitates suspension of one liquid in another

Emulsion Mixture in which one liquid is suspended in minute globules in another liquid without either losing its identity (e.g., oil in water)

Encapsulated formulation Herbicide enclosed in capsules (or beads) to control rate of release of active ingredient and thereby extend period of activity

Epinasty Increased growth on one surface of plant organ or part, causing it to bend

Exchange capacity Total ionic charge of the adsorption complex

Field capacity Percent water remaining in soil after free drainage has ceased

Floating plant Free-floating or anchored aquatic plant adapted to grow with most vegetative tissue above the water surface; plants rise and fall with water level

Flowable Two-phase formulation containing solid herbicide suspended in liquid that forms a suspension when added to water

Formulation (1) Pesticide preparation supplied by manufacturer, (2) process of preparing pesticides for commercial use

Fumigant Volatile liquid or gas to kill insects, nematodes, fungi, bacteria, seeds, roots, rhizomes, or entire plants; usually applied in an enclosure of some kind or to covered soil

Germination   Activation of metabolic sequences culminating in renewed growth of the seed embryo; morphologically observable as radicle protrusion through the seed coat

Granular   Dry formulation consisting of discrete particles usually < 10 mm^3, designed to be applied without water

Growth regulator   Substance effective in minute amounts for controlling or modifying plant processes

Growth stages of cereal crops   (1) Tillering—when additional shoots are developing from the crown; (2) jointing—when stem internodes begin elongating; (3) boot—when leaf sheath swells due to the growth of developing spike or panicle; (4) heading—when seed head emerges from the sheath

Hard water   Water that contains minerals, usually calcium and magnesium sulfates, chlorides or carbonates, in solution, to the extent of causing a curd, or precipitate, rather than a lather, when soap is added; hard water may cause precipitates to form in some herbicidal sprays

Harvest index   The amount (weight) of grain vs total plant foliar dry weight

Herbaceous plant   Vascular plant without woody tissues

Herbicide   A chemical used for killing or inhibiting the growth of plants; phytotoxic chemical (from Latin *herba*, plant, and *caedere*, kill)

Hormone   Growth-regulating substance occurring naturally in plants or animals; refers to certain manmade or synthetic chemicals with growth-regulating activity, more correctly called synthetic regulators, not hormones

Hypocotyl   Portion of the stem of a plant embryo or seedling below the cotyledons

Incorporation   Mixing or blending of herbicides into soil

Interference   Total adverse effect that plants exert on each other when growing in a common ecosystem; includes competition and allelopathy

Invert emulsion   One in which water is dispersed in oil; oil forms the continuous phase with water dispersed therein; usually a thick, mayonnaise-like mixture results

Kairomone   Allelochemical of favorable adaptive value to the organism receiving it

K_d   Ratio of sorbed to dissolved pesticide at equilibrium in a water/soil slurry

K_{oc}   Soil organic carbon sorption coefficient; K_d divided by the weight fraction of organic carbon in soil

Label   Directions for herbicide use created by the manufacturer and approved by federal or state regulatory agencies

Lay-by   Refers to the stage of crop development (or the time) when the last regular cultivation is done

LD_{50}   Dose (quantity) of a substance that causes 50% of test organisms to die; usually expressed in weight (mg) chemical per unit of body weight (kg)

Leaching Usually refers to movement of water through soil, which may move soluble plant foods or other chemicals

Lodge (lodging) Of plants, to be beaten down by action of rain and wind; often encouraged by high rates of nitrogen fertilizer

Mechanism of action Precise biochemical or biophysical reaction or series of reactions that create a herbicide's final or ultimate effect; many herbicides have primary and secondary mechanisms of action

Mesocotyl Elongated portion of the axis between the cotyledon and coleoptile of a grass seedling

Miscible liquids Two or more liquids capable of being mixed, which will remain mixed under normal conditions

Mode of action Sequence of events that occur from a herbicide's first contact with a plant until its final effect (often plant death) is expressed

Monocot Abbreviation of monocotyledon, the preferred term to describe grasslike plants

Mulch Material (grass, straw, plastic, plant residue) spread on soil to cover or protect soil

Necrosis Death of tissue

Nonselective herbicide Herbicide used to kill plants generally without regard to species

No-till = No-tillage Planting without prior soil disturbance

Noxious weed Plant arbitrarily defined by law as being especially undesirable, troublesome, and difficult to control

Perennial Plant that lives from year to year; in most cases, in cold climates, stem and foliage die, but roots persist (e.g., field bindweed, dandelion, Canada thistle, johnsongrass, leafy spurge)

Pesticide Any substance or mixture of substances intended for controlling insects, rodents, fungi, weeds, and other forms of plant or animal life considered to be pests

Phenology Study of naturally occurring phenomena that recur periodically, such as flowering and time of seed germination

Phloem Living plant tissue that transports metabolic compounds from site of synthesis to storage or site of use

Phytotoxic Poisonous or inhibitory to growth of plants (from Greek *phyton,* plant and *toxikon,* poison)

Plagiotropic Used primarily for roots, stems, or branches to describe growth at an oblique or horizontal angle

Postemergence After emergence of a specified weed or crop

Preemergence After a crop is planted but before it emerges

Preplant Before planting

Preplant incorporated Herbicide applied and blended into soil prior to planting

Proherbicide A precursor to a herbicide; example bialophos

Radicle Part of the plant embryo that develops into the primary root

Rate and Dosage Synonymous terms: Rate is preferred and usually refers to the amount of active ingredient applied to a unit area regardless of percentage of chemical in the carrier

Registration Process of gaining approval to sell an herbicide or other pesticide from the U.S. Environmental Protection Agency (US/EPA) as governed by the amended Federal Insecticide, Fungicide, and Rodenticide act

Residual To have a continued effect over a period of time

Resistance Decreased response of a population to a herbicide (*see* Tolerance)

Rhizome Underground stem capable of sending out roots and leafy shoots

Ribonucleic acid (RNA) Polymeric constituent of all living cells, consisting of a single strand of alternating phosphate and ribose units with the bases adenine, guanine, cytosine, and uracil bonded to the ribose, the structure and base sequence of which determine the proteins synthesized

Safener Substance that reduces phytotoxicity (*see* Antidote)

Selectivity Property of differential tolerance, some plants are affected others are not; essential attribute of many herbicides

Sink Plant site with a high rate of metabolic activity where food resources are used

Soil incorporation Mechanical mixing of herbicides in soil

Soil injection Mechanical placement of herbicide in soil

Soil persistence Refers to the length of time a herbicide remains in soil; it may refer to effective life (i.e., the time during which plants are killed), or it may refer to total soil residence time

Soil sterilant Herbicide that prevents growth of all plants; effects may be temporary (a few months) or long-term (years)

Solution Homogenous mixture of two or more substances

Solution concentrate Liquid formulation that forms a solution when added to water

Soluble powder Dry formulation that forms a solution when added to water

Spike stage Early emergence stage of corn in which leaves are tightly rolled to form a spike, usually before corn is more than 2 in. tall.

Spot treatment Application of herbicide to localized or restricted areas as opposed to overall, broadcast, or complete coverage

Spray drift Movement of airborne liquid spray particles

Stolon Aboveground creeping stem that can root and develop new shoots (e.g., bermudagrass)

Stunting Retardation of growth and development

Submersed plant Aquatic plant that grows with all or most vegetative tissue under water

Surfactant Material added to pesticide formulations to impart spreading, wetting, dispersibility, or other surface-modifying properties

Suspension Liquid or gas in which very fine solid particles are dispersed, but not dissolved

Sympatry Occurring in one area (*see* Allopatry)

Symplast Functionally integrated unit consisting of all living cells of a multicellular plant

Synergism Action of two or more substances that creates a total effect greater than the sum of independent effects; achievement of an effect by two substances that neither is capable of achieving alone

Systemic herbicide Herbicide translocated in plants that has an effect throughout the entire plant system

Teratogenic Produces birth defects

Tolerance Natural and normal variation that exists within a species and can evolve quickly (*see* Resistance); or (legal definition) amount of pesticide chemical allowed by law to be in or on the plant or animal product sold for human consumption (legal definition)

Tolerant Capable of withstanding effects (*see* Resistance)

Toxicity Potential to cause injury, illness, or undesirable effects

Trade name A trademark or other designation of a commercial product

Translocation Transfer of photosynthate or other materials such as herbicides from one plant part to another

Turion Scaly shoot developed from a bud on a subterranean or submerged rootstock

Volatility Measure of the tendency to change state from liquid to gas

Weed Any plant that is objectionable or interferes with the activities and welfare of man (definition accepted by the Weed Science Society of America)

Weed control (1) Process of limiting weed infestations so crops can be grown profitably or other operations can be conducted efficiently; (2) process of reducing weed growth or weed infestation to an acceptable level

Weed eradication Complete elimination of all live plants, plant parts, and weed seeds from an area

Weed management Relatively new term in the lexicon of weed science that has several definitions; a synthesis of definitions follows: Rational deployment of appropriate technology to minimize weed impacts, provide systematic management of weed problems, and optimize intended land use (it is likely that the evolving definition will incorporate determination of an economic threshold)

Weed prevention Stopping weeds from invading an area

Wettable powder Powder that forms a suspension (not a true solution) in water

Wetting agent Chemical that when added to a spray solution causes the

solution to contact (wet) plant surfaces more thoroughly

Winter annual   Plant that starts germination in the fall, lives over winter,
and completes its growth, including
seed production, the following season
(e.g., downy bromegrass); many summer annuals can behave as winter annuals, if they germinate in fall and
live over the winter

Xylem   Nonliving plant tissue that con-
ducts water and solutes from roots
to shoots

Index